CONSTRUCTION FAILURE

CONSTRUCTION FAILURE

SECOND EDITION

Jacob Feld (deceased) and Kenneth L. Carper

A WILEY-INTERSCIENCE PUBLICATION

JOHN WILEY & SONS, INC.

New York / Chichester / Brisbane / Toronto / Singapore / Weinheim

Library of Congress Cataloging in Publication Data:
Feld, Jacob.
    Construction failure / by Jacob Feld and Kenneth L. Carper.—2nd
ed.
        p.   cm.
    Includes index.
    ISBN 0-471-57477-5 (alk. paper)
    1. Building failures.   2. Structural engineering.   I. Carper,
Kenneth L.   II. Title.
    TH441.F43   1996
    624.1'71—dc20   96-33425

Printed in the United States of America

10 9 8 7 6 5 4 3 2 1

CONTENTS

Preface   xi

Preface to the First Edition   xv

1   Failures: Causes, Costs, and Benefits   1

     1.1   Definition of Failure / 2
     1.2   Historical Notes / 3
     1.3   Recent Publications and Professional Society
         Activities / 8
     1.4   Failure Causes: Technical and Procedural / 13

         1.4.1   Fundamental Errors in Concept / 14
         1.4.2   Site Selection and Site Development Errors / 16
         1.4.3   Programming Deficiencies / 16
         1.4.4   Design Errors / 17
         1.4.5   Construction Errors / 18
         1.4.6   Material Deficiencies / 20
         1.4.7   Operational Errors / 21

     1.5   Forensic Engineering: Lessons from Failures / 21
     1.6   References / 22

**2   Natural Hazards and Unusual Loads: Effect on the
Built Environment**   24

     2.1   Gravity / 25
     2.2   Seismic Events / 28
     2.3   Extreme Winds / 39
     2.4   Flood / 43
     2.5   Fire / 44
     2.6   Unusual Loads: Blast, Vibration, and Collision / 48
     2.7   Deterioration / 52
     2.8   Summary / 53
     2.9   References / 57

v

3 Earthworks, Soil, and Foundation Problems   **59**

3.1 Problem Soils / 61
3.2 Below-Surface Construction / 65
    3.2.1 Deep Foundations: Piles and Caissons / 65
    3.2.2 Culverts, Pipelines, and Tunnels / 71
    3.2.3 Rock as an Engineering Material / 78
3.3 Foundations of Structures / 83
    3.3.1 Undermining of Safe Support for Existing
           Structures / 83
    3.3.2 Load Transfer Failure / 85
    3.3.3 Lateral Soil Movement / 87
    3.3.4 Unequal Support / 88
    3.3.5 Downdrag and Heave / 91
    3.3.6 Flotation and Water-Content Fluctuations / 93
    3.3.7 Vibrations and Seismic Response / 95
3.4 At-Surface Construction / 97
    3.4.1 Slabs-on-Grade and Pavement Failures / 97
    3.4.2 Slopes and Slides / 99
    3.4.3 Subsidence / 104
    3.4.4 Retaining Walls and Abutments / 105
3.5 Summary / 108
3.6 References / 109

4 Dams and Bridges   **111**

4.1 Dam Failures / 112
    4.1.1 Failures of Completed Dams / 117
    4.1.2 Dam Failures during Construction / 128
4.2 Bridge Failures / 130
    4.2.1 Failures of Completed Bridges / 133
    4.2.2 Bridge Failures during Construction / 150
4.3 Summary / 157
4.4 References / 158

5 Timber Structures   **162**

5.1 Timber as a Structural Material / 162
5.2 Importance of Connections / 167
5.3 Protection from Deterioration / 173
5.4 Repair and Rehabilitation / 177
5.5 Ponding Failures and Drifting Snow / 178
5.6 Proprietary Systems / 180
5.7 References / 183

6 Steel Structures   **186**

6.1   Steel: The "Ideal" Structural Material / 186
6.2   Stability Problems in Steel Structures / 188

    6.2.1   Erection Failures / 188
    6.2.2   Stability Failures in Completed Structures / 193

6.3   Steel Connections / 205

    6.3.1   Critical Details / 205
    6.3.2   Failure Case Studies Involving Connections / 209

6.4   Corrosion / 228
6.5   References / 231

7 Reinforced Concrete Structures (Cast-in-Place)   **235**

7.1   Reinforced Concrete as a Structural Material / 235

    7.1.1   Advantages of Reinforced Concrete / 236
    7.1.2   Undesirable Characteristics of Reinforced
           Concrete / 237
    7.1.3   Evolution of Concrete Design and Construction
           Standards: Early Failure Examples / 239

7.2   Formwork Problems and Failure of Temporary
      Structures / 242
7.3   Quality Assurance or Quality Control Problems / 243
7.4   Shrinkage, Expansion, and Plastic Dimensional
      Changes / 251
7.5   Abrasion and Surface Deterioration / 262
7.6   Shear Failures: Flat-Plate Structure / 265
7.7   Detailing for Ductility: Lessons from Earthquakes / 274
7.8   Corrosion in Reinforced Concrete Structures / 281
7.9   References / 287

8 Precast and Prestressed Concrete Structures   **290**

8.1   Prestressed Concrete: The Modern Structural
      Material / 291
8.2   Prefabrication: Reliable Member Production
      Techniques / 292
8.3   Connections: Critical Details / 298
8.4   Importance of Quality Control at the Construction
      Site / 302
8.5   Transportation and Erection Failures / 306
8.6   Performance in Extreme Winds and Seismic
      Events / 315
8.7   Corrosion of Prestressing Tendons / 319
8.8   References / 324

9   Masonry Structures

327

9.1   Characteristics of Masonry: Traditional and
Contemporary Construction / 327

9.2   Failures due to Aging / 332

9.3   Construction Errors and Workmanship
Deficiencies / 335

9.4   Settlement, Expansion, and Contraction / 337

9.5   Incompatibilities with Other Materials / 343

9.6   Masonry Cladding, Curtain Walls, and Facades / 348

9.7   Corrosion / 360

9.8   Seismic and Wind Performance / 363

9.9   Interior Partitions and Decorative Screens / 367

9.10 References / 368

10   Nonstructural Failures

371

10.1   Building Envelope Problems: Facades, Curtain Walls,
and Roofs / 372

    10.1.1   Roofs / 373

    10.1.2   Facades / 375

    10.1.3   Failure Case Study: John Hancock Mutual Life
Insurance Company Building, Boston,
Massachusetts / 382

10.2   Heating, Ventilating, and Air-Conditioning
Problems / 392

10.3   Suspended Ceilings / 396

10.4   Equipment Failures / 399

10.5   Nonstructural Component Repair Costs Following
Recent Seismic Events / 402

10.6   References / 406

11   Construction Safety and Failures during Construction

409

11.1   Construction: A Dangerous Occupation / 409

11.2   Construction Failures due to Design Errors / 412

11.3   Excavation, Trenching, and Foundation Construction
Accidents / 413

11.4   Construction Loads May Exceed Design Loads / 420

11.5   Materials and Assemblies Not Yet at Design
Strength / 422

11.6   Stability Problems with Incomplete Structures / 429

11.7   Renovations, Alterations, and Demolitions / 438

11.8   Other Construction Hazards / 442

11.9   Strategies to Improve Construction Safety / 444
11.10   References / 447

**12 Responsibility for Failures: Litigation and   451
     ADR Techniques**

12.1   Responsibility: Ignorance, Negligence, and the
        Standard of Care / 453
12.2   Risk Management and Dispute Avoidance / 461
12.3   Traditional Litigation of Construction Disputes / 464
12.4   Alternative Dispute Resolution Techniques / 466

        12.4.1   Arbitration / 467
        12.4.2   Mediation / 471
        12.4.3   Mediation/Arbitration / 473
        12.4.4   Minitrials / 473
        12.4.5   Private Judges / 475
        12.4.6   Dispute Review Boards / 475
        12.4.7   Other ADR Techniques / 477

12.5   Trends in Dispute Avoidance and Dispute
        Resolution / 478
12.6   The Architect or Engineer as Expert Witness / 480
12.7   References / 481

13 Learning from Failures   483

13.1   Why Are There So Many Failures? / 484
13.2   What Can We Do to Reduce the Frequency and
        Severity of Construction Failures? / 486

        13.2.1   Definition and Assignment of
                 Responsibilities / 489
        13.2.2   Structural Integrity / 489
        13.2.3   Project Peer Review / 490
        13.2.4   Dissemination of Failure Information / 490
        13.2.5   Other Failure Mitigation Developments / 491

13.3   The Critical Role of Education / 492
13.4   Concluding Remarks / 493
13.5   References / 494

Index   495

PREFACE

It gives me a great deal of pleasure to present this second edition of Jacob Feld's classic text, *Construction Failure.* This revision has been six years in the writing and production. The original 1968 edition was the first title in the highly successful Series of Practical Construction Guides edited by M. D. Morris, P.E., and published by John Wiley & Sons, Inc. The Wiley Series of Practical Construction Guides now numbers 61 volumes.

The 1968 edition of this book and a 1964 monograph by Jacob Feld, *Lessons from Failures of Concrete Structures,* published by the American Concrete Institute, were among the first major works to discuss openly the results of failure investigations in the contruction industry. Many lessons were learned from the case studies presented in these two books, and other professionals were encouraged by Feld's example to acknowledge failures and disseminate information about them.

Several books on construction failure have been written since the original edition of this work appeared, and a number of professional societies have made great strides in enhancing the discussion of failures. These efforts have not eliminated failures from the industry, however, nor is such a task achievable. Indeed, the principal causes of failures remain unchanged since Feld wrote his books and are likely to exist well into the future. Most of Feld's writings are as timely today as they were 30 years ago.

While preparing this edition I have attempted to preserve as much of the original work as possible both in content and in style, although approximately 80 percent of the book is new material. The intent of the book is to improve practice as a result of learning from failure. The presentation is aimed at practitioners and the references listed at the end of each chapter are also practice oriented.

Recent case studies are included that illustrate the principles outlined in the first edition. Redundant examples from the earlier edition have been deleted to make room for these new examples. Design techniques and tools have evolved, as have construction procedures and approaches to project delivery. Many new materials and proprietary systems have been introduced to the industry. Buildings have become more complex than they were in Feld's generation and the expectations of users have become more demanding. The construction industry has been plagued by escalating litigiousness since Feld's day, and this has encouraged the development of alternative dispute resolution techniques, new ways of settling arguments. Any revisions or additions to the original text are intended to reflect these changes in the industry. The basic principles underlying Jacob Feld's classic text continue to be relevant.

I first read Jacob Feld's books in 1972, my final year as an architecture student. These books greatly influenced my career, as I was caused to contemplate the serious implications of professional responsibility and the need for technical competence. Partly as result of reading *Construction Failure,* I returned to graduate school and obtained a graduate degree in structural engineering. My interests have continued to focus on the subject of learning from failure.

In my teaching at Washington State University I often integrate failure case studies into my lectures. These case studies help to reinforce principles of structural design and also introduce important nontechnical aspects of professionalism into our discussions: the value of clear communication, the contractual and traditional relationships among the various parties in the construction industry, the importance of defining roles and responsibilities for all parties, and techniques for avoiding and resolving construction disputes.

I am grateful to my many colleagues, both in engineering and architecture, that have contributed to this book. The encouragement I have received from my fellow members on the Technical Council on Forensic Engineering of the American Society of Civil Engineers (ASCE) is especially appreciated. Several of the photographs were given to me by ASCE colleagues, as noted in the figure captions. Many of the recent case studies discussed here are condensed from articles in the ASCE *Journal of Performance of Constructed Facilities.* It was my great fortune to be the founding editor of this journal in 1986, a position that continues to be one of the most rewarding aspects of my professional career.

There are a number of people who deserve acknowledgment. Among these are editors: M. D. Morris, Daniel Sayre, Robert Argentieri, and Charles Schmieg. All are true professionals, providing constant encouragement and guidance. Neal FitzSimons, practicing consulting engineer in Kensington, Maryland, and Donald Vannoy, University of Maryland, contributed valuable insights, especially at the outset of the project. Christopher Stoski, architecture student at Washington State University, prepared the line drawings. Most particularly, I am indebted to Robert A. Rubin of the law firm, Postner and Rubin, New York City, for his thoughtful review and comment on a draft of Chapter 12.

Since Jacob Feld's writings had such an influence on me as a young professional, I was delighted to respond to the request from the publisher to work on a new edition. I trust that my respect for the original volume and its author are evident in this work.

KENNETH L. CARPER, PROFESSOR

School of Architecture
Washington State University

PREFACE TO THE FIRST EDITION

In some 40 years of consulting civil engineering practice and, to a smaller extent in the previous 10 years of work in engineering design and construction, I have seen many instances of construction failure, both collapse and distress, which seemed inexcusable. In return for a half century of satisfying and profitable occupation I offer this collection of experiences to the construction industry, not as a castigation or indictment but rather as a warning in a series of lessons to be learned from unfortunate or ill-advised procedures which led to these failures. The building of structures is the contribution made by this industry to the advancement of the living standards of humanity. It can be done better, however, with less grief and fewer troubles. A review of what has not worked so well as it might have can help to provide the necessary improvements in technique and control in every facet of the construction industry.

It would be impossible to give the names of the hundreds of people associated with the investigations and reports of these incidents. Even a list of references became too voluminous for inclusion in this book. Thankful credit is due to Professor Seymour S. Howard, Jr., of Pratt Institute, Brooklyn, New York, for his critical comments on the final text and for the preparation of the diagrams used as illustrations. The horrendous typing job of the several preliminary texts was performed by Mrs. Esther Whitton. The final version was produced by Ruth Wales of Typing Unlimited.

JACOB FELD (1899–1975)

New York, New York
June 1968

Jacob Feld
1899–1975

1

FAILURES: CAUSES, COSTS, AND BENEFITS

This volume is a compilation of experience, that of the authors and of many others. Experience is a great asset to professionals practicing in any discipline, whether that experience comes from success or failure (Addis 1990). Experience is of particular value to those practicing in the construction industry, where failures can result in death, injury, and significant economic damages. The construction industry, from ancient times, has benefited from trial-and-error and trial-and-success experience.

The construction industry is very complex, with many specialized disciplines contributing to the project delivery system. Professionals in each of these specialized disciplines learn from their own mistakes, and many other errors could be avoided if each knew about the mistakes made by others. In the hope that this compilation of experience will tend toward that end, this book is presented to the construction industry. Because many failures are not explained and often not even reported publicly, no complete coverage of the subject is claimed or even attempted. Lessons can be learned from the material presented here, and other lessons are available in the many excellent references given at the end of each chapter. Since the publication of the first edition of this book, it is gratifying to observe the increased willingness among construction industry professionals to discuss failures openly, thus advancing the art and science of practice. The many recent publications and professional society activities aimed at disseminating the results of failure investigations have come despite increased construction litigation. This sharing of critical performance information is a testament to the seriousness with which construction professionals view their responsibilities to public safety and welfare.

1.1 DEFINITION OF FAILURE

If we define failure as catastrophic structural collapse, there are few failures. But if nonconformity with design expectations is defined as failure (and this is the more logical and honest approach) and if one takes the trouble to measure the shape, position and condition of completed structures, then there are many failures, far more than the list of incidents that are covered by news media, both technical and public. This statement is more applicable to complicated statically indeterminate structures than to simple pin-connected structures. Unwanted displacements and unexplainable deformations often occur, and users of the facility may question whether they are failures or normal, acceptable (but unexpected) strains.

For the discussion of failures to be most helpful, performance problems that are less than catastrophic or life threatening must be included. Other industries have encouraged the discussion of less spectacular performance problems by describing them in words other than *failure*. Dam designers and nuclear engineers, for example, prefer to use words such as *incidents* and *accidents* to describe shortcomings that are not catastrophic. The Technical Council on Forensic Engineering of the American Society of Civil Engineers has adopted the definition of *failure* first advanced by Leonards (1982): "Failure is an unacceptable difference between expected and observed performance."

In addition to structural collapse, this definition includes serviceability problems such as distress, excessive deformation, premature deterioration of materials, leaking roofs and facades, and inadequate interior environmental control systems. The relationship between expectations and performance given in this definition is useful, for it helps to explain the large number of claims filed each year stemming from both realistic and unrealistic expectations.

Almost all failures or perceived failures generate controversy. Long and expensive argument and litigation usually follow, during which the experts are quizzed by their clients' lawyers and cross-examined at great length by the opponents' counsel in an attempt to pin down the "proximate" cause of the failure or incident, as if the legal fiction that responsibility stems from a single and exclusive cause makes it possible to determine such cause by either observation or deduction. Sometimes there is a single simple explanation for a failure, but failure usually results from a combination of conditions, mistakes, oversights, misunderstandings, ignorance, and incompetence, or even dishonest performance. Often, there is no single item that can be pinpointed as the sole cause of failure. Yet each individual deficiency can be said to be the responsible straw that broke the camel's back, and if one straw had not been added, the camel's back would not be broken. The dimension between success and failure in a constructed project may be that of the thinnest hair.

It is the intention of this book to improve the construction industry. Included are lessons describing errors in judgment, in design, in detail, in control, and in performance.

1.2 HISTORICAL NOTES

The science of archeology is dependent on the uncovered debris of engineering failures. If the Code of Hammurabi (about 2000 b.c.) is typical of ancient regulations, few engineer/architect/builders had the opportunity to learn from their own mistakes; there was no such thing as a second chance. There was, however, a clear incentive to learn from the errors made by their late colleagues. The five basic rules from the Code of Hammurabi covering construction failures are reproduced in original, transliteration, and translation in Figure 1.1. It is not likely that these rules stopped all failures, but they certainly must have been a deterrent to shoddy construction practice. In addition, these harsh guidelines would have discouraged many innovations. The Code of Hammurabi also eliminated the probability of repetitive malpractice, at least by the same practitioner.

It is of passing interest to note that the symbol for "builder" is the framework of a house, consisting of a main center post with four corner posts all connected and braced by roof beams, and the center post is the symbol for "person." The fundamental concept of providing safety for human occupancy is embodied in this ancient symbol, along with an implicit suggestion that the builder is placed in the most vulnerable position to suffer the consequences should the design or construction prove to be inadequate.

Some warnings in the Hebraic literature against building on shifting sand may stem from experiences encountered when the Hebrews were the construction battalions in Egypt. History and fable report instances of summary execution in ancient civilizations of the designers/builders of successful monuments as a guarantee that competing or better structures would not be erected to rival the architectural gem. If these stories are based on fact, this practice, like the Code of Hammurabi, would have discouraged innovation, but this time to avoid resounding success. There would have been little incentive to explore novel or innovative designs unless one had an overriding desire to conclude one's career suddenly at the top, achieving immortality in the prestigious architectural journals yet to be published.

That entire ancient cities collapsed from normal aging, aided by enemy attacks, is found in the deep excavations at city sites with clearly separate layers of rubble. In the exploratory trenches dug at the Walls of Jericho, for example, Major Tulloch, the retired chief engineer of the Allenby invasion of Palestine in World War I, uncovered at least six distinct collapsed walls before coming to the wall that Joshua felled. Tulloch's explorations presented evidence that the Jericho wall fell from the undermining of the foundation stones. His view was that these operations were undertaken by the attackers while the defenders were distracted by the blowing of horns recorded in the scriptures.

In Greco-Roman times the construction industry was in the hands of trained slave artisans. Successful work was rewarded by gifts of substance as well as freedom. With little in the way of competitive financial or time constraints to

A. If a builder build a house for a man and do not make its construction firm and the house which he has built collapse and cause the death of the owner of the house — that builder shall be put to death.

B. If it cause the death of the son of the owner of the house — they shall put to death a son of that builder.

C. If it cause the death of a slave of the owner of the house — he shall give to the owner of the house a slave of equal value.

D. If it destroy property, he shall restore whatever it destroyed, and because he did not make the house which he built firm and it collapsed, he shall rebuild the house which collapsed at his own expense.

E. If a builder build a house for a man and do not make its construction meet the requirements and a wall fall in, that builder shall strengthen the wall at his own expense.

Translated by R.F. Harper.
"Code of Hammurabi" p. 83 - seq.

Jacob Feld 1922.

Figure 1.1 Code of Hammurabi (2200 B.C.).

worry the builder, important work was well done and with apparent great success, as evidenced by the many examples still existing. Similarly, in the Medieval period, time was no object, craftsmanship was highly valued, and the great successful structures built for the church or state have lasted for centuries. One wonders whether our modern works will last to comparable age (Chiles 1984).

In the common law developed in England and written in the fifteenth-century court records from the reign of Henry IV, the rule was ". . . if a carpenter undertake to build a house and does it ill, an action will lie against him."

Of course, by "ill" is meant "not well," and if all work were done well, there would be few records of failure. Historically, then, the burden rests on the construction industry to see that work is done "not ill" and each partner in the industry, from the architect who conceives the project to the superintendent who directs the performance of a part thereof, must lend every effort to avoid every foreseeable possible cause of "ill" structures. Avoidance of failure has always been at the heart of engineering decisions.

In North America, the Napoleonic code became the basis of the common law wherever the original settlers were French. In many ways this code places a greater responsibility on the designer and the professional in charge of the work as agent of the owner to safeguard the investment and guarantee proper and adequate performance than is expected under English legal procedure.

In the period 1870–1900, construction became a large industry in the United States, chiefly influenced by the expansion of the transportation system with its necessary bridges, storage facilities, and industrial plants. Comparatively little is recorded of failures in buildings, most of which were heavy masonry wall-bearing structures with good timber floors.

Bridge spans, however, soon exceeded the capability of timber trusses, and fierce competition in the sale of iron spans resulted in many failures. These were spectacular and were almost daily newspaper headlines. Even foreign technical journals commented on the great number of unsuccessful designs of bridges in the United States. Engineering magazines from 1875 to 1895 contained as many reports of railroad accidents and bridge failures as today's daily newspapers contain reports of automobile traffic accidents. Even as late as 1905, the weekly news summary in *Engineering Record* regularly described the most serious railroad wreck of the week, usually related to a bridge failure.

The *Railway Gazette* in 1895 published a discouraging summary of iron bridge failures resulting from railway traffic, listing 502 cases in the period 1878 through 1895, noting that the first 251 occurred in 10 years, whereas the second 251 occurred in only eight years. In the years 1888–1891 there were 162 such accidents. These reports were widely published and discussed. They must have had considerable influence on the design engineers and bridge vendors of those days.

Nearly every new development or innovation in the construction industry has the potential to create a concentration of failures. If these failures are openly discussed and the warnings are heeded, repetition stemming from the same causes can be prevented. When reinforced concrete was first gaining acceptance as a satisfactory and economic structural material, the editors of *Engineering News* wrote in the April 9, 1903 issue:

> One of the special advantages usually claimed for concrete work is that it can be safely built by unskilled labor, but in view of some recent accidents which occurred it appears well to point out that this principle is not of universal application. For concrete in large masses such as abutments, foundations, retaining walls, etc., there is no doubt that unskilled labor should be employed under proper supervision. But for certain other classes of work, girders and floors in concrete buildings, it appears that some degree of skilled labor should be employed, or at least the entire work should be under strict and constant supervision by skilled foremen, architects or engineers. In the carpentry work for forms and falsework especially, there is frequent evidence that the weight of the mass to be supported and the hydrostatic pressure of very wet concrete in columns are not realized by the men who are entrusted with the construction of this part of the work. When concrete has once thoroughly set it will stand very hard service and even overload or abuse, but it should be very strongly impressed upon the men engaged in concrete construction that the wet mass is simply a dead weight to be supported, having absolutely no supporting power in itself. If steel bars, rods, etc. are used, these simply add to the weight of the wet mass. The fall of a concrete floor at Chicago, noted in our issue of December 4, seems to have been due entirely to an ignorant man blindly following out instructions which were probably extremely indefinite. He was told to go in and knock out some of the shoring, and he proceeded to knock out every bent, until the unsupported concrete gave way, and its fall broke through other completed concrete work below. . . .
>
> In another case which came under our personal observation a laborer was found knocking away some of the struts and braces under a green concrete floor, simply because (as he told the superintendent when discovered) a carpenter told him to get some lumber. When informed that he stood a good chance of killing himself and other men in the work, as well as wrecking the building, he simply became surly and appeared to think that the superintendent was making a fuss about nothing. . . .
>
> There is no doubt that a great deal of concrete work is built by men who are really not competent to undertake it, and that much more work is done without sufficiently strict and continuous expert supervision to ensure the best and safest results. In view of the enormous increase in the use of concrete, and in the variety of purposes for which it is used, it is well for engineers to bear these facts in mind.

This editorial should be required reading for everyone in the construction industry at least once a year. Only one revision is necessary to bring this editorial up to date: the acknowledgment that women, as well as men, are now present on the construction site. The warnings given in 1903 are otherwise relevant today.

In 1918 the American Railway Engineering Association published an editorial article on "Study of Failures of Concrete Structures" with the subtitle "A Compilation of Failed Concrete Structures and Lessons to Be Drawn Therefrom." The study covered a period of 25 years and concluded with this statement:

> The one thing which these failures conclusively point to is that all good concrete construction should be subjected to rigid inspection. It should be insisted upon that the Inspector shall force the Contractor to follow out the specifications to the most minute details. He must see that the materials used are proper and are properly mixed and deposited, also that the forms are sufficiently strong and that they are not removed until after the concrete has set. It is believed that only by this kind of inspection is it possible to guard against the failure of concrete structures.

So the 1903 editorial advice still held in 1918; the lesson was not learned and is still not learned, as evidenced by examples discussed in Chapters 7 and 11. Consistent repetition and publication of the consequences of neglect are needed to combat the inertia of ignorance.

In 1924, Edward Godfrey, a consulting structural engineer well known for his frank and tireless criticism of improper design techniques, published a book on *Engineering Failures and Their Lessons* (Godfrey 1924), consisting principally of discussions and letters written in the years 1910–1923. Every phase of civil engineering is covered in the 20 chapters of Godfrey's book. The discussion of concrete failures is somewhat compromised by two of Godfrey's incorrect beliefs: that stirrups in concrete beams are of no value as shear resistance and that "hoop-tied" column bars are a detriment to the performance of a concrete column.

In 1952, Henry Lossier published a small book, *La Pathologie du Beton Arme,* that cites many examples of failures, mostly in concrete frames and special structures. The book included warnings that errors in the choice of framing and in design details are the most important factors causing failures. A revised and expanded edition of the book was published in 1955, and in 1962 the work was translated into English and published by the National Research Council of Canada as *Technical Translation 1008* (NRCC 1962). This work covers reports on a great variety of structural and foundation failures both in design and execution.

Under the title *Engineering Structural Failures,* Rolt Hammond published a book in 1956 detailing the causes and results of failures in modern structures of various types (Hammond 1956). In an interesting foreword Sir Bruce White quotes comments made in 1856 by Robert Stevenson, then president of the British Institution of Civil Engineers:

> [N]othing is so instructive to the younger Members of the Profession as records of accidents in large works, and of the means employed in repairing the damage.

A faithful account of those accidents, and of the means by which the consequences were met, is really more valuable than a description of the most successful works. . . . Older Engineers have derived their most useful store of experience from the observation of those casualties which have occurred to their own and to other works, and it is most important that they should be faithfully recorded in the archives of the Institution.

During the 1960s, three influential books were published detailing construction failure case studies. Thomas McKaig's *Building Failures* is a compilation of over 200 failure cases from all aspects of construction (McKaig 1962). Two books were written by Jacob Feld: *Lessons from Failures of Concrete Structures* (Feld 1964), published by the American Concrete Institute, and the first edition of this book, *Construction Failure* (Feld 1968).

Perhaps the most prominent building failure to occur in the 1960s was the collapse of a 22-story precast concrete building in London, the Ronan Point housing project. A gas explosion on the eighteenth floor started a progressive collapse that continued down to the first floor of the building, killing four occupants. The failure led to code requirements for structural redundancy in England and continues to have an influence on designers in the United States, with regard to structural integrity, provision of adequate connections, and concerns for prevention of progressive collapse (RCPS 1973). The Ronan Point collapse is discussed further in Section 8.4.

Recording and disseminating the results of failure investigations has been an important component in the advancement of engineering design and construction practices throughout the history of construction. That process has continued in the past couple of decades and has been enhanced by several significant professional society initiatives.

1.3   RECENT PUBLICATIONS AND PROFESSIONAL SOCIETY ACTIVITIES

A number of catastrophic structural failures occurred in the 1970s and 1980s. These failures were widely published, both in the public media and in professional journals. Public attention was focused intensely on the construction industry and several new strategies aimed at mitigating construction failures were introduced within the industry.

The safety record of the construction industry was called into question due to the large number of deaths and injuries occurring on construction project sites. Many of these casualties were associated with equipment failure and isolated excavation and trenching accidents. However, the past two decades have also witnessed several catastrophic structural collapses during construction that have generated wide publicity. These collapse incidents include:

- *Commonwealth Avenue, Boston, January 1970.* A 17-story reinforced concrete apartment building collapsed, killing four construction workers and injuring 20 others.

- *Bailey's Crossroads, Virginia, March 1973.* Progressive collapse of a 26-story reinforced concrete residential tower killed 14 construction workers and injured more than 30.
- *Cooling Tower Scaffold Collapse, Willow Island, West Virginia, April 1978.* Premature loading of cast-in-place concrete resulted in loss of life for 51 construction workers, the most costly construction accident since the first collapse of the Quebec bridge over the St. Lawrence River in 1907. (The Quebec failure killed 74 people.)
- *Rosemont Horizon Arena, Chicago, August 1979.* Glue-laminated timber roof arches spanning 89 m (290 ft) collapsed, killing five workers and injuring 16.
- *Harbor Cay Condominium, Cocoa Beach, Florida, March 1981.* A five-story cast-in-place reinforced concrete building collapsed due to design and construction deficiencies, killing 11 construction workers and injuring 23 others.
- *East Chicago, Indiana Highway Ramp Accident, April, 1982.* Thirteen construction workers were killed and 17 injured when the falsework collapsed during construction.
- *L'Ambiance Plaza, Bridgeport, Connecticut, April 1987.* A 16-story post-tensioned prestressed concrete lift-slab apartment project collapsed suddenly during construction, resulting in 28 deaths.

These construction tragedies and lessons learned from them are discussed in Chapters 7, 8, and 11.

The construction process is inherently dangerous, but many recurring accidents are avoidable, such as those resulting from unprotected falls and improperly shored foundation and trenching excavations. The National Safety Council (NSC), the Occupational Safety and Health Administration (OSHA), and other public oversight agencies have focused attention on the poor safety record of the industry, publishing alarming statistics and introducing new regulations and enforcement policies. Their efforts and increased awareness within the industry have brought about some improvement.

The American Society of Civil Engineers (ASCE) and other professional societies have responded with activities and publications aimed at improving the safety of construction, including provisions for more integrity in the design of temporary structures. In 1987, the Center for Excellence in Construction Safety was established at West Virginia University with funding from the National Institute for Occupational Safety and Health (NIOSH) and widespread support from all segments of the construction industry.

In addition to the foregoing listed failures during construction, several significant collapses of completed buildings occurred in the period 1978–1981, alarming the public and generating a congressional investigation of the industry. These failures included:

- *Civic Center Coliseum, Hartford, Connecticut, January 1978.* Total collapse of a steel space truss roof spanning 110 by 92 m (360 by 300 ft)

Figure 1.2   Hartford Civic Center Coliseum, Hartford, Connecticut (collapsed January 18, 1978). (This photograph was taken by permission from *Toward Safer Longspan Buildings,* published by the American Institute of Architects, 1981.)

occurred (Figure 1.2). No one was in the building at the time of collapse, so there were no injuries.

- *C. W. Post Center Auditorium, Greenvale, New York, January 1978.* A steel and aluminum dome spanning 52 m (171 ft) collapsed under snow load. There were no injuries, because the building was unoccupied at the time of collapse.
- *Crosby Kemper Jr. Memorial Arena, Kansas City, Missouri, June 1979.* A coliseum roof collapsed partially during a rain and windstorm (Figure 1.3). No one was injured, again because there were no occupants.
- *Hyatt Regency Hotel Walkways, Kansas City, Missouri, July 1981.* When these pedestrian walkways in the hotel convention center atrium collapsed, 113 people were killed and more than 180 were injured.

These dramatic structural failures, coming so close together, were understandably responsible for a great deal of public alarm. The U.S. Congress conducted a set of hearings regarding structural failures during August 1982. Hearings were held under the direction of the House Subcommittee on Investigations and Oversight, a subcommittee of the House Committee on Science and Technology. Quite a few experts from the construction industry were interviewed and a number of recommendations emerged from the hearings.

Figure 1.3 Crosby Kemper Jr. Memorial Arena, Kansas City, Missouri (collapsed June 4, 1979). (© Wilborn & Assoc., Photographers, Kansas City, Missouri.)

Some of these recommendations have been implemented, particularly with respect to increased efforts to disseminate failure information and improve the project delivery system (Carper 1987).

Several important books have been written during the last 10 years on the subject of construction and engineering failures. These include *Structural and Foundation Failures* (LePatner and Johnson 1982), *Construction Disasters* (Ross 1984), *To Engineer Is Human* (Petroski 1985), *Building Failures* (Addleson 1987), *Design and Construction Failures* (Kaminetzky 1991), and *Design Paradigms* (Petroski 1994).

Professional societies within the construction industry renewed efforts during this period to communicate about failures and seek new solutions. Some disciplines have traditionally been more active than others in disseminating information about failures. The Association of Soil and Foundation Engineers (ASFE), in particular, has been a leader in this regard, with a long list of published failure case studies (Gnaedinger 1987). The American Concrete Institute (ACI) has published a number of failure case studies. Other societies began to discuss failures openly for the first time. For example, the American Institute of Architects (AIA) appointed a panel and published a small brochure, *Toward Safer Long-Span Buildings* (AIA 1981).

Some of the more promising professional activities have brought together professionals from various disciplines. For some time the construction industry has been fragmenting, with professionals specializing to the detriment of cross-disciplinary communication. Bringing these specialists together to discuss common concerns has stimulated some creative ideas.

In March 1986, a Construction Industry Roundtable meeting was held in Kansas City, Missouri, where 113 people were killed five years earlier in the Hyatt Regency Hotel walkway collapse. Attending were 40 industry leaders representing all segments of the industry, including engineers, architects, constructors, insurers, attorneys, material suppliers, fabricators, owners, and building officials. Recommendations emerging from this interdisciplinary conference centered on organizational peer review, independent project peer review for projects exceeding a defined threshold, on-site inspection by the engineer of record, better delineation of roles and responsibilities, shop drawing review, competitive bidding for design services, professional registration, communication among parties, and public education on the positive accomplishments of the construction industry.

Another interdisciplinary project developed at that time was the Architecture and Engineering Performance Information Center (AEPIC). Established at the University of Maryland in 1982, AEPIC serves as a repository for performance information.

The Engineering Foundation sponsored two conferences on Structural Failure, one in Santa Barbara, California, in November 1983, and one in Palm Coast, Florida, in December 1987. Recommendations addressed structural integrity, life safety assurance, peer review, definition and assignment of responsibility, and unified risk insurance (Bell, Kan, and Wright 1989).

In 1982 the American Society of Civil Engineers (ASCE) established the Technical Council on Forensic Engineering (TCFE). TCFE maintains six committees, all dedicated to the reduction of failures in the construction industry: Committee on Dissemination of Failure Information, Committee on Practices to Reduce Failures, Committee on Forensic Engineering Practice, Committee on Education, Committee on Research, and Committee on Publications.

With National Science Foundation support, TCFE sponsored an interdisciplinary workshop on "Reducing Failures in Engineered Facilities" in Clearwater, Florida, in January 1985. Several specific projects and failure reduction strategies were identified, many consistent with those proposed by the Kansas City Roundtable and the Engineering Foundation conferences (Khachaturian 1985).

One of these proposals, the need for enhanced dissemination of failure information, has been implemented in the form of a new interdisciplinary journal. The TCFE Publications Committee is responsible for the quarterly *Journal of Performance of Constructed Facilities,* published by ASCE and cosponsored by AEPIC and the National Society of Professional Engineers, Professional Engineers in Private Practice (NSPE/PEPP). Established in 1986, the journal is dedicated to interdisciplinary discussion about the causes and costs of failures and performance deficiencies in the construction industry. Case studies are contributed by practicing professionals and researchers, and generic problems within the industry are also discussed. Now in its tenth year of publication, the journal has met with considerable success, indicating that expanded discussion about failures is both acceptable and useful to practicing professionals.

The ASCE Technical Council on Forensic Engineering has been active in presenting technical sessions and has participated in projects related to other emerging failure reduction strategies, such as project peer review, constructability reviews, and alternate dispute resolution (ADR) techniques. Most recently, the TCFE Committee on Education has published a book of important historic failure case study summaries for use in undergraduate professional education (Shepherd and Frost 1995).

It should be noted that buildings were not the only constructed facilities that experienced dramatic structural failures during the 1970s and 1980s. Among the more prominent failures of other civil structures were:

- *Teton Dam, Idaho, June 1976.* Failure of a 93-m (305-ft)-high dam cost over $1 billion in property losses, loss of 11 lives, and over 2000 injuries.
- *Mianus River Bridge, Connecticut Turnpike, June 1983.* Three persons were killed when a connection gave way, allowing a section of the 26-year-old bridge to collapse into the river.
- *Schoharie Creek Bridge, New York State Thruway, April 1987.* Ten persons died when a pier failed due to scour of the pier footing, a situation that might have been detected by a periodic inspection program.

These failures and others led to enhanced dam safety programs and greater emphasis on inspection and maintenance of existing civil structures. Failures of dams and bridges are discussed further in Chapter 4.

In addition to structural collapse, other costly performance problems in constructed facilities are receiving much attention. The facade problems experienced by high-rise structures such as Boston's John Hancock Tower and the Amoco Tower in Chicago are well-known examples. Some new materials and systems, such as corrosion-resistant steels, certain types of metal stud/brick veneer facade systems, certain mortar additives, and fire-resistant plywood products, have introduced unexpected performance problems or unanticipated accelerated maintenance costs in buildings. These deficiencies, although generally not life threatening, have had significant economic consequences. This type of performance deficiency is perhaps more responsible for the increase in construction litigation and the related escalation in insurance premiums paid by design professionals than is the infrequent structural collapse.

1.4 FAILURE CAUSES: TECHNICAL AND PROCEDURAL

Failures may result from a single error. It is more common, however, for a failure to be the result of several interrelated contributing factors. These may involve technical problems and unexpected deficiencies in material performance. Procedural deficiencies may result from human errors in judgment or from the human tendencies toward ignorance, incompetence, negligence, and greed. The constructed project may be subjected to environmental conditions or loads that are unpredicted by the designer or by accepted standards of practice.

Procedural causes of failure, involving human deficiencies, are often interdisciplinary and may be the result of communication deficiencies or unclear delineation of roles and responsibilities.

For this discussion the causes of failures in constructed facilities are classified in seven categories:

1. Fundamental errors in concept
2. Site selection and site development errors
3. Programming deficiencies
4. Design errors
5. Construction errors
6. Material deficiencies
7. Operational errors

Of course, a particular failure may be the result of a combination of two or more of the factors listed. These contributing factors may be impossible to isolate and quantify. When a failure occurs during the service life of a facility, contributing factors from every item in the list may be found. Sorting out an equitable distribution of responsibility for such a failure presents a significant challenge.

1.4.1 Fundamental Errors in Concept

Some failed projects may be described as fundamental errors in basic concept. The project may be unique, an original attempt to build something beyond available technology. The scale of the project may be outside the envelope of past experience. The project may have been located in an unusual environment, where the prediction of environmental effects was unreliable. Some failures of this type are not engineering failures at all, but rather, economic failures. Those in control of the project may discover that their concept is flawed technically or that the resolution of evolving problems will require far more economic investment than originally anticipated. Such projects may be abandoned and will be considered failures by most observers.

Most of the failures discussed in this book do not fall into this classification, but a few examples will be given here. Perhaps the first recorded example of a failure due to a concept error is Biblical: the collapse of the Tower of Babel presented in Genesis 11. The proposed concept, an unreinforced masonry tower "whose top would reach to heaven," was fundamentally flawed. The inevitable collapse generated a great deal of confusion and the first recorded incidence of breakdown in communication.

The story of the U.S. Navy "Big Dish" project of 1948–1962 is the record of another failure due to error not in design or in construction but in concept (Figure 1.4). The science of radio-wave reception required tolerances impossible to achieve with construction technology available at the time.

In 1948 the Naval Research Laboratory suggested the construction of this radio telescope for scientific work. Congress approved a budget of $20 million in 1956 for construction of the 180-m (600-ft)-diameter parabolic shell. By the end of 1957, feasibility studies reported that such a structure could be

Figure 1.4 Model of 1948 proposal for U.S. Navy "Big Dish" radio telescope project.

built but at a cost of $52.2 million. Soon thereafter, the Navy decided to combine military functions with the instrument's scientific capabilities, increasing the cost projections to $79 million. These cost estimates were subsequently revised to $126 million and then to over $200 million. In September 1961, however, some years after construction had begun, and with designs not yet fully complete, Congress set a ceiling of $135 million for the project.

In 1962, with expenditures of $63 million, work was stopped and the project was abandoned. The Controller General's report justified termination of the project as follows: "Our belief is based upon reports prepared in 1960 and earlier by scientists within and outside the government who reviewed specific [technical] problem areas and indicated serious doubts that the instrument, if completed, would have the desired capabilities."

More recently, in 1983, financial markets were shocked by the largest bond default in history, when the Washington (State) Public Power Supply System (WPPSS) abandoned an ambitious multiple nuclear power plant construction project. Of five reactors to be constructed simultaneously by the consortium of 115 public utilities, only one was completed. The failure of this project and the resulting $2.25 billion municipal bond default occurred not from design or construction errors but as a result of cost overruns, changing societal attitudes, and power demand projections that did not materialize.

To these failures involving unique or large-scale projects should be added those projects that may be considered failures due to changing societal expectations. Changing environmental regulations, for example, are expected to cause certain projects that meet current expectations to be redefined as unacceptable solutions in the future. Although these projects may be considered successful today, the "moving target" of societal expectations may eventually classify them as failures, especially if they require replacement before their projected useful life has been realized.

1.4.2   Site Selection and Site Development Errors

Failures often result from unwise land-use or site-selection errors. Certain sites are more vulnerable than others to failure. The most obvious examples are sites located in regions of significant seismic activity, in coastal regions, or in floodplains. Other sites pose problems related to specific soil conditions, such as expansive soils or permafrost in cold regions.

Recognition of the characteristics of particular site conditions through appropriate geotechnical studies can lead to decisions about site selection and site development that reduce the risk of failure. Unnecessary exposure to natural hazards is an unfortunate consequence of historic patterns of human settlement. Examples of failures associated with unwise site selection are included in Chapter 2.

1.4.3   Programming Deficiencies

Failure has been defined as an unacceptable difference between expected and observed performance. This definition implies that the expectations of the client must be clearly understood by the designer, and that they must be realistic. When the project does not perform as expected, even if the expectations are unrealistic or unachievable within given economic restraints, the client is likely to define the project as a failure.

A considerable volume of construction litigation results from unclear or unrealistic expectations. This type of failure could have been avoided through communication during the programming phase of a project. A program should clearly define the scope and intent of a project at the outset, so that general agreement can be reached on a way to measure the success of the completed project.

1.4.4 Design Errors

Design errors have contributed to many of the failure case studies presented in this book. These include:

- Errors in design concept
- Lack of structural redundancy
- Failure to consider a load or a combination of loads
- Deficient connection details
- Calculation errors
- Misuse of computer software
- Detailing problems, including selection of incompatible materials or assemblies that are not constructable
- Failure to consider maintenance requirements or durability
- Inadequate or inconsistent specifications for materials or expected quality of work
- Unclear communication of design intent

Deficiency in the basic design of a structure, such as amount of reinforcing steel at points of maximum moment or incorrect dimensions of concrete or steel sections to provide sufficient resistances for normal loading, is a rare cause of failure. One case was caught in 1925, just before placing concrete in forms for beams spanning 19.5 m (64 ft) over a school auditorium in Yonkers, New York. About 4 in^2 of reinforcement steel had been specified and placed, although the design should have required 40 in^2. At that time, girders of such size were unusual and the error was discovered by an inquisitive young engineer on the contractor's staff. The girders would undoubtedly have failed if the error had not been corrected.

In 1961, another young construction engineer noticed a significant error at the almost completed upper-level addition to the Birmingham, Alabama, Legion Field Stadium. A structural steel addition to provide 8600 seats was being rushed to meet a deadline for the Alabama–Tennessee football game. The upper deck consisted of sloping plate girders on two columns, with cantilevers in each direction. A mezzanine was added in the center portion only, and the load of the mezzanine was suspended from the girders. During a final inspection of the steel work, a junior employee of the contractor asked the design engineer why the girders were all identical, whether with or without the hanging mezzanine load. A quick check of the design calculations revealed that the computations had omitted the hanger reactions on the affected girders. Additional steel angles were immediately welded to the top and bottom chords of the girders to correct the deficiency.

On March 27, 1981, a cast-in-place concrete condominium project collapsed while under construction in Cocoa Beach, Florida, killing 11 construction workers. Several factors contributed to this collapse, discussed in Chapter 7. The most important factor was a design error. The designer never performed any calculations to check punching shear, the most common failure mode associated with this type of structure.

The failure of the Hartford, Connecticut, Civic Center Coliseum space truss roof in January 1978 (see Figure 1.2 and Chapter 6) has been attributed to design assumptions that were not executed in design details. The 110 by 92 m (360 by 300 ft) roof experienced a total collapse when compression chords of the space truss buckled. Bracing, assumed at midspan of the members for input to the sophisticated computer analysis, was not provided in the final design.

The Kemper Memorial Arena roof failure in Kansas City, Missouri (June 1979), involved a connection detail. This collapse is also a good example of the need to provide structural redundancy if the design is to have resistance to progressive collapse (see Figure 1.3 and Chapter 6). The Kansas City Hyatt Regency Hotel pedestrian walkway failure (July 1981) is attributed to a poorly designed connection, made more unsafe by a change during construction. This collapse, which killed 113 people, is reviewed in Chapter 6.

Some failures result from unclear communication of design intent. One famous example is the 1950 collapse of a 7.3-m (24-ft)-high reinforced concrete retaining wall in Manhassett, New York. The design called for $1\frac{1}{4}$-in.-diameter vertical reinforcing bars, spaced at regular intervals. Unfortunately, the drafter placed the "1" of the "$1\frac{1}{4}$ in." on a dimension line (Figure 1.5). Incredibly, $\frac{1}{4}$-in.-diameter smooth wires were provided and installed. Total collapse started as soon as the wall was backfilled. This type of failure (and many others recorded in this book) could be prevented if a representative of the design engineer is retained to provide field inspection services.

1.4.5 Construction Errors

Failures can result from construction errors. Such failures may involve:

- Excavation accidents
- Construction equipment failure

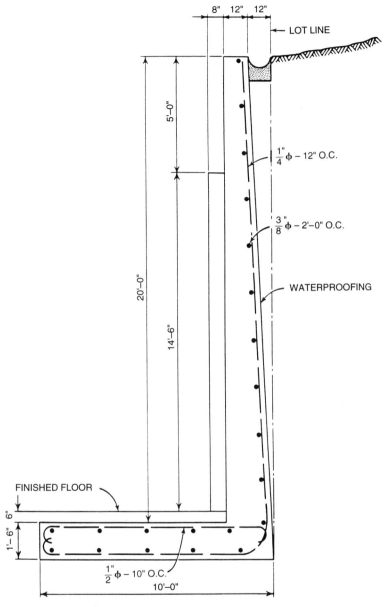

Figure 1.5 Drawing of a retaining wall that failed because of misinterpretation of reinforcement bar size due to a drafting error.

- Improper construction sequencing
- Inadequate temporary support
- Excessive construction loads
- Premature removal of shoring or formwork
- Nonconformance to design intent

Construction is a dangerous occupation. Many failures of cast-in-place concrete structures occur during construction due to inadequate temporary support, premature removal of shoring, and premature loading of concrete. Examples include the 1973 failure of a 26-story residential tower in Virginia (14 killed) and the Willow Island, West Virginia, cooling tower scaffold collapse of April 1978 (51 killed). These cases are reviewed in Chapters 7 and 11.

Precast concrete and steel frame structures often experience stability failures when temporary bracing is inadequate. Improper construction sequencing is also a source of failure. In 1985, a Denver, Colorado, highway structure collapsed during construction when eight girders were placed on incomplete piers. One worker was killed and four were seriously injured. Construction sequence is absolutely critical for certain construction types, such as post-tensioned prestressed concrete. The designer makes assumptions regarding sequencing, and if these are not communicated clearly to those responsible for field operations, the results can be catastrophic.

Some failures are the result of gross violations of the design documents. The July 1983 collapse of the Magic Mart Department Store in Bolivar, Tennessee, is one example. Following the collapse, investigators were able to find very few similarities between the design documents and the as-built construction. Structural steel members were not the sections specified, and the structural grid was even rotated 90 degrees from that shown on the drawings (see Chapter 6).

1.4.6 Material Deficiencies

Some would claim that materials do not fail; people fail. Although it is true that most materials problems are the result of human errors involving a lack of understanding about materials or the ignorant juxtaposition of incompatible materials, there are failures that can be attributed to unforeseeable material inconsistencies.

Designers have come to rely on modern structural materials. However, manufacturing or fabrication defects may exist in the most reliable structural materials, such as standard structural steel sections or centrally mixed concrete. In 1980, for example, over 130 buildings in California's San Francisco Bay area experienced serious structural defects as a result of poor-quality aggregate that was inadvertently used by four concrete suppliers. Spalling of the concrete was attributed to several tons of expansive brick that was accidentally dumped onto an aggregate pile at a cement plant. Stone facade panels or glass curtain wall units may contain undetected critical flaws. Although these examples may, in fact, derive from human errors, they can hardly be considered design or construction errors.

1.4.7 Operational Errors

Certainly, failures can occur after occupancy of a facility as the result of owner/operator errors. These may include alterations made to the structure, change in use, operational judgment errors, negligent overloading, and inadequate maintenance.

Cases exist where shear walls have been removed by owners, reducing the lateral force resistance of the structure, and shifting the center of rigidity so that unanticipated torsional moments were experienced under lateral loads. Roof membrane deterioration can cause water damage to structural members, leading to failure. Conversely, adding roof membranes without removing existing materials increases the dead load on structural members.

Sometimes, failures occur because maintenance or operation personnel do not have the requisite knowledge or skills to operate the facility properly. This is particularly true for technically sophisticated systems. For example, a number of claims involving deficient interior air quality and interior environmental comfort in buildings stem from the technical complexity of modern sophisticated environmental control systems. Unless facility personnel are specially trained to operate the systems, the results will not be satisfactory.

Not all maintenance-related failures are the fault of the owners and operators of facilities. Designers and constructors share the responsibility for durability of constructed projects, through appropriate selection of materials, design for maintainability, and quality of construction.

The June 1983 failure of the Mianus River bridge in Connecticut due to a corroded connection, and the April 1987 collapse of the scoured piers of the New York State Thruway bridge over Schoharie Creek, are examples of operational errors involving deficient inspection and maintenance provisions. These are discussed in Chapter 4.

1.5 FORENSIC ENGINEERING: LESSONS FROM FAILURES

Forensic engineering is the name given to the activity of failure investigation. A literal interpretation of forensic engineering implies presentation of the findings of an investigation in litigation or mediation. Forensic engineers practicing in disciplines outside the construction industry, such as in traffic accident reconstruction or product liability disputes, may spend a great deal of time in the courtroom. Forensic engineering in the construction industry, however, is not always related to litigation. In fact, only a very small percentage of cases actually proceed all the way to the courtroom, although most investigations are conducted with that possibility in mind.

While the examples presented in this book are derived from forensic investigations, it is not the purpose of this book to serve as a guideline for conducting such investigations. Readers who are interested in the techniques and procedures for forensic engineering practice, including the preparation of reports and guidelines for providing expert witness testimony, are referred to two recent publications (ASCE 1989, Carper 1989). The principal contribution of the forensic engineer is the introduction of rational interpretation of technical factors into the dispute settlement process. However, forensic engineering also serves to advance the art and science of engineering design as the results of failure investigations are disseminated to practicing design professionals. Forensic engineers are the pathologists of the construction industry, and their efforts are sometimes compared to those of their counterparts in the medical professions. A forensic medical investigation, although often related to litigation, may result in recommendations that effect improvements in medical practice (Carper 1986).

The failures discussed in this book are presented with this intent. There is no intended criticism of the unfortunate parties involved with these cases. Surely, hindsight provides a favored perspective, and the unknowns confronting the designer are far more complicated and diverse than those facing the failure investigator. Successful design in construction, as in all fields of engineering endeavor, presents the formidable task of anticipating, predicting, and mitigating all possible modes of failure.

In his delightful book *To Engineer Is Human,* Henry Petroski writes:

> Failure analysis is as easy as Monday-morning quarterbacking; design is more akin to coaching. However, the design engineer must do better than any coach, for he is expected to win every game he plays. That is a tough assignment when one mistake can often mean a loss. And when defeat occurs, all one can hope is to analyze the game films and learn from the mistakes so that they are less likely to be repeated the next time out. (Petroski 1985)

1.6 REFERENCES

Addis, W., 1990. *Structural Engineering: The Nature of Theory and Design,* Ellis Horwood Series in Civil Engineering, Ellis Horwood Ltd., Chichester, West Sussex, England.

Addleson, L., 1987. *Building Failures: A Guide to Diagnosis, Remedy and Prevention,* Van Nostrand Reinhold, New York.

AIA, 1981. *Toward Safer Long-Span Buildings,* Long-Span Building Panel, American Institute of Architects, Washington, DC.

ASCE, 1989. *Guidelines for Failure Investigation,* Technical Council on Forensic Engineering, American Society of Civil Engineers, New York.

Bell, G., F. Kan, and D. Wright, 1989. "Project Peer Review: Results of the Structural Failures II Conference," *Journal of Performance of Constructed Facilities,* American Society of Civil Engineers, New York (November).

Carper, K., ed., 1986. *Forensic Engineering: Learning from Failures,* American Society of Civil Engineers, New York.

Carper, K., 1987. "Failure Information: Dissemination Strategies," *Journal of Performance of Constructed Facilities,* American Society of Civil Engineers, New York (February).

Carper, K., ed., 1989. *Forensic Engineering,* Elsevier Science Publishing Co., Inc., New York.

Chiles, J., 1984. "Engineers Versus the Eons, or How Long Will Our Monuments Last?" *Smithsonian,* Smithsonian Institution Press, Washington, DC (March).

Feld, J., 1964. *Lessons from Failures of Concrete Structures,* American Concrete Institute, Detroit, MI.

Feld, J., 1968. *Construction Failure,* John Wiley & Sons, Inc., New York.

Gnaedinger, J., 1987. "Case Histories: Learning from Our Mistakes," *Journal of Performance of Constructed Facilities,* American Society of Civil Engineers, New York (February).

Godfrey, E., 1924. *Engineering Failures and Their Lessons,* Superior Printing Co., Akron, OH.

Hammond, R., 1956. *Engineering Structural Failures,* Odhams Press Ltd., London, UK.

Kaminetzky, D., 1991. *Design and Construction Failures: Lessons from Forensic Investigations,* McGraw-Hill, Inc., New York.

Khachaturian, N., ed., 1985. *Reducing Failures of Engineered Facilities,* American Society of Civil Engineers, New York.

Leonards, G., 1982. "Investigation of Failures," *Journal of the Geotechnical Engineering Division,* American Society of Civil Engineers, New York (February).

LePatner, B., and S. Johnson, 1982. *Structural and Foundation Failures: A Casebook for Architects, Engineers and Lawyers,* McGraw-Hill, Inc., New York.

McKaig, T., 1962. *Building Failures: Case Studies in Construction and Design,* McGraw-Hill, Inc., New York.

NRCC, 1962. *La Pathologie du Beton Arme* (1955 by H. Lossier), English translation by National Research Council of Canada (Tech. translation no. 1008). Ottawa, Ontario, Canada.

Petroski, H., 1985. *To Engineer Is Human,* St. Martin's Press, Inc., New York.

Petroski, H., 1994. *Design Paradigms: Case Histories of Error and Judgment in Engineering,* Cambridge University Press, New York.

RCPS, 1973. *Structural Failures: Modes, Causes, Responsibilities,* Research Council on Performance of Structures, American Society of Civil Engineers, New York.

Ross, S., 1984. *Construction Disasters: Design Failures, Causes, and Prevention,* Engineering News-Record, McGraw-Hill, Inc., New York.

Shepherd, R., and J. D. Frost, 1995. *Failures in Civil Engineering: Structural, Foundation and Geoenvironmental Case Studies,* Committee on Education, Technical Council on Forensic Engineering, American Society of Civil Engineers, New York.

2

NATURAL HAZARDS AND UNUSUAL LOADS: EFFECT ON THE BUILT ENVIRONMENT

Prior to a discussion of specific failure case studies, it is important to review the forces and agents that are destructive to the built environment. We often consider the natural environment to be peaceful, but in fact there are violent forces constantly at work in the natural environment. It is only through understanding the character of these forces, and with respect for their magnitude, that the designers and builders of structures are able to provide facilities safe for human habitation and use.

The natural environment is extremely complex and is constantly changing, involving a multitude of interrelated variables. The natural environment is not static. It is a dynamic system, one in which changes are occurring continually. These changes can be dramatic, even cataclysmic. Natural disasters such as earthquakes, floods, mud slides, tornados, wildfires, and hurricanes are examples of the destructive dynamics of the natural environment.

There is nothing like a violent natural phenomenon such as a typhoon or earthquake for bringing to view the errors, omissions, and shortcuts in construction performance. Every discontinuity, every weakness, every incompatibility of deformation becomes the focus of a structural failure, with immediate loss in total strength and reduction in available resistance.

Even well-built objects, constructed by human beings and placed in the environment, are temporary. This is because they are static objects with a limited capacity to adapt to sudden changes in the natural environment that surrounds them. Apart from sudden violent change, there is a natural cycle of aging and disintegration of all assemblies into their primary elements, which change under the forces of climate into different assemblies. Human actions may accelerate both the generation and degeneration cycles.

In his poem "Mending Wall," Robert Frost writes:

Something there is that doesn't love a wall,
That sends the frozen-ground-swell under it
And spills the upper boulders in the sun . . .
Something there is that doesn't love a wall,
That wants it down. . . .   (Frost 1915)

In this chapter we review briefly the natural hazards and unusual loads that may act on structures to bring about failure. The fact that failures do occur from time to time is easily explained when the cumulative effects of these destructive agents are considered. Indeed, other questions may arise: Why aren't failures more common? Why do most facilities survive their expected lifetime and provide satisfactory service for many years? The answers can be found only in the dedication of design and construction professionals throughout the industry, who worry a lot about failure and do their best to understand and mitigate the effects of these forces. Often, there is more than a little luck involved, as well.

2.1 GRAVITY

Most loads are directed vertically, because most load forces are the result of gravity. Even when other forces trigger a failure, such as seismic or wind events, gravity is involved in the ultimate collapse of the displaced structure.

Since gravity is always present and usually predictable, failures occurring under the influence of gravity alone, especially when loads are predicted by experience, result in extreme embarrassment for the professionals involved. Such was the case following the failure of the Hartford, Connecticut, Civic Center Coliseum roof under a modest snow load in 1978, and the failure of the Hyatt Regency Hotel pedestrian walkways in Kansas City in 1981 under the gravity load of people observing a dance. The public outcry resulting from such failures is understandable. The triggering loads were ordinary, predictable forces, well within established design guidelines and applicable codes.

Among environmental gravity loads, snow loads have been particularly troublesome. Design snow loads are based on experience. Sometimes weather patterns produce extraordinary conditions, with snow loads exceeding accepted standards of practice. It has also become apparent that the uniform snow-load distribution pattern assumed in the past is inadequate. The configuration of the structure and of adjacent structures can produce nonuniform load distributions from drifting snow.

In the winter of 1977–1978, more than 130 major roof collapses occurred in the northern United States. During the following winter, more than 200 roofs collapsed in the Chicago area under record snow loads. In January 1979, Chicago had 1194 mm (47 in.) of snow with no melting; there were over 100 reported roof failures in one week. Other northern counties in Wisconsin, Michigan, and all the New England states experienced similar problems due to record snowfalls in 1978 and 1979. These failures brought about the development of new standards, codes, and design guidelines for considering nonuniform or drift loads.

Figure 2.1 shows one of these failures, collapse of a new junior high school roof in Waterville, Maine. The collapse occurred on February 9, 1978, opening day for the new school (Zallen 1988). Wind and the configuration of the structure contributed to the formation of deep drifts that exceeded the capacity of the open-web roof joists.

With level roofs getting lighter and their spans getting longer, several cases of roof failure from water ponding during rainstorms should warn designers to take proper precautions. In most of these designs the roof drains are located at the columns, where there is no deflection of the framework. Dead weight

Figure 2.1   Failure of a new junior high school roof (1978) in Waterville, Maine, due to snow drift loading. (Courtesy of R. Zallen.)

of conventional sheet-metal decking, roofing, steel joists, and beam framing is seldom over 720 Pa (15 lbf/ft^2). Live loads anticipated by the design may vary from 1000 to 1900 Pa (20 to 40 lbf/ft^2). Normal dead-weight deflections can introduce ponds in each structural bay that do not drain. Sudden heavy rains add water that increases the deflection so rapidly that failure of a critical detail may occur, producing partial or total roof collapse.

In July 1967, parts of lightweight roof constructions failed in both Edison, New Jersey, and outside Boston when steel joists gave way during heavy rains. High-strength steel joists may experience large deflections and are more susceptible to such failures when the design is based on strength considerations only. Careful check for stiffness is required in the design and adequate provisions for camber, and positive drainage should be included. Scuppers should be provided for emergency drainage when water is trapped behind high parapets.

Failures can occur due to loss of stability under gravity loads. These failures are particularly dangerous, since they are sudden, giving little or no warning to occupants. For example, older wood-floor buildings with beams embedded into lime mortar brick walls are vulnerable to the twisting action of overloaded beams, which can buckle the wall and cause instantaneous collapse. Many ponding failures during rainstorms are sudden, involving loss of stability at critical locations, especially in steel-framed structures.

A number of technical articles on failure case studies illustrate the importance of providing stiffness in critical framing members and connections to avoid such a collapse. Stiffness of the beam–column connection is an essential link in the resistance to such failures, especially for conditions of negative bending moment in continuous beams. Where the beam is subjected to negative moment, the roof deck provides no lateral support to the compression flange (the bottom flange). If the design does not increase the size of the beam to accommodate this situation, or include a detail that provides lateral support, local buckling is very likely.

Total collapse of the roof of a department store in Bolivar, Tennessee, in July 1983 was caused by ponding and insufficient stiffness of a steel beam–column connection (Figure 2.2). The design drawings included a detail that would have provided support to the connection, but the detail was not executed in the field construction (Bell and Parker 1987). This failure is discussed further in Chapter 6, along with other cases involving stability problems in structural steel buildings.

Many of the failures presented in the remaining chapters of this book were caused by gravity loads. In some of these cases, design or construction defects were responsible for the collapse. Sometimes, however, the failures were simply the result of carelessness and negligence of the users. Figure 2.3, for example, shows the collapse of a bridge in Washington State under a load that greatly exceeded the posted capacity. Although some misuse of facilities is to be expected, anticipation of such gross negligence is not the responsibility of the designer.

Figure 2.2 Collapse of a department store roof in Bolivar, Tennessee, in 1983. Ponding of rainwater caused buckling of a beam-to-column connection. (Courtesy of Simpson, Gumpertz & Heger, Inc.)

2.2 SEISMIC EVENTS

Earthquakes are among the most terrifying natural hazards. The magnitude of natural forces associated with earthquakes, and our limited ability to predict their timing, location, and severity make them especially threatening to human life. Much has been learned from experience with earthquakes, and engineered buildings are now much better able to cope with the effects of ground shaking than they were just a few decades ago.

Each new earthquake produces new knowledge and confirms knowledge gained in the past. The September 1985 Mexico City earthquake can be seen as an immense and very costly laboratory. Lasting only 4 minutes, this earthquake destroyed over 250 buildings in the downtown area and killed over 10,000 people (Figure 2.4). The only positive effect of such an event is the knowledge that can be gained to mitigate the effects of future catastrophes of this magnitude.

There is nothing like a major earthquake to illustrate the limitations of human engineering in relation to the potential destructive power of the natural environment. Although a great deal of progress has been made in predicting the effects of earthquakes on structures, there is still much to be learned.

Henry Degenkolb, a structural engineer who practiced in California and contributed much to advances in earthquake engineering, wrote and lectured

Figure 2.3 Negligent overloading of a county bridge in Washington State. (Courtesy of Washington State Department of Transportation.)

Figure 2.4 Failures of nonductile reinforced concrete buildings caused many of the deaths and injuries in the 1985 Mexico City earthquake. (Courtesy of the National Geophysical Data Center.)

extensively about the important contributions of trial-and-error experience to engineering knowledge in general, and to earthquake engineering in particular. The idea that a builder could predict ahead of time, through mathematical theory, what would happen to a beam under load is a very recent development in the history of construction. Before such theories were available, designers had to base their designs on the inventory of past experience. Even now, when the input data are uncertain, as for seismic events, the inventory of experience provided by past earthquakes is at least as reliable as untested theory. Degenkolb noted that the ultimate test of a mathematical design theory is whether or not it agrees with the trial-and-error solution (Degenkolb 1980).

Engineering judgment is required when dealing with uncertainties, and mathematical theories must be tempered with empirical evidence. Earthquakes are humbling events. James Amrhein, structural engineer and executive director of the Masonry Institute of America, has given the following relevant definition of *structural engineering:*

> Structural Engineering is the art and science of molding Materials we do not fully understand;
>
> into Shapes we cannot precisely analyze; to resist Forces we cannot accurately predict;
>
> all in such a way that the society at large is given no reason to suspect the extent of our ignorance (Amrhein 1985).

The science of geology has advanced our capability to predict the locations where most seismic activity is likely to take place. The development of the theory of plate tectonics has given us a better understanding of the geologic origin of earthquakes. However, we also know from experience that no region is completely free from the threat of earthquakes. In fact, very large seismic events have occurred far from identified plate boundaries. Examples include the 1886 Charleston, South Carolina, earthquake and a series of three earthquakes in New Madrid, Missouri, in 1811–1812. A repeat of the 1811–1812 Missouri earthquakes (estimated Richter magnitude 8.5) could cause $50 billion damage over a 500,000-km^2 (200,000-mi^2) area.

The damaging effects of earthquakes can often be mitigated through land-use planning. Certain sites, by virtue of their location or their specific geology, are particularly susceptible to seismic effects. For the purpose of this discussion, seismic effects can be classified in four categories: (1) ground rupture, (2) ground failure, (3) tsunami, and (4) ground shaking. Of these four, only ground-shaking effects can be reduced significantly through high-quality structural engineering. Land-use planning is a more effective tool for addressing the other three categories.

Ground rupture occurs in the fault zone. Fissures and permanent relative displacements are likely in these regions. If the zone of expected surface faulting can be identified, this region should simply be avoided for the placement of constructed facilities. The 1906 San Francisco earthquake, for example, produced horizontal relative displacements along the San Andreas fault in excess of 6 m (20 ft). Clearly, it is not possible to design structural systems to accommodate such displacements. Structural engineering, no matter how sophisticated, cannot provide safety in the fault zone where ground rupture is likely.

Ground failure can be induced in unstable soils by ground shaking. Such failure phenomena include landslides, large-scale settlements of loosely compacted soils, and liquefaction. These effects can occur far from the epicenter of the earthquake. It is possible to identify sites that are susceptible to ground failure (Figure 2.5). The best strategy for addressing these problems is to avoid development on the vulnerable sites, again through wise land-use planning ordinances. Ground modification techniques may be applied, but these are costly and require site-specific solutions. Liquefaction is a particularly difficult problem. When the soil liquefies it loses its shear strength, and a shallow foundation, no matter how well it is constructed, can no longer provide adequate stability. Only those sites where loosely compacted soils are found in conjunction with high groundwater tables are susceptible to liquefaction, but a great deal of development has occurred in the past on such sites, and substantial damage is expected in the future.

Figure 2.5 The 1964 Anchorage, Alaska, earthquake produced landslides in the Turnagain Heights area, a condition that had been predicted by geologists. (Courtesy of the National Oceanic and Atmospheric Administration.)

The seismic sea wave, or *tsunami,* is a threat to coastal communities everywhere. In the United States, the west coast and Hawaii are particularly vulnerable. Tsunamis can be very damaging, and their effects cannot be resisted by structural engineering and construction quality. Hawaii has been zoned for tsunami potential, but the best defense is simply through warning and evacuation. Damage to the built environment due to seismic sea waves will continue to occur, but the dramatic loss of life historically associated with these events has been greatly reduced with the development of prediction and warning capability. The most destructive tsunami recorded accompanied the eruption of Krakatoa in 1883. Waves over 40 m (130 ft) high swept over the island of Java, killing more than 36,000 people. Entire villages disappeared. Japanese art depicts a profound respect for the damage potential of the dynamic seacoast. In 1896, a tsunami killed 27,000 Japanese and destroyed 10,000 houses. A seismic sea wave can cause damage far from the epicenter of the earthquake. The 1964 Anchorage, Alaska, earthquake produced a tsunami that flooded 30 city blocks in Crescent City, California, over 2500 km (1500 mi) from Anchorage.

Most structural engineering solutions focus on mitigation of the effects of *ground shaking.* Fortunately, there is much that can be done to improve the performance of constructed facilities when subjected to earthquake-induced vibrations. Since the forces exerted on the structure are related to its mass, we know that lightweight materials will reduce the damaging effects of the earthquake. We have learned that architectural configuration plays an important role in earthquake resistance. Regular building floor plans, with rigid elements distributed symmetrically, help to reduce the potential for torsional stresses. A flexible structural system provides the ability to store energy temporarily through deformations and prevents total collapse in a major earthquake. Conversely, deformations must be limited in moderate earthquakes to prevent damage to nonstructural components. Problems with costly nonstructural component damage in moderate earthquakes have received considerable attention since the 1989 Loma Prieta and 1994 Northridge earthquakes in California.

We know from experience that structural discontinuities, such as setbacks and abrupt changes in stiffness at a floor level, will produce damaging stress concentrations. We have learned that the integrity of connections is critical and that systems that provide structural redundancy have a greater opportunity for survival. We know that falling nonstructural components (precast concrete panels, parapets, and masonry curtain walls) are potentially as hazardous to human life as collapse of major structural components, and that these components must be designed and detailed with care (Lagorio 1990).

These principles have been derived through experience—learning from failure and success. They are reflected in modern building codes, and these evolving codes have been responsible for significant reduction in property loss, deaths, and injuries where they have been applied.

One of the great misfortunes of the earthquake problem, however, is that so many large-scale seismic events occur in regions of the world that are least prepared to deal with them: socially, economically, and technically. Although we now know the value of strong, ductile, lightweight, reliable materials and the importance of engineering and construction quality, the appropriate materials, economic resources, and engineering expertise simply do not exist in many locations. The large loss of life associated with earthquakes in many regions, for example, is clearly related to the limited availability of construction materials. Where heavy, nonductile, unreinforced masonry is the only material available for residential construction, the effects of even moderate earthquakes can be devastating.

Some of the more costly earthquakes, as measured by loss of life, are:

Earthquake	Richter Magnitude (Approximate)	Number of Deaths
1908 Italy	7.5	75,000
1915 Italy	7.0	30,000
1920 China	8.6	180,000
1923 Japan	8.3	143,000
1927 China	8.3	200,000
1932 China	7.6	70,000
1935 India	7.5	30,000
1939 Chile	8.3	30,000
1970 Peru	7.8	67,000
1976 China	7.8	250,000
1990 Iran	7.7	50,000

In most of these earthquakes, people were crushed by heavy masonry (stone or clay) and poorly detailed reinforced concrete buildings. Other recent damaging earthquakes have occurred throughout the world. The loss of life is more a reflection of building materials, the quality of construction, and the degree of preparedness than the magnitude of the geologic event. In 1960, an earthquake in Morocco (Richter 5.6) killed 12,000 people. The 1980 earthquake in Italy (Richter 6.8) killed 3000 and caused $24 billion in damage. In 1982 a Richter magnitude 6.0 earthquake killed 2000 residents of Yemen. Over 1000 people lost their lives in the El Salvador earthquake of 1986 (Richter 5.4). In 1988 the Armenian earthquake (Richter 6.9) killed 25,000 people and caused $16 billion damage to marginally engineered and poorly constructed unreinforced masonry and precast concrete buildings. Most of the 50,000 Iranians killed in the 1990 earthquake (Richter 7.7) were crushed by unreinforced masonry buildings and adobe residential construction with heavy mud roofs.

Mexico City has suffered repeatedly from the effects of earthquakes. In 1957 an earthquake damaged over 1000 buildings and killed 125 people. The earthquake vibrations were greatly magnified by the soft soils in the central city area. Tall buildings with periods up to 2 seconds experienced resonance. On September 19, 1985, a much more costly but similar event killed over 10,000 people and caused $5 billion damage. The 1985 earthquake registered 8.1 on the Richter scale, the strongest earthquake ever to have its characteristics recorded in detail. This was also the first set of strong motion records on soft soil. Larger accelerations were recorded than were previously thought possible on soft soils. This very large shock was accompanied by severe amplification of vibrations in the central city area, due to the poor local soil conditions. Over 800 buildings collapsed.

Lessons are still being learned from the 1985 Mexico City earthquake. Recommendations have included greater attention to provision of symmetry in architectural design; there were a number of failures due to torsion. The need for careful detailing of reinforcement to ensure confinement of damaged concrete was also evident, as was the need for connection integrity in general. It was recognized that detailing alone is not sufficient; the designer must also inspect the construction to verify that the details are in fact executed during construction. New guidelines for separation between structures were developed, as there were a number of buildings that experienced failures due to pounding. These failures were particularly dramatic when the floors of adjacent buildings were at different elevations, so that stiff floors repeatedly impacted on columns of the neighboring buildings at midheight.

These lessons came at great cost. The Mexico City event was accompanied by profound hardship and suffering among the 15 million residents. It is difficult to imagine the burden of such a catastrophe on a population that is already struggling each day with inadequate shelter, food, and basic necessities that people in the industrialized world take for granted.

None of the earthquakes in the United States have yet been marked by this degree of human suffering. One reason is the use of timber light-frame construction in residential construction. Nevertheless, a large seismic event in a populated region of the United States is potentially a catastrophe of enormous economic and social impact. The sudden loss potential for a single earthquake in California has been estimated at over $100 billion, with over 20,000 lives lost. And California is not alone; as many as 70 million people in 39 states face significant risk from earthquakes (ACIDNHR 1989).

The recent events in Northridge, California (1994), and Kobe, Japan (1995), showed that the economically developed and technically advanced nations are not immune from severe damage and human suffering caused by earthquakes. Over 5400 people perished in Kobe, a region that was not as well prepared as assumed before the earthquake.

It has been noted that the model building codes have evolved through trial-and-error experience. There is perhaps no clearer connection between failures and the development of codes and standards than is seen in the seismic codes. The *Uniform Building Code* is recognized as one of the best seismic design codes in the world, and its provisions have evolved primarily as a result of studying empirical evidence at the sites of major earthquakes throughout the world. The prodigious efforts of members of the Structural Engineers Association of California (SEAOC) and others have brought improved safety into modern engineered facilities and have most certainly saved many lives.

One can easily trace the development of seismic building code provisions in relation to significant earthquakes in the United States and elsewhere. Each major earthquake has been followed by refinement of codes and standards as knowledge has evolved through trial-and-error experience (Ross 1984).

The 1906 San Francisco earthquake and the Tokyo earthquake of 1923 led to increased awareness of the damage potential of earthquakes and to the understanding that it is the breaking of rocks that releases accumulated strain energy in the form of seismic vibrations. Engineering societies began to document their observations and to seek the incorporation of their findings into legislation. In 1925, an earthquake in Santa Barbara, California, resulted in adoption of the first earthquake design codes. These provisions were aimed at improving the performance of certain types of construction, mainly brick masonry and concrete.

In 1933, some of the new ideas were tested in the Long Beach, California, earthquake and proved that improvements in construction standards could reduce earthquake losses. The failure of several schools in the 1933 earthquake was especially alarming. Fortunately, the earthquake occurred in the evening when the school buildings were not occupied. Again, many of the failures were of unreinforced masonry bearing wall buildings with timber joists or roof trusses. A number of poorly detailed reinforced concrete buildings also experienced failure, and improvements were quickly incorporated into the California Seismic Code, a state law that served as a model for earthquake codes around the world.

Each earthquake since Long Beach has been studied and has produced further refinements in design and construction practices. This process continues to be reflected in improvements to the seismic provisions in the *Uniform Building Code*. The 1964 Anchorage, Alaska, earthquake (Richter 8.4) lasted 3 minutes and killed 114. This earthquake was accompanied by a large tsunami that caused substantial damage to California coastal cities. The lesson learned in this event was that high-quality construction, combined with good structural design and stable sites, could allow buildings to survive intense and prolonged ground shaking. Several failures of concrete structures illustrated the need for attention to connection details and confinement of reinforcement. The importance of land-use planning was evident, especially in the large number of failures associated with landsliding and ground settlements.

Insofar as code development is concerned, one of the most influential seismic events was the 1971 San Fernando, California, earthquake. Although this was not an especially significant geologic event (only 6.6 on the Richter scale) it provided a great deal of evidence on structural performance and generated a political climate that supported adoption of new design standards. The earthquake killed 58 and caused damage of $1 billion, much of this to highway structures. Important lessons were learned about the detailing of reinforced concrete structures, both in bridges and in buildings. New provisions for enhancing the ductility and overall stability of structures emerged from this earthquake. Since substantial damage was done to several hospitals and other emergency facilities, the code was revised to reflect the importance of certain building occupancies. The failure of the Olive View medical center, a brand-new facility, brought about widespread recognition of the importance of architectural configuration (Figure 2.6).

In 1983, an earthquake in Coalinga, California (Richter 6.5), destroyed over 40 older unreinforced masonry buildings and cost $25 million. Collapse of an unreinforced parapet wall killed two children in the 1983 Challis, Idaho, earthquake (Richter 7.3). These events provided little new knowledge but confirmed the need for upgrading the seismic performance of existing buildings.

In 1987 an earthquake occurred in the Los Angeles area. Known as the Whittier earthquake, this event had a Richter value of 6.1. It lasted only 5 seconds, but six people were killed and there was $358 million damage to property. This earthquake provided the first opportunity to test some of the new concepts for upgrading existing masonry buildings. Tilt-up concrete structures were also tested for the first time, and much was learned about the critical need for attention to connection details in precast concrete construction.

Figure 2.6 Olive View medical center, a new facility, was destroyed by the 1971 San Fernando, California, earthquake. Inadequate reinforced concrete connections and an irregular architectural configuration contributed to the collapse. (Courtesy of the Federal Emergency Management Agency.)

On October 17, 1989, the Loma Prieta earthquake (Richter 7.1) caused $6 billion damage to facilities in the San Francisco–Oakland area, killed 62 people, and injured 3800. Perhaps the most dramatic failure was the collapse of a portion of the elevated Interstate 880 Nimitz Freeway (Figure 2.7). This failure illustrated some of the same problems with deficient reinforced concrete detailing that were observed following the 1971 San Fernando event. The I-880 collapse made evident the painful realities of economic limitations: There simply is not enough public money to replace the inventory of existing structures that are known to be deficient according to current design standards. The Loma Prieta earthquake also illustrated the dramatic effect that local site conditions can have on earthquake performance. Soft soil sites experienced significant amplifications, and major failures were associated with liquefaction.

Several large earthquakes occurred in California in 1992, to remind residents of that state that they live in earthquake country. Two of these were in northern California in April 1992 (Richter 6.9 and 6.5). There were no deaths, but damage was $50 million. Two southern California earthquakes killed one person and did $92 million damage in June 1992 (Richter 7.1 and 6.5).

The January 17, 1994, earthquake in Northridge, California (Richter 6.8), caused 61 deaths and over 9000 injuries. Approximately 112,000 structures were damaged, and the urban infrastructure was also affected severely. The Northridge earthquake proved to be the most costly natural disaster ever

Figure 2.7 The reinforced concrete Cypress Street Viaduct collapsed in the 1989 Loma Prieta, California, earthquake. The structure lacked several important seismic details that have been incorporated into recent building codes, focusing attention on the need to upgrade existing facilities for seismic performance. (Courtesy of the Portland Cement Association.)

experienced in the United States, with estimated losses of $30 billion. Older nonductile concrete bridges and buildings failed, as expected. Some poorly detailed precast concrete buildings also failed dramatically, while other well-designed precast and cast-in-place concrete facilities performed admirably. The alarming costs associated with nonstructural component damage in buildings were especially disconcerting, as were the surprising brittle fractures in moment-resisting steel frames. All of these issues are currently under study, and significant revisions are expected in the seismic provisions of the building codes. These lessons are discussed further in Chapters 6, 7, 8, and 10.

The Northridge earthquake and the Kobe, Japan, event exactly one year later served to underline the vulnerability of the modern urban environment to earthquake damage. Clearly, there is a need for preparation for disaster, even in cities where earthquake probability is relatively low. The implications for Seattle, Boston, and even New York City are profound (Carper 1995).

At the present time, efforts are under way to bring greater uniformity of seismic design into the various model building codes used throughout the United States. These national standards are the product of a decade of work by the Building Seismic Safety Council (BSSC) under the National Earthquake Hazards Reduction Program (NEHRP).

In the past, building codes have been directed primarily at new construction. It is recognized, however, that there is a substantial seismic risk associated with existing buildings, bridges, and other facilities that do not conform with current design standards. Much work is now being directed at the rehabilitation of unreinforced masonry buildings and nonductile concrete frame structures. The magnitude of the problem is quite impressive, as many of the deficient reinforced concrete structures were constructed less than 20 years ago. The inventory of unreinforced masonry buildings is vast; over 11,000 such buildings exist in San Francisco alone. Many of these marginally engineered structures are currently serving a valuable function as low-cost housing, so the demolition of this building stock would bring about severe social problems. Some of these buildings are historically significant, and the upgrading of such structures requires a great deal of creativity. In many cases, the dangerous components (parapets, balconies, and other ornamental features) are the very reasons for the aesthetic value placed on the building.

The retrofit of reinforced concrete bridge connections and columns is also expected to be an important area of construction activity as public funds become available. Retrofit of existing structures to improve seismic resistance requires unique, site-specific solutions for each individual structure. The 1989 Loma Prieta, 1994 Northridge, and 1995 Kobe events proved that such efforts are effective and needed desperately.

Emerging concepts of base isolation, where the structure is mechanically separated from the impact of seismic vibrations, show promise for retrofit of existing structures as well as for new construction. These concepts were pioneered in Japan and New Zealand. The first new U.S. building to rest on base isolators is the Foothills Communities Law and Justice Center in Rancho Cucamonga, California.

The 1894 historic landmark Salt Lake City and County Building in Salt Lake City, Utah, was retrofitted in 1989 by excavating for the installation of 447 steel and rubber bearings under the building foundations. The forces exerted by an earthquake will be reduced by these isolators, mitigating some of the problems with the irregular configuration of the structure. Base isolation also shows some promise for retrofitting the column–girder connections in box-girder concrete bridges with spans of less than 90 m (300 ft). These solutions are also site specific, and application of base isolation systems may not be appropriate for tall structures, where overturning is a problem, or where large displacements are expected.

2.3 EXTREME WINDS

Windstorms cause 350 deaths and $4 billion to $5 billion damage to the built environment annually in the United States. Although the loss of life has been reduced by warning and evacuation planning, economic losses related to wind are escalating each year. Major damage is caused by the extreme winds associated with hurricanes and tornados. In addition, turbulent wind patterns in the dense urban environment, or in the vicinity of large buildings of complex configuration, hinder the use of public urban spaces and result in costly damage to the curtain walls of buildings.

Structural damage resulting from storms of hurricane, tornado, or typhoon intensity is to be expected, although these forces can be resisted by proper and well-known design procedures. Far too much damage to buildings and their components occurs due to winds that are normal or predictable. Resistance to such forces requires appreciation of the fact that wind forces are not uniform horizontal pressures. Uplift, suction, and torsional effects have caused most of the damage, and most of the damage is due to poor-quality construction. The most common source of failure is the inadequacy of connections in nonengineered or marginally engineered light-frame construction.

Six hurricanes hit the Atlantic and Gulf coast states in 1985, costing a total of $4 billion. Hurricane Hugo (September 1989) killed 27 people in the United States and caused $10 billion damage. This was the highest economic cost of any U.S. hurricane in history to that point. Hurricane Andrew (August 1992) caused over $20 billion damage and was followed one month later by Hurricane Iniki, which caused $1 billion damage in Hawaii.

One of the reasons for escalating wind damage costs is that the regions of the United States most susceptible to the potential for hurricanes (the Gulf coast and the southeastern coastal states) are the regions that have developed most rapidly in the past few decades. In many cases this development has been right on the seacoast. In the two decades preceding Hurricane Andrew, thousands of suburban homes and condominiums were built in Florida. Lessons learned from past great hurricanes were forgotten by many in the rush to provide housing for the influx in population. As a result, over 60,000 homes were destroyed by Hurricane Andrew in Florida alone, and over 250,000 people were left homeless. The $20 billion damage experienced by Florida and Louisiana made this the most costly single natural disaster ever experienced in the United States to that date. (This loss was exceeded in the 1994 Northridge, California, earthquake.)

Florida residents once had a greater consciousness regarding hurricanes. In 1926 a storm killed more than 200 and damaged or destroyed every building in Dade County. In 1935, 408 people were killed in the Florida Keys. From 1944 through 1950 the state experienced an average of one hurricane each year. This regular experience caused the residents to prepare for the hurricane season and construction techniques were refined.

However, between 1965 and 1992, there was only one major hurricane in Florida. This period was one of dramatic growth in the state. The population increased sixfold between 1950 and 1992. Many more people live at risk, but at the time of Hurricane Andrew it was estimated that only 10 percent of the people living in Florida had any personal experience with hurricanes. Fortunately, prediction and warning skills have improved. The minimal loss of life (only 38 deaths) is due entirely to accurate prediction and timely evacuation. Nevertheless, the threat to lives is very real, and significant property damage is likely to occur due to coastal storms in the southeastern and eastern United States. Some meteorologists predict that the cycle of more frequent severe storms is returning, with associated risks for over 44 million people living in coastal communities from Maine to Texas. This prediction gained credibility with the extremely active 1995 hurricane season: Many coastal regions are unprepared for a disaster the scale of Hurricane Andrew and will be simply impossible to evacuate completely, given little time and congested transportation corridors.

Tornados are another source of costly damage. These small-scale, extremely violent winds are among the least predictable of all natural phenomena, despite much research conducted by the Nuclear Regulatory Commission and others. Wind speeds may reach a velocity of 480 km/h (300 mi/h) in the most severe tornados. Although it is possible to design critical facilities for such wind velocities, it is not cost-effective to build most structures to resist tornados in their entirety. However, certain portions of residential-scale structures may be economically designed to provide a reasonable degree of safety to the occupants. When a severe tornado hits an urban location, the results can be quite dramatic (Figure 2.8).

Some of the more interesting and instructive wind-related failures have involved bridges. The collapse of the Tay Bridge in 1879 on the railroad line between Dundee and Edinburgh took 75 lives. The bridge was probably designed to resist a horizontal force of 575 Pa (12 lbf/ft^2) on all exposed surfaces, with no allowance for the exposed surface areas of the railroad coaches. This figure came from recommendations by John Smeaton in 1759

Figure 2.8 Damage to downtown Kalamazoo, Michigan, caused by a 1980 tornado. (Courtesy of D. Jennings.)

as the pressure of a storm or "tempest." Following the Tay Bridge failure, design wind loads were increased greatly.

Perhaps the most dramatic bridge failure from wind action was the 1940 collapse of the Tacoma Narrows suspension span in Washington State. The Tacoma Narrows bridge was the third longest span in the world at the time of its construction. During its four months of use, the bridge became known as "Galloping Gertie," due to its large oscillating deformations. A photographic record of the torsional oscillations of this bridge made by F. B. Farquharson did more to prove the necessity for aerodynamic investigation of structures than all the theoretical reports written previously. The film of this collapse is still used in college physics courses throughout the world to illustrate principles of wave motions. Both the Tay Bridge and the Tacoma Narrows failures are reviewed in Chapter 4.

The effects of extreme winds on structures are quite well established. Evolving codes and standards now recognize that pressures are not distributed uniformly. Concentrations of pressure exist at the leading edges of walls and roofs, and special considerations must be given to locations where uplift or suction may combine with internal pressure to create critical conditions. These conditions occur at eaves or overhangs and at the leeward wall when openings exist in the windward wall. Uplift on long-span, lightweight roofs often causes failure of entire roof systems, following the failure of portions of the windward wall, which allows internal pressure to add to the uplift.

Failure of roofing and other finish materials is common due to connections that are incapable of developing sufficient resistance to negative pressures. In 1958 the three-year-old circular coliseum auditorium in Charlotte, North Carolina, lost half the aluminum roofing cover when a wind with 120-km/h (75-mi/h) gusts passed over the area. The 20-gauge metal roofing was nailed to expansion sockets set in the fiberboard insulation of the concrete dome roof, but there was insufficient uplift resistance.

Failures to nonstructural curtain walls are also common, and these can be particularly life threatening. Since such walls are often considered architectural elements rather than structural components, there is some confusion over responsibility for their engineering design. While collapse of a curtain wall may not precipitate collapse of the entire building, the falling materials involved may themselves be extremely hazardous. In 1989 a small freestanding masonry curtain wall at a New York elementary school cafeteria failed in a windstorm. The wall was only 6 m (20 ft) wide and 3.8 m (12.7 ft) high, but when it fell, nine children were killed and a dozen others were injured (Figure 2.9).

Certain types of structures are more susceptible to wind failures because of their lightweight character. These include pneumatic (air-supported) struc-

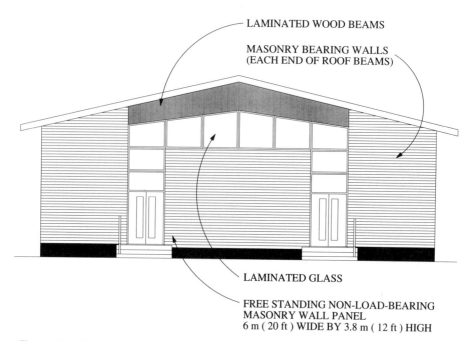

Figure 2.9 Nine elementary school children were killed in New York when the freestanding masonry curtain wall of this cafeteria collapsed in a 1989 windstorm.

tures, cable structures, tents, and other long-span, lightweight forms of roof construction.

Wind forces cause proportionately more damage to structures that are under construction than to completed structures. Incomplete structures without diaphragms, shear walls, or permanent connections are particularly vulnerable to sudden failure. Construction site safety requires care in bracing incomplete structures to guard against this type of failure. During construction, concrete block bearing walls, with the roof bracing not yet installed and with no temporary support, present large surface areas and may fail under wind loads far below those used for design. Most people will be surprised to learn that small amounts of wind load can cause failure of a concrete block wall 6 m (20 ft) high. Steel frames and precast concrete buildings under construction are vulnerable to collapse in moderate winds when lateral bracing is insufficient and the floors and walls are not yet present to provide mass and lateral stability. Such construction-related wind failures are discussed in Chapter 11.

In addition to the damage done by extreme winds, much damage results from turbulent wind patterns in dense urban environments around tall buildings or buildings of unusual configuration. Curtain wall components may suffer costly damage from the fluctuating positive and negative pressures associated with such turbulence, and urban public spaces may be rendered unusable. A boundary-layer wind tunnel is an extremely useful design tool to predict wind patterns, although future development on adjacent building sites may affect wind patterns significantly. Case studies involving curtain wall problems related to wind are included in Chapter 10.

2.4 FLOOD

There are approximately 5 million kilometers (3 million miles) of streams in the contiguous United States and about 6 percent of the land area in the contiguous United States is subject to potential flooding. By the year 2000, the annual loss due to floods is expected to be about $5 billion. Increasing costs associated with flooding are due to the continuing urbanization of agricultural or otherwise undeveloped riparian land (FEMA 1986). Flooding was especially severe in 1993 in the midwestern and central regions of the United States, with damage along the Mississippi and Missouri rivers approaching $10 billion.

Three characteristics that have major effects on flooding potential are elevation, drainage, and topography. Floods and hurricanes often come together in coastal regions, and the effects are difficult to separate for statistical studies.

Although most federal flood control projects in the United States have helped to mitigate damage in frequent smaller floods, the risk of catastrophic damage in large floods has actually been increased by some projects. The building of dams has encouraged development on the floodplains below the dams. These regions are still located in floodplains, and overtopping or failure of the protective dams can have disastrous results. Some of the failures discussed in Chapter 4 show the potentially catastrophic consequences of dam incidents.

Besides property losses, floods have been responsible for major loss of life. The famous Johnstown, Pennsylvania, flood due to the failure of an earthfill dam killed over 2200 people in 1889. Over 2600 Italians were killed in 1963 when the Vaiont Dam was overtopped by a landslide-induced wave. The development of modern prediction, warning, and evacuation capabilities has greatly reduced the threat to human life of coastal and riverine floods. However, occasional flash floods associated with mountain rainstorms or dam failures continue to claim human lives each year.

Current flood protection projects under way include a massive multiple-dam and channel improvement project in southern California in the Santa Ana River valley. When completed by the year 2000, the project will provide protection to more than 3 million people now living under one of the most critical flooding threats in the United States. The Santa Ana River has flooded 14 times this century. The most severe flood was in 1938, before much development had occurred. Damage was confined primarily to agricultural land. A similar flood today could kill 3000 people and result in over $18 billion damage to property (ASCE 1991).

Floods can be caused by human error. On April 13, 1992, 1.5 million cubic meters (350 million gallons) of water poured into scores of buildings and underground parking garages in Chicago. The Chicago River had broken through a 6-m (20-ft) breach into the network of 90-year-old freight tunnels beneath the downtown area. The damage was caused by a pile replacement project at the Kinzie Street Bridge. Piles had been driven too close, or directly into, the tunnel located 6 m (20 ft) below the riverbed.

The damage to the tunnel wall was done six months before the flood, but lagging repair efforts permitted the breach to occur. Property damage was estimated at more than $1 billion in the downtown Loop area. Over 300 buildings were affected. Some buildings had as much as 11 m (35 ft) of water in their basements. Fish were reported swimming in the lobbies of the City Hall and the State of Illinois Building. The flood was followed by three weeks of furious plugging and pumping efforts by the U.S. Army Corps of Engineers, Chicago city staff, and private contractors. The cleanup effort alone cost over $37 million.

2.5 FIRE

While earthquakes represent the greatest potential for sudden loss in a single catastrophe, fire is by far the costliest natural hazard each year. The annual damage to facilities and the yearly loss of human life related to fire are more severe than for any other hazard. The United States has a poor fire safety record compared to other nations, despite efforts to make improvements. We have grown accustomed to these losses and somewhat complacent about them, just as we have come to accept the large number of automobile accident deaths each year. Fire losses in the United States can be reduced, however, and the planners, designers, and constructors of facilities can make a difference.

Public fire departments in the United States respond to over 2 million fires annually, including about 650,000 fires in structures. Approximately 75 percent of structure fires occur in residential construction. Every 15 seconds a fire department responds to a fire somewhere in the nation. Fires occur in structures at the rate of one every 49 seconds; a residential fire occurs every 66 seconds. About 5000 people are killed in fires each year in the United States, 78 percent of these in homes. Nationwide, there is a fire death every two hours.

The magnitude of these statistics provided by the National Fire Protection Association is difficult to comprehend (Karter 1992). However, they should be compared with statistics from 1971, when twice as many people died due to fire. The reduction in deaths in the last 20 years is largely the result of improved prevention and containment and better firefighting techniques and equipment. The increasing use of fire warning systems, including smoke and heat detectors in residential construction, has contributed to a reduction in the annual loss of life, as has the mandated use of sprinklers and smoke control systems in commercial construction.

In 1991, about $9.5 billion in property damage occurred as the result of fire. Of this, $8.3 billion, 88 percent of the total, was from structure fires. This statistic includes the Oakland, California, wildfire of October 1991, which killed 25 people and caused $1.6 billion in property damage, making it the single most costly natural disaster in the United States during 1991. The Oakland fire destroyed 2449 single-family residences, 437 apartments and condominiums, and other structures. This fire demonstrated the role land-use planning can contribute to reducing fire losses. The congested, narrow, and circuitous streets that served the exclusive residential area proved to be an insurmountable obstacle to the movement of firefighting equipment and hindered evacuation of the residents.

In 1910, the Chamber of Commerce Building in Cincinnati, Ohio, burned with an intense fire. The steel roof trusses deformed and fell to the cellar, collapsing all floors. There are many similar examples of dramatic failures in unprotected structural steel buildings that occurred prior to the widespread recognition of the need to provide adequate fire protection to structural steel. In contrast, in 1946, the 22-story La Salle Hotel in Chicago lost all the wood trim and fixtures, although the fire-protected steel frame came through without damage. That same year the Winecoff Hotel in Atlanta, Georgia, a 15-story steel fire-protected building with one open unprotected stairway, experienced similar loss of nonstructural materials in a fire, with minimal damage to the structural steel.

In 1965, the top floor of the Chicago Civic Center building was gutted by fire. Two years later, just before the opening of the National Houseware Show, the McCormick Place Chicago Exhibition Hall had a fire that reduced to ashes displays valued at $100 million, collapsed a 60-m (200-ft) section of precast concrete outer wall, and destroyed the entire structural steel roof framing covering 32,000 m² (8 acres) of floor area. The facility, insured for almost $30 million, was destroyed entirely within an hour. The applicable building code permitted roof steel located 6 m (20 ft) or more above a "fireproof floor" to be exempt from fire protection. Such large losses, together with the awareness of what could have happened if the fire had come when thousands of people were inside such a large facility, usually bring about a reevaluation of fire protection requirements. Criteria for the permissible volume of combustible materials within any one enclosure were also called into question following the McCormick Place fire. Since the contents in the exhibit did not differ much from those in any large wholesale product warehouse or retail sales space, and the large capacity available with the Chicago firefighting equipment could not cope with the fire, a serious question arises as to whether there will ever exist such a thing as "fireproof construction."

Fire protection of structural members is important. Such protection provides structural integrity during a fire for the safety of occupants and firefighters. Recently, however, greater importance is being placed on critical protection for *people* during the early stages of a fire, and for those elements that will facilitate evacuation, such as provision of adequate exits, smoke containment, and effective warning systems. Crowded facilities with poorly located, deficiently maintained, and insufficiently marked exits are disastrous in an emergency.

In modern fire codes, ensuring the safety of people takes precedence over protecting buildings and their contents. The lessons leading up to these codes have been costly, and carelessness continues to circumvent the intent of life safety provisions in the codes. A brief review of some of the more important historic fires is given here, and there are many other lessons that have contributed to our understanding of the effect that burning contents can have on the occupants of buildings and on "fire-resistant" construction itself. The rapid development of smoke and the interaction of toxic products of combustion associated with modern synthetic materials are only now beginning to be understood. Further research into the biological effects of combustion of the plastic materials used in furnishings and finishes in buildings is necessary. When assessing the threat of fire to human life, the contents of buildings need at least as much consideration as the construction materials.

In 1905, Chicago's Iroquois Theater burned, killing 602 people when the stage curtain ignited. Unmarked and blocked exits contributed to the disaster. The Triangle Shirt Waist Company in New York burned six years later, with 145 employees trapped and killed behind locked exits.

In 1942, 492 people died in the Boston Coconut Grove nightclub fire due to inadequate exits. The Beverly Hills Supper Club fire in Southgate, Kentucky, in May 1977 killed 165 people. The building had a convoluted plan that had evolved through many additions and alterations. Combustible interior finishes combined with poorly marked and blocked exits to trap occupants in the intense fire.

Stauffer's Inn in Harrison, New York, just north of New York City, burned in December 1980, killing 26 people. In March 1990, 87 people died in a New York City social club fire when an arsonist set a blaze near a main exit stair.

These multiple-death fires are reported in the national news media. It should be noted, however, that far more people are killed and far more property is destroyed in the large number of small fires that occur each year in poorly constructed and marginally maintained light-frame residential buildings.

Older buildings require special consideration and retrofit measures can be costly. However, the record of fires in older buildings is sufficient to show that retrofit efforts are justified. In December 1958, Our Lady of the Angels School in Chicago burned, killing 90 children and three nuns in a one-hour fire. The school had been constructed in 1910 and remodeled in 1951. Flames and dense smoke moved rapidly through open corridors and an open stairway that acted as a flue. There were no sprinklers. Rubbish piled under wooden stairs ignited to start the fire. Ironically, the building had passed an inspection only two months prior to the fire.

In the past two decades, the high-rise fire has been recognized as a serious problem, and new theories about containment and rescue have been developed specifically for tall buildings. The potential for catastrophe was first realized in the November 21, 1980 fire in the MGM Grand Hotel in Las Vegas, Nevada. Eighty-four persons were killed and 500 injured, principally because of the rapid spread of smoke throughout the building; at least 70 were killed by asphyxiation. The fire started in the seven-year-old 26-story building due to improperly installed wiring. There were no sprinklers and no smoke alarms, both now required by code. The manual alarm system failed. Many code violations were discovered in the subsequent investigation, mostly penetrations in walls and floors that contributed to rapid smoke spread. This fire developed very rapidly, consuming the 19,000-m^2 (200,000-ft^2) gambling area in 10 minutes.

Other fires in high-rise buildings have occurred, both before and since the MGM Hotel fire, emphasizing the unique fire hazards associated with tall buildings. Five floors of the 110-story World Trade Center in New York City were damaged in February 1975 due to improperly installed wiring. In June 1980, a fire on the sixteenth floor of a 42-story Park Avenue office building in New York trapped many occupants above that floor. There were only six injuries and no deaths, a demonstration of the value of compartmentalization. However, if a sprinkler system had been in the building, it might have detected and stopped the blaze. It took the all-out effort of New York City firefighters over three hours to bring the fire under control.

The Dupont Plaza Hotel fire in San Juan, Puerto Rico, killed 96 people on New Year's Eve 1986. There were no sprinklers and no smoke alarms on hotel room floors in this 22-story hotel, despite the lessons of the earlier, similar MGM Grand Hotel fire in Las Vegas.

In May 1988, a fire burned the twelfth through sixteenth floors of the 62-story First Interstate Bank Building in Los Angeles. Over 300 firefighters worked for three hours to control this fire, an experience that gave impetus to sprinkler retrofit ordinances. A 10-story Atlanta, Georgia, office building burned in July 1989, killing five occupants. There was no sprinkler system. In 1991, a fire in the One Meridian Plaza Building in Philadelphia burned eight floors and killed three firefighters before sprinklers on the thirtieth floor stopped the fire. These fires taught valuable lessons that improved material selection, encouraged greater use of sprinklers, and enhanced understanding of the importance of smoke spread control.

Current building codes and life safety codes include new considerations to protect atriums, large rooms, and tall buildings. There are new guidelines and standards for smoke control, sprinkler systems, finishes, and furnishings. Los Angeles, New York City, and other municipalities have passed ordinances for the retrofitting of existing buildings to meet critical current standards. But these changes have been adopted slowly and sporadically, generally following major local fires. The ordinances are the result of public outrage immediately following a disaster rather than a carefully considered incremental evolution toward greater levels of fire safety.

Of course, the diligent enforcement of codes is as critical as the codes themselves. Poor code enforcement was a significant factor in the Las Vegas MGM Grand Hotel fire (Ross 1984). Alterations that allowed fire and smoke to spread more easily throughout the building were clear code violations.

Industrial fires can be especially intense. In 1953, the General Motors transmission plant in Livonia, Michigan, was a complete loss from fire. The entire building was of noncombustible construction, framed in steel with brick and steel walls. The contents were not considered to be especially dangerous, but the four-year-old plant burned and all 140,000 m^2 (1.5 million square feet) was totally destroyed. The damage was estimated at $55 million, but fortunately only three of the 4200 employees who worked in the building were killed. There were no fire separation walls, not enough sprinklers, and no roof vents. These factors prevented containment and hindered access. The steel frame was not fire protected. The fire spread quickly, in part due to accumulations of oil from industrial processes. This fire and similar industrial fires illustrate the need for good fire prevention practices, including tidy storage of industrial materials. Prevention practices are especially critical for industrial facilities, since many good fire-resistant design details are difficult to achieve given the functional requirements for uninterrupted open manufacturing spaces.

2.6 UNUSUAL LOADS: BLAST, VIBRATION, AND COLLISION

Impact loads or vibrations resulting from accidental explosion or from operating equipment can cause sudden damage to structures, often with loss of life.

Sometimes these incidents can be anticipated and proper precautions can be built into the operation or necessary resistances included in the construction.

Various kinds of gas explosions have caused major damage. There have been a great number of buildings affected by improper gas heating elements, leaky pipes, and containers. In 1932 the State Office Building in Columbus, Ohio, lost areas of some 1000 m^2 (10,000 ft^2) of the first and basement floors. Gas accumulation under the lowest floor of this 12-story steel-framed building exploded and lifted the slabs, which fell as rubble.

During the 1965 reconstruction of a Titan II missile silo near Searcy, Arkansas, an explosion, apparently of a tank of diesel fuel serving an engine for auxiliary power, started a fire that knocked out the electrical power and killed several people. The missile warhead and fuel had been removed, but the 445-Mg (490-ton) cover door was closed. Loss of electricity prevented operating the door, and very high pressures developed from the blast.

In 1966, a trailer truck loaded with 19 m^3 (5000 gallons) of asphalt was crossing a two-lane steel bridge in Atlanta, Georgia, when it struck the steel center divider and broke through the floor. The span collapsed and fell to the railroad tracks below. The fall snapped a suspended 150-mm (6-in.) gas line and the explosion started a fire visible 30 km (20 mi) away. The steel span and truck were demolished completely, but the driver escaped with some burns and minor injuries. That same year, a 109-Mg (120-ton) cylindrical high-pressure autoclave used for curing concrete blocks at a Hamilton, Ontario, plant suddenly exploded. The 32-m (106-ft)-long and 2.6-m (8.5-ft)-diameter vessel slid 46 m (150 ft), cutting through a truck and stopping just short of the highway. The 3.6-Mg (4-ton) door tore through several buildings, including the laboratory, and flew 60 m (200 ft).

There are well over 1 million kilometers of pipelines for long-distance transport of gas and liquid fuels in the United States. Some of these lines operate at very high pressures; only a small portion of them are located in populated areas. In 1965, the Federal Power Commission issued a report on pipeline failures. For the 15½-year period ending June 1965, 2294 incidents were reported. The worst was a gas explosion on March 4, 1965, near Natchitoches, Louisiana, which killed 17 people and formed a crater 8 m (27 ft) long, 6 m (20 ft) wide, and 3 m (10 ft) deep. Publicity attendant on such incidents led to the development of state and federal regulations on pipeline design, installation, and operation.

On February 13, 1981, a blast in Louisville, Kentucky, caused by leaking solvent from a processing plant destroyed sewers serving a 12-block area, causing structural damage to 27 buildings. Miraculously, only four people were injured.

Mexico has experienced several disastrous gas explosions. In November 1984, 334 people were killed in an explosion at a gas storage area in Mexico City. On April 22, 1992, 200 people were killed and 1470 were injured in a violent explosion in Guadalajara, Mexico. Leaks from a corroded 300-mm (12-in.)-diameter gasoline pipeline saturated the soil and contaminated the sewer system. Eleven separate blasts destroyed 1124 houses and ripped up 8 km (5 mi) of streets.

Explosions other than those caused by gas occur frequently in constructed facilities. Certain industrial operations generate fine particles of dust or other explosive materials. Immense explosions of this nature have been experienced in grain storage elevators and other agricultural buildings.

Damage to structures from blasting vibrations and pile-driving operations during construction has been the subject of much litigation. This type of failure is discussed in Chapter 11.

Even a small explosion can be extremely damaging to a structure, depending on the degree of structural redundancy present in the system. When there is little redundancy and the connections are poorly detailed, a local failure can lead to catastrophic progressive collapse. Such was the case in the Ronan Point housing project failure in London in 1968. A small gas explosion on the eighteenth floor of the 22-story building blew out a precast concrete load-bearing wall panel. This precipitated a progressive collapse all the way down to the ground level. There were four deaths in this failure, which is discussed further in Chapter 8. As a result of this collapse, building codes in the United Kingdom were revised to require consideration of potential for progressive collapse. Designers are required to consider the loss of load-bearing structural members and provide alternative load paths so that failures will be localized.

The "structural integrity" provisions in recent U.S. model building codes are also motivated by the desire to include greater redundancy in structures. The U.S. Department of State has been particularly interested in structural integrity and progressive collapse in the design of buildings that may be subjected to terrorist activities. State Department buildings must now include provisions to prevent progressive collapse (Yokel, Wright, and Stone 1989). Intentional explosions can be catastrophic in structures that lack continuity, particularly when key structural elements are targeted. These State Department concerns have been motivated, in part, by several disasters involving suicide terrorist bombings. On April 18, 1983, 63 people were killed in a progressive collapse of the U.S. Embassy building in Beirut, Lebanon. A truck loaded with explosives had been driven into the building. Six months later, on October 23, suicide terrorists set off explosives at U.S. and French barracks in Beirut, killing 241 American and 58 French members of a peace-keeping force. Again, the victims were buried under tons of concrete in a progressive collapse. The only defense against such intentional acts of sabotage, other than reliable military intelligence, is a barrier to prevent access and provision of structural redundancy.

The problem of intentional destructive action by domestic terrorists is receiving more attention in the United States since the bombings of the World Trade Center in New York City on February 26, 1993, and the Alfred P. Murrah Federal Building in Oklahoma City on April 19, 1995. Six people died in the World Trade Center bombing; the incredible Oklahoma City tragedy claimed 168 lives. There are a few things that technical experts and urban planners can do to guard against such attacks, but the options are limited indeed. Unfortunately, a free society must remain somewhat vulnerable to acts of terrorism. The dynamic and destructive forces already existing in the natural environment are a sufficient challenge for designers and builders. It is truly disheartening to contemplate the additional damage and suffering that can be intentionally inflicted by misguided or insane human beings.

Collision between a large moving vehicle, such as a truck, barge, or ship, and a structure usually results in damage to the structure. The Pontchartrain Causeway in Louisiana was struck by barges seven times between 1956 and 1966. Extra precast concrete elements were kept available on site to permit rapid replacement. In 1958, a 46-m (150-ft) concrete span at Topeka, Kansas, was put out of commission by a barge. The next year the Hood Canal Bridge in Washington State was damaged and some of its precast boxes sunk by a similar collision, with consequent serious delay in completion of the project. In 1964, the Maracaibo Bridge in Venezuela was damaged by a seagoing vessel that missed the main channel and hit a pier.

In 1966 the Raritan River railroad bridge was hit by a barge, with serious effects on New Jersey commuter traffic, and the Poplar Street Bridge over the Mississippi River, under construction, had some of its falsework carried away by a string of barges.

Physical contact can do catastrophic damage to the moving equipment as well as the structure. In 1959, a 160-m (525-ft) grain ship bumped into a 61-m (200-ft) lift bridge over the Buffalo River, wrecking the span. Falling water after the crash lowered the stern of the boat, which became enmeshed with the bridge steel. Fuel and cargo had to be removed to lighten the boat so that it could be pulled free.

The Sunshine Skyway Bridge spanning Tampa Bay, Florida, was heavily damaged in a tragic accident during a rainstorm in May 1980. A phosphorus-carrying freighter collided with one of the main piers of the bridge, causing collapse of the southbound span and the deaths of 35 people whose cars fell into the water below. After the accident, the Florida Department of Transportation determined that the bridge could not be repaired, and a new bridge was built. Construction of the new bridge began in 1982 and was completed seven years later at a cost of $244 million. Demolition of the old bridge was an impressive engineering and environmental challenge, costing over $8 million. The demolition was not completed until 1992 (Terpening and Irwin 1992).

Many other examples exist where trucks, ships, and other moving vehicles have collided with critical structural elements of buildings and bridges, sometimes with catastrophic results. Provision of structural redundancy is the best design defense against unusual collision loads. For certain structures, provision of protective barriers is another alternative, such as the use of fenders to protect bridge piers from collisions.

Of course, there will be an occasional collision resulting from equipment failure that simply cannot be resisted, even by a well-designed structure with built-in redundancies. In October 1992, an El Al 747 cargo plane crashed into a 10-story apartment building in Amsterdam, the Netherlands. The structure was demolished and more than 80 people in the building were killed. Such unusual impact loads cannot be predicted by designers, although a similar crash of a B-25 bomber into the seventy-ninth floor of the Empire State Building on a foggy morning in 1945 proved the value of redundant structural systems.

2.7 DETERIORATION

Since structures do not have the capacity to resist or adapt indefinitely to all natural forces and hazards, including the time-related degradation of materials, regular maintenance and repair are required. The purpose of maintenance is to prevent failure and to extend the useful life of a facility.

There is a tendency among design professionals to classify maintenance as the owner's responsibility. Unfortunately, the owners, operators, and users of facilities tend to view constructed facilities as passive objects, requiring no regular attention. Furthermore, many decisions that affect durability and maintainability are made early in the programming or design phase of a project. Durability, together with aesthetics, functionality, strength, stiffness, and stability, deserve thoughtful consideration in every design decision (Carper 1991).

All structures and facilities require maintenance and repair. Even brilliantly engineered and well-constructed buildings and bridges require both repair following large-scale seismic events and periodic retrofit to comply with changing standards and expectations. The most carefully designed and thoughtfully sited structures will ultimately fail or deteriorate without maintenance.

Lack of regular inspection and maintenance can be the cause of catastrophic structural failure. Far more common, however, is a gradual loss of attractiveness or usefulness and an increase in operational expenses. Most problems with aging buildings, for example, result from degradation of the envelope membrane materials (roofs and walls). Similarly, the deterioration of a highway bridge deck and other highway surfaces is accompanied by increasing costs for all users of the highway, even though structural failure may not occur.

Causes of deterioration include aging of materials, overloading, excessive use, and severe environmental conditions. Dampness, water penetration, fatigue, and progressive corrosion are common recurring problems.

Failure of the bridge over the Mianus River on the Connecticut Turnpike in June 1983 was attributed to inadequate maintenance and nonredundant design. Three persons were killed when a corroded pin-hanger connection gave way, allowing a section of the 26-year-old bridge to collapse into the river.

Design detailing can either enhance or hinder the adequate inspection and maintenance of a facility. Most owners and operators will act to correct life-threatening deterioration if it is visible. Poor detailing, however, can permit the unseen degradation of critical structural elements in inaccessible locations.

In 1985, the ceiling over a swimming pool in Switzerland collapsed suddenly, killing 12 people. The failure was traced to stress corrosion in stainless steel hangers, exposed to the corrosive environment in an unseen location. The necessity for periodic inspection of structures, even where special access must be provided, is too often forgotten.

Sometimes, renovation projects uncover unpredicted structural deterioration. In such cases, corrective repairs must be undertaken before other work can proceed. In the 1983 renovation of the historic University of Illinois Memorial Stadium, corroded structural steel columns were discovered. This problem was unrelated to the purpose of the renovation. Nevertheless, it was a critical condition that required remedial repairs (Wilkinson and Coombe 1991).

Some types of facility have traditionally had more maintenance problems than others. Parking garages, for example, have an alarming proclivity for deterioration and structural failure. Parking garages are among the most vulnerable facilities for material degradation. Open to the weather, they experience extreme temperature variations. In winters of the northern United States, vehicles constantly deposit ice-melting salts on the concrete decks. The decks are not periodically rinsed by rainfall as bridge decks are. The accumulation of these salts and temperature variations gradually degrade the concrete decks and accelerate corrosion of the reinforcing steel.

Maintenance and repair projects occupy an increasingly significant proportion of overall construction activity for several reasons. Among these are interest in conserving historic structures, concern for deterioration of the infrastructure, and desire to upgrade the structural integrity of buildings and bridges built prior to the implementation of current standards. In addition, the durability of constructed facilities has suffered from the current emphasis on least initial cost. This emphasis has produced an inventory of newly constructed works requiring costly repairs to correct premature deterioration and other performance deficiencies.

Improvements in modern analytical methods have led to reductions in factors of safety for strength. While improving efficiency, speed of construction, and low initial cost, these reductions have in many cases diminished the capacity for long-term performance (Moncarz, Osteraas, and Wolf 1986; Kuesel 1990). As attention is focused on the high costs associated with premature deterioration of newly constructed works, it is expected that the design professions will begin to place more emphasis on durability and maintainability.

2.8 SUMMARY

In this chapter we have reviewed the forces arising from natural hazards and the unusual loads that act on constructed facilities. Some of these hazards involve disasters that claim many lives and produce widespread, dramatic damage to the built environment. These disasters exact such a toll in human suffering throughout the world that the United Nations declared the 1990s the International Decade for Natural Disaster Reduction. By making this declaration, the member nations have recognized both the severity of risk presented by natural hazards and the promise that international cooperation and scientific and technical progress hold for understanding the hazards and mitigating their effects (ACIDNHR 1989).

Consideration of the many destructive agents at work in the natural environment should instill a certain amount of humility in the attitude of designers and builders. The objects they are building are temporary; they will not last forever. All structures are in the business of resisting forces that will ultimately destroy them. As James Gordon states in his book, *Structures: Or Why Things Don't Fall Down:*

> The entire physical world is most properly regarded as a great energy system: an enormous marketplace in which one form of energy is forever being traded for another form according to set rules and values. That which is energetically advantageous is that which will sooner or later happen. In one sense a structure is a device which exists in order to delay some event which is energetically favored. It is energetically advantageous, for instance, for a weight to fall to the ground, for strain energy to be released, and so on. Sooner or later the weight *will* fall to the ground and the strain energy *will* be released; but it is the business of a structure to delay such events for a season, for a lifetime, or for thousands of years. All structures will be broken or destroyed in the end, just as all people will die in the end. It is the purpose of medicine and engineering to postpone these occurrences for a decent interval. (Gordon 1978)

While Gordon is certainly correct in his assessment of the nature of physical forces and their destructive consequences for the works of humankind, there are many things that builders can do to increase the potential for survival. Thoughtful site selection is an extremely important aspect of planning for success.

In the natural environment, some locations are more susceptible than others to violent dynamic changes. The tools of modern technology provide the ability to identify those locations that are most likely to experience the hostile effects of change. Those sites that are most sensitive to the effects of strong winds, flooding, or seismic events can often be identified. Design guidelines, land-use planning ordinances, and other public policies can be implemented so that these sites are avoided and conflicts between static objects and the dynamic environment are minimized.

Unfortunately, political concerns and economic considerations have often taken precedence in design and land-use decisions. Coastal regions, susceptible to hurricanes and tsunamis, may be the most dynamic environments of all. Yet these regions have been developed, even very recently, to an extremely dangerous degree. Many schools, hospitals, and other essential facilities in

California have been located in close proximity to clearly defined seismic faults. The coastal regions of Florida and the Gulf coast states have experienced dramatic increases in population and property investment, with very little regard for the potential for hurricane activity. The prediction capability of modern technology has been largely ignored in these decisions, at least with respect to the potential for large-scale property loss. As a result, Hurricane Andrew, in 1992, became the most costly natural disaster in the history of the United States, even though lives were saved through timely warnings.

Inappropriate land-use decisions and settlement patterns have given rise to many problems and conflicts throughout the world. These conflicts add unnecessarily to those already inherent to the design process and bring about an unnecessary risk of loss of life and property due to natural hazards. Those responsible for unwise land-use development have sometimes belatedly asked engineers to provide solutions to these conflicts. At best, technology can deliver only temporary relief from these problems, often with damaging impact on the natural environment (Carper 1993).

Figure 2.10 is an example of poor site selection. The photograph is of a residential roof failure in Peru. A large boulder, dislodged in an earth tremor, has rolled down a hillside and through the residence, destroying the roof. If this failure had occurred in the litigious United States, it might be expected that legal action would be brought against the architect, the structural engineer,

Figure 2.10 Destruction of a residence in Peru caused by a displaced boulder. (Courtesy of R. Hamblin.)

the roof truss manufacturer, and perhaps even the roof shingle supplier. A more rational consideration of the failure, however, would suggest that it is simply not cost-effective to design residential roofs successfully to resist loads of such magnitude. There are forces in the natural environment that are simply too large to resist.

However, this failure should not be casually dismissed as an "Act of God" for which no one is to be blamed. Technology has provided the tools to permit planners to locate such boulders, to predict earth tremors and landslides, and to plot the likely trajectories of dislodged objects. This failure was the result of poor land-use planning or site selection rather than of inadequate design or construction.

Such errors are made over and over again in the siting of structures. One example is the development of the Turnagain Heights area of Anchorage, Alaska, prior to the 1964 earthquake. Geologists had long warned of the dangers of building in this area, yet there were few ordinances established to discourage the type of development that was destroyed in the earthquake (Figure 2.11). This failure cannot be dismissed as an unavoidable casualty. It should be recognized as a planning failure, not a design or construction deficiency. Incidentally, less than 30 years after the earthquake, new homes are currently being constructed in the immediate vicinity. Unfortunately, memo-

Figure 2.11 Turnagain Heights landslide caused by the 1964 earthquake in Anchorage, Alaska. (Courtesy of the Federal Emergency Management Agency.)

ries are limited, and wisdom is all too often displaced by shortsighted economic considerations.

The design professional who is motivated by a desire to mitigate the loss of life and property due to natural disasters can make a substantial contribution through involvement in the formation of public policy, particularly in the land-use planning arena. There is a far greater opportunity to reduce such losses through wise settlement patterns than though the technology of structural engineering for individual buildings. Those facilities that are located in the least vulnerable locations and whose design is founded in thoughtful consideration of the dynamic character of the natural environment will have the greatest opportunity for survival for the longest useful life.

2.9 REFERENCES

ACIDNHR, 1989. *Reducing Disasters' Toll: The United States Decade for Natural Disaster Reduction,* Advisory Committee on the International Decade for Natural Hazard Reduction, National Academy Press, Washington, DC.

Amrhein, J., 1985. "Reinforced Masonry Seminar," University of Idaho, Moscow, ID, September 11, 1985.

ASCE, 1991. "Corps Ready to Avert Future Flood Threat," *Civil Engineering,* American Society of Civil Engineers, New York (September).

Bell, G., and J. Parker, 1987. "Roof Collapse, Magic Mart Store, Bolivar, Tennessee," *Journal of Performance of Constructed Facilities,* American Society of Civil Engineers, New York (May).

Carper, K., 1991. "Facilities Maintenance and Repair of Damaged Structures," in *Advances in Construction Materials, Techniques and Management,* S. Somayaji and M. S. Mathews, eds., Sri Venkatesa Printing House, Madras, India.

Carper, K., 1993. "Building as a Response to Technological and Creative Processes," in *The Built Environment: Creative Inquiry into Design and Planning,* T. Bartuska and G. Young, eds., Crisp Publications, Inc., Los Altos, CA.

Carper, K., 1995. "The January 17 Earthquakes in Northridge, California (1994) and Kobe, Japan (1995): Old Principles—New Lessons," published in Japanese and English by Nihon University, Tokyo, Japan.

Degenkolb, H., 1980. "Failures: How Engineers Learn—Or Do They?" Paper Preprint 80–017, ASCE Convention, Portland, OR, April 14–18, American Society of Civil Engineers, New York.

FEMA, 1986. *Multi-hazards and Architecture,* TR 20 (Vol. 1B), Federal Emergency Management Agency, Washington, DC.

Frost, R., 1915. "Mending Wall," in *North of Boston,* Henry Holt & Co., New York.

Gordon, J., 1978. *Structures: Or Why Things Don't Fall Down,* Penguin Books, New York.

Karter, M., 1992. *Fire Loss in the United States During 1991,* National Fire Protection Association, Quincy, MA.

Kuesel, T., 1990. "Whatever Happened to Long-Term Bridge Design?" *Civil Engineering,* American Society of Civil Engineers, New York (February).

Lagorio, H., 1990. *Earthquakes: An Architect's Guide to Nonstructural Seismic Hazards,* John Wiley & Sons, Inc., New York.

Moncarz, P., J. Osteraas, and J. Wolf, 1986. "Designing for Maintainability," *Civil Engineering,* American Society of Civil Engineers, New York (June).

Ross, S., 1984. *Construction Disasters: Design Failures, Causes, and Prevention,* Engineering News-Record, McGraw-Hill, Inc., New York.

Terpening, T., and M. Irwin, 1992. "Out with the Old," *Civil Engineering,* American Society of Civil Engineers, New York (September).

Wilkinson, E., and J. Coombe, 1991. "University of Illinois Memorial Stadium: Investigation and Rehabilitation," *Journal of Performance of Constructed Facilities,* American Society of Civil Engineers, New York (February).

Yokel, F., R. Wright, and W. Stone, 1989. "Progressive Collapse: U.S. Office Building in Moscow," *Journal of Performance of Constructed Facilities,* American Society of Civil Engineers, New York (February).

Zallen, R., 1988. "Roof Collapse Under Snowdrift Loading: Snowdrift Design Criteria," *Journal of Performance of Constructed Facilities,* American Society of Civil Engineers, New York (May).

3

EARTHWORKS, SOIL, AND FOUNDATION PROBLEMS

The necessity for a proper foundation support for any structure is an accepted axiom in engineering. All loads must be transferred to the underlying soils in such a way that the resulting movements can be tolerated by the structure and stability will be maintained over the life of the structure. Introduction of a foundation into a soil mass causes a new set of physical conditions, changing the size and geometry of the soil mass, as does the later addition of loadings during and after construction. The construction of foundations near an existing structure may undermine an existing satisfactory support system. Sometimes, foundation construction changes site drainage conditions, creating serious problems for neighboring properties.

Each construction site is unique. The soil is the first material encountered by the constructor, upon which all subsequent materials are placed. It is essential to understand the characteristics of the soil to have confidence in the safety of that which is placed within and upon it. Soil characteristics vary, sometimes considerably, on a given construction site. Sufficient preconstruction soil testing requires careful consideration of the number of borings, their proper locations, and the depth to which borings should be taken. Numerous foundation failures have occurred due to unforeseen soil conditions that were, in fact, not unforeseeable had adequate numbers of borings been specified and taken to adequate depth. LePatner and Johnson (1982) provide examples that illustrate the need for soil borings that penetrate below the bearing area, all the way to the base of the zone of influence, where the soil is unaffected by the imposed loads. The engineering literature contains innumerable examples of costly failures that could have been avoided had qualified geotechnical

59

engineering consultants been retained, both to provide preliminary information to designers through soil exploration and testing and to provide guidance during construction relative to interpretation of actual conditions (ASFE 1990).

In below-grade construction, the unexpected often happens. Costly surprises and construction schedule delays are common. The question then arises: "Who should pay for the unexpected?" This dilemma has taken on a new dimension in modern construction. In the past, the unexpected was limited to unforeseen natural conditions. Today, the surprises include encountering hazardous wastes. The costs of hazardous waste disposal and remedial measures to rehabilitate a site are extremely high. In many cases these costs force abandonment of the site, the project, or both. Traditionally, unexpected site conditions have been a risk assumed by the project owner unless it can be shown that the condition encountered during construction could have been detected by ordinary exploration and testing procedures.

Foundation failures may bring about catastrophic collapse of an entire structure, such as the 1987 failure of a 13-story reinforced concrete apartment building in Brazil that killed 50 persons. What is far more commonly associated with foundation litigation, however, is deficient performance in the form of unacceptable deformations, leading to cracks, moisture penetration, and other serviceability problems.

Geotechnical failures related to difficult sites appear to be increasing. Part of the explanation for this apparent increase lies in the fact that more challenging sites are now being developed. Much of the prime real estate with few inherent risks has already been used. What generally remains in certain regions of the country is hilly or mountainous terrain subject to landslide activity, low-lying land subject to flooding and/or subsidence, alluvial fan areas subject to debris flows and mudflows, and land encumbered by other geologic and environmental hazards (Shuirman and Slosson 1992).

Below-grade construction in the increasingly congested urban environment also entails difficult challenges. Special problems in urban construction include the presence of buried cables and utilities, existing tunnels, and adjacent structures, all of which complicate the construction. The process of dewatering a construction site to facilitate construction, for example, presents unique problems on the urban site (Gould, Tamaro, and Powers 1993).

Modern foundation design is a combination of engineering principles tempered by records of successful and unsuccessful experiences. It is far from an exact science. Perhaps more than any other field of construction engineering, geotechnical engineering relies on experience-based professional judgment. Foundation construction involves skilled procedures that maintain stable conditions during all construction activities until the final design is incorporated into the ground. Several below-grade construction accidents are discussed in Chapter 11. Others are included in this chapter, together with soil and foundation problems in completed projects.

3.1 PROBLEM SOILS

Certain sites and specific soil conditions present extraordinary challenges. The more difficult sites include those with expansive soils, organic soils, and contaminated soils. Fills frequently present significant problems, requiring diligence during placement. Numerous costly failures involve nonengineered fills, poorly engineered fills, fills containing organic materials, and fills that are not carefully monitored during construction.

Other challenging sites are those with sinkhole potential, permafrost conditions in cold regions, steep slopes, and sites with large seasonal water content fluctuations or varying water table conditions.

The bearing capacity of the soil is an important consideration, but settlement potential is equally critical. The soil bearing capacity controls the extent of loading that soil can support before it experiences shear failure. However, for many lightly loaded foundations, unacceptable differential settlements are the most serious problem, leading to costly performance or serviceability deficiencies. For lightly loaded construction, settlement considerations may be the dominant factor in a successful design (Greenfield and Shen 1992).

The importance of retaining a qualified geotechnical engineer cannot be overemphasized. Unfortunately, some jurisdictions do not require geotechnical investigations or observation during construction, and if budgets are tight, developers and owners tend to expend only the amount required by the governing agencies. James Slosson, a prominent consulting geologist, has postulated *Slosson's law,* which he describes as a corollary to Murphy's law. Slosson's law states that "the quality of professional work will sink to the lowest level that government will accept" (Shuirman and Slosson 1992). There is no question that many costly soil-related failures are avoidable with a modest expenditure for competent geotechnical information at the outset of design.

Another source of foundation problems resulting from false economy is the misuse of a preliminary soils report. Often, a preliminary report is made prior to determining the actual location of buildings, and the report is then used without subsequent verification of conditions at the building locations. Similarly, a soils report intended to establish preliminary bearing pressures for foundation design may be misused to design a septic system, without conducting critical percolation tests. The purpose and scope of a geotechnical investigation should be noted clearly by the consultant. The safest strategy is to retain the geotechnical engineer throughout foundation construction and site development.

Proper soil tests include complete information about the behavior and engineering properties of the soil: plasticity, potential for volume change (settlement and expansion), and moisture content. It is also important to explore for the presence of groundwater, the degree of soil compaction, and slope stability. Soil density information can be obtained from penetration tests at the site. Interpreting the data from these explorations and others requires a great deal of judgment by a geotechnical engineer.

As noted earlier, contaminated soils can be extremely costly to remediate. Such soils exist on abandoned waste disposal sites or are associated with past industrial occupancies involving heavy metals or other chemical contaminants. Certain industrial processes have generated large quantities of waste materials that have been used for unstable fills (Raghu and Hsieh 1989). Remediation measures are all costly: excavation and replacement, containment, or chemical treatment. The presence of organic deposits may be a significant problem. Organic material is continually decaying. Construction of structures on these deposits is very risky.

Planning, design, and construction of engineered fills should be undertaken only by a qualified geotechnical engineering consultant. Continuous compaction testing is necessary for successful structural fills. It should be noted that deep fills may develop problems even if placement has been carefully monitored. Deep fills and fills with varying depths have problems due to differential long-term settlements from their own weight. Soil is heterogeneous even under the most favorable conditions.

Fills should be carefully examined for organic content. LePatner and Johnson (1982) note that "organic soils are potential sources of problems; fills are potential sources of problems; fills with organic inclusions are double sources of problems."

Water content of the soil is an important factor to consider. A high groundwater table in conjunction with loose soil deposits indicates a potential for vibration-induced liquefaction in a seismic event or due to construction vibrations. Such soils can be densified by controlled blasting or by deep compaction to increase seismic resistance. Fluctuating water content causes volume changes in the soil mass. Depletion of an aquifer, desiccation from tree roots, or dewatering of a construction site will result in shrinkage of the soil mass, while added water will weaken the soil structure and cause certain soils to expand. Sloping sites can become unstable when the water content is increased, due either to excessive rainfall or poorly managed development that permits excessive irrigation or uncontrolled discharge of domestic or industrial water into the soil. Increasing the water content in the soil increases its unit weight, decreases its cohesion, and decreases the angle at which slopes are stable.

The costly damage to structures built on expansive clay soils is an impressive amount. F. H. Chen has estimated that by the year 2000, the cost of damage to buildings constructed at sites with expansive soils will exceed $4.5 billion (Greenfield and Shen 1992). Expansive soils do more damage to single-family residential structures annually in the United States than the total combined residential damage from floods, hurricanes, earthquakes, and tornados. Expansive soils cover approximately one-fifth of the United States (Figure 3.1). Local precipitation patterns (cyclical moisture changes) influence the expansion/contraction potential of these soils; damage potential varies greatly between regions.

Sites that contain expansive soils can be developed successfully, but the soils must be explored carefully and the appropriate remedial measure adopted

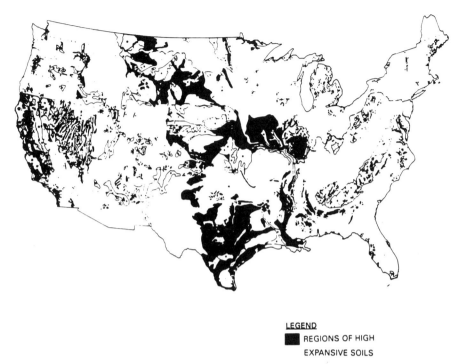

Figure 3.1 Expansive soil deposits in the United States. (Courtesy of the J.H. Wiggins Company.)

for the specific site and the specific type of facility. Greenfield and Shen (1992) note that these sites can cause costly damage in light construction. Budgets may be tight and owners/developers may not recognize the need for soil investigation, as would be the case for more heavily loaded construction and larger projects. Expansive clays present some of the most challenging soil conditions on which lightly loaded structures are constructed. Some clays exert very large uplift pressures when they expand. Lightweight structures do not have enough weight to counteract these large uplift forces. Furthermore, the expansion is not uniform, resulting in large differential movements (Meehan and Karp 1994).

Potential solutions for expansive soils include drilled shafts, post-tensioned or heavily reinforced structural slabs with stiffening beams, and heavily reinforced continuous wall footings. The best but most costly solution may be the use of deep foundation systems, founded in nonexpansive soils, and protected from seasonal moisture variations. If the active expansive soil layer is not too deep, it may be possible to remove the expansive material to a depth protected from seasonal moisture variation and replace it with engineered fill using nonexpansive materials. Subgrade treatments are also common. These include water pressure injection to increase the moisture content and preswell the soil; chemical injection to alter the soil composition chemically, rendering it nonexpansive; lime slurry pressure injection to inhibit moisture migration; or direct mixing of lime with the soil. Site drainage is also an important aspect of construction in regions with expansive soil problems (Jones and Jones 1987).

The problem of settlement of a soil mass has two components: the initial settlement caused by deformation of the soil particles due to the weight of construction, and consolidation settlement due to the expelling of fluid from the soil. Consolidation settlement is time dependent. This can occur over considerable time in clay deposits of low permeability.

To mitigate the potential for large consolidation settlements on sites with compressible clays, a deep foundation system may be required. Other remedial solutions include soil removal and replacement, and preloading the site. Preloading may require considerable time. It should be noted that preloading is not a workable solution for soils containing organic material, since long-term decay of the material will continue to take place. Wick drains are sometimes used to accelerate the surcharge-consolidation process. Dynamic deep compaction (dropping a large weight on the surface) can also be used.

Some sites, mostly in the southeastern states, are susceptible to the formation of sinkholes. Sinkholes are depressions usually associated with acidic groundwater that dissolves limestone strata, resulting in gradual subsidence or sudden collapse of the overlying material. Several techniques are available for detecting the presence of sinkhole formation and mitigating their effects (Sputo 1993).

Frost action can increase the soil volume considerably. Ice lenses can continue to grow due to capillary action with the groundwater in soils with small pore space. When thawing occurs, large settlements may follow. Silty soils are more susceptible to frost heave than are other soil types. Permafrost sites (perennially frozen ground conditions in cold regions) require particular care and specialized engineering expertise (McFadden and Bennett 1991). Ordinary foundation methods, when imposed in permafrost, can cause severe structural distress and related environmental damage.

In recent years, several soil-modification techniques have been developed to improve the character of problem sites. These include the use of artificial soils, geosynthetic materials, geotextile fabrics, and reinforced earth using soil nailing (grouted steel bars). Chemical grouts are now used to stabilize soils for tunnel construction and repair. New equipment for deep compaction and preloading of compressible or loose soil deposits is available.

Compaction of a soil mass involves forcing air out of the soil to increase its density. Compaction of clayey soils is accomplished by rolling or kneading the soil. For nonclayey soils, vibration is usually effective. Compaction of engineered fills is best accomplished under controlled conditions by placing soil in layers.

Artificial fills, such as expanded polystyrene material, have been used successfully. The weight of artificial materials may be much less than the weight of compacted natural soil. This will result in less settlement when the underlying soils are highly compressible. The negative skin friction on a driven pile or drilled shaft will also be reduced when this fill material is used. Polystyrene fills are quite costly, however, and their use may not be acceptable when the groundwater table is high (Greenfield and Shen 1992).

This brief discussion of a few of the more common problem site conditions should be sufficient to conclude that accurate information about the character of the soil is a prerequisite for satisfactory performance of facilities constructed upon it. Any soil can become a problem soil if its properties are not understood. The competent geotechnical engineer is a valuable member of the design and construction team.

3.2 BELOW-SURFACE CONSTRUCTION

3.2.1 Deep Foundations: Piles and Caissons

Piles and caissons are structural elements used to transmit loads to the soil. When the required structural value of a pile or a group of piles is not provided, catastrophic failure can result. Piles are inserted in the ground to transfer loads to soil layers through direct bearing, friction, or both. If the pile cannot be inserted to the depth desired, a different type of foundation or even abandonment of the site is the proper solution.

Timber piles eventually deteriorate in a soil layer subject to alternate wet and dry conditions or where they are exposed to certain marine animals and fungi. Creosote or other pressure-treatment deterrents are necessary for the long life of timber piles. Timber continuously submerged in fresh water will last indefinitely, but it is weakened rapidly when the water level recedes. The danger of lowering water tables to timber pile foundations has been recognized for many years. In 1918, the Chicago River Warehouse suffered cracked walls when the river level dropped. The almost 50-year-old building was jacked up and the piles repaired by concrete caps replacing the rotted tops.

As areas become heavily populated, groundwater recedes because of sub-surface utility and transit system construction and the increased demand for domestic and industrial water. For this reason, untreated timber piles should never be considered as permanent support. Numerous expensive residences in Boston's historic Beacon Hill neighborhood, for example, have required costly pile replacement due to loss of timber piles from receding water tables.

Wood borers are normally expected along brackish waters, but they are also active in inland waters, as was found in 1933 at St. Paul, Minnesota. Timber piles under a four-story industrial building were eaten away completely. The original pile installation was in 1886 for a two-story building, but when two stories were added in 1912, the piles had been inspected and found to have sufficient integrity.

Borers remove very small amounts at a time, and eventually the remainder of the pile cannot withstand the loading, causing sudden failure, such as happened in 1958 when a 30-m (100-ft) section of a Brooklyn, New York, pier collapsed under a load of cargo being unloaded from a ship. Since that time, procedures and instrumentation have been developed to permit nondestructive evaluation of existing piles through ultrasonic and other techniques. Regular inspection and repair can prevent such catastrophic surprises.

Structural design of columns is most efficient and least troublesome when the column is loaded axially. Walls supported on piles in a single line should similarly be positioned to avoid eccentric loading. Many wall-bearing buildings are placed on piles located at the center of the foundation wall width. When the exterior wall face is at the property line, the building loads may be applied considerably eccentric to the single line of piles. The walls will have a tendency to rotate, with major cracking near the return walls. Many examples of such failures exist in historic masonry buildings. Correction requires extensive shoring, underpinning, and addition of new piles to provide stability.

Steel pipe piles driven to high bearing value should be checked for continuity and integrity before being filled with concrete. Unexpected intermediate resistances in the form of boulders or other buried obstructions may be encountered. In the fills placed above a swamp crossed by the Long Island Expressway at Alley Ponds, Queens, New York, such obstructions accordioned 7.3 m (24 ft) of pipe into 1.2 m (4 ft) of crushed steel with practically no change in diameter. In the glacial moraine hillside not far distant from that bridge location, 18.3-m (60-ft)-long H-piles driven for sheeting of a subway excavation were diverted by the boulders and became structurally useless. One pile section actually bent into a horseshoe shape while driving, and the tip came out of the street surface without warning. Special techniques are also required for driving steel piles into permafrost. Frozen ground can be virtually impenetrable. Preparation of the ground prior to driving the piles is essential (McFadden and Bennett 1991).

Much has been learned about pile foundations from actual historic experience with this form of construction. The experience with mandrel-driven shell piles for a hanger at the San Francisco airport in 1963 indicates some of the troubles to be guarded against when installing high-capacity piles in soft layered soils interspaced with granular soil lenses. Piles were 27 m (90 ft) long with several segments of tapered interlocked shells and a bottom closed-end section of 270-mm (10.75-in.) pipe. With a hammer of 44,000 N-m (32,500 ft-lb) rated energy and a 24.4-m (80-ft) mandrel, extended to the bottom of the pipe, driving was performed to obtain 623-kN (70-ton) bearing values. Driving lengths required to develop the desired resistance did not agree with lengths computed based on soil test values, but load tests confirmed the driving resistance criteria. In some of the piles, with apparent satisfactory penetration resistance and electric light examination showing straight and clean piles to the bottom, measurement showed loss in length up to 6 m (20 ft). The visible "bottom" was not the bottom of the pipe but a squeezed or bent section. Satisfactory piles left overnight before concrete filling were found squeezed almost completely at the bottom of a 9-m (30-ft)-deep soft soil layer. About

5 percent of the driven piles required replacement because of shell or pipe failure. Outer collars were added at the critical depth over the new shells before driving, to prevent squeezing.

Because of improper installation of extrusion piles for the foundation of a five-story concrete office building near Montreal, Canada, in 1958, local failure in the pile shafts resulted in settlement and tipping of the almost completed structure. The building was 40 by 23 m (130 by 75 ft) in plan, of slab and girder design. Piles were uncased concrete shafts with extruded bulbs to carry 1.1 MN (125 tons). When about 90 percent of the dead load was in place, a worker noted a "bang" and the elevator installers reported that the shafts were out of plumb sufficiently to make the cars inoperative. An immediate inspection revealed extensive cracking in the basement floor, a settlement of 200 mm (8 in.) at the north end, and a lean of the north wall 100 mm (4 in.) outward. Within 10 days, 45 steel-tube piles were jacked to rock refusal, and with steel grillage horseheads the building load was picked up.

The first assumption was that the rock below the piles had failed. Core holes were made and 280 m^3 (10,000 ft^3) of slurry was injected, but settlement continued at the rate of 13 mm (0.5 in.) per week. H-beams with brackets were installed along the north wall and 12 more tubes were jacked within the building, all within five weeks of the first settlement. Settlement reached a maximum a few weeks later and the building was 460 mm (18 in.) average out of level, without a single sign of distress in the concrete floors or columns. Borings to check the rock showed no change in thickness or deformation of the sedimentary rock sheets and no grout in the seams. Most of the grout had become a cover layer on top of the rock.

Careful check of design and materials used gave no clues. The total weight of the building, some 4500 Mg (10,000 kips) was located with less than 150 mm (6 in.) eccentricity from the center of gravity of the pile locations. Piles of this design had been used successfully for large loads, and the dead loads averaged only 76 Mg (167 kips) per pile with little variation. Rock depths agreed closely with pile lengths and there was no lateral movement to indicate slippage along rock seams. Elastic squeeze of the rock seams could not explain the magnitude or the differential amounts of settlement. There was no unbalanced loading that could shear the piles. Groundwater and soil tests, including concrete made up with groundwater (which unexpectedly tested higher than concrete made with city water), showed no unusual potential for corrosion failure.

Nine of the piles were then completely exposed for inspection within split-pipe hand-dug caissons. Piles had been specified as 457 mm (18 in.) in diameter constructed of 20.7-MPa (3000-psi) concrete with a cage of five No. 6 (19-mm) vertical bars within a 6-mm ($\frac{1}{4}$-in.) spiral on a 203-mm (8-in.) pitch. The pile casing was driven to a resistance of 10 blows to the last 13 mm ($\frac{1}{2}$ in.) of penetration, and concrete placed therein was compacted by a 27-kN (3-ton) hammer as the casing was retracted. Exposed failures in the piles were all similar, at 1 to 4 m (3 to 13 ft) below the pile cap. The concrete was crushed, with seams of mud enclosed and the vertical bars bent into the shape of ears. The deformed length of the bars exactly matched the settlement at that location for each pile (Figure 3.2).

The explanation for failure was then evident. The casing had been retracted above the concrete being placed, and mud had squeezed into the shaft to

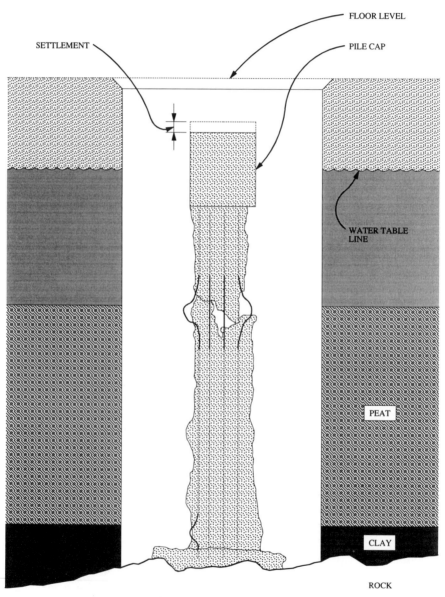

Figure 3.2   Section through hand-excavated inspection caisson.

seriously weaken the strength of the pile. As soon as one pile failed, probably near the north end, adjacent piles became overloaded and failed, at the section of concrete that had been necked most by the squeeze of the mud. The building was completely demolished, new caisson piles with metal liners left in place were installed, and the building was reconstructed. A large economic loss resulted from a small lack of proper supervision in the pile installation.

A somewhat similar incident was found during expansion of the LaGuardia Airport in New York in 1961. After the concrete-filled steel-tube piles had been completed, some of the piles had to be cut down to permit installation of added fuel tanks. Some of these piles were found not to be completely filled with concrete. An investigation was undertaken to explore the condition of all the piles. A random sampling of piles was drilled with rock drills, and when found suspect (as indicated by a nonuniform drill resistance and change in dust color) the piles were also cored. Some were extracted to correlate the drilling and coring information with visible exposure after the steel shell was removed. Very good correlation was proven. Enough improper piles were discovered in the random sampling to warrant the decision that the piles were not usable. Two were found with voids over 9 m (30 ft) in length. All pile caps already in place were removed, and new piles were added to each cluster, designed to carry the full loadings. Since all work had been done by an experienced contractor under continuous supervision, it makes one wonder how many such conditions exist and have not been discovered (or discovered but not reported).

Uncased concrete cast-in-place piles have typically given trouble when installed in plastic soils. In 1925, piles driven in soft clay at Plainfield, New Jersey, were found sometimes to be necked or even completely sheared from the lateral pressure set up by driving adjacent casings. Uncased piles must be installed with special control to guarantee the structural integrity of the structural shaft. In soils that squeeze readily, a precored hole must be provided within the plastic layer to avoid lateral displacement and vertical heave of neighboring piles.

One common error in pile driving in plastic soils is the disregard of internal lateral soil displacement. Heaved and buckled or collapsed pile shells can be avoided by thicker shells or a proper sequence of installation with allowance of sufficient time for the induced high pore pressures to be dissipated. All piles should be driven from one general level, even where this entails more expensive later excavation of partial cellars, pits, and trenches. Excavation of lower areas, even if sheathed and braced for normal earth pressures, will not stop the lateral drift of piles driven adjacent to such excavations. The cost of corrective measures at a hospital in Chicago, where the boiler room area was first excavated, from pile cap redesigns and added piles, exceeded the cost of the total original pile installation. This type of failure is such common knowledge that its repetition is completely inexcusable.

Cored and hand-dug caissons are of little value unless the bottom of the caisson is at the resistant layer of soil chosen for bearing and the shaft is constructed of solid continuous concrete. After the holes are excavated, they must be inspected not only for determination of concrete requirements but also for conditions at the bottom of the excavation. When the bearing condition is not equivalent to that assumed in the design, modifications must be made. Sometimes, loose fill can fall into the excavation, voiding the expected quality of the caisson.

Some historic examples illustrate potential problems with caisson foundations. In 1966, three incidents of seriously faulty caissons were discovered in Chicago. At the 2.5-m (8-ft)-diameter core-drilled caisson foundations for a 100-story office building, four units were found to contain earth intruded into the concrete. In one caisson, over a 4.3-m (14-ft) length of earth was discovered starting 15 m (50 ft) down. Steel erection was stopped when a caisson showed distress under the weight of the lowest column section, about 11 Mg (12 tons), necessitating investigation of all caissons by coring.

In a 25-story concrete building a 1.2-m (4-ft)-diameter caisson settled 44 mm (1.75 in.) before the frame was topped out, causing the column to drop and several floor slabs to crack. One would not normally expect such lack of quality control on such an important item of a building assembly, particularly when repair costs are many times the cost of properly built caissons. In each defective caisson determined by drilled cores, 3700 lineal meters (12,000 lineal feet) of 50-mm (2-in.) cores in the case noted, all defective concrete was removed and replaced. Column loads required temporary shoring before concrete replacement.

In a similar incident in 1961, at a building in the Chicago Loop area, an inspector's question as to proper contact between caisson and the rock was settled by coring the full depth, only to find hollows in the concrete near the bottom of the shaft. All caissons were then cored and 20 percent were found unsatisfactory. The corrective measure ordered by the architect was to add two 457-mm (18-in.) caissons along the side of each caisson in question, which were 1.2 and 1.5 m (4 and 5 ft) in diameter. Almost five years of litigation resulted in an agreement to settle all claims out of court to avoid adverse publicity. With such examples in the record, test coring of all caissons became an added expense in such foundation work.

Design and construction techniques for pneumatic caissons have long been established and continuous control is usually provided, resulting in many successful foundations. Failures have been rare in recent years, yet a single omission in the control sequence can lead to a mishap, such as the mysterious explosion in the caisson for a bridge foundation in the St. Lawrence River at Three Rivers, Quebec, in 1965. The explosion heaved workers and materials 30 m (100 ft) into the air and tore the caisson in half, floating one 4-m (13-ft) cell 5 m (15 ft) above the others. Personnel were working at 275 kPa (40 psi) air pressure in the 15.8 by 40.2 m (52 by 132 ft) caisson, built pneumatically to control position more accurately. The bottom had penetrated 10.7 m (35 ft) into the clay bottom, with 15 m (50 ft) of water in the channel. The blast came at 4:00 P.M., as the shifts were changing after a 90-minute work period. The foundation work inside the 24 cells would have been completed in a few days. The force of the explosion threw the entire foundation out of line and at least six of the cells filled with water. The immediate area was known to contain gas pockets in the subsoil, and some local residents had even tapped such gas pockets for heating and cooking. Explosions in underground construction are less common in modern construction, but they continue to occur from time to time, usually associated with tunnel construction (Section 3.2.2).

Deep foundation construction in congested urban areas presents special challenges, including coordination of the work with existing foundations and subsurface construction (Lin and Hadipriono 1990). The placement of concrete in drilled piers or belled caissons requires skillful work by competent workers, especially when the shafts are flooded. Pumping water while placing concrete disturbs the soil and the concrete, producing contaminated concrete. Experienced deep foundation contractors exercise caution to keep the bottom of the underwater concrete pipe (tremie) inserted into the fresh concrete. When these precautions are not followed, the result can be economic disaster (Kaminetzky 1991). Of course, dewatering a site to avoid such problems requires constant vigilance to prevent damage to adjacent structures and below-grade utilities in the urban context. Some of these concerns are discussed further in Section 11.3.

3.2.2 Culverts, Pipelines, and Tunnels

Despite extensive research and wide dissemination of experience, pipelines, both water supply and sewer, continue to fail in service. Premature failures from design and construction deficiencies become the object of much public scrutiny, since they are often funded locally and since they affect the general public directly. Of course, many infrastructure failures also occur due to lack of maintenance. The problem is encountered everywhere but especially in metropolitan areas.

For small, premature pipeline failures, the cause is often that the trench was dug too wide and lack of proper bedding provision for the pipeline. Improper control of backfilling is also often cited.

Buckling of a 1.7-m (66-in.) corrugated pipe sewer outfall line in Salinas, California, in 1958, affecting 610 m (2000 ft) of the 2300-m (7500-ft) length, was blamed on faulty backfill. The pipe of 10- and 12-gauge metal was laid through sandy clay, river sand, and muck, in a trench kept dry by wellpoints. Crushed rock was placed 150 to 600 mm (6 to 24 in.) below the flow line for stability, with native soil above as backfill, there being no special specification on this important item.

Grade changes at the surface must be done with consideration of the effect on buried structures. An old brick 2.4-m (8-ft) circular sewer in Atlanta, Georgia, designed for 5-m (15-ft) cover was loaded with 12 m (40 ft) of new highway embankment. The shape changed to elliptical, with large longitudinal fissures in the crown. Some 180 m (600 ft) had to be shored and rebuilt as a gunited 1.8-m (6-ft) pipe with two layers of mesh reinforcement. Smaller pipe under such changed conditions of loading requires complete replacement.

Pile-supported rigid pipelines crossing roadways in soft soil areas will locally stop the natural soil consolidation under the pavement, as happened at the Long Island Expressway in New York City in 1962. Combined with high temperature, a section of 230-mm (9-in.)-thick double reinforced roadway slab heaved up 760 mm (30 in.), closing all traffic. Compatibility between the rigid pavement on a flexible base and the rigid sewer below was established by using a concrete arch cap on timber sills spanning the sewer, with open space above the sewer to allow for future settlement of the pavement.

In recent years, emphasis has been placed on repair of the infrastructure. Several significant pipeline failures have reminded the public of the need for diligent maintenance and repair. For the aging infrastructure, rehabilitation is sometimes preferable to replacement, even if the cost is greater. Rehabilitation is often the desirable solution in the dense urban environment, since easements and access for replacement construction are difficult to obtain. Repairs to operating pipeline systems are difficult, but several methods are available that do not require extensive shutdowns. One technique, called *sliplining,* uses a new pipe that is slipped inside the existing pipe, with the new pipe then grouted in place. Increasing concern for the seismic vulnerability of lifelines (oil pipelines, water supply lines, and waste lines) is also stimulating funding for pipeline rehabilitation projects.

Several explosive failures in spirally prestressed concrete pipelines illustrate that new materials are not immune from problems. One such pipeline, a 2.14-m (7-ft)-diameter water supply line in Tampa Bay, Florida, ruptured suddenly in 1990. In 1992, concerns were expressed regarding the integrity of precast prestressed pipe sections that were installed in the 1970s as part of the 540-km (336-mi) aqueduct that supplies Colorado River water to central Arizona. There have been no ruptures yet in this pipeline, but extensive corrosion damage was discovered in sections of the 10-km (6-mi) portion of the aqueduct that involved prestressed pipe. The sections are 6.4 m (21 ft) in inside diameter with a wall thickness of 38 mm (1.5 in.). These problems and many similar failures involving prestressed concrete pipe serve as reminders that there are no miracle materials. Modern materials, like their predecessors, require careful attention to specifications, design, control in manufacture, and competent installation. When prestressed concrete members are damaged during transportation or installation, corrosion can easily become a catastrophic problem, even in well-designed and manufactured pipelines.

Of course, pipelines using the older materials continue to experience failures, some with dramatic consequences, such as the multiple failures of a buried 2.74-m (9-ft)-diameter steel water supply line in Colorado in 1980–1982 (Moncarz, Shyne, and Derbalian 1987; Eberhardt 1990). In 1991, operational errors led to a central California accident when workers at a power plant inadvertently shut a turbine gate and a bypass gate simultaneously. This action caused a shock wave that blew open a 6-m (20-ft) section of penstock. The resulting sudden pressure loss destroyed a 305-m (1000-ft) section of 2.4-m (8-ft)-diameter steel pipe, requiring several million dollars for repairs.

One of the most highly publicized pipeline failures in modern times was the massive rupture of the 30-year-old Point Loma sewage outfall system off the coast of San Diego, California, in February 1992 (Prendergast 1992). This failure caused 685,000 m³ per day (180 million gallons per day) of primary-treated effluent to be discharged into the Pacific Ocean, 300 m (1000 ft) offshore, for two months, a total of 45 million cubic meters (11.5 billion gallons) of effluent. Round-the-clock repairs in harsh winter weather involved an unprecedented amount and size of pipe and volume of flow. The 150-m (500-ft) broken section of the 3450-m (11,300-ft)-long outfall was 960 m (3150 ft) offshore in 11-m (35-ft)-deep water. The pipe sections were reinforced concrete 2.74 m (9 ft) in diameter. Emergency repairs cost approximately $14 million. Very little corrosion or joint deterioration was found in the broken pipe. One forensic investigation determined that the most probable cause of the failure was 77,000 m³ (100,000 yd³) of trapped air in the outfall line, which caused the pipe to become buoyant and more susceptible to movement in underwater currents.

Large corrugated steel culverts are distressed and may fail when the desired uniformity of soil pressure is not obtained. Causes are nonuniform compaction of backfill and application of eccentric loads, as happened in Crestwood, New York, in 1963. A wheel load from a concrete truck caused failure of a 910 by 1200 mm (3 by 4 ft) corrugated drain with a 900-mm (3-ft) cover. Immediately after the pipe was reconstructed under most careful control, identical failure resulted from the next concrete truck crossing.

Large culverts built in nonporous soils must be designed for possible uplift from hydrostatic pressure. One such concrete box culvert built in San Juan, Puerto Rico, in 1964 in a soil of poorly stratified red to white silty clays and fine sands failed during a heavy rainfall. At the time of failure there was 1.8 m (6 ft) of water in the 9.1-m (30-ft)-wide 2.9-m (9.5-ft)-deep rigid frame structure. The floor buckled for about a hundred meters (several hundred feet), while the vertical walls remained substantially fixed in position.

This failure was similar to that of a concrete four-pass aeration tank at the Jamaica Bay sewage treatment plant in New York in 1965. The soil was a hydraulic fill sand on top of meadow mat. The tanks were designed on the assumption that an earlier construction program had removed the impervious meadow mat. Adjacent 35-year-old operating tanks were known to permit some raw sewage leakage into the subsoil. Sufficient accumulation of this waste liquid had eliminated all filter capacity of the soil. Upward pressure developed to the extent that the new tank floors were heaved and displaced and the intermediate walls were crowned up a maximum of 800 mm (32 in.) in the 61-m (200-ft) length. End walls and sidewalls had been backfilled and were held by the soil. When the floor was broken by the distortion, blows developed, the major geyser being almost 600 mm (24 in.) in height. With the tanks filled to that level, the floors receded to almost the original position and the walls returned to a straight line. A few months later when the tanks were pumped out, a similar but much smaller distortion occurred, and this again receded upon flooding. To correct the work, pressure relief valves were installed in the floor, an added pile foundation was provided for the support of the tank walls, and a new floor was constructed, since all metal wall stops had been sheared.

Of all engineering structures, tunnels are most vulnerable to small troublesome and large catastrophic failures. The construction process removes macroscopic volumes of soil or rock from the internal structure of a stable mass and upsets the state of equilibrium. Sudden release of internal strains causes rock movements which cannot be predicted precisely by preconstruction geological and geotechnical investigations. Careful monitoring of the conditions within the soil mass with modern geotechnical instrumentation has improved the safety of such construction significantly, but accidents are still all too common (see Section 11.3).

Rock falls can come near the portal or, more often, under the highest rock cover. In the diversion tunnel for the Green Peter Dam in Oregon, a mass of rock broke through the steel roof supports when the heading had advanced only 18 m (60 ft). In the tunnels carried into the Cheyenne Mountain in Colorado, thorough geological investigations did not reveal the cross-bedding of seams in the center of the mountain. A large cavern formed, requiring extensive structural support where cross-drifts met, and $1 million was spent on rock bolting to stabilize incipient spalling.

Successful completion of a tunnel despite unpredictable obstacles is a monument to the stamina and determination of the workers in a recognized hazardous occupation. Some tunnels suffer more than an average share of troubles. The 13-km (8-mi)-long Shimizu Railway tunnel in Japan was holed through in August 1966 after three years of hardships endured by three contractors who divided up the work. This is a tunnel parallel to the 10-km (6-mi)-long earlier Shimizu Tunnel, completed in 1931 after nine years of work at the cost of nearly 50 lives. The work on the center section of tunnel, which required an inclined access shaft and had 1200 m (4000 ft) of rock cover as it crossed the mountain ridge, was exposed to the greatest hazards. Water came from the rock seams in one 45-m (150-ft) stretch at 1.5 MPa (215 psi) and up to 0.42 m³/s (15 ft³/s), requiring six 300-kW (400-hp) and three 375-kW (500-hp) pumps to eject the water through the access shaft. Sealing the joints with silicate and epoxy resin grouting finally stopped the flows, with water pressure buildup in the rock to 4.2 MPa (615 psi). Under the highest peak elevations, rock pressures caused squeeze of the walls and falling rock, requiring the installation of steel bents at 1.5-m (5-ft) centers with full lagging cover. At 240 m (800 ft) from the tunnel portal a hot spring at 50°C (122°F) required boring a ventilation shaft to permit work even on a shift reduced in length to 4 hours. Complaints of nearby hotel resorts that their supply of natural hot springs was being depleted then required chemical sealing of the flowing seams. An unusual problem was electricity leakage from the operating tunnel, making it hazardous to activate electrical detonators and requiring electric insulation for all ducts and rails carried into the tunnel.

Lebanon also had an eight-year job to hole through a 16-km (9.9-mi) 3-m (10-ft)-diameter water tunnel in 1965 to connect the Awali and Litani rivers. Construction was stopped in 1959 when the heading, advancing in limestone, entered loosely cemented sandstone, and a flow of 1.0 m³/s (265 gal/s) brought down the roof, filling up almost 3.5 km (2 mi) of excavated tunnel. The water flow increased to a maximum of 5.7 m³/s (1500 gal/s), and then leveled at 0.2 m³/s (50 gal/s), after some 100,000 m³ (130,000 yd³) of sand was deposited in the tunnel. Work was completed after treating the problem as a soft ground excavation.

Another seven-year-long tunnel job was the original twin-bore Wilson Tunnel through the Koolau Mountain near Honolulu, driven through the well-known geological volcanic ash. A recount of events during the construction of this tunnel illustrates the uncertainties and political implications of such projects. The tunnels each carry two lanes of traffic and are 7.9 m (26 ft) wide. Total length is 846 m (2775 ft), originally shown in the 1953 contract documents as 587 m (1925 ft) in rock and 260 m (850 ft) in earth, on a 6 percent grade. The work started on one tunnel in rock with a three-platform jumbo on which nine drills were mounted as a full-face operation. By June 1954 the rock section changed to earth. Work continued as a full-face method with 200-mm (8-in.) I-beam arch ribs spaced 0.6 to 0.9 m (2 to 3 ft), excavating the sides ahead so that the posts could be set first.

On July 10, 1954, several tunnel supports collapsed 18 m (60 ft) back of the heading; soil fell in, burying the equipment and filling 52 m (170 ft) of the tunnel. A crater formed in the mountain 37 m (120 ft) above the tunnel. Water flow increased, a second mudflow started 40 m (130 ft) back from the heading, followed by a third flow on August 14, when 60 m (200 ft) of tunnel was filled and five workers were killed. Work was stopped on the earth section and continued in the rock section while the city and the contractor searched for consultation advice.

With an election close at hand and the mayor's position at stake, concerted efforts were made to clear the city of any responsibility. Some but not all of the tunnel consultants contacted by the authorities signed a report placing the blame on the contractor's methods, but the electors defeated the mayor in November. With 550 m (1800 ft) of the 846 m (2775 ft) clear, the 460-mm (18-in.) concrete was placed in the rock section and probings were made by a pipe 60 m (200 ft) into the cave-in dirt. The contractor claimed that a tunnel could not be driven in the earth section and offered to complete the work as either a wide-open cut or else a concrete tunnel in open excavation to be filled over later to avoid possible slides in the deep open slopes. The proposed open cut would have been 60 m (200 ft) deep maximum, 15 m (50 ft) wide at the bottom, and 200 m (650 ft) at the top.

The contractor claimed that the extra costs of the experts' recommendations would pay for the open cut for two tunnels (the city contemplated a parallel tunnel in the future) and the lined second tunnel in the rock section. To avoid further soil flow, the cave-in section was bulkheaded with a 0.9-m (3-ft) concrete wall and some 1150 m³ (1500 yd³) concrete was force-pumped behind the bulkhead to fill the space between the roof and the debris of the earth section tunnel. The new mayor felt that the contractor should continue the work despite the consultants' recommendation that the contractor be terminated. The resulting litigation must have been welcomed by the legal profession: claims and cross-claims between city and contractor and a $1.5 million damage suit by the contractor against the consultants for malicious and false statements ". . . made in bad faith and in reckless disregard of the truth and facts."

In May 1955, the city ordered the contractor to complete the work, but he insisted on an agreement for extra costs and new limitations on liability, claiming that the design was deficient. Some nine months later an agreement was reached with some interesting clauses:

1. The contractor would assume the cost of reexcavating the 150 m (500 ft) of cave-in.
2. The city would assume the extra cost for a heavier design of temporary and permanent tunnel lining.
3. There would be no penalties for delay in completion of the project.
4. Both the city and the contractor would replace their tunnel superintendents.
5. The city would purchase any steel ribs on hand that did not fit the enlarged tunnel section.
6. All cross-claims were waived except those arising from damage suits of the families of workers killed in the cave-ins.

One year later an exploratory drift had been carried through the cave-ins to the end of the tunnel, approach road contracts were signed, and more than $1 million in damage suits were filed against the city for the death of the five workers. The earth section was redesigned to incorporate a curved invert, and the original 460-mm (18-in.) wall thickness was changed to 1.2 m (48 in.) at the maximum cover and 812 mm (32 in.) elsewhere in the earth section. Excavation was changed to side drifts, with very close spacing of steel arch ribs. Progress was slow, and concrete was placed in 0.9-m (3-ft) segments immediately behind excavations. Initial progress was only 150 mm (6 in.) per day as the saturated soil was being removed. The first bore was completed in May 1959 at a cost of $1 million above the original contract of $4.5 million. Soon after resuming work, the same contractor received a contract to build the second parallel tunnel and complete it one month after the first was completed, at a cost of $4 million. The tunnels are open today and in heavy use. The normal increase in traffic required the twin bores. Since different facets of the construction industry will draw quite different conclusions as to the cause of these problems (both technical and political), no attempt is made to point to the lesson to be learned from this incident, except to illustrate the complexity of such projects.

Tunnels in sand usually require fairly simple repetitive steps of protection and excavation. But a sewer tunnel in Brooklyn going through coarse dry sand was suddenly confronted with live free-flowing sand when the excavation crossed under a series of railroad tracks. The vibration of the passing trains converted the stable sand into a fluid mass, requiring a revision in excavation methods.

In 1979, silty sand oozed into a 3-m (10-ft)-diameter Cleveland, Ohio, sewer tunnel. A boring machine was trapped in the accident. The remaining 790 m (2600 ft) of the 1800-m (6000-ft) tunnel was completed under compressed air conditions at a reported cost of $11 million, three times the initial estimate.

Natural gas in shale rocks and human-caused fumes have caused serious explosion and poisoning accidents in tunnel work. At the El Colegio Hydro-electric project in Colombia, pockets of methane gas in the shale were released into the free air pressure in the tunnel and caused five explosions; the most serious came after the tunnel was lined and grouted. In a Japanese irrigation tunnel at Monomura, 25 workers died of carbon monoxide poisoning after their foreman brought in a gasoline-powered generator because an electric extension cord was not available. He ordered the workers to continue even after they complained of dizziness, and he was therefore charged with negligence leading to the deaths.

A sewer tunnel was rocked by a series of explosions in 1962 in Philadelphia. Although the 1.8-m (6-ft)-diameter tunnel was being dug in ground known to be impregnated with petroleum wastes and the contractor did check the exhaust air periodically, sufficient seepage gas accumulated to be set off by an electric spark from an open switch or a defective connection. Sparkproof lighting and additional ventilation was ordered before work was continued.

Recent underground construction accidents show that tunneling work is still extremely hazardous, despite modern regulations and improved techniques. Examples include a 1990 tunneling accident in Tokyo and 10 fatalities from a variety of accidents related to a major sewer excavation project in Chicago, leading to a full-scale investigation in 1987 (see Section 11.3). A catastrophic explosion in a Milwaukee, Wisconsin, tunnel killed three workers in 1988 (see Section 11.8).

The incidents described here are only a few cases to indicate the types of trouble encountered in tunneling work. There have been many such cases and they have spurred on the development of tunnel coring and boring machines to guard against a sudden change in material and to provide protection during excavation of the heading. Japanese and European construction researchers have pioneered developments in robotic tunneling equipment. Also, new materials and techniques are available to improve site conditions for tunneling work. Chemical grouts are used to fill voids and stabilize incompetent rock material in mine and tunnel excavations. In 1985, a major rehabilitation project used foamed chemical grouts to extend the useful life of the 81-year-old Mt. Washington tunnel in Pittsburgh, Pennsylvania. Microfine cement grouts are routinely injected into loose soils and fine sands to improve the soils prior to excavation for new tunnels. When problem cohesive soils are encountered, ground freezing is an option in modern tunneling projects.

3.2.3 Rock as an Engineering Material

Failures in work involving natural rock nearly always stem from the false assumption that "rock" is a homogeneous continuous solid. The high cost of flawless precious stones should serve as an indication that geologic materials vary. Continuity and homogeneity are very unusual conditions in natural rock structure.

Normally one expects to find the softer, seamier rocks on top of a more massive bedrock. However, this is not always true, as demonstrated in the northwestern corner of Westchester County, New York, where the cap rock is an emery of great hardness and toughness with no seam structure but the underlying rocks are seamy and brittle granites. Cavity formations in sedimentary rocks have given considerable trouble. In the New York subway contract along St. Nicholas Terrace near 155th Street, where borings indicated dense schists and all buildings were required to be underpinned to rock bearing, excavation of rock in the subway cut uncovered an unusual condition. For a length of over 60 m (200 ft) there was a mud-filled seam in the rock up to 6 m (20 ft) in height within the bedrock. All along this exposure, the rock had to be underpinned to the lower rock as a protection for the buildings and for the subway structure, which had been designed on the assumption of a structural section in rock. Preconstruction borings had been located 90 m (300 ft) apart and had missed the discontinuity completely.

Tunnels planned as rock excavations may encounter discontinuities in "solid" mountains. One of the most carefully explored mountains prior to tunneling operations was the location of the North American Air Defense NORAD command station in Cheyenne Mountain, already mentioned in Section 3.2.2. It was chosen because of the rock density and continuity exhibited by the core tests. At the intersection of the galleries near the middle of the mountain, however, the rock fell apart and a domed cavern formed twice the desired height. Numerous permanent steel supports and a close pattern of rock anchors were necessary to provide internal stability.

A different difficulty resulting from misinterpreted borings is the case of a shield-driven tunnel near the East River in New York, designed on the basis of river depth soundings to locate rock levels. The profile was chosen so as to get a full-earth section except for the shore approaches, and the contract indicated that this would be the condition to expect. Construction soon proved that the borings were inaccurate. Rock protruded above subgrade, irregularly and in varying amounts, making progress difficult and costs almost double that associated with an all-earth section excavation under air pressure. When some of the abandoned test boring casings were encountered in the heading, the mystery was solved. The ends of the casings were curled into spring spirals where they had hit sloping rock surfaces. The depth to rock was incorrectly assumed to be the length of embedded pipe at ultimate resistance.

Bedrock may be mineralogically continuous but have open internal seams, often filled with water-deposited soft muds which seriously affect its engineering value. An example of such a condition was exposure of the fine-grained gneiss and schist at the Manhattan approach to the New York Battery Tunnel. Two roadways were built with an intended rock core separation about 9 m (30 ft) wide. Exposure showed the rock to be dense and hard but completely broken into blocks with open gaps between the blocks. The sections of rock were lying on a sloping bed, so that considerable steel bracing was needed to salvage the lower blocks for embedment in a concreted core wall. The existence of such seams places an indefinite load-bearing value on the rock and practically eliminates the use of uplift or lateral anchors. Even rock bolts are of questionable value.

Natural and human-made cavities in bedrock can develop into large-scale surface settlements or local sinkholes. Drifts from a 40-m (130-ft)-deep coal mine carried under Spring Creek near Indianapolis caused settlements up to 400 mm (16 in.) in a center pier and abutments of a bridge crossing the creek, with serious damage to the deck. To avoid interruption of use, piles were jacked to rock at each support and the concrete spans, each weighing 200 Mg (220 tons), were jacked to a level condition and underpinned.

When some column footings of a large one- and two-story factory located in the southwestern corner of Virginia, founded on spread footings on soil and on limestone rock and some on drilled caissons to rock, settled and moved laterally during a cement slurry grouting of cavities, reevaluation of the foundation program resulted in a complete design change. Borings showed that the grout was not consolidating the surface soils or closing the rock seams but was disappearing into lower cavities. The solution and pressure effect of the liquid grout had simply broken the skeleton framework of the rock, causing loss of support.

Grouting was stopped, and certain floor areas where spans were favorable were designed as supported slabs on the column footings. The rest of the area was covered with a harrowed-in skin of limestone dust and fragments, which was consolidated before placing the floor. All piping under the floor was encased in concrete to protect against leakage. The perimeter was surrounded by a concrete-surfaced trench to prevent infiltration of rainfall into the ground. Columns that had settled were shimmed level, and in the areas of movement, column-base anchor bolts were left without nuts to prevent future footing settlement from placing downward loads on the columns and distorting the frame. Subsequent use of the building proved that the remedial work was sufficient protection against the ground conditions.

A spectacular sinkhole accident was reported by J. E. Jennings at the 1965 International Conference of Soil Mechanics and Foundation Engineering. A three-story crusher plant at a gold mine some 80 km (50 mi) west of Johannesburg, South Africa, located in dolomite rock collapsed completely into a sinkhole without warning. Groundwater was lowered 120 m (400 ft) and surface settlements of 6 m (20 ft) have been measured. The largest recorded sinkhole in these dolomites is 90 m (300 ft) in diameter and 37 m (120 ft) deep. Any human-made change in conditions of the subsurface should be a warning of possible subsidence, and in limestone-dolomite rocks, of change in water movements that will dissolve the supporting rock.

Internal seams in the rock volume, often altered by the release of the confining pressures resulting from excavations at upper levels and by later modification through change in water pressure and flow, are the controlling factors in rock strength and stability. Tests on core samples that do not exhibit the extent of the seams, or tests that do not take into consideration future changes in weathering, can mislead a designer into serious error. Even a minor earth tremor can liquefy a saturated mud seam and cause a major rockslide.

Tree roots have tremendous expansive power. A number of rock cliffs in the northern part of New York City have failed from ice and water pressure after the roots have opened natural seams. One such fall 2300 to 3050 m^3 (3000 to 4000 yd^3) in volume crushed a one-story factory building. The fall pulled out part of the rear yards of the apartments on top of the cliff but, fortunately, did not undermine the foundations.

Tunnel excavation is also vulnerable to rock movements along internal seams that are cut as excavation progresses. In some rocks there is a warning of popping or shear creaking. Modern geophysical instruments prewarn of the possibility of rock movements. These valuable instruments supplement the less reliable visual and audible sensing and the judgment of experienced tunneling workers.

Rock is an aggregate of fused or compressed discrete mineral particles. Elastic properties determined from core samples are reliable only for low stresses. For high-stress conditions, the seam structure and anisotropy of the rock must be considered, as well as the plastic creep potential from crack and crystal adjustment with time. Comparison of test results obtained on core samples and in large-scale tests in tunnels and in deep dam excavations are available. Such data should be reviewed carefully before accepting strength and elastic modulus values of rocks as determined from core samples in the laboratory.

Even in the best granites, such as the quarries on Vinalhaven Island in Maine, plastic flow has been found in the rock at some 46-m (150-ft) depths. In one case, the solid sheets buckled upward, with serious pop-outs. The fractures made further quarry excavation wasteful since so little usable stone was recovered. Similar rock movements—elastic rebound followed by continuing plastic deformation—were reported in the shale exposures at the Oahe Dam in South Dakota when some 1.1 MPa (12 tons/ft^2) of overburden pressure was removed. Release of large vertical loads over a limited area in deep excavations alters the physical characteristics of the bedrock, causing change in shape, opening of seams, and creating zones for potential water infiltration and softening.

Rock movements of the dolomite cuts made in the 1890s for the wheel pits to harness Niagara water power required continuous addition of bracing at the bottom of 55-m (180-ft)-deep cuts. With the advent of electric power replacing the belt-transmitted water-turned shafts, plants were constructed on the banks of the Niagara gorge. Some rock movements were reported in 1958 in the construction of tunnels and open trenches for the Ontario Hydro development on the Canadian side.

After the Schoellkopf power plant on the New York side of the Niagara River was damaged by a slide following an earth tremor, the Niagara Power Authority built twin concrete conduits 14 m (46 ft) wide by 20 m (66 ft) high set into open-cut trenches with 15 m (50 ft) of rock between the trenches. The cuts were up to 50 m (165 ft) deep in horizontally layered limestone and dolomite. To contain the vertical sides during general excavation, the rock was first "pre-slit" by blasting within closely spaced line holes. Within the line holes, the drilling pattern was on a 2.1-m (7-ft) grid, loaded to break the rock within 1.8 m (6 ft) of the walls, and the outer 1.8 m (6 ft) on each side was slashed by light charges. Lifts were about 3.7 m (12 ft), with a final cut of 1.8 m (6 ft) to reach subgrade. A rectangular pattern of holes was drilled in the bottom for embedment of rod anchors to resist hydrostatic uplift on the empty conduit.

Considerable difficulty was found at the greater depths from continuous rock movement (Figure 3.3). Line holes went out of position, anchor holes were distorted, and the formwork for the floor was squeezed enough to crush the 300 by 300 mm (12 by 12 in.) timber sills. Rail supports for the sidewalls were continuously displaced, the subgrade heaved and cracked, and even the 1.5-m (5-ft)-thick slab built with a centerline construction joint heaved to open the joint a full 50 mm (2 in.). Measurement of movements indicated an initial 75 mm (3 in.) at the centerline, which increased in one month to 216 mm (8.5 in.) with a 64 mm (2.5 in.) heave along the sidewalls. Local areas displaced even more, accompanied by loosening of the subgrade sheets and raveling of the sidewalls. Replacement of the load by the construction of the conduits and backfill to the surface reestablished stable conditions. However, this experience indicates the need to consider and compensate for movements during the construction period.

Squeeze of rock in deep tunnel construction is an expected phenomenon, as is the accompanying flaking, popping, and spalling of the rock exposures. In the French half of the Mont-Blanc Tunnel, more than 100,000 rock anchors were inserted to prevent rock bursts in the overstressed granite. About 20 percent of these were placed in the face of the tunnel as a safety measure before progressing. Rock bolting can be successful in the prevention of rock separation and fall, but only after the exposure has been made, when the rock

Figure 3.3   Subgrade upheaval in deep rock cut.

structure has already been altered. If the rock could have been sewn together before exposure, much of the rock separation could have been avoided. Modern soil grouting techniques have been used successfully to stabilize certain incompetent materials prior to tunneling.

The properties of rock can change over the years. Subway excavation in previously developed areas provides an opportunity to examine rock conditions previously tested and used to support loadings. Some rocks, such as mica schists, may change radically in just a few years. This is probably not the case with igneous rocks unless seams permit internal weathering. In the construction of New York's Sixth Avenue Subway, several startling examples were uncovered of change in rock even when it was sealed by concrete footings. At Forty-Second Street, an office building was constructed in 1917 with all foundations reportedly placed on "hard ringing rock." However, this structure required underpinning to sound rock when it was exposed in 1936, to as much as 7.6 m (25 ft) in height. The design engineer for the original building construction personally examined the condition and brought his original field notebook, showing that each of the piers had been examined and tested at the time of construction. The maximum live full load bearing was 1.9 MPa (20 tons/ft^2). The rock had softened at a rate of about 300 mm (12 in.) in depth per year; removal was by pick and shovel.

The Queensboro subway crossing Sixth Avenue was built as a twin tunnel in rock, using timber supports later encased in concrete and grout. Construction reports showed hard and dense but seamy and blocky schist. Ten years after completion, the roof was exposed in the Sixth Avenue excavation, and no hard rock of any kind was found. The grout-encased timbers were removed after excavating the dense but soft disintegrated mica schist, using a small power shovel with no blasting or chipping.

Other rocks also "misbehave" on exposure. These include expansive materials, discussed in Section 3.1. In Denver, Colorado, a shale formation excavated for a concrete reservoir was noted to swell as much as 100 mm (4 in.) following heavy rains. Similarly, east of Cleveland, Ohio, the shale contains as much as 4.5 percent pyrite which expands upon oxidation and heaves footings, floors, and walls unless corrective measures are taken.

Natural "rock" can be the best foundation material, but problems of design and construction must always be approached with an open mind. In construction, all rocks are not equal. Nor, after construction, are they as stable as before construction started. When one considers that soils come from the rock of mountains, and mountains continuously wear away, it follows that the term *stability of solid rock* should not be blindly accepted.

3.3 FOUNDATIONS OF STRUCTURES

3.3.1 Undermining of Safe Support for Existing Structures

Careful, thorough soil investigation beyond the boundaries of a proposed foundation or earth structure is a necessity. The fact is that there will be a stress change in the soil when excavated material is removed. The designer should always assume that this stress change may cause ground movement outside the boundaries of the excavation.

When designers ignore the presence and condition of foundations of adjacent existing structures, the results can be tragic, as illustrated by a few historic examples here and more recent examples in Section 11.3. Excavation adjacent to or below an existing wall or column clearly will reduce the support value of the foundations. Yet such evident error has often been committed.

In Wheaton, Illinois, in 1926, excavation about 1 m (3 ft) below an adjacent wall footing in stiff blue clay caused a 300-mm (12-in.) settlement of the neighboring building. Excavation for a deep cellar near a five-story wall of a residential building in New York in 1960 caused total collapse of the 90-year-old building.

Again in New York, prior to the excavation for a three-level garage below street grade for a multistory apartment building, an adjacent five-story wall-bearing building was underpinned to rock level. During excavation of the bedrock using controlled blasting because of the adjacent buildings along two lot lines, a major slip of the rock carrying the underpinning removed all foundation support under the rear part of the adjacent building. Part of the wall fell and part hung to the stable portion, with the masonry corner remaining plumb. The cost of providing a new foundation exceeded the value of the building. None of the borings within the project limits indicated such a loose condition within the bedrock. Seamy rock must be provided with temporary shoring until covered and braced by the permanent construction.

In some soils, excavation need not be carried even down to adjacent footing level and yet cause trouble. In 1924 the soft clay excavation for the Lafayette Hotel in Buffalo, New York, caused the adjacent wall footings to settle 150 to 200 mm (6 to 8 in.), and displace laterally the same amount, even before the old footings were exposed. Similar lateral clay squeeze during the 1905 extension of the cellar for the Meyers Department Store in Albany, New York, caused partial wall collapse when only 1.5 m (5 ft) of the old wall footing was exposed.

In 1893, after its conversion to an office building used by the War Department, the Ford Theater Building in Washington, D.C., was involved in an undermining failure. An excavation in the cellar under the piers of the front wall caused some 12 m (40 ft) of the wall to collapse below the third floor, hurling occupants and equipment into the cellar. In 1964, a foundation for a two-story building in Ottawa, Canada, was being placed in a trench in clay soil about 450 mm (18 in.) deeper than the footing of an adjacent building. Part of the new wall was concreted, but the remainder of the trench was left exposed to the weather. After two days, the neighboring wall collapsed as the clay below its footing softened.

Alongside a three-story government building in Brussels, Belgium, in 1962, an uncontrolled excavation reportedly caused "hollow sounds" in the sidewalk. As the decision was being considered whether to evacuate the government building, it collapsed, causing 10 deaths and 20 serious injuries.

Overloading stable foundations can cause failure, as was proven empirically in Cairo, Egypt, in 1964. A two-story concrete building built in 1948 with 125-mm (5-in.)-thick concrete walls was raised successfully to three, four, and then five stories. The five-story building contained 15 dwelling units. The original foundations had not been upgraded and they overloaded the soil to the point of failure. The apartment building collapsed suddenly, taking 31 lives.

Even rock excavation below structures resting on rock but with unfavorable sloping seams can cause loss of support and failure of the wall. In one such case in New York, the rest of the building was reinforced by cable ties introduced for all footings and anchored to the rock at the opposite side of the building. The undercutting of the sloping rock sheets disturbed the continuity of support when the rock sheets moved slightly to reach a new position of stability.

Excavation for drains and pipe changes alongside existing buildings, being considered minor work, often is undertaken without regard for the effect of undermining. One failure occurred in 1965 at a new shopping center in Atlanta, Georgia, where the footing was undermined by a sewer excavation. The sewer excavation was 4.6 m (15 ft) below the footing. It permitted a shear failure in the firm sandy micaceous silts, dropping the column, together with the 18-m (60-ft) prestressed concrete double-tee roof beams.

In the Chicago clays, a tunnel excavation in the street caused the 16-story Unity building to lean 150 mm (6 in.) in 1900, which increased to 680 mm (27 in.) within 10 years. Expensive foundation underpinning and correction of the column connections were required to avoid collapse.

In many regions, coal mining removes natural support, sometimes with catastrophic results. As early as 1923, partial failure of the Spring Creek bridge near Springfield, Illinois, was diagnosed as being caused by coal mining. Suddenly, the center pier and one abutment of the two-span concrete bridge settled from 200 to 400 mm (8 to 16 in.), even though all foundations were placed on rock when the bridge was constructed in 1921. Inspection of a coal mine 40 m (130 ft) below the surface showed that the roofs over seven rooms directly below the bridge had collapsed, filling the 1.5 m (5 ft) of excavation height. The bridge was jacked to grade and underpinned with brick masonry as an emergency measure, with permanent reconstruction postponed until all subsequent consolidation of the bedrock had been completed. In 1963, the Zeigler–Royalton High School in southern Illinois was closed when the building sank 100 mm (4 in.), with damage limited to falling plaster. The settlement was blamed on the collapse of an abandoned coal mine.

Deficient control of the flow of domestic water and storm water can erode a supporting soil structure, undermining the foundations of structures. Plans were announced in the late 1980s to demolish about 1000 row houses in Philadelphia, many with severely cracked and uneven foundations. The houses were built between 1910 and 1920 on 3 to 12 m (10 to 40 ft) uncompacted fill made of ash and cinders. The area was originally a valley containing a creek. The fill was eroded by leakage from water lines and broken downspouts.

3.3.2 Load Transfer Failure

A properly designed frame will usually adjust its load somewhat to compensate for differential foundation settlement, thereby resisting total collapse. When the state of stability is disturbed by partial support loss through differential settlement, the reactions are redistributed among the available supports. While the structure system may still achieve overall equilibrium, the load transfer may result in significant stress increases in certain supports.

In a 13-story commercial building in New York supported by four continuous pile caps and timber piles, pumping operations from a neighboring excavation for a depressed roadway lowered the water table and caused one line of piles to settle. The settlement pulled one pile cap down slightly (Figure 3.4). This released the exterior wall support, and the load transfer to the interior line of columns overloaded their steel beam grillages. The 600-mm (24-in.)-deep beams collapsed to a 300-mm (12-in.) depth, with consequent 300-mm (12-in.) settlement of every floor in the building. The floor beams rested on girders between columns and had only a loose bearing, so that they took

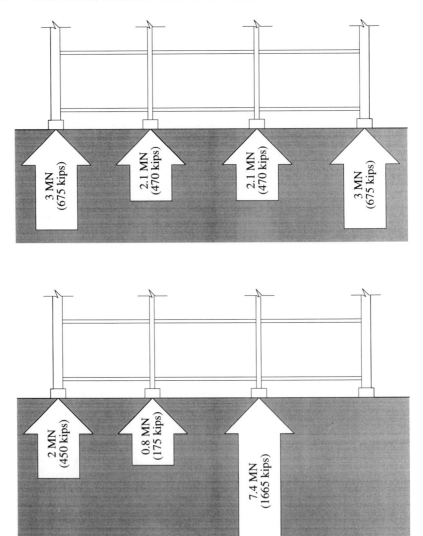

Figure 3.4 Significant change in reaction forces caused by uneven settlement.

the new trough shape without failure. After the entire building was shored internally, new shallow grillages were provided and each floor was releveled. The small settlement of the exterior wall was neutralized and its bearing restored on the existing piles.

When foundation disturbance results in masonry distress, investigators seeking the cause of the distress should consider the possibility of a load transfer. The actual foundation failure may not be located at the same position where the masonry distress is evident.

3.3.3 Lateral Soil Movement

In the foundation construction fraternity, it is usually conceded that a small amount of lateral movement in a structure causes much more damage than an equivalent amount of vertical settlement. Lateral movement is the result of imbalanced horizontal pressure, usually caused by removal of a resisting force component. Sometimes, however, lateral movement can be caused by a temporary or permanent increase in active soil pressure without providing a corresponding added resistance. In the design analysis, consideration must be given to possible changes in both active pressure and passive resistance as water content changes. Saturation increases the active pressure and may decrease the passive.

A heavy double-deck concrete viaduct built in 1966 into a clay hillside in Cleveland, Ohio, suffered shear cracks in the partly embedded columns when lateral pressures built up from saturation. The columns were 15 m (50 ft) high and 1.8 m (6 ft) in diameter set on 1.8-m (6-ft)-thick caps with 21-m (70-ft)-long piles. The piers were built prior to the installation of an adjacent embankment. Normal procedure is to construct the embankment first, so as to consolidate any underlying compressible materials before inserting the structure, but time constraints caused a reversal of the sequence. The amount of consolidation would have been the same under normal procedures, but the effect on the piers would have been much less severe. The columns were jacketed to a 2.4-m (8-ft) diameter, and a larger pile cap was built on top of the original cap.

Dredging of the waterway of the Mississippi River in 1962 removed the lateral support of the grain elevator foundation at Jackson, Mississippi, and seriously damaged the bins. Excavation for roadbuilding has regularly been responsible for lateral soil movement and related damage to existing structures.

In 1943, lateral flow of the soil under a 24-story office building in São Paulo, Brazil, caused total collapse. A similar subsurface slide at Altena, Germany, in 1960, under an eight-story office building was stopped by inserting 44-mm (1.75-in.)-diameter rods in 150 holes drilled 6 m (20 ft) deep.

Several instances of collapse have been reported when adjacent buildings were demolished, leaving an open cellar lower than the foundations of the remaining building. In 1957, an 11-story residence building in Rio de Janeiro, Brazil, collapsed completely after excavation adjoining the wall removed the lateral support of its building piers. Underpinning operations were not ordered until too late, after settlements and rotation were noted for almost two years. Even restricted excavations can cause similar failures. This was proven by the collapse of the Boston Club building in 1925, with the loss of 44 lives, caused by digging seven underpinning pits, 1.2 m (4 ft) wide and 2.1 m (7 ft) deep below the foundations, without lateral bracing. The result was 4.3 m (14 ft) of earth pressure with no provision for lateral support.

3.3.4 Unequal Support

All footings will settle when loaded. Structural problems result when the footings do not settle equally. Unequal settlements will occur unless soil resistances and load distributions are equal. Differential settlements can cause load transfer and tipping of the structure. When the framework is not structurally rigid, brittle materials such as masonry walls will crack along the shear surfaces bounding the volume that is supported by the less yielding structural elements. There are a great number of such structures, with foundations partly on rock and partly on soil, or with piles of unequal bearing value, or with foundations of more than one structural type but resting on soils of unequal compressive support. Numerous swimming pools in California, for example, have been built on sloping sites, partly on fill. Many of these have suffered total collapse due to inadequate design for this dissimilar bearing strata condition (Kovacs 1992).

Correction of these support deficiencies usually requires underpinning of the weaker foundation to equalize support, or in some cases, the addition of a subsurface enclosure to increase the bearing value of the weaker soil. An early example of the latter method, in 1919, was the addition of 2600 pinch piles to enclose the foundation of a grain elevator in Portland, Oregon. The 222-kN (25-ton) piles on which the structure had been built exhibited progressive unequal settlement. The added piles were 6 to 8 m (20 to 25 ft) longer than the original piles.

Aging of the support must be considered, as was shown by the settlement of the Baltimore, Maryland, pumping station, founded on large concrete piers carried to rock. Inspection of the pier bottoms in 1908 showed that some of the rock bearings were no longer on rock but on material of clay consistency. Certain metamorphic rocks, such as schists and shales, are known to lose their solid characteristics when loaded, especially if water saturated, and more rapidly in unclean waters.

There are some classical examples of tilting buildings due to nonuniform foundation conditions. The Tower of Pisa is probably the best known and researched leaning example, and it is still in use with (or perhaps because of) its 5-degree list. Another remarkable example of large settlements is the Palace of Fine Arts, built on the compressible clays of Mexico City. The building is still in use, with settlements measured in meters. Settlement of the Palace of Fine Arts has been quite uniform, causing little structural distress. The structure has actually settled and risen as a rigid body, resulting from the hydraulic effect of new large structures that have been built on adjacent sites.

The Pisa Tower is 54.6 m (179 ft) high and has a wall thickness of 4 m (13 ft) at the base. It is now 4.3 m (14 ft) out of plumb, with a rate increase in lean of 25 mm (1 in.) in 25 years. The foundation mat, a 18-m (60-ft)-

diameter plate with a hole in the center, rests on a fine volcanic silt. The foundation has rotated so that it is 2.1 m (7 ft) out of level, with the highest base pressures at the low side. Tilting became noticeable, causing the work to cease, in the latter part of the twelfth century, when the tower had been built to only three tiers. A 60-year cooling-off period elapsed, and since the tower was not moving, another story was added. Work stopped again when additional tilting was noted. Another 100 years went by, and the tower was completed with some changes in design in an attempt to counterbalance the tilt tendency. In 1932, an attempt to stabilize the soil by grout injection proved unsuccessful. The added 900 Mg (1000 tons) of grout added weight but little stability to the soil. A 1965 report, following an exhaustive investigation, recommended limitations on the pumping of groundwater in the vicinity. Another council was formed in 1983, and an international commission in 1990, to monitor the condition of the tower continuously. The tower continues to increase its lean and the lower portion has been strengthened. Plans for stabilization of the tower are frequently in the news. However, no one has proposed righting the tower. The Leaning Tower of Pisa adds about $2 million to the local economy from tourism. A vertical tower would probably remove Pisa as a tourist attraction, but so would a collapsed tower.

In 1906, at Tunis, Tunisia, two warehouses on concrete grid mats tilted some 25 degrees but were righted by loading the high sides. The first building was leveled and became stable after settling 4.6 m (15 ft). The second building had tipped so that the overhang increased to 5.2 m (17 ft) in 17 hours. It also leveled, but only after settling 5.5 m (18 ft). In 1913, the Transcona grain elevator in Manitoba, Canada, consisting of 65 concrete bins, 29 m (95 ft) high on a concrete mat 23.5 by 53 m (77 by 175 ft), settled 300 mm (12 in.), then tipped 27 degrees when about 85 percent full of grain (Figure 3.5). The structure weighed 18,000 Mg (20,000 tons) and the grain weighed an additional 20,000 Mg (22,000 tons). By an extraordinary engineering feat, the structure, which was 10 m (34 ft) out of level, was righted by underpinning with piers to rock support (Morley 1996). In 1955, a smaller one-year-old grain elevator, consisting of 20 bins, 37 m (120 ft) high on a 15.8 by 65.8 m (52 by 216 ft) mat in Fargo, North Dakota, tilted, but the bins broke up and the structure was a total loss.

A dramatic failure by rotation due to unequal support was a nine-story apartment building in Rome, Italy, that settled at each end and split into two sloping halves. The separation into two buildings, with a triangular gap about 1 m (3 ft) wide at roof level, occurred in 1959, when living space was at a premium and habitable space could not legally be kept vacant. Nineteen of the 28 apartments remained in use after stiffening beams were added in the lobby. Glass rods were inserted across the break in walls and partitions, with the intention that their breakage would warn of added movements.

Buildings placed partly on earth and partly on rock have too often been designed on the assumption that the bearing values permitted by local building codes will ensure equal settlement. If such soil conditions exist, a separation

Figure 3.5 Transcona grain elevator, Manitoba, Canada, 1913. (Courtesy of The Foundation Company, Inc., Toronto, Ontario, Canada.)

joint must be provided and each portion should be designed as an individual structure. Such buildings, especially if wall bearing, crack almost immediately when nature provides the omitted joint. Underpinning and repair costs, together with the economic loss of delayed or vacated use, far exceed the cost of properly separated foundation and structure design. Some instances have been investigated where soil shrinkage due to local dewatering triggered the differential settlement.

Foundations on soils with large water-level fluctuation are also vulnerable. In 1949, at Corning, New York, a hospital addition on loose gravel soon showed greater settlement of the exterior walls than at the interior columns. The differential movements caused considerable distress in the interior partitions and finish surfaces.

New York City had two multistory building collapses in the early days of such construction, due to improper foundations. In 1895, the Ireland Building at Fifth Avenue and Third Street was almost completed to its eight-story height as a "fireproof iron frame" with masonry bearing walls when almost half of the 30-m (100-ft) length collapsed completely, killing 16 workers. An interior column was set on a 0.92-m² (10-ft²) footing with one corner above a backfilled brick-lined well only 460 mm (18 in.) deep below the normal footing bottom. The collapse was triggered by the shearing of this footing, resting on very fine wet sand, and the column with its base punctured through the break and penetrated the sand over 1 m (3 ft).

In 1904, the collapse of the Darlington Apartment House, 13 stories of steel frame and concrete floors with cast-iron columns in all wall lines, caused the death of 25 construction workers. Collapse was total, coming without warning. Foundations at the front of the building were on rock, which shelved off rapidly to the rear, where clayey soils covered some natural springs. Since this design was then considered the most modern type of structure and many similar multistory residence buildings were planned and occupied, the failure investigation was conducted in great detail. Although the cast-iron columns and connections to the beams were admittedly vulnerable to distortions, it was the inequality of foundation support that caused the collapse.

A 91-m (300-ft)-high, 24-story concrete office building in São Paulo, Brazil, was found to be over 600 mm (24 in.) out of plumb when the elevators were being installed in 1941. One corner leaned over the street 650 mm (25.5 in.). The building was pile supported, but it was assumed that lenses of fine wet sand had been punctured by an adjacent excavation and that its flow had affected the piles. Cement grout injection was unsuccessful, as was injection of aluminum salt solution intended to stabilize the fine sand. The soil was then frozen to a depth of 18 m (60 ft) by circulating brine through 160 double-walled pipes over a period of eight months. Holes were then excavated for 1.2-m (4-ft)-diameter shafts adjacent to the building columns and filled with concrete. Using these piers as reactions, the columns were then raised by 40 jacks, ranging in size from 9 kN to 8 MN (100 to 950 tons) each, so that the building was within 6 mm (0.25 in.) of plumb. The underpinning and repair of cracks, especially around the elevator shafts, took almost two years and cost over $1 million, about 50 percent of the original total estimated construction cost. The additional payment was made willingly by the owner, with no attempt to recover from any party involved in the design and construction. Because of the rapid inflation in the economy, legal rents more than doubled during the two-year delay, so the increase in building value exceeded the repair costs.

3.3.5 Downdrag and Heave

Once a footing, caisson, or pile foundation is loaded and the bearing soil compresses to provide resistance, the structure remains stable since the foundation no longer settles. This condition is reached rapidly in the case of granular soils and over considerable time for clays.

As an uncontrolled or loose fill settles, there is increased potential for downdrag forces to cause a driven or auger cast pile or a drilled shaft to fail. The settlement creates friction forces that act downward on any piles or shafts that pass through the fill. The downdrag forces may be large enough to cause the piles or shafts to settle excessively or fail because of overstressing (Greenfield and Shen 1992).

Downdrag and heave actions can also be activated by subsequent construction nearby. Stability depends on fringe areas as well as the soil directly below the footing or in the immediate vicinity of pile tips. If the soil below the footings is removed or disturbed, settlement or lateral movement is induced. If the fringe area is removed, the soil reaction pattern will change.

Even if there remains sufficient resistance without measurable additional settlement, the center of the resistance is shifted to a new position and a moment is introduced, affecting the stability of the support. When the fringe area is loaded by a new structure, causing new compression in the affected soil volume, the old area can be required to carry some of the new load through internal shear strength in the soil mass. In that case, there will be a new, unexpected settlement of a previously stable building.

If the new building is attached to the existing construction, the soil compression settlement from the new load will cause a partial load transfer to the existing structure, with possible overloading of previously stable footings. In plastic soils, these new settlements are often accompanied by upward movements (heaves) some distance away. Liquid soils cannot change volume, so every new settlement produces an equal volume heave. Disregard of such elementary considerations is the cause of most of the litigation stemming from foundation construction in urban areas, where neighboring structures are suddenly placed in danger of serious damage or even complete failure. Settlements and upward movements of the soil can occur late in the life of a structure, due to changing conditions around the site.

Downdrag from soil shrinkage with consequent differential settlements is a phenomenon observed in many countries, when groundwater tables recede or from desiccation by tree growth. These problems are especially evident in clay soils. The 1942 failure of a theater wall in north London was correlated with the growth of a line of poplar trees. In Kansas City, Missouri, 65 percent of the homes in one residential area were found to be affected by soil desiccation from vegetation growth. After some 15 years of satisfactory behavior, structures at the Selfridge Air Force Base in Michigan suddenly started settling and tipping from a volume change in the underlying clays as the water table receded, aided by tree evaporation. A long-time study of Ottawa, Canada, homes that similarly correlated soil desiccation with such damage was reported in 1962.

Piles embedded in soil layers that consolidate from dewatering or from surcharge may become overloaded when the greater soil density increases surface friction and greater soil loads hang on the piles. The added load may cause considerable increase in settlement and may even pull the pile out of the pile cap, or pull the pile cap free of the column or wall that it was intended to support. Several New York public buildings are located on an island in the East River that has some 9 m (30 ft) of rubbish fill over 15 to 23 m (50 to 75 ft) of soft silt and clay above a decomposed rock. Normal aging and desiccation, aided by tidal action suction of the soft silts, has shrunk the fills. Downdrag on the piles nearest the shore line has caused serious damage. In one steel-framed building the corner pile cap settled away from the column and a 50-mm (2-in.) gap was found under the column base plate. Fortunately, no nuts had been placed on the anchor bolts; otherwise, much more damage would have resulted. This was an unusual condition in which a construction oversight turned out to be a help rather than a detriment.

Dewatering of the ground can result from addition of new subsurface construction. In 1945, a 20-year-old four-story hospital in Buffalo, New York, started to settle when a sewer was constructed 60 m (200 ft) away. The piles, which had been located in a filled creek bed, were found to be moving. Holes were drilled in the pile caps and new steel piles were driven to rock support as a corrective measure. The Old Criminal Building, built in New York in 1890, was affected by adjacent subway construction in 1909. The mat foundation settled, causing the exterior walls to drop 100 to 125 mm (4 to 5 in.).

In 1960, the four-year-old Justice Building in Little Rock, Arkansas, showed buckling and cracking of walls, floors, and ceilings. An investigation found that the design had not provided footings that would resist the heave of the high-porosity clay soils used for support. A calculated risk had been taken, governed by concerns for an inadequate budget.

Soil bearing capacity can be neutralized by frost action, which can heave the contact surface. At Fredonia, New York, the frost from a deep-freeze storage facility froze the soil and heaved the foundations upward 100 mm (4 in.) in 1953. A system of electrical wire heating was installed to maintain soil volume stability. During an extremely cold winter in Chicago, frost penetrated below an underground garage and broke the buried sprinkler line. This caused an ice buildup which heaved the structure above the street level and sheared a number of the supporting columns.

As noted previously, when exposed to air and moisture, certain natural rocks will expand as the minerals change in composition and volume. In a strip along the shores of Lake Erie east of Cleveland, Ohio, pyrite content in the shales runs to almost 5 percent, and oxidation of the yellowish iron sulfide forms a product that has 10 times the original volume. Structures founded on this shale can suffer significant damage from localized heaving. In the Cleveland area, several sewer lines have broken, and irregular levels appear in floors and stairways. Expansive soils, and corrective measures for dealing with them, are discussed in Section 3.1.

3.3.6. Flotation and Water-Content Fluctuations

Except for well-consolidated granulated soils, a change in water content will modify the dimensions and structure of the supporting soil, whether from flooding or from dewatering. There are incidents where localized settlements within an existing building resulted from cooling water pumping by the occupants. A 12-story concrete building in Brooklyn, New York, supported on spread footings on medium sand, suddenly experienced settlement of several interior columns when large groundwater pumps were installed to provide industrial water. Groundwater levels had receded about 9 m (30 ft). The building owner was forced to purchase city water and stop pumping. Several columns had to be underpinned.

A large two-story steel-framed factory for television assembly in the Syracuse, New York, area experienced similar problems. Groundwater had been pumped into the building for air-conditioning requirements for two years. The factory was built on an apparently good base—thin layers of limestone and shale—but directly above deep layers of salt. The floors and roof deflected downward directly around the pump walls. Borings showed that the fast flow of water had dissolved the salt seams, causing underground settlement. Columns were jacked up and the floors resurfaced. An outside source of water was obtained for further operational requirements.

Pumping from adjacent construction excavations for the purpose of intentionally dewatering the construction site affects the stability of spread footings and can cause dragdown on short piles. Such damage often results in expensive litigation and has brought about requirements in some jurisdictions to recharge the soil to preserve the original groundwater level outside sheathed excavations. As noted earlier and in Section 11.3, dewatering of the soil around untreated timber piles can initiate decay and cause pile collapse.

Continuous pumping from deep wells for construction of the Brooklyn approach to a subaqueous tunnel in lower Manhattan in 1941–1945 to maintain the depressed water table, even with construction closed during World War II, dropped the adjacent street by 230 mm (9 in.), affecting all the structures facing the street. Similar pumping for a subway in Brooklyn constructed in fine and medium sands was done under conditions that sheathed the excavation in steel sheeting. All the water pumped was recharged into the ground continuously on each side of the excavation. There was practically no measured settlement of the adjacent two- and three-story buildings, even with excavations 6 m (20 ft) below groundwater level, and no underpinning of the shallow foundations.

Typical of the lawsuits arising from ground dewatering by construction operations is that filed by the Boeing Company in 1963 against a sewer contractor for $190,000 damage to a Renton, Washington, building. The two-story pile-supported plant suffered settling of the ground floor and cracking of partitions. The trench was sheet piled, but pumping by deep wells lowered the groundwater 9 m (30 ft) at a distance 400 m (1300 ft) away from the Boeing building. Another sewer nearer to the building was then built by excavating into the water table and tremie-concreting a base seal to reduce the amount of pumping.

Since water-content changes can cause detrimental changes in soil, owners should be advised of foundation maintentance techniques, including proper landscaping and drainage. Desiccation from vegetation growth can present problems. Mature trees planted close to the foundation can lower soil moisture levels as their roots grow beneath the foundation. This will cause shrinkage of the soil and differential settlements.

Problems associated with expansive soils have been discussed previously. Clay heave from oversaturation must be expected, and in such soils the structures must either be designed to tolerate upward displacement or the supporting soil must be protected against moisture. Many parts of the world have trouble from heaving bentonitic clays. With no change in water content, these soil volumes will remain stable.

Saturation of soils on sloping sites will diminish resistance to sliding. Such failures are discussed in Section 3.4.2. Saturated soils also exert large lateral forces on basement walls and retaining walls (Diaz, Hadipriono, and Pasternack 1994). The lateral pressure of saturated clay can be greater than hydrostatic pressure. In a large housing development built in the early 1940s on a sloping terrain in Hartford, Connecticut, the concrete cellar walls were backfilled with the local clay. Before the site had been surfaced and drains provided, a heavy rain caused a large flow of water to saturate the backfills on the high side of many buildings and the walls collapsed (Figure 3.6). During reconstruction of the walls, the builder decided that the high cost of bringing in granular fill was not warranted and again used the local soils. After a second "unusual" storm caused a repetition of the failures, granular fill was purchased and used. The buildings were in use for many years without further failure of the cellar walls.

3.3.7 Vibrations and Seismic Response

Earth masses that are not fully consolidated will change volume when exposed to vibration impulses. Vibration sources include construction equipment (especially pile drivers), operating equipment in a completed building, traffic on a rough pavement, and shock from blasting. Examples of failures from construction vibrations and blasting operations are given in Section 11.3.

Foundations must also be designed with consideration for the shock spectra provided by nature, including the severe ground shaking that accompanies seismic events (see Section 2.2). The effect of earthquakes on foundations is usually less than on the superstructures, especially in earthquakes of short duration. Nevertheless, the character and integrity of a foundation system will significantly influence the degree of damage in the superstructure. Foundation design must be coordinated with the expected action of the superstructure. For example, if the superstructure contains seismic separation joints to allow portions of the structure to move independently, the foundation system must also be designed to accommodate such independent relative motion.

Long-duration earthquakes, such as the Niigata, Japan, and Anchorage, Alaska, earthquakes of 1964, can severely alter the character of subsoils. In this case a great deal of the damage may be the direct result of foundation failure because of soil bearing collapse, liquefaction, or severe settlement.

In the Anchorage earthquake, vertical and lateral movement due to the instability of the soil destroyed many buildings, regardless of whether they were properly built or designed for seismic resistance. In the Turnagain resi-

(a)

(b)

Figure 3.6 Wall failures from liquefied clay backfill.

dential area, 75 fine houses were destroyed by the large consolidation and lateral flow of the underlying clays (see Figure 2.11). Five years earlier the U.S. Geological Survey warned of the incipient danger from the sensitive clays underlying the sand-gravel beds in the Anchorage area. At four other discontinuous areas, similar soil failures dropped buildings as much as 4.5 m (14 ft) vertically and caused total failure of two blocks of small commercial buildings while similar buildings on the opposite side of the street were unaffected.

3.4 AT-SURFACE CONSTRUCTION

3.4.1 Slabs-on-Grade and Pavement Failures

Pavement failures, although not usually catastrophic or life threatening, represent a significant cost to the public. Despite the long history of road and pavement construction, engineers and paving contractors continue to design and build pavements that fail (Rollings and Rollings 1991).

Progressive damage to pavements is often the result of deficient provision for drainage. Failure to consider climatological effects on the subsoil has resulted in many pavement failures. In most cases, lack of moisture control is the proximate cause. Regional soil conditions and rain intensity variabilities demand quite different solutions on various sites. Copying a successful design from one locality is hardly a guarantee of sufficiency in another location with different soil composition and rainfall intensity.

In 1955, the New York State Thruway 3 km (2 mi) west of Albany was washed out by the erosion of some small springs at the foot of a moisture-saturated fine sand embankment. Soil loss was progressive and the entire pavement under two lanes was eaten away. The soil is stable under normal rains but cannot withstand steady contact with flowing water.

At the Florida Sunshine State Parkway, the lime rock formation used as the base near Ft. Pierce had become oversaturated before the pavement was placed. The asphalt surface prevented drying out and the rock became greasy. The pavement broke up under traffic movement. About 18 km (11 mi) of roadway required extensive repair. The saturated rock was scraped off and replaced with dry material. Similar subgrade disintegration affected parts of the London-to-Birmingham expressway in England, where asphalt surface cracking, local settlements up to 150 mm (6 in.), and shoulder failures required closing the road to traffic in 1959. The underlying soils were clays and limestones. Heavy rains during the construction period brought compaction problems and saturated the base.

Continuous disintegration under traffic of the easterly end of the New York State Thruway (Berkshire section) after two years of use required extensive reconstruction in 1960. The native soil was glacial outwash layered clayey silt and fine sand covered by harder till. In the saturated state, excavation was very difficult and actually could not be done until a dry summer period permitted equipment to operate on it. The rock surface just below this layer prevented drainage, and excessive pumping action by the traffic ruined the 230-mm (9-in.) reinforced concrete slabs. Some 11,000 m³ (14,000 yd³) of pavement was removed, and many other areas were strengthened by pumping asphalt emulsion to fill the gaps between concrete and subbase. The pumping of the slabs accumulated water under pressure which bubbled out when grouting holes were drilled through the pavement. The fine-grained soils had traveled upward and sealed the porous base layer.

Roads on expansive clay soils, found in many parts of the world, require special attention to moisture control. Serious failures are reported in Mysore State, India, in parts of South Africa, and in North America, where excessive differential swelling and shrinkage of the subgrade soil causes bumps at all cross-drains and cracking and heaves under rigid slabs. Lime stabilization is a possible remedy, with strict control of moisture content before covering the base.

Concentration of water drainage will ruin the best-made fills in fine soils of nonuniform porosity. In arid western regions of the United States where valley-alluvial loose silts dry out and form interior seepage channels, the first rain melts away the clay and silt particles along the channel faces. Soon thereafter interior cavities form and the grain structure collapses, opening up gullies in the slopes and sinkages under pavements.

This same type of internal structural change resulted in large-scale erosion and collapse of the slopes and shoulders of many miles of railroad trackage built in 1949 at an ammunition storage depot located in the marls of north central New Jersey. The marls and the local silty sands have large percentage of content of very fine flat particles. The trackage was an extension of existing facilities built in 1942 with similar soils and identical geometric design. Erosion of the 1942 construction was minor in nature. Further study, however, revealed a significant difference in construction method. The 1942 fills had been placed by small truckloads and were bulldozer-positioned in horizontal layers with fairly uniform compaction. The 1949 fills were placed by 19-m³ (25-yd³) scraper pans on rubber tires. The loaded tires compacted grooves in the fill, which became buried drainage ditches for rainwater runoff. Where a weak spot existed in the side or bottom of these grooves, the water flow piped out through the fill and the material above fell down to form open gullies.

Since it was impractical to remove the drainage troughs in the fills, a full-scale research program was carried out to determine the best material for covering the slopes as a protective measure against the erosion. After adjacent sections were covered with materials of varying density and grain size, from straw to crushed rock, the test track was covered with artificial rain at the rate of 25 mm (1 in.) per hour. Physical and photographic observations soon eliminated most of the test materials. A lightweight graded blast-furnace slag was found to be the best cover since it held moisture and had some tensile restraint to blow-out forces, and it was not so heavy that it sank through the saturated silts.

Another case of a fine soil base failing under concentrated water flow is a somewhat different use of the material. A large tank farm of volatile fuels consisted of steel tanks with hemispherical bottoms. Each tank was surrounded by a concrete ring wall, within which a hemisphere was carefully shaped and lined with a fine uniform-grained sand as a base for the steel tank. The installation in central New York State was completed in 1952, with some 40 separate tanks connected by valved piping. It was soon noted that the piping fittings were straining under outside forces and that all the tanks were gradually and continually leaning to the east. The prevailing winds being westerly, the tanks were rolling on their bases as a pneumatic bearing, after the rain runoff from the tanks had saturated the fine sand and formed a liquid lubricant. The highly volatile contents of the tanks required a repair with minimum fire risk, and the cause of the difficulty eliminated any liquid form of grouting. Any addition of water would only aggravate the condition, and cement or chemical solidification not uniformly covering the entire base surface might cause local plate buckling and tank failure. The accepted and successful cure was to stabilize the soil cushion by intrusion, forcing dry limestone dust into the wet mass by wooden rods activated by hand power and, after some consolidation was accomplished, by the aid of an air hammer at the end of a long broomstick. The blend of dry limestone dust and saturated fine sand, after three or four passes around each tank perimeter, provided a stable base. The exposure at the perimeter was then sealed and waterproofed against further infiltration. All tanks remained slightly out of plumb but in the same direction, so that the piping required little adjustment.

3.4.2 Slopes and Slides

Surface topography is never naturally stable. Gravity, aided by wind and rain, moves the hills into the valleys and the valleys into the oceans. Continuous tectonic and intermittent seismic action forms new hills and the cycle of change continues. Construction workers locally interrupt the natural cycle of change by excavating into the ground with slopes in the natural soil, by filling on top of the ground with slopes outlining the embankment, by covering the ground with impervious blankets, by stepping the surface and holding the upper levels with retaining walls, by modifying the groundwater content or level, and by storing soil-like materials within bins. In some of these constructions, natural soil strengths are degraded or the necessary strength of the soil in the relocated position is not provided, and a failure results.

Since slides are often the result of continuing natural processes, historic structures that once were founded on relatively safe sites may eventually become vulnerable to geologic hazards. For example, a number of historic houses built on the bluffs along the Mississippi River in the vicinity of Natchez,

Mississippi, are now threatened by continuing erosion, gradually causing the 60-m (200-ft)-high bluffs composed of wind-deposited loess soils to recede.

A study of aerial photographs of a site will often disclose prior massive earth movements, in the form of displaced contours and areas of light-colored newer growth cover that surface surveys and borings may not detect. The lack of earlier slides does not constitute insurance that new excavations or loadings will be stable. The existence of prior movements, however, is a warning that special precautions are necessary. The "self-healing" of slip surfaces seldom restores the original strength of the soil mass.

Greenfield and Shen (1992) note that stability of a slope depends on the soil type and strength, the character of the groundwater table, the strength of weak subsurface layers, and the geometry of the slope. Slopes can collapse from a variety of reasons, including:

- Their own weight
- The influx of water reducing the shear resistance of the soil (from irrigation or from leakage from swimming pools, water pipes, or other water sources)
- Dynamic loading from an earthquake
- Erosive action
- Change in geometry of the slope
- Undercutting
- Creep (slow moving of the soil downslope)

The state of California suffers regularly from the effects of landslides, rock slides, and mudflows. Here, many of the factors influencing landslide potential are found in the extreme: steep terrain, unstable sites containing a wide variety of soil types, frequent seismic activity, and the legendary firestorm and mudflow cycles in heavily developed coastal residential communities. Maintenance of the coastal highways in California is a constant battle against nature, requiring large expenditures of federal and state funds to remove debris and repair slide damage. The terrain in the Big Sur area, Malibu, Palos Verdes Peninsula, and other sections of the Pacific Coast Highway in California is breathtakingly scenic, but the same rugged coastline that provides the scenery is extremely vulnerable to the destructive effects of landslide activity.

Typical of the extensive damage that can result from the development of areas with prior landslide activity, usually for residential use, is the case of the Palos Verdes Peninsula in the County of Los Angeles, California. Beginning most recently in 1956, the hilly coastal slopes became active, ultimately destroying many costly homes, despite extensive measures taken to save them. Following long, drawn-out, and horrendously expensive litigation, the County of Los Angeles imposed strict controls on future development of similar slide-prone land. These new controls were imposed not only as a precaution against increased liability exposure on the part of the county, but also as a protection against revenue loss from shrinkage of the real property tax base resulting from casualty deductions.

Another well-known example is the costly Big Rock Mesa landslide in Malibu, California, that occurred in 1983. Malibu is a long, narrow community along the Pacific Coast Highway, consisting of expensive homes populated by a number of celebrities. There is one narrow row of homes along the beach, bordered by a narrow section of the Pacific Coast Highway, then steep, rugged cliffs topped by mesas. Some of these mesas have been heavily developed for residential use. The steepness of the bluff east of the highway is the result of wave erosion and undercutting of the toe of the slope over centuries. The highway in the vicinity of Malibu is often closed by slides. In addition to the obvious geologic hazards of the region, Malibu is frequently subjected to brushfires in the dry fall season that denude the slopes, increasing the potential for subsequent mudslides in the rainy season.

For years, while development on the mesas increased, residents voted against the installation of a sanitary sewer system, choosing instead to use individual septic systems. Thus development became a significant contributor to the eventual instability of the slope, as heavy rainfall, increased irrigation, and discharge from the septic systems raised the groundwater level. Too much groundwater was acknowledged by all parties as the triggering mechanism for the massive 1983 slide. Prior to development, the mesas were used for cattle ranching. Water pumped from the ground for this use helped to maintain a stable groundwater level, counteracting rainwater infiltration.

The entire bluff is visibly broken into various scarps and slumps resulting from the 1983 slide. Litigation arising from this slide amounted to claims of about $300 million. If the principal case had gone to trial, an estimated 300 attorneys would have been present, requiring taxpayers to rent and remodel a temporary courtroom at a projected cost of $2 million. However, this particular lawsuit was settled prior to signing the lease for the temporary courtroom. Nevertheless, other suits continued in the courts for years, with successive appeals that eventually exonerated the county at the appellate court level, citing numerous deficiencies by the developers and homeowners that contributed to their own unfortunate but predictable losses.

A thorough review of the Malibu landslide and eventual resolution of related litigation is beyond the scope of this book. The reader is referred to Shuirman and Slosson (1992) for an extremely informative account. This landslide presents a clear illustration of the vulnerability of ill-conceived human endeavors, and how development can increase the potential for "natural" disaster.

Massive natural slides have occurred around the world. In 1961 at Kiev, USSR, a landslide moved houses and barracks into a ravine and killed 145 people. In June 1961, some 8000 slides hit terraced suburban developments around Tokyo, Yokohama, and other cities in Japan after heavy rains, killing 270 people. The 2500-year-old city of Agrigento, Sicily, lost 200,000 m^2 (50 acres) of developed homes and apartment houses from a slide of a thin layer of rock-overlying clay. The blame was placed on overloading the hillside with buildings up to 10 stories high. Only a month before the slide, after years of discussion, the city passed an ordinance banning new apartment construction. Complaints of water flooding the city from unknown sources were being studied, but no report had been issued as to its source or effect on the hillside.

Slides have wrecked power plants located on steep hillsides. In 1956, probably triggered by a slight earth tremor, 45,000 Mg (50,000 tons) of rock demolished two-thirds of the Schoellkopf station at the bottom of the Niagara gorge. Some warning was given when water started to leak into the building, but the rock suddenly sheared vertically in the almost 60-m (200-ft) height. Similarly, an earth slide, coming about a month after a 1965 earthquake shook the Seattle, Washington, area, damaged the lower Baker powerhouse, then 40 years old.

Melting snow in 1967 saturated a Japanese mountainside, and the resulting slide dammed the river, flooding the Odokoro power plant in Niigata. A 180-m (600-ft)-long barrier of earth and rock 40 m (130 ft) high formed the temporary dam. In July 1983, a catastrophic landslide in the Andean slopes near Bogota, Colombia, killed 128 workers at the Guavio hydroelectric project. The project under construction was a 250-m (820-ft)-tall rockfill dam, the tallest in the Americas. With such massive earth movements comes the danger of damming streams and rivers with unstable rock and soil deposits—an immediate threat to downstream developments should a dam give way suddenly, releasing the impounded water.

At the open-pit Beattie Mine in northwestern Quebec in a glacial-varved clay area, the excavation in 1943 was 300 m (1000 ft) long, from 12 to 60 m (40 to 200 ft) wide and 90 m (300 ft) maximum depth. Excavation consisted of 26 m (85 ft) of clay overlying 5 m (15 ft) of sand-till before exposing the rock. Some small slides had occurred in 1937 and in 1942. On the night of June 15, 1943, 760,000 m³ (1 million cubic yards) of clay filled the open excavation and flowed into the underground tunnels. The flow was initiated by the collapse of a pillar in the pit, followed by a rock fall. All the clay on the north and east went into the pit. By 1946, some 1.5 million cubic meters (2 million cubic yards) of material had been removed, including several smaller slides. Investigation of soil properties showed the clay to be normally consolidated and easily subject to earth flows.

Landslides in clays have been studied for many years. In 1846, Alexandre Collin wrote about his experiences in analyzing and trying to control slips in clays, marls, and sandy clays. Fifteen slides are described in the constructions for the Canal de Bourgogne during the period 1833–1846. Slides were from 3 to 12 m (10 to 40 ft) in height. Also in this period, Collin describes the investigation of eight other slides in France, one 18 m (60 ft) high. An English translation of Collin's book by W. R. Schriever of the Division of Building Research, National Research Council, Canada, was issued by the University of Toronto Press in 1956 (Schriever 1956).

Canal operation often requires continuous maintenance in the form of slide removal. The intermittent side-slope failures in the Panama Canal, especially at the Culebra Cut, have necessitated removal of soil volumes that may have exceeded the original canal excavation.

Rock exposures are weakened by weathering, often aided by tree root expansion from landscaping plantings. Moisture accumulation in the expanded seams freezes and expands further in the winter, eventually loosening the rock. In February 1957, some 900 Mg (1000 tons) of rock fell out of the slope along the New York State Thruway, closing all three southbound lanes north of Yonkers. During construction of an interstate highway in western North Carolina completed in 1966, repeated rock slides in a sidehill cut added 610,000 m³ (800,000 yd³) to the 3 million cubic meters (4 million cubic yards) of contract excavation. One slide of 190,000 m³ (250,000 yd³) took a slash almost 150 m (500 ft) high into an apparently solid-rock benched slope cut into the Smoky Mountains.

Small slides are quite common in foundation construction and are often the result of carelessness or an attempt to save the cost of temporary shoring. A clay bank collapsed in 1965 at a post office construction in Columbia, South Carolina. Of eight construction workers building formwork for a 6-m (20-ft)-high retaining wall, seven were killed. The original soil bank sloped at 60 degrees from the horizontal. To clear space for the wall footings, the toe of the slope was cut back 3 m (10 ft) and protected by 100-mm (4-in.) lagging against some timber piles. The slide snapped the piles and smashed the form-work. A similar accident in Seattle, Washington, occurred in 1960 when workers trimmed the toe of a 11-m (36-ft)-high sandbank for wall footings. The slope of $\frac{3}{4}$: 1 had been stable without shoring until the footing pits were cut into the toe.

Massive avalanche slides of human-made mountains of debris are examples of poor control and regulation. In 1965, an avalanche of stored cinders buried a housing development in Kawasaki, Japan. Torrential rains had saturated a 15-m (50-ft)-high mound of fly ash, 90 m (300 ft) long and 70 m (230 ft) wide, that flowed down a slope and demolished 15 homes. Storage of cinders and fly ash in the critical area had caused some alarm, and the trucking company responsible for the mound had erected a barrier of sandbags and wooden poles. The entire barrier sheared and was carried by the slide.

In 1966, a tragic coal waste slide took 200 lives at Aberfan, Wales. A conical pile of coal waste slurry had accumulated from a coal mine some 6.5 km (4 mi) away and had become 18 m (60 ft) high and 140 m (450 ft) in diameter. Heavy rains saturated the mass and some 1.8 million megagrams (2 million tons) broke out and slipped 450 m (500 yd) into the town within a few seconds, engulfing the school and 18 houses. Despite several minor slips, the warning of danger was not heeded. Some drainage channels cut into the storage pile could have averted the disaster, but a less hazardous location for the storage would have been the safer solution.

Geotechnical engineers and geologists have developed numerous techniques for slope protection and erosion control. Geosynthetic materials are available, and procedures for using vegetation to assist in establishing slope stability are well known. These techniques, however, must be used with the knowledge that natural processes will continue, changing the character and perhaps even the stability of the site. Poorly constructed new slopes and nonengineered modifications to existing slopes can only be expected to enhance the destructive capacity of these natural processes.

In unstable mountainous regions where slides simply cannot be eliminated, the best that engineers can do is provide instrumentation that will warn of impending slides. With this instrumentation, lives can be saved through timely evacuation of the area. Automated data acquisition systems, developed for use in dams, are now being used to monitor precursors to landslide activity.

Such instrumentation, including extensometers, microseismic sensors, inclinometers, piezometers, and hydrometeorologic devices, has been installed in the Italian Alps. This system was installed following a 1987 catastrophe when 32 million cubic meters (42 million cubic yards) of soil fell from a height of 1280 m (4200 ft), killing 27 people and forming a fragile dam and lake in the Valtellina Valley. The unstable dam was 80 m (260 ft) high and over 24 km (15 mi) long. The dam impounded approximately 20 million cubic meters (700 million cubic feet) of water. Another 230,000 cubic meters (8 million cubic feet) of unstable material still existed precariously on the slope above the site. Potential overtopping and subsequent failure of the dam was a clear danger to downstream inhabitants.

3.4.3 Subsidence

Even without earthquake impulse, land masses are always rising or falling, usually at a very slow rate. Land subsidence is sometimes generated by human endeavors that change natural conditions. Heave also results from change in water content of the soil or exposure to deep frost. When these changes take place in built-up areas, damage and even collapse of surface structures can entail serious losses. The long-time subsidence of Mexico City correlated with the pumping depletion of the water table over many years is a classic example. Pumping is now carefully controlled, and water supply sources beyond the basin of the city are replacing the local wells, but subsidence continues.

Subsidence is a very serious problem in coastal regions, where the risk of flooding is increased. A number of the world's great cities, including London, Tokyo, Bangkok, and Shanghai, are vulnerable to coastal flooding problems due to subsidence. In the United States, the cities of Houston and New Orleans are facing significant subsidence problems. All of these cities are founded on soft sand, silt, or clay, and the subsidence is principally the result of depletion of the groundwater for domestic and industrial use.

In Long Beach, California, subsidence has been responsible for considerable damage to structures. The subsidence has been monitored continuously since the early 1940s and is the result of changes in the land surface from filling and dredging to provide space for industrial facilities, pumping of water from the subsurface for industrial and domestic use, lowering of the groundwater table during dewatering for construction of dry docks, and pumping of oil from the sand layers bounded on top and bottom with impervious beds of weak shales.

Alviso, a historic Hispanic residential section of San Jose, California, was severely damaged by flooding in 1983 (Shuirman and Slosson 1992). The community is sited on tidal flats between two rivers with inadequate upstream flood control. During the past few decades, the average elevation of the ground was lowered from 0.6 m (2 ft) above sea level to 1.2 m (4 ft) below sea level by groundwater extraction. Under such conditions, the increased exposure to flooding from severe rainstorms or unusually high tides is obvious.

Another dramatic subsidence was the sudden split and settlement of a 670-m (2200-ft) length of the fill placed across the Great Salt Lake in Utah to carry the Southern Pacific Railroad. Design of this fill was based on thorough soil investigations, yet the load of some 1.5 million cubic meters (2 million cubic yards) of material, partly placed in water, caused a 3-m (10-ft) layer of salt to give way. Failure came six days after completing the fill. Reconstruction in 1958 was on the basis of a modified design with wider base and flatter slopes. The rest of the 21-km (13-mi) fill had a few minor slides, but stability was soon reached when the underlying clays compacted and gained strength under load.

3.4.4 Retaining Walls and Abutments

The history of soil mechanics is in the ancient attempt to build stable retaining walls. Until the twentieth century, the greatest part of the literature in soil mechanics was concerned with the determination of earth pressures and the analysis of retaining wall failures. Recent theoretical and empirical developments have brought about safer designs and relatively fewer failures, but some will continue to plague the succession of false prophets who announce the "final, complete solution" to the earth pressure problem.

Historically, in the upper hilly part of New York City, where many retaining walls separate the stepped levels of adjacent yards, the annual spring thaws were accompanied by retaining wall failures. Many of these walls were not engineered, and some did not even have a construction drawing to guide the builder. Controls have been tightened somewhat and time has eliminated the most vulnerable cases. Fatalities, narrow escapes, and masonry cascading into adjacent occupied property were the source of much litigation before the retaining wall was recognized as a structure requiring investigation for safety, even though it was not itself occupied by people.

A few of the more spectacular wall failures are described as early examples of the errors in such work. In 1925, the city of Los Angeles built a multiple-arch retaining wall, 137 m (450 ft) long and 18 m (60 ft) high, to hold fill for the relocated and widened Cahuenga Avenue leading from Hollywood to Lankershim, California. The design was chosen over that of a bridge to span Indian Gulch, since it provided space for disposal of fill from an adjacent excavation and the arch design would impose smaller loads on the soil level chosen for bearing, about 5 m (15 ft) below the ground surface. Buttresses were spaced 9 m (30 ft) apart and the arches were 460 mm (18 in.) thick at the top and 610 mm (24 in.) at the spread footings. Arches sloped $4\frac{1}{2}$ horizontal to 10 vertical and were capped with a concrete beam and railing. The Santa Barbara earthquake on June 29, 1925, came at the time the fill was half-height. One of the end buttresses, founded on rock, moved about 50 mm (2 in.), with only slight changes in the buttresses on soil. The movement thereafter unexpectedly continued and was 200 mm (8 in.) after 18 months. Six months later it was 380 mm (15 in.) downward and 430 mm (17 in.) outward. Analysis indicated that slippage was along a steeply sloping buried rock surface 12 m (40 ft) below the ground level. Adjacent property was purchased and the arched wall was filled in completely, to the very top. Even with the downhill side backfilled, movement continued.

A stone masonry gravity wall built in 1890 to retain a steep hillside and Geohring Street in Pittsburgh, Pennsylvania, extending up to 11 m (35 ft) above Itin Street, by 1940 was leaning 760 mm (30 in.) at its greatest height. At this time reconstruction was essential, and a scheme was devised that permitted maintaining traffic on the streets. Sheathed pits on 7-m (23-ft) centers were cut into the high street and concrete counterforts built therein to subgrade levels of the wall. The wall was then cut away in pieces and replaced by a stone-faced reinforced concrete wall connected to the counterforts. All new work was carried to shale rock bearing.

In 1957 the Union Pacific Railroad built a 21-m (70-ft)-high concrete retaining wall along the Albina Yards at Portland, Oregon, to make room for more tracks. The wall was 2.1 m (7 ft) thick at the base and tapered to 0.5 m (1.5 ft) at the top. On April 4, 1958, a 94-m (310-ft) section of the wall fell over, releasing the backfill over ground completely prepared for the new tracks. The wall cost $450,000, but the cost of resulting damages was even higher. The project was abandoned, an additional 25,000 m^2 (6.2 acres) of private land was purchased to permit a $1\frac{1}{2}$:1 slope to be cut into the hillside, and some 300 m (1000 ft) of a four-lane highway was moved back from its damaged position and rebuilt. During reconstruction a steel sheetpile protection was installed to control the hillside before 126,000 m^3 (165,000 yd^3) was excavated.

In 1962, a retaining wall consisting of cast-in-place drilled-in piles carried to the top of a rock cliff, the concrete face making up 6 m (19 ft) of the 23-m (75-ft)-high enclosure of the One Main Place project in Dallas, Texas, fell into the plaza. This collapse occurred after a fissure cave-in 6 m (20 ft) wide and 5 m (15 ft) deep developed in the upper street. The sequence of events was first a report of leaking gas, a small drop in the pavement, the street cave-in two hours later, and the fall of 23 m (75 ft) of wall 18 hours later. The wall consisted of 760-mm (30-in.) drilled cores spaced 1.5 m (5 ft) on centers, carried about 1.5 m (5 ft) into the rock. A continuous concrete base beam at the bottom of the piles was anchored to the limestone rock by prestressed diagonal tiebacks 4.6 m (15 ft) long and spaced 1.5 m (5 ft) apart between the piles. The space between piles was covered by a sprayed 90 by 90 mm (3.5 by 3.5 in.) mesh-reinforced mortar. To permit street and utility restoration and to protect the existing buildings, a similar concrete wall was built in the opposite side of the street but with cores carried 4.6 m (15 ft) into the rock, tiebacks eliminated, and the street backfilled. A subsequent geotechnical investigation disclosed some faults in the rock, which were then crossed by drilled anchors, and a new wall was built to enclose the five cellar levels.

A most interesting wall failure from lateral pressure occurred in 1942 at Chambersburg, Pennsylvania. A building for the cold storage of apples had been in service for many years. Concrete walls 200 mm (8 in.) thick were covered with brick facing and tied to the floors with the slab steel extended. Normal storage was in crates, which became expensive and hard to get because of war conditions. In 1944, apples were stored on the floors uncrated and the lateral "apple pressure" forced a 15 by 30 m (50 by 100 ft) section outward, with a grand spillage of apples.

Bridge abutments have their share of failures. Maintaining bridge profiles when an abutment settles (or rises) often requires some ingenious solutions. In 1957, one end abutment of a three-span deck girder bridge carrying the Connecticut Turnpike over railroad tracks settled progressively during construction, eventually reaching 430 mm (17 in.) total settlement. Settlement of the piles due to an unforeseen compressible substratum was noted before the bridge abutment was completed. Rather than drive new piles about 30 m (100 ft) to rock, a jacking bracket was added to the abutment face and the bridge deck was kept level by successive wedging. A battery of nine hydraulic jacks lifted the deck, and steel plates were added under the sole plate. Total settlement was almost 600 mm (24 in.) before the conditions stabilized and the piles reached a gravel layer.

In Chicago, one abutment of an old bridge failed in 1959 and the approach street dropped 150 mm (6 in.). Temporary piles and shoring were inserted under the steel bridge and a pile bulkhead was placed next to the abutment to hold the street. The abutment had developed a crack 80 mm (3 in.) wide and the settled part was rebuilt on new pile supports.

There have been some reports of bridge abutments, apparently carefully designed with long pile supports running through soft clay or silt soils to a proper bearing, that moved more laterally than vertically. The effect on a bridge structure is serious, involving expensive repair and alteration. The reason seems to be the error in the design assumption that loads in the piles (the toe piles are usually battered) are purely axial. The true condition is some bending in the piles, possibly from the lateral pressure of the earth on the high side, causing deformations, and with weak soils there is not enough passive resistance to maintain the structure in a fixed position. Investigation of the stability of the abutment as a retaining wall is of course necessary, but the foundation piles cannot be treated as an independent element. The combination of wall and piles must be checked for stress and deformation under the variation in loadings that exist in a bridge abutment, first without bridge load, then with dead load only, and then with a reasonable live-load allowance. The factor of safety against failure is usually greater under the condition of full specified live load than under any of the prior conditions that must be sustained.

A number of new techniques and materials for the retaining of earth have been introduced to modern construction, including soil nailing and the use of geogrid reinforced backfills. Despite the development of theory, sophisticated computer software, and new materials and methods, retaining-wall failures continue to occur (Leonards, Frost, and Bray 1994). Recent failures confirm the continuing need for quality design and construction. This quality control must include observation of actual field conditions to verify design assumptions and diligence in monitoring the construction to ensure that the work accomplished in the field conforms to design intent. These have always been the guiding principles for failure avoidance; their importance has not been diminished by modern theories, materials, and techniques.

3.5 SUMMARY

Structural distress from foundation insufficiency causes, at best, continuous high-cost maintenance and sometimes results in total collapse. All structures have some tolerance for unequal settlement of the foundation elements, but when the support is stressed beyond the elastic limit, eventual failure is unavoidable without immediate strengthening of the foundation.

The numerical percentage of structures with foundation failures is very small, but every single incident itself is a serious indictment of the capabilities of the design and construction talent. Foundation failure has the unique property that adequate adjacent structures are often seriously affected, causing third-party litigation and expensive liability for damages.

There have been a number of large catastrophic incidents which have been well publicized in technical journals as well as the general press. Many lessons can also be learned from the results of the less publicized, smaller cases. In fact, the size of a project has very little to do with the size of the problems it can cause. Some of the most costly litigation involves small-scale residential construction, where insufficient geotechnical information was available prior to construction. This is illustrated by many examples discussed in the geotechnical literature (ASFE 1990). These examples suggest rules for what *not* to do, something that cannot be learned from the most minute study of the theories of soil mechanics or from the glorious record of successful work. Sharing the experience of such information is of unlimited value as a warning of pitfalls to be avoided if the list of foundation failures is not to increase. This warning is of special value in the design and construction of foundations in congested urban areas and in the use and control of unusual problem soil conditions.

3.6 REFERENCES

ASFE, 1990. *The Real World of Engineering: Case Histories 1–40,* ASFE: The Association of Engineering Firms Practicing in the Geosciences, Silver Springs, MD.

Diaz, C., F. Hadipriono, and S. Pasternack, 1994. "Failures of Residential Basements in Ohio," *Journal of Performance of Constructed Facilities,* American Society of Civil Engineers, New York (February).

Eberhardt, A., 1990. "108-Inch Diameter Steel Water Conduit Failure and Assessment of AWWA Practice," *Journal of Performance of Constructed Facilities,* American Society of Civil Engineers, New York (February).

Gould, J., G. Tamaro, and J. Powers, 1993. "Taming the Urban Underground," *Civil Engineering,* American Society of Civil Engineers, New York (February).

Greenfield, S., and C. Shen, 1992. *Foundations in Problem Soils,* Prentice Hall, Upper Saddle River, NJ.

Jones, D., and K. Jones, 1987. "Treating Expansive Soils," *Civil Engineering,* American Society of Civil Engineers, New York (August).

Kaminetzky, D., 1991. *Design and Construction Failures: Lessons from Forensic Investigations,* McGraw-Hill, Inc., New York.

Kovacs, G., 1992. "Swimming Pools Supported by Dissimilar Bearing Strata," *Journal of Performance of Constructed Facilities,* American Society of Civil Engineers, New York (May).

Leonards, G., J. Frost, and J. Bray, 1994. "Collapse of a Geogrid Reinforced Retaining Structure," *Journal of Performance of Constructed Facilities,* American Society of Civil Engineers, New York (November).

LePatner, B., and S. Johnson, 1982. *Structural and Foundation Failures: A Casebook for Architects, Engineers, and Lawyers,* McGraw-Hill, Inc., New York.

Lin, H., and F. Hadipriono, 1990. "Problems in Deep Foundation Construction in Taiwan," *Journal of Performance of Constructed Facilities,* American Society of Civil Engineers, New York (November).

McFadden, T., and L. Bennett, 1991. *Construction in Cold Regions,* John Wiley & Sons, Inc., New York.

Meehan, R., and L. Karp, 1994. "California Housing Damage Related to Expansive Soils," *Journal of Performance of Constructed Facilities,* American Society of Civil Engineers, New York (May).

Moncarz, P., J. Shyne, and G. Derbalian, 1987. "Failures of 108-Inch Steel Pipe Water Main," *Journal of Performance of Constructed Facilities,* American Society of Civil Engineers, New York (August).

Morley, J., 1996. " 'Acts of God': The Symbolic and Technical Significance of Foundation Failures," *Journal of Performance of Constructed Facilities,* American Society of Civil Engineers, New York (February).

Prendergast, J., 1992. "Perils of Point Loma," *Civil Engineering,* American Society of Civil Engineers, New York (November).

Raghu, D., and H. Hsieh, 1989. "Performance of Some Structures Constructed on Chromium Ore Fills," *Journal of Performance of Constructed Facilities,* American Society of Civil Engineers, New York (May).

Rollings, R., and M. Rollings, 1991. "Pavement Failures: Oversights, Omissions, and Wishful Thinking," *Journal of Performance of Constructed Facilities,* American Society of Civil Engineers, New York (November).

Schriever, W., 1956. (English translation of 1846 geotechnical publication by the French engineer, Alexandre Collin.) University of Toronto Press, Toronto, Ontario, Canada.

Shuirman, G., and J. Slosson, 1992. *Forensic Engineering: Environmental Case Histories for Civil Engineers and Geologists,* Academic Press, Inc., San Diego, CA.

Sputo, T., 1993. "Sinkhole Damage to Masonry Structure," *Journal of Performance of Constructed Facilities,* American Society of Civil Engineers, New York (February).

4

DAMS AND BRIDGES

Dams and bridges are utilitarian structures whose purpose is to withstand natural forces or connect two points. Beyond their utilitarian function, however, they can be objects of great beauty—especially certain bridges—which have become symbols of a time or place. Indeed, some of the great bridges have become destination points in themselves and have provided valuable cultural qualities to the communities they serve. Who could imagine New York City without the Brooklyn Bridge (Figure 4.1) or San Francisco without the Golden Gate Bridge?

The evolution of the technology of dam and bridge construction is a fascinating story, involving ambitious and successful projects undertaken by the most creative and talented men and women of their times. Yet this evolution has not come without occasional failures and human suffering. In this chapter we review a few of the more influential failures and their contributions to advances in construction technology.

Bridges and dams appear to be static objects, and generally, rather substantial objects. There is a tendency to assume that they are permanent, passive objects that need little attention. However, like all other constructed facilities, they interact continuously with their environment and with their users. There is no guarantee of permanancy, even when these objects are designed and maintained in accordance with accepted standards. Bridges expand and contract with seasonal and daily thermal cycles. They interact with wind, and are often subjected to abuse, including abrasion, overloading, and corrosive deicing salts. Dams are influenced by changing temperatures, seismic events, material degradation, unusual weather conditions, and operational errors. In time

Figure 4.1 Brooklyn Bridge, New York City, completed in 1883. From a rendering drawn by bridge designer John Roebling's chief draftsman, Wilhelm Hildenbrand (ca. 1867). (From the Roebling Collection, No. MC 4. Archives and Special Collections, Rensselaer Polytechnic Institute.)

of war, these structures may be the object of intentional destruction, since they are essential elements of the infrastructure.

Materials used in the construction of dams and bridges have the same problems and potentials that they have when used in other constructed facilities (see Chapters 5 to 9). These structures, however, do require special consideration because of their scale and unique environmental exposures. For both bridges and dams, a critical concern is restraint of movement, or provisions to accommodate movement. Problems often arise with deficient expansion joints and support conditions that do not perform as expected. Foundation considerations are extremely important, as will be illustrated by reference to some catastrophic dam failures. The cyclic loading of bridge members and connections introduces the need for fatigue considerations. Also, in certain bridges, corrosion can become a serious problem, especially where structural elements are of small cross section to begin with and subject to very high unit stresses, as for cable structures (Robison 1988, Watson and Stafford 1988).

4.1 DAM FAILURES

Spectacular and devastating collapses of dams, sufficient to unleash the force of large volumes of contained water, are the subject of much publicity and discussion. Of course, the number of safe and harmless dam installations is predominant. Relatively few of these safe facilities are ever discussed in the media—and then usually only at the time of dedication, when some public figure takes part.

Failure of a large dam produces direct property losses to buildings and their contents, highways, bridges, railroads, agricultural land, and to the dam itself. In catastrophic cases there can be a large number of injuries and deaths. In addition, there are indirect costs involving loss of income or business (especially for hydropower facilities), transportation delays, and inconvenience.

The potential for loss of life in a dam failure is to some extent a function of the type of dam. For example, a thin arch concrete dam may be more susceptible to instantaneous structural failure, while seepage failure of an earth-fill dam may be more gradual, giving warning, so that downstream areas can be evacuated prior to catatrophic collapse.

Failures have occurred in masonry and concrete dams, but most of these were either foundation problems or due to lack of strength in the built soil or rock embankment. Failures often have been traced to piping between contact surfaces. Shrinkages of fills, chemical and physical erosion, damage from rodents, and loss of soil and rock along the contacts between concrete and other natural or artificial materials open channels for water seepage through gaps of increasing dimension. Eventually, the flows may become large and fast enough to erode an appreciable opening, with serious results.

Concrete dams have failed due to inadequate foundations, poor construction quality, uneven settlement, concrete deterioration, and overtopping. The case of downstream damage or dam failure due to overtopping should not always be classified a structural engineering or construction failure; it is most likely due to inadequate spillway capacity decisions or extreme unexpected floods.

Earth and rock dam failures are usually the result of foundation problems, excessive seepage or piping, overtopping due to inadequate spillway or failure of the control gates, erosion in the spillway area, erosion of the embankment, or liquefaction of the soil.

Several catastrophes have occurred due to sliding of rock and earth into reservoirs behind both large and small dams. This causes a shock wave and potential overtopping of the dam. In 1960, a small dam in Spain near the city of Biscay collapsed from such a flood wave, resulting in the deaths of 20 people in the village downstream. A massive slide in Italy in 1963 behind the Vaiont Dam caused overtopping of the dam and a disaster of immense proportions, including the loss of more than 2600 lives, even though the dam itself did not fail (see Section 4.1.1.c).

Seismic events have been responsible for dam failures. In some cases, the seismicity of an entire region has been increased due to the hydrostatic pressures induced by large reservoirs.

Just as is the case for other construction projects, the quality of construction is critical to a dam's performance. If anything, diligent monitoring of construction quality is more important for a dam, since the consequences of failure are so dire. Qualified full-time independent inspectors should observe field conditions throughout the placement of concrete, especially monitoring the placement and compaction of fill material in the construction of earth dams. A very common deficiency in earth dams is undercompaction of fill material, particularly in confined areas where large equipment is difficult to operate. Inspectors must monitor this activity continuously; tests alone cannot verify the adequacy of compaction, since tests are confined to small specific areas. A small zone of material that does not conform to contract requirements at a critical location can cause total failure of a dam. Finding such a zone through random test sample collection is unlikely. A representative of the design engineer is the most appropriate person to monitor the work. Karl Terzaghi once noted that "the design of a dam is not finished until the construction is complete" (Johnson 1980).

In 1962, as part of one of his last lectures, Karl Terzaghi discussed the basic reasons for the failures of the Malpasset Dam and Wheeler Lock from foundation inadequacies that eluded detection despite extensive site investigations, considerable testing, and the application of the most advanced analysis and theory. He stated:

All foundation failures that have occurred in spite of competent subsurface exploration and strict adherence to the specifications during construction have one feature in common: The seat of the failure was located in thin layers or in weak spots with very limited dimension. . . . If the failure of a structure would involve heavy loss of life or property, the structure should be designed in such a way that it would not fail even under the worst foundation conditions compatible with the available data. If no such consequence would ensue, the cost of the project can be reduced considerably by a design involving a calculated risk. . . . With increasing height and boldness of our structures, we shall find out more about this interesting subject, but always by hindsight, as we did in connection with these failures. (Jansen 1983)

These statements are not very encouraging to one looking for a reliable factor of safety in the design. Twenty years earlier, Terzaghi wrote:

Since about 1928 the attention of research men has been concentrated almost exclusively on the mathematics of stability computations and on refining the technique of sampling and testing. The results of these efforts have been very useful and illuminating. At the same time, they created a rather indiscriminate confidence in laboratory test results. In some incidents confidence is warranted, but in many others it can lead to erroneous conclusions because the attention of the investigator is diverted from what is really essential. (Terzaghi 1943)

At the closing session of the Eighth Congress of the International Commission on Large Dams (ICOLD) in 1964, president-elect J. Guthrie Brown said:

In the past, dam builders have acquired very substantial knowledge about all the factors that contributed to the safety of the dam structures themselves, but they have had very little reliable or accurate information about the strength and nature of the underlying strata on which the dams are supported. . . . All knowledge gained in the studies of dam design is of no avail if the foundations on which a dam is to rest are in doubt. Recent failures have all been concerned with foundation movements, or geological slips. It is from the geologist and the combined science of soil mechanics and rock mechanics that the engineer must obtain the details and the scientific information he requires to confirm his own experience and judgment of the foundations on which rest the safety of his dam. (ICOLD 1965)

The past few decades have seen the continued failure of dams, despite progress in soil investigation and control, improvements in theoretical analyses of internal stress, modern geotechnical instrumentation, and computer-aided design methods. These failures have occurred throughout the world, often with catastrophic consequences to communities and the environment. In this chapter a few of the important failures are discussed, together with some potential disasters that were averted through maintenance, repair, and alert operational decisions. First, a brief chronological account is given, then some prominent case studies are reviewed in more detail. Only a few of the notable historic dam failures are discussed here. An interesting list of such failures between 1883 and 1928 appears in the *Engineering News-Record* (ENR 1928).

The failure of a privately owned dam caused the Johnstown, Pennsylvania, flood of 1889, the worst civil disaster ever experienced in the United States. More than 2200 people were killed in this catastrophe, illustrating the magnitude of loss that can accompany a dam collapse (see Section 4.1.1.a).

In 1926, a concrete dam near Dolgarrog in North Wales failed through scouring out of a 9-m (30-ft) section of the clay under the footings. The flood overtopped and caused complete failure of an earth-filled dam with concrete core wall some 3 km (2 mi) downstream. Large masses of earth and boulders cascaded into the town, causing extensive property damage and the loss of 16 lives. The flood entered the plant of the Aluminum Corporation, submerging the generators and causing an explosion of the reduction furnaces. The concrete dam was only 5.5 m (18 ft) high and was planned to have a solid concrete foundation 4.4 m (14.5 ft) wide and 2.4 m (8 ft) thick, embedded into the clay a minimum of 1.8 m (6 ft). The water depth was only 4.6 m (15 ft). The failure took out the bottom part of the dam, the footing, and a large mass of the underlying glacial clay. Subsequent investigation showed that the footing had been built only 450 mm (18 in.) into the clay and that the concrete contained large pieces of granite rock with little mortar covering. The preceding year the reservoir was dry for a considerable period, allowing the clay to shrink and crevices to form. Sudden rains filled these gaps in the clay and started the undermining of the shallow footing.

The catastrophic failure in 1928 of the St. Francis Dam of the Los Angeles Aqueduct in California was the result of a weak foundation, but one section was also affected by a hillside slide. Although arched in plan, the design was a concrete gravity section, with a maximum 62.5-m (205-ft) height and a 53.6-m (176-ft)-wide base. The center 30-m (100-ft)-long section did not move; the west side went downstream much farther than the east side. Foundations rested on rock: schist at the left bank and in midstream and a red conglomerate on the other side. Excavation into the rock was 9 m (30 ft) deep at midchannel and to lesser depths toward the higher ground. The flood moved down the San Francisquito Canyon at 25 km/h (15 mi/h), scouring out the powerhouse and other structures and taking several hundred lives. Fragments of the dam, some with rock adhering to the base, were moved 0.8 km ($\frac{1}{2}$ mi). The debris came from 180 m (600 ft) of the 213-m (700-ft)-long dam. The bedrock had not been grouted; the foundation rock was simply unsuitable for a dam at this location, and insufficient engineering judgment had been applied in its design. An official report concluded that the St. Francis Dam collapse could be attributed "wholly to the unsuitability of the materials on which the dam was built." The California Dam Safety Law of 1929 was passed as a result of this failure.

An expected failure under the earth dike of the Pendleton Levee in Arkansas was carefully instrumented and observed in 1940. Knowing that the foundation was a soft clay with marginal stability, the work was scheduled so as to measure and record the actual failure, which came when the triangular section had reached a height of 9.8 m (32 ft). The width of the levee base was 97.5 m (320 ft). The failure came on the land side of the fill as a vertical subsidence of 2 to 3 m (6 to 10 ft), 15 to 21 m (50 to 70 ft) wide, a 3-m (10-ft) horizontal movement of the rest of the fill toward the landslide, and an upheaval 1.2 to 1.5 m (4 to 5 ft) high just beyond the land-side toe. Failure was explained by the theory that all maximum shear resistances did not mobilize simultaneously along the interior slide surface; failure began where the local strength was not sufficient and progressed successively along such surface. Normal design procedure, based on the physical characteristics of the soils as determined in the laboratory, indicated a factor of safety of 1.3 at the point of failure if all maximum resistances were mobilized. In the discussion section of the Pendleton Levee report it was noted that the cause of failure was a lubricated horizontal shear surface in the clay. Terzaghi referred to four similar failures in earth dams that had occurred at horizontal clay surfaces: the Lafayette Dam in California (1928), the Clingford II water supply embankment near London (1937), the Marshall Creek Dam in Kansas (1937), and the flood-protection dike at Hartford, Connecticut (1941). Loading such a clay layer introduces pore water pressures that degrade the physical strength of the virgin soil.

In 1959, a 34-m (112-ft)-high masonry and concrete dam built two years previously across the Tera River in northwestern Spain failed when 17 of the 28 masonry buttresses gave way as the crest was topped by heavy rains. The flood destroyed 125 of the 150 buildings in the downstream village of Rivadelago and killed 140 people. Litigation resulted in the conviction of four engineers on the staff of the power company for neglect and insufficient supervision of construction.

The dramatic collapse of the Malpasset concrete dome-shaped dam near the French Riviera in 1959 was caused by a rock shift under the left bank foundations. The structure was 66 m (218 ft) total height above rock trench bottom, with 57 m (186 ft) of head. More than 400 people were killed as a result of this collapse, discussed in Section 4.1.1.b.

The 1963 catastrophic overtopping of the Viaont Dam in Italy due to a massive earth slide into the reservoir was a disaster of almost inconceivable scale. Yet the influence of this experience contributed to the saving of lives in the United States that same year when the Baldwin Hills Reservoir in Los Angeles, California, failed (see Sections 4.1.1.c and 4.1.1.d).

In 1965, the 50-m (163-ft)-high concrete arch dam across Maltilija Creek in Ventura County, California, was found to suffer from exterior cracking, interior swelling, and general disintegration of concrete in the upper 6 m (20 ft) of the 16-year-old structure. The defects were discovered by the state inspection of dams that followed the Baldwin Hills Dam collapse in 1963 (see Section 4.1.1.d). The concrete was removed for a height of 9 m (30 ft) and over a length of 85 m (280 ft) to reduce maximum stress conditions as a precaution against collapse.

Late in 1966, the 12-m (40-ft)-high concrete Ambursen Buttress Dam at Mill Brook near Pittsfield, Massachusetts, built in 1909, failed by disintegration of the concrete slab between three buttresses. Ice buildup between buttresses had been a constant source of damage to the concrete. Only local road and bridge damage resulted from the 34,000 m^3 (9 million gallons) of water released.

In 1966, a 18-m (60-ft) length of the spillway of a homogeneous clay dam near Mercer, Pennsylvania, collapsed due to piping erosion through the embankment next to the left wing wall of the spillway. The spillway was 94 m (308 ft) long, and the dam was 13 m (43 ft) high. A similar seepage closed down the Fontanelle Dam in Wyoming when a hole developed near the right abutment of the 48-m (157-ft)-high irrigation and water supply earth-fill dam. The flow was about 4.25 m^3/s (150 ft^3/s). It washed out a hole in the downstream face that was 24 m (80 ft) across, 46 m (150 ft) high, and 18 m (60 ft) deep.

The failure of the Teton earth dam in Idaho in 1976, due to inadequate design, caused massive environmental damage, over \$1 billion property damage, and the loss of 11 lives (see Section 4.1.1.e).

4.1.1 Failures of Completed Dams

In this section a few of the more prominent dam failures are discussed briefly. Some of these failures have become legends in the history of technology and are the subjects of volumes of literature. References are given for readers who wish further information.

4.1.1.a Johnstown, Pennsylvania, Flood of 1889

(*References:* McCullough 1968, Daniels 1982, Frank 1988, ENR 1989)

On May 31, 1889, a three-story wall of water destroyed most of the town of Johnstown in southwestern Pennsylvania, killing more than 2200 of the 30,000 inhabitants. This event has been described as the worst civil disaster ever experienced in the United States.

Cause of the disaster was the failure of a privately owned earth and rock dam located 23 km (14 mi) to the northeast and 140 m (450 ft) higher than the city, on the South Fork of the Conemaugh River. The dam, built in 1840–1852, was competently designed and carefully constructed to provide dry-season water storage for a state-operated canal system. Successive rolled layers of clay and earth 600 mm (24 in.) thick were placed, and each layer was allowed to sit under water for a few days to create a watertight barrier. The downstream face was built with heavy rock. The dam was 284 m (931 ft) long and 22 m (72 ft) high. At the top it was 3 m (10 ft) thick, and it was 83 m (272 ft) thick at the valley floor. An arched stone culvert at the center of the dam was used to discharge flow into the canal through five 600-mm (24-in.)-diameter cast-iron pipes controlled by valves. A spillway 26 m (85 ft) wide and 2.75 m (9 ft) deep was also cut through the rock near the eastern end of the dam to handle water flows during heavy rains. At the time of its construction, the dam was one of the largest dams in the United States, and the reservoir behind the dam was the largest human-made lake in the United States.

When operation of the canal system became unprofitable due to competition from the railroads, the dam was purchased from the state by the Pennsylvania Railroad.

The South Fork Dam experienced a partial failure on June 10, 1862, when the upstream portion of the stone culvert running under the dam collapsed. There was not much damage to downstream property in this event, but a large section of the dam over the damaged portion of the culvert collapsed. The reservoir was reduced to a lake 3 m (10 ft) deep.

In 1879, the property was sold to Benjamin F. Ruff for $2000. He planned to repair the dam, making it and the reservoir the centerpiece of a private summer resort, the South Fork Hunting and Fishing Club. He intended to fill in the gap resulting from the 1862 failure and to rebuild the dam to its original height.

Reconstruction of the dam was conducted without regard to quality, either in design or construction. Fill of any composition was purchased from local landowners, including mud, brush, tree stumps, hay, and even a few loads of horse manure. Apparently, no attempt was made to compact any of this fill material properly. On December 25, 1879, the repairs were swept away by a rainstorm. Work was resumed in April 1880 using better methods and supervised by a man who had some experience building railway embankments.

However, the discharge system had been destroyed in the 1862 failure, and no provisions were made to replace it. Water rising in the lake seeped through the dam to the stone embankment side and started washing away any earth that was mixed with the stone on the downstream side of the dam. Hay, straw, and cut brush were placed across the upstream face to stop the seepage. Then earth and clay were dumped over the hay and straw to create the watertight section of the dam.

The project was completed in the summer of 1881, including some revisions to the top of the dam, lowering it so that a roadway could be constructed. This decreased the reservoir capacity and restricted the spillway. The spillway was further restricted by barriers to contain the game fish stocked in the lake before the club was opened.

For eight years, the resort operated without incident. The dam continued to impound a lake with a perimeter of 11 km (7 mi) and a volume of 18 million cubic meters (20 million tons) of water. There was some evidence of leakage, but since there was no workable discharge system, the water level in the lake could not be lowered to make repairs. Some permanent settlement had also taken place at the center of the dam.

On May 28, 1889, a storm struck the Johnstown area with the worst sudden rainfall that had ever been recorded in the area. Approximately 150 to 250 mm (6 to 10 in.) of rain fell in 24 hours over a large region. The amount of water entering the lake behind the dam was 285 m^3/min (10,000 ft^3/min), and the spillway capacity was only 170 m^3/min (6000 ft^3/min). Debris washed downstream contributed to restriction of the spillway.

The lake rose 600 mm (24 in.) overnight on May 30. By 10:00 A.M. the next morning the water was less than 300 mm (12 in.) from the top of the dam. About 11:30 A.M., despite frantic efforts to create a new channel, the water began to overtop the dam. The "resident engineer" at the dam rushed to inform a telegraph operator to warn the inhabitants of the valley below that the dam might collapse. This message, unfortunately, was not taken seriously. At 3:10 P.M. the dam collapsed in the center, leaving a gap of more than 120 m (400 ft).

A 12-m (40-ft)-high wall of water traveling at 25 km/h (15 mi/h) surged down the Conemaugh River valley. The destruction is legendary, with numerous accounts given in poetry and art as well as in the technical literature. There are many moving accounts of the flood recorded by witnesses and many photographs and lithographs that attest graphically to the immensity of the event (Figure 4.2). Perhaps most terrifying of all was a catastrophic fire that engulfed the debris as it collected at the Pennsylvania Railroad's massive stone bridge below the town. The debris behind the bridge was a tangled mass rising 10 m (30 ft) above the water and extending 30 city blocks upstream. Hundreds of Johnstown residents, many still alive, were trapped in the burning debris. The final death toll was 2209. Among the survivors were 565 children who had lost parents. Ninety-nine families ranging from two to 10 members were gone entirely.

Figure 4.2   Flood of 1889 in Johnstown, Pennsylvania (Kurz and Allison print, 1890).

Levy and Salvadori (1992) note that the dam as originally designed, and even as built, would have safely handled the flow into the reservoir from the heavy rainfall. The modifications that were made to satisfy the comforts of a select few were the direct cause of the disaster. No real engineer was involved in the planning or implementation of the reconstruction and modification of the facility. In court actions undertaken after the catastrophe, the event was deemed to be a "providential visitation" or act of God. Ironically, the only plaintiff against the Pennsylvania Railroad requested compensation for the loss of 10 barrels of whiskey. One can only marvel at the progress in liability law in the intervening century.

4.1.1.b Malpasset Dam, France, 1959

(*References:* Ross 1984, Levy and Salvadori 1992)

While the gravity dam relies on bulk for stability and an unyielding foundation to support its weight, the thin arch dam achieves its stability through shape. To develop its strength, the thin arch dam requires solid and stable rock at the abutments. However, even the thin arch dam can be destroyed by a weakness in the foundation rock, as was tragically illustrated in the mysterious sudden failure of a magnificent structure in France.

Malpasset Dam, on the Reyran River, was the world's thinnest arch dam at the time of its construction. Its location seemed ideal because of the excellent condition of support for the abutments. The wall was 66 m (217 ft) high, 1.5 m (4.9 ft) thick at its crest, and only 6.9 m (22.6 ft) thick at the base. It measured 223 m (732 ft) between abutments. The graceful dam was built between 1952 and 1954, and the 6.5-km (4-mi)-long reservoir began to fill.

The dam served satisfactorily for five years as the water in the reservoir rose, but shortly after 9:00 P.M. on December 2, 1959, after five days of heavy rain, the dam collapsed completely, with no warning. The failure released 50 million cubic meters (40,000 acre-ft) of water into the narrow Reyran Valley. Destruction was total for 5 km (3 mi) downstream, including road and railway bridges. The town of Frejus, 8 km (5 mi) downstream was demolished and 421 inhabitants were killed.

The subsequent exhaustive investigations could find no fault with the design concept or calculations. Construction quality was also excellent. The 48,000 m^3 (62,500 yd^3) of concrete was placed with the utmost care and control by the most experienced technical talent available. All potential causes were investigated, including the possibility of sabotage and even the unlikely chance that the dam had been struck by a meteorite.

Eventually, however, a thin clay seam about 38 to 40 mm (1.5 in.) thick was found in the rock below the arch abutment. Some engineering experts reported their opinion that this seam could have been detected from a proper preconstruction boring program. Lack of thoroughness in gathering of foundation information was cited by these investigators as the direct cause of the failure. It is now generally accepted that the unidentified thin, clay-filled seam in the rock adjacent to the left bank of the dam acted as a lubricant, permitting the foundation to move. This displacement destroyed the dam.

Legal action was begun by the French government against the design engineers, on the charge of involuntary homicide through negligence. Three years of litigation ended with their acquittal. Responsibility was never firmly established, but one investigator suggested that the brilliant and accomplished designer, André Coyne, had been "misled by his own genius." He had trusted his intuition rather than guaranteeing the solidity of the dam through sufficient geotechnical investigations.

4.1.1.c Vaiont Dam, Italy, 1963

(*References:* ENR 1963, USCOLD 1979, Ross 1984)

At the time of the Malpasset Dam collapse, another concrete arch dam was nearing the end of construction north of Venice, Italy. The Vaiont Dam was the highest thin arch dam in the world when it was completed. Located in a deep, narrow river gorge between limestone walls in the Italian Alps, the dam rises to the lofty height of 262 m (858 ft). The length of the arc at the crest of the dam is 191 m (625 ft). The Vaiont Dam is of the double-curved type, with its upstream face convex in both profile and plan at any section. It is 3.5 m (11 ft) thick at the crest and 22 m (73 ft) thick at the base.

The dam was completed in 1960. Its principal purpose was for hydroelectric power generation, as it is one of a series of dams, reservoirs, and powerhouses in the Piave River Valley system. Flood control was a secondary purpose for the Vaiont Dam.

On October 9, 1963, an entire mountainside slid into the Vaiont Reservoir during a severe rainstorm. The slide contained over 240 million cubic meters (314 million cubic yards) of rock, and almost filled the complete reservoir in a period of 15 to 30 seconds. The rockslide traveled at speeds of up to 30 m/s (100 ft/s). The rock movement was so massive that it registered on seismographic instruments throughout central and western Europe.

The slide displaced the reservoir water into an immense wave that climbed to 140 m (460 ft) above the reservoir level in one place. A 100-m (330-ft)-high wave overtopped the dam and smashed into the canyon below. Two minutes after the rock slide began, a 70-m (230-ft)-high wall of water leveled the town of Longarone, 1.5 km (1 mi) downstream, killing nearly all the inhabitants, then swamped three villages in the wider valley below. More than 2600 lives were lost. It was all over in less than 15 minutes.

The air shock wave preceding the wave of water caused complete destruction of the dam installations. The force on the dam was calculated to have been 36,000 MN (4 million tons). Remarkably, the dam itself did not fail, except for some minor crumbling along a small section of the crest near the left abutment. *Engineering News-Record* described the experience as a "disaster without structural failure" (ENR 1963).

The slide's immediate cause was attributed to erosion from water in the reservoir. Engineers had recognized the potential for slides in the reservoir walls, and the stability of the rocky enclosure was being monitored. However, no added resistance against sliding had been provided.

A slide had been observed moving toward the reservoir about 400 mm (16 in.) per day for about 10 days prior to the catastrophe. Engineers had been expecting a slide of about 19 million cubic meters (25 million cubic yards). The reservoir had been drawn down six days before the disaster in anticipation of the smaller slide. When it came, the water level was about 23 m (75 ft) below the crest of the dam. The actual slide, however, was of a magnitude beyond imagination.

The material involved in the slide comprised a volume slightly larger than the working volume of the entire Vaiont Reservoir itself. Ross (1984) notes that the amount of earth and rock that slid into the reservoir was more than twice the volume of the largest earth dam ever built (Fort Peck on the Missouri River in Montana). Moving this mass out of the reservoir would have been a monumental task, especially since there was no place to put it in the narrow river valleys in the region. As a result, the slide was never cleared from behind the dam; it remains there today, making the dam useless for the generation of hydroelectric power. The Vaiont Dam structure stands as a monument to the strength of a well-designed and constructed thin arch dam (Figure 4.3).

Figure 4.3 Vaiont Dam, Italy, (Courtesy of the McGraw-Hill Companies, Inc.)

The lesson of the Vaiont Dam is the importance of understanding not only the geology in the immediate vicinity of the dam but also completely around the reservoir. This requires a major effort in the preconstruction investigation. In a 1964 report on the Vaiont Dam, Claudio Marcello, retiring president of the International Commission on Large Dams (ICOLD), stated that "borings . . . may yield but a discrete sampling, and nature is not, unfortunately, so obliging as to comply with any continuity rule which could allow interpolation of results in between."

In his report on the 1961 Wheeler Lock failure in the Tennessee River, Karl Terzaghi similarly argued that no method of exploration can always be expected to reveal the existence of minor geological details, such as weak spots in the rock structure. The Vaiont Dam is a sobering reminder that meticulous engineering calculations are not sufficient to prevent failure entirely.

A number of Italian engineers were charged with manslaughter in connection with the Vaiont Dam catastrophe; three were eventually convicted.

4.1.1.d Baldwin Hills Reservoir, California, 1963

(*References:* ENR 1964, USCOLD 1975, Ross 1984)

Memory of the wave that swept over the Vaiont Dam caused engineers in Los Angeles, California, to order the evacuation of a section of the city just before the Baldwin Hills Reservoir failed in 1963. This action was no doubt responsible for saving many lives.

The Baldwin Hills Dam in southwestern Los Angeles was built in 1948–1949 as a balancing reservoir. About 1.5 million cubic meters (2 million cubic yards) of fill was used to provide the basin and the 200-m (650-ft)-long, 47-m (155-ft)-high dam for the 77,000-m^2 (19-acre) reservoir, which was lined with a membrane seal, a porous drainage course, a compacted earth blanket, and a layer of asphaltic concrete to guard against leakage. Some minor interconnected faults were found during construction but were not believed to be a hazard.

By 1951, however, these minor faults had moved sufficiently to rupture the reservoir's asphalt membrane floor. This allowed drainage from the reservoir into the soil. The break went undiscovered, even during a routine maintenance operation in 1957, when the reservoir was drained. By then the slow seepage of water had begun to weaken the soil below the reservoir and dam.

In December 1963, a fault movement enlarged the cracks in the lining, increasing the water flow and putting pressure on the dam's tile and gravel drains. On the morning of December 14, the dam failed, releasing over 1 million cubic meters (300 million gallons) of water onto a residential hillside and causing $50 million damage to property. Only five people lost their lives, thanks to quick action on the part of Department of Water Resources engineers.

Once empty, the reservoir revealed a large crack running its entire length. Probable cause of the crack was subsidence related to a 3-m (9-ft) settlement in the area from oil and water withdrawals over the preceding 40 years.

This failure served as a warning to the state of California. As noted previously, California had already passed a tough dam safety code following the St. Francis Dam failure in 1929, which killed several hundred persons and caused extensive property damage. The Baldwin Hills failure stimulated renewed interest in dam safety, and the state embarked on an ambitious dam inspection program. Inspections throughout the state identified a number of dams at risk and led to additional dam safety legislation.

4.1.1.e   Teton Dam, Idaho, 1976

(*References:*   ENR 1980a, Daniels 1982, Ross 1984, ENR 1985b)

In 1972 at Buffalo Creek in West Virginia, a nonengineered tailings pile from a coal mine collapsed following a heavy rainstorm, killing 125 persons. After this and a series of several other dam failures in the 1970s both in the United States and abroad, the federal government began to take an interest in dam safety on a national scale. The failure that gained the most attention and eventually led to significant action was the dramatic collapse in 1976 of the earth-fill Teton Dam in Idaho. This was the first catastrophic failure ever experienced by the U.S. Bureau of Reclamation in over 75 years of successful dam construction.

The Teton River presented a very difficult site geologically and a controversial site from an environmental point of view. State-of-the-art techniques were used in the dam's construction. The 93.5-m (307-ft)-high earth-fill dam involved a 930-m (3,050-ft)-long embankment comprised of 7.6 million cubic meters (10 million cubic yards) of earth and rock fill. A thick curtain of concrete 915 m (3000 ft) long and as much as 90 m (300 ft) deep was also constructed beneath the trenches and extended 300 m (1000 ft) beyond the abutments. Keyways 21 m (70 ft) deep and 9 m (30 ft) wide at the bottom were excavated into relatively stable rock in both abutments. A core trench across the canyon floor linked the keyways. Special outlet works were included in the design to prevent too-rapid filling of the reservoir.

But several oversights diminished the factor of safety against failure. According to the design specifications, a tight seal required grouting of cracks wider than 12 mm (0.5 in.); smaller cracks were ignored. The outlet works needed to maintain a moderate reservoir fill rate were still unfinished at the time of the collapse, even though the dam was essentially complete and the reservoir had been filling since October 1975.

There were many small cracks in the dam, and these, combined with the swiftly rising water pressure, brought about the collapse. Most investigators believe that water eroded cracks in the seam between the trench system and the curtain of concrete, scouring the earth fill to begin a piping action.

Some seepage was first noted on June 3, 1976, two days before the collapse. On the early morning of June 5, 1976, a muddy trickle appeared on the face of the right downstream embankment. This trickle grew to a steady flow within a few hours as the piping widened to a tunnel. Evacuation of the towns below the dam, Newdale, Teton, and Rexburg, was ordered. A last-minute effort to plug the leak with bulldozers failed at about noon. The roof of the tunnel collapsed in a rush of water and mud. The bulldozers and 3 million cubic meters (4 million cubic yards) of the dam were gone.

A 5-m (15-ft)-high wall of water, rushed down the river valley, flooding 780 km^2 (300 mi^2), with the 370 million cubic meters (300,000 acre-ft) of water stored behind the dam. By the end of the day, 11 people were dead, more than 2000 were injured, and 25,000 residents of the downstream communities were left homeless. Property damage was over \$1 billion.

It was the conclusion of two independent investigations and a 10-member panel of engineers appointed by the Secretary of the Interior and the Governor of Idaho that the principal cause of the Teton Dam failure was deficient embankment design and inadequate field inspection by the Bureau of Reclamation. The contractor was exonerated by the panel's report. Although some minor construction deficiencies were found, the panel found that construction conformed to the design in all significant aspects.

The cause of failure was attributed to a combination of unique geologic factors existing at an inappropriate dam site and related design decisions. The embankment material was not adequately protected from erosive seepage. One aspect of the design that was singled out for criticism was inadequate provision for drainage. The report noted that the designers went to great lengths to keep water from seeping through the dam but expended almost no effort to ". . . render harmless whatever water did pass." The panel determined that piping eroded the base of the embankment's impermeable core material in the keyway that was cut into the right abutment. Water then burst through the dam's downstream face because it was inadequately drained. However, because the washout carried away the failed section, the panel noted that the specific cause of the piping could probably never be determined with certainty. The panel also cited the lack of sufficient instrumentation to warn of impending failure.

Investigations of catastrophic dam failures of the scale of the Teton incident are quite difficult since so much of the structure and its immediate environment is destroyed in the failure. The cause of such a collapse may never be established with certainty; only a general consensus may emerge. At an International Workshop on Dam Failures held at Purdue University in 1985, nine years after the Teton disaster, at least seven theories for the probable cause of the failure were presented:

- Water leaked through the grout curtain under a grout cap at the right abutment key trench.
- The fill was cracked by hydraulic fracturing or differential settlement.
- Upward hydrostatic pressure lifted the fill off the base of the key trench.
- Water entered the key trench through an open rock joint, flowed over the grout cap, and exited through a downstream rock joint, eroding the fill.
- Wetted key trench fill sloughed off into an open rock joint, opening a gap in the fill.
- A layer of soil was compressed too much during the dam's construction, resulting in a dry seam. As water flowed through the permeable seam later, the seam collapsed and created a flow path.
- A seam of wet soil existed at the right abutment because of frost action, heavy rain, too high a moisture content when the soil was placed, or some other reason. Water seeping through this seam eroded the dam.

In reviewing the results of the conference, *Engineering News-Record* editors wrote: "As the vast number of successful dams attests, [engineering] judgment is generally pretty good. But, as the perpetually unexplained details of the spectacular failures remind us, any engineer's job description should list *humility* as an essential requirement" (ENR 1985b).

The Teton Dam collapse generated considerable discussion at the federal level and several proposals for funding national dam safety programs. These programs were given added incentive, and finally implemented one year later, in 1977, following the Toccoa Dam failure in Georgia, which killed 39 students and teachers at a bible college in Georgia.

4.1.1.f *Spokane River Dam, Washington State, 1986*

(*Reference:* Hokenson 1988)

A final example will illustrate the need to plan for unusual circumstances and the damage that can occur when operators rely on mechanical equipment for safety of a dam facility. On May 20, 1986, lightning struck an interconnection electrical tie line to the Upriver Dam hydropower facility in Spokane, Washington. The lightning triggered a sequence of events that did extensive damage to the facility and revealed design and operation flaws.

The facility includes a 12-m (40-ft)-high concrete dam and gated spillway with two power plants and a power canal. The dam and one powerhouse were built in 1936. The second powerhouse was constructed in 1984, along with other modifications.

The lightning strike caused a fault on the tie line, which shut down the generators by locking out all the circuit breakers. Although power was restored within 6 minutes, power to the spillway gate controls was not restored soon enough. This caused a surge of rejected turbine discharge. The surge washed over the closure embankment and onto a transformer, shorting it out. Both manual and remote attempts to open the spillway gates failed. The Spokane River began flowing unchecked through the project area, breaching the embankments and the dam abutments. Finally, six and a half hours later, a portable generator made it possible to open the spillway gates.

The dam itself escaped major damage except for a 30-m (100-ft)-long section of the parapet wall. There was substantial erosion of 150 m (500 ft) of canal bank, and 60 m (200 ft) of canal lining was washed away. The most costly damage was to the foundations of a powerhouse. The 12,000-Mg (13,000-ton) masonry structure was undermined, displaced, and tilted on its foundations.

An investigation by the Federal Energy Regulatory Commission found deficiencies in the electrical systems, mechanical systems, design of the reservoir freeboard system, the emergency action plan, operator training and staffing, equipment maintenance, management, and regulatory agency review.

Much of the damage could have been avoided had the reservoir freeboard system been adequately designed. Not enough freeboard was provided when the second powerhouse was added to the system in the 1984 renovation.

A fast-track repair project stabilized the powerhouse building and put the generating plant back on line. As part of the repair work, more freeboard was provided and provisions were incorporated to reduce the effects of surges. Mechanical and electrical system modifications were also made. Costs of the repair were over $11 million, including a $3 million project to lift and move the displaced powerhouse structure back into position (Figure 4.4). In addition

Figure 4.4   The Upriver Dam powerhouse, Spokane, Washington, was displaced by an accident in 1986.

to the direct damage cost, the city of Spokane lost considerable revenue during the power outage.

There were no deaths or injuries in this incident and no significant damage other than to the facility itself. This case study is included not because it was of tragic proportions, but simply to illustrate the multiple weaknesses that can be uncovered in a particular chain of events.

4.1.2   Dam Failures during Construction

Large earth-fill dams built before 1930 were normally hydraulic-fill, slowly depositing graded materials. The development of better compaction procedures permitted dry-fill mechanical placing and consolidation. Both methods have experienced failures during construction. Hydraulic fills often broke through the edge dikes. Dry fills are prone to slides during construction, sometimes of very large volumes. A few of the notable historic examples are given here.

The Calaveras Dam in California was discussed in the engineering news media for a decade. It was completed in 1926 as an earth and rock dam 67 m (220 ft) high above bedrock with a volume of 2.7 million cubic meters (3.5 million cubic yards). The lower half was built by hydraulic filling, but in 1918 some 610,000 m^3 (800,000 yd^3) of the upstream embankment and part of the clay core slid into the reservoir. Repairs were made by adding rolled fill and repairing the cutoff trench fill. Additional clay fill was rolled between broken rock berms to complete the dam.

The Fort Peck Dam on the Missouri River in Montana was scheduled for completion in 1939. It was then the world's largest earth dam, requiring more than 75 million cubic meters (100 million cubic yards) of earth and gravel to be placed by hydraulic methods. After an exhaustive geological investigation, a steel sheet pile cutoff had been driven to the top of firm shale.

In 1939, a major slide occurred suddenly near the right abutment, after which a reevaluation of design assumptions placed the blame on a buried bentonite seam in the shale beds. The addition of fill load on top of the normal weight covering the saturated seams had caused a release of water that could not readily escape through the overlying impervious shale. This provided a lubricant for the sudden movement of the overlying soil. The reconstruction and completion of the dam involved a modified design, with flatter slopes and reduction of shear stresses within the dam. The new core and the topping off of the dam were constructed by rolled-fill methods after the additional upstream berm and the fill from the slide area were replaced by hydraulic-fill techniques.

In 1956, some 1.5 million cubic meters (2 million cubic yards) of fill slid out of the east abutment of the Oahe Dam at Pierre, South Dakota. The slide developed slowly, moving 2 to 4 m (8 to 12 ft) in six months with vertical displacement of up to 9 m (30 ft) in the rolled-fill embankment. The area of the slide was 760 m (2500 ft) wide by 610 m (2000 ft) in the direction of movement, but the volume that had to be removed and then replaced with flatter slopes was only a small part of the 60 million cubic meters (78 million cubic yards) embankment. A combination of existing fault planes and bentonite seams in the underlying shales, triggered perhaps by nearby excavation operations along the face of the abutment and aggravated by subsurface water pressures, was blamed for the slide occurrence.

Rolled-fill earth dams containing clay or materials that can alter into a plastic mass develop high-pore pressures that provide temporary artificial supporting strength. When the volcanic ash hydraulic fill for the Alexander Dam on Kauai, Hawaii, was about 80 percent completed in 1930, a trapezoidal section having reached 29 m (95 ft) of the 37-m (120-ft) total height, half of the material sloughed out. The explanation was that the soft grains had crushed under the weight of the fill and reduced the expected porosity, slowed drainage, and the side was pushed out by the liquid pressure of the added fill (ENR 1930).

The Waco Dam in Texas was almost topped out to its 43-m (140-ft) height in 1961, when longitudinal and transverse cracks appeared in a 460-m (1500-ft) section of the embankment. The crest of the dam sank 6 m (19 ft) and the downstream slope shifted 8 m (26 ft). There was little movement at the upstream slope or toe. The fill was rolled local soils over a bed of cretaceous shales and limestones. An evaluation of rock conditions after the slide revealed some weak shale layers with intrusions of clay and the capability of built-up pore pressures under the new loading. The repair consisted of strengthening the fill with added berms and sealing the cracks with a sand–bentonite–cement grout.

In the West Branch Dam near Warren, Ohio, a 30-m (100-ft) stretch of a multipurpose embankment 3010 m (9900 ft) long and 28 m (93 ft) high sank about 300 mm (12 in.) in 1965, when a saturated layer of clay some 9 to 27 m (30 to 90 ft) below the streambed compressed under the new applied loading. To remove the water rapidly, electro-osmosis was used, with anodes introduced into some 1000 drilled holes.

One additional comment should be included regarding the economics of dam construction and the potential for costly disputes during the course of the project. In Chapter 3 it was acknowledged that a true determination of subsurface conditions is not always possible. In dam construction, the economic consequences of discovering unforeseen conditions can be immense. Unexpected conditions that may be encountered in such work should be anticipated by contract agreements at the outset of a project. Otherwise, the resulting construction delays and the complicated litigation will consume valuable resources that could be better spent on completing the project.

4.2   BRIDGE FAILURES

The history of iron bridge failures during the nineteenth century is a fascinating story and one that continues to impress engineers today with their professional responsibilities. The frightening number of railroad bridge collapses between 1875 and 1900 was discussed in Section 1.2. This was a time when the railroads were expanding. The economics of the expansion and railroad equipment technology demanded bridges to support heavier and heavier loads, and the bridges had to be constructed in irregular, unusual environments.

These bridges were subjected to large gravity loads, to dynamic vibrations, and to the changing positions of the train (Figure 4.5). Petroski (1985) has noted that the railroads brought together the machines of the mechanical engineer and the stationary structures of the civil engineer, and the demands of each provided stimulus for the evolution of the other. The trial-and-error experience of the railroad bridge builders provided much of the foundation for modern fracture mechanics and other material science knowledge.

Complete collapse of bridge structures is not common now, but there are still difficulties. Most bridge failures in this century have been due to design methods that were not sophisticated enough to account for subtle conditions and secondary loads, such as wind-induced excitations and thermal effects (Ross 1984). There have also been a number of failures caused by detail deficiencies: joints, bearings, supporting corbels, poorly welded connections, and so on. Other failures were caused by impact loading from collisions. Finally, a few catastrophic collapses have been due to inadequate maintenance.

Figure 4.5 A railroad bridge must resist dynamic loads and gravity loads in changing positions. (Courtesy of *Lewiston Morning Tribune,* Lewiston, Idaho.)

A trapezoidal piece of elevated highway structure in Chicago shifted laterally in 1966, after five years of use, sufficiently to displace the bearings away from the base plates. The span dropped 80 mm (3 in.), to rest on the concrete pier. The 7.6-m (25-ft)-wide section was 10.7 m (35 ft) long on one side and 24.4 m (80 ft) on the other. One end was fixed to a steel bent. At the other end, the beams rested on sliding bearings 100 mm (4 in.) wide, which dropped when the supports had bent sufficiently. The lateral force responsible for such distortion came from an accumulation of braking traction from vehicles on the roadway, unequal thermal changes in the length of the deck because of the large discrepancy in dimension of the two sides, and possible locking of the expansion joints by filling of the space with debris. When the span was jacked up, taking the load off the pier, the support moved somewhat toward the fixed end, indicating that some lateral movement had occurred. After raising the span to the proper elevation, larger base plates were installed and shimmed tight to the pier. To avoid future separation of the two supports, cable ties were added to connect the piers.

Sometimes insufficient consideration of details in bridge design causes unacceptable performance short of structural collapse. Such serviceability problems can require large expenditures for corrective repairs or excessive maintenance. For example, the assembly of carefully detailed steel expressway frameworks often does not produce the smooth riding pavement planned, when camber built into the stringers and floor beams is not reduced sufficiently by the dead load. With slab and pavement thicknesses fitted between prefabricated steel expansion dams, sitting directly on steel floor beams, a series of scallops may result.

Such a rough-riding expressway in Brooklyn, New York, was closed to traffic in 1960, six months after dedication, when the asphalt surface became too rough for traffic, even under restricted speeds. Correction required the removal of the asphalt, disregard of thickness and grade levels called for in the contract documents, and an empirical placement of a wearing surface to connect the tops of the expansion dams, located 18 m (60 ft) apart. The concrete deck had been placed with reference to the top of the stringers, in a series of curved surfaces that were expected to flatten under the weight of the concrete. Possibly the rigidly welded end connections of the stringers prevented the mechanically bent stringers from giving up their camber, and the hardened concrete fixed the beams in the curved position.

However, should the designer not provide camber in the stringers, the result will usually be an irregular profile with the expansion dams creating high ridges. In any case, the thickness of the wearing surface can be varied to correct for deck irregularities. On New York's Tappan Zee Bridge, where the concrete slab was the wearing surface, formwork for each panel had to be tailor-fit to the surveyed steel elevations with a variation in slab thickness to provide an acceptably smooth roadway. Here, too, the camber did not reduce sufficiently under dead load.

4.2.1 Failures of Completed Bridges

In this section a few of the more influential and interesting failures of completed bridges are presented briefly, along with references for further study. In some of these cases, wind and storm were the cause of collapse. Other wind failure examples are given in Section 2.3.

Collision damage to bridges is discussed in Section 2.6. The best known recent cases are the 1980 Sunshine Skyway Bridge collapse in Tampa Bay, Florida, and the September 1993 fog-related barge collision into a bayou bridge near Mobile, Alabama, which claimed 47 lives in a subsequent Amtrak railroad accident. There are many other examples resulting from the impact of trucks, train derailments, barges, ships, and debris in river channels. To mitigate such accidents, warning devices are helpful, and a structural pier protection system is recommended. This may take the form of fenders, sand-filled cofferdams, horseshoe-shaped islands, and so on. Sliding cofferdams are recommended where barges contain toxic or environmentally damaging materials. In these cases, the impact must be absorbed without catastrophic damage to the vessel, which may be of greater concern than damage to the bridge.

4.2.1.a *Tay Bridge, Scotland, 1879*

(*References:* Ross 1984, Levy and Salvadori 1992, Brown 1993)

The collapse of the Tay Bridge in Scotland is one of the most published engineering disasters of all time. Completed in 1877, the bridge was 1.5 km (1 mi) long. It was comprised of 85 iron lattice-truss spans supported 27 m (88 ft) above an inlet of the North Sea on the east coast of Scotland, known as the Firth of Tay.

The bridge was one of two proposed by Sir Thomas Bouch, engineer for the North British Railway Company. The Tay Bridge, and another over the Firth of Forth, were inspired by fierce competition between rival railroad companies and the desire to eliminate slow ferry crossings over the large estuaries. Both bridges were nearly the same overall length, but the Firth of Tay was more shallow, permitting the economical construction of multiple spans.

Construction began in 1871. The six-year project was difficult and there were several accidents, causing the deaths of 20 workers. At the time of the opening of the Tay Bridge, on September 26, 1877, Thomas Bouch had completed the design for the companion Firth of Forth Bridge. For the deeper estuary, he proposed a dramatic and daring suspension bridge.

Most of the spans of the Tay Bridge were deck trusses, with trains traveling above the deep trusses. However, to provide clearance for the shipping channel, the 13 center spans were raised above the railway level so that trains ran inside them: a through-truss arrangement, or lattice tunnel.

On Sunday evening, December 28, 1879, two years after the completion of the bridge, a severe gale struck the Firth of Tay. At about 7:00 P.M., with the violent windstorm still building, a mail train with six coaches containing 75 crew and passengers stopped at the south end of the bridge. The night watchman at the southern approach to the bridge gave a harrowing account of the events that transpired subsequently.

The watchman discussed the weather conditions with the engine driver, but the driver decided to risk the crossing. (Reflecting on the consequences of this unwise decision should surely cause impatient modern-day air travelers to appreciate a pilot's decision to delay a departure for weather reasons.) The train never reached the north end of the bridge.

When the watchman did not receive confirmation of a safe crossing, he began to crawl out on the bridge on his hands and knees, creeping forward while clinging to the bridge in the howling gale. Just before giving up and turning back, he noticed that the horrible shrieking of the wind in the trusses seemed to be diminishing. He crawled on to the next pier and found the structure completely sheared away. To his front and immediately below was black emptiness. The entire train, all 75 passengers, and 13 spans of the bridge had fallen into the Firth of Tay. In the morning, it could be seen that nearly a third of the bridge was gone (Figure 4.6).

Several theories were proposed for the disastrous collapse. Some investigators held that the wind against the side of the train buckled the track bracing, causing a lateral failure of the spans. Others proposed that the bracing system for the cast-iron towers supporting the bridge failed under the severe wind

Figure 4.6 Tay Bridge failure, Scotland, 1879. (From the *Illustrated London News;* courtesy of the Hulton Getty Collection.)

load, and that the towers and trusses collapsed together. Still others believed that the train had simply been blown off the rails, tangling into the lattice through-trusses, leading to a progressive collapse of adjacent spans and supporting towers.

The British Board of Trade investigation uncovered many deficiencies in design and construction quality control. Criticism of the designer was intense, and Sir Thomas Bouch died in disgrace less than a year after the collapse. Ethical questions were raised regarding the fact that Bouch had a financial interest in the firm that built the bridge; however, this was not an uncommon practice at the time. (Similar questions arise from time to time regarding modern "design-build" or "turnkey" projects.) Cast-iron columns contained voids that had been filled with beeswax and iron filings so that they were not visually apparent. Foundation design was inadequate, based on incomplete borings.

The investigation report was especially critical of the design's lack of provision for wind loading. Bouch had been very complacent in his consideration of wind effects, relying on tables produced by John Smeaton nearly a hundred years earlier that suggested a static design load of only 575 Pa (12 lbf/ft^2) for a "tempest."

As a result of the inquiry, British engineers came to appreciate the character of dynamic forces. The immediate effect on design was an overreaction, increasing the design wind pressure requirement for new bridges to 2700 Pa (56 lbf/ft^2). This law resulted in many bridges that were overdesigned and inefficient. Wind loading remained a point of argument for many years.

Understandably, Thomas Bouch's design for a suspension span over the Firth of Forth was dropped after the Tay Bridge collapse. In its place, Sir John Fowler Benjamin Baker designed a massive landmark structure that was completed in 1890 (Figure 4.7). The Firth of Forth Bridge is perhaps the stiffest, strongest, and safest railroad bridge ever constructed. It remained the longest span bridge until 1918, when the Quebec Bridge was completed in Canada.

Like the Quebec Bridge, the Firth of Forth Bridge is a cantilever truss design. Cantilevers form the two main spans, which are supported on massive piers 45 m (150 ft) above the water. Each of the main spans is 521 m (1710 ft). There are more than 45,000 Mg (50,000 tons) of steel tubes and girders in the bridge.

The Firth of Forth Bridge was a remarkable achievement, one that established many new world records, including the volume of masonry in the piers; the height, length, and depth of its cantilevers; the scale of its free spans; and the volume of steel in the whole structure. Some observers acclaim the structure for its aesthetic qualities as well, although others see it as a grossly inefficient overreaction to the failure of the Tay Bridge. William Morris, the nineteenth-century poet, artist, craftsman, and socialist, expressed his opinion of the Firth of Forth Bridge articulately when he called it ". . . the supremest specimen of all ugliness . . ." (Brown 1993).

Figure 4.7   Firth of Forth Bridge, Scotland, completed in 1890. Reed International Books.)

4.2.1.b   Ashtabula Bridge, Ohio, 1876

(*References:*   MacDonald 1877, Gasparini and Fields 1993)

On December 29, 1876, during a winter snowstorm, an iron railroad bridge collapsed at Ashtabula, Ohio, after 11 years of service. A train pulled by two locomotives was crossing the bridge, heading west. The first locomotive was almost across when the bridge began to fail. That locomotive barely made it safely to the west abutment, but the second locomotive and 11 cars fell 20 m (65 ft) into Ashtabula Creek. The fall and the fires caused by the coal stoves in the cars caused over 80 deaths.

The most probable cause of the failure was fatigue and brittle fracture at a flaw in an iron casting. The Ashtabula Bridge was built between 1863 and 1865, using prestressing methods developed for timber Howe trusses. At the time, structural analysis was more an art than a science. Determining the capacity of slender compressive members was an unresolved issue. Standard design and material specifications did not exist. The concept of fatigue was not commonly understood.

The Ashtabula Bridge collapse was investigated extensively by engineers. Public reaction to the tragedy added credibility to the call for independent consulting bridge engineers and for standard design specifications. For material scientists, the failure illustrated the unreliability of iron castings; 10 years later, specifications explicitly forbade the use of cast iron in any part of a bridge structure.

Construction and fabrication quality was found to be generally good. Numerous criticisms of the design emerged from the investigations, leading to the suicides of two people involved in the design. Unfortunately, much of the criticism was not founded in fact.

The cause of the failure generally accepted today was identified initially by MacDonald (1877). A fatigue crack, originating at a flaw in an iron casting, propagated under repeated stress cycles over the decade of use. On the day of the collapse, low temperatures had reduced the fracture toughness of the iron. The stress from the train load caused the crack to become unstable and a brittle fracture occurred. It is likely that if the flaw had not been there, the bridge would have served its full expected life. At the time of this failure, however, material science was in its infancy. Investigation of cracking potential due to fatigue was not yet part of standard engineering practice.

4.2.1.c Hackensack River Bascule Bridge, New Jersey, 1928

(*References:* ENR 1929, Ross 1984)

On December 15, 1928, the eastern leaf of a double-leaf bascule bridge in Hackensack, New Jersey, fell while it was being lowered into place. The bridge was less than two years old. Design deficiencies, principally the failure to account properly for dynamic effects in the moving structure, were blamed for the failure.

The bridge was adequately designed for static loads, but it was actually an operating piece of machinery. The collapse was caused by dynamic distortions and oscillations involving friction and inertia. Large oscillatory motions and cyclic stress reversals, while the bridge was in motion, produced excessive stresses in the counterweight tower and progressive fracture. Failure to consider the bridge as a machine was directly responsible for the collapse.

This was an extremely important structural accident, for it showed that a moving bridge must be designed as a dynamic object. The final investigative report noted that ". . . structural dynamics (and the science of vibrations) will occupy a position of marked importance in the future development of bridge engineering" (ENR 1929).

In the aftermath of the Hackensack Bridge failure, vigorous debate took place in the professional and technical journals regarding the proper consideration of dynamic effects. This debate involved some of the most prominent people in modern structural engineering history, including D. B. Steinman and Hardy Cross. Developments in analytical theory and physical modeling procedures were inspired by this accident.

4.2.1.d   Tacoma Narrows Bridge, Washington State, 1940

(*References:* Petroski 1985, 1993; Levy and Salvadori 1992; Puri 1994)

One of the most spectacular failures in the history of engineering occurred on November 7, 1940, when the Tacoma Narrows Bridge over Puget Sound in Washington State collapsed due to wind-induced excitations. The 853-m (2800-ft) main-span suspension bridge failed dramatically four months after completion in a wind of only 68 km/h (42 mi/h).

Lightweight suspension bridges had experienced failures previously. In 1826, one year after completion, Thomas Telford's Menai Straits Bridge in eastern England was partially destroyed in a hurricane. The deck of the 170-m (550-ft)-long span had undergone 5-m (16-ft) vertical undulations before breaking. The Menai Straits Bridge was repaired and strengthened, but wind deformations continued to be large and unacceptable. In 1854, the 308-m (1010-ft) span suspension bridge over the Ohio River at Wheeling, West Virginia, experienced a similar failure.

John Roebling, designer of the Brooklyn Bridge (completed in 1883), achieved numerous successful suspension spans at a time when other designers were experiencing failures. Roebling's designs were based on an understanding that stiffness of the deck was absolutely critical to wind resistance. Nevertheless, those who followed Roebling continued to attempt longer and longer spans with more economical slender structures. Suspension bridges evolved through the first part of the twentieth century into longer and sleeker designs, making them more flexible. These structures were experiments outside the envelope of experience.

Built between 1938 and 1940, the Tacoma Narrows Bridge was opened to traffic on July 1, 1940. At the time of construction, the 853-m (2800-ft) main span was the third longest suspension span in existence. The span of the Golden Gate Bridge in San Francisco was 1280 m (4200 ft), and the George Washington Bridge span in New York was 1070 m (3500 ft). The slender Tacoma Narrows Bridge presented a dramatic, graceful silhouette suspended from its two 128-m (420-ft)-high towers. It was designed without stiffening trusses, for aesthetic reasons.

Indeed, the Tacoma Narrows Bridge was the most slender suspension bridge ever built, both in width and in depth. The width of the deck was only 12 m (39 ft) and the depth of the stiffening girders was only 2.4 m (8 ft). The bridge supported two lanes of traffic, as compared with six lanes for the Golden Gate Bridge and 12 on the George Washington Bridge. In terms of span/width ratio, the Tacoma Narrows Bridge had less than one-third the stiffness of the Golden Gate and George Washington spans. Not only was the Tacoma Narrows Bridge so slender, but it was also extremely lightweight. With only two traffic lanes, the dead load was one-tenth as much as that of any other major suspension bridge yet constructed. In addition to these unique features, the structure exhibited very little damping capability. Once oscillations began, they continued for a long time.

Yet the Tacoma Narrows Bridge was designed by Leon Moisseiff, one of the world's leading authorities on suspension bridges. Moisseiff was a consulting engineer for the Golden Gate Bridge, the Bronx–Whitestone Bridge in New York City, and the San Francisco–Oakland Bay Bridge. His design theories and equations were respected and used internationally. The Tacoma Narrows structure was similar to that used for the Bronx–Whitestone Bridge, although Bronx–Whitestone is much heavier, wider, and stiffer. The Bronx–Whitestone Bridge experienced uncomfortable vibrations and vertical movements, but none that compromised its structural integrity.

From the day the Tacoma Bridge was completed, disturbing vertical oscillations occurred in strong winds. The Puget Sound region is prone to high winds, and the structure survived several windstorms without structural damage. Vertical displacements had reached as much as 1.25 m (4.2 ft) amplitude, much more than the small displacements experienced at Bronx–Whitestone. The bridge quickly earned the nickname "Galloping Gertie."

Scale models of the bridge were constructed at Princeton University and at the University of Washington to search for corrective measures. Several ideas were attempted, but none were implemented in time to save the bridge. Tie-downs were installed, but they broke during moderate winds. Streamlining the profile of the edge girders was found to have a positive effect in the model, but the bridge failed before this idea could be applied.

Until November 7, 1940, the vertical waves on opposite sides of the bridge had not gotten out of phase, even though the motion was quite severe. On the morning of November 7, however, torsional oscillations began to develop. Failure occurred shortly before noon, when almost the entire suspended span was torn loose. The collapse was initiated at midspan when the stiffening girders buckled. The suspender cables failed and flew high into the air above the main cables, while large sections of the deck fell progressively toward the towers. The 335-m (1100-ft) side spans remained intact, bending the towers backward 4 m (12 ft) under their pull (Figure 4.8).

No lives were lost in the collapse; there were only a few people on the bridge. Among these was F. B. Farquharson of the University of Washington, who had been conducting experiments on a model of the bridge. At the time of the collapse, he was on the main span taking motion pictures of the abnormal twisting motions. Farquharson's marvelous photographs have subsequently been studied by physics and engineering students throughout the world, even to this day.

Farquharson described his experience on the bridge deck during the final minutes. He reported that the deck tilted from side to side more than 45 degrees, that the edges of the deck had an up-and-down movement of 8.5 m (28 ft), and that the acceleration of the deck at times exceeded that of gravity.

The valuable documentation of the motions of the Tacoma Narrows Bridge inspired a great deal of research on lightweight and tension structures. As

Figure 4.8   Tacoma Narrows Bridge, Washington State, 1940. (From the Special Collections, University of Washington Libraries, Farquharson No. 12.)

noted, this was not the first suspension span to fail due to dynamic wind-induced vibrations, but its measured and photographed performance was extremely influential.

Clearly, the bridge was the victim of its narrow deck and lack of torsional stiffness. Federal investigators noted that structural members exhibited remarkable strength. Their report emphasized the need for rigidity in both the vertical and horizontal planes. The need for greater attention to dynamic response and provision of structural damping was also noted.

As for the strength of the materials and members, the report stated that behavior of the structure during the final hour was a severe test. Under the violent dynamic motions recorded by the photographs, structural members and connections were subjected to stresses far beyond the safe limits for which these members were designed. The fact that they did not fail earlier attested to the excellence of their design and the quality of materials and workmanship.

The fundamental weakness of the Tacoma Narrows Bridge was its great flexibility, vertically and in torsion. The comparatively narrow width gave adequate lateral rigidity, but combined with the extreme vertical flexibility, the narrowness made the structure extremely sensitive to torsional motions created by aerodynamic forces.

The final report suggested further that the behavior of the bridge could only be explained by its response to aerodynamic forces. The investigators pointed out that ". . . a complete analysis of these forces and of the response of a suspension bridge thereto is not possible with present knowledge; further observations, experiments and mathematical analyses are required. . . ."

The report did not criticize the choice of a suspension bridge for the Tacoma Narrows, noting that a suspension span was the most suitable and economical that could have been chosen and that the particular site was satisfactory. Nor was the quality of materials and workmanship blamed in any way for the collapse. The principal conclusion was that further studies were needed for advancing the science of suspension bridge design.

Many learned from the Tacoma Narrows failure, particularly because of its extensive documentation. Eventually, another suspension bridge was built on the same site. Construction started in 1946. In the meantime, much analytical work and structural modeling was undertaken, including improved sophistication in wind-tunnel modeling. These developments were accompanied by vigorous and heated discussion in the technical literature, involving such distinguished bridge designers as Othmar Ammann and D. B. Steinman, who argued the merits of various stiffening schemes. The new bridge, however, was of more conventional design, with four lanes of traffic instead of two and with deep stiffening trusses, 10 m (33 ft) deep rather than the 2.4-m (8-ft) depth of the stiffening girders in the original bridge.

Othmar Ammann, who designed the George Washington and other monumental bridges, once wrote:

> [T]he Tacoma Narrows bridge failure has given us invaluable information. . . . It has shown [that] every new structure which projects into new fields of magnitude involves new problems for the solution of which neither theory nor practical experience furnish an adequate guide. It is then that we must rely largely on judgment and if, as a result, errors or failures occur, we must accept them as a price for human progress. (Ammann, Von Kármán, and Woodruff 1941)

When considering Ammann's comments, Petroski (1985) notes that extending the envelope of experience does not necessarily equate to failure. Several bridges are successful testaments to leaps in engineering progress. Petroski responds:

> For a hundred years now the Eads Bridge has crossed the Mississippi at St. Louis in three chrome steel arches while the Brooklyn Bridge has been suspended over the East River in New York from four steel-wire cables. These two bridges represent some of the earliest uses of pneumatic caisson technology and steel in bold new designs; both stand as unequaled achievements of their era. They both involved several simultaneous leaps of engineering and hence invitations for something to go wrong. (Petroski 1985)

4.2.1.e Silver Bridge, Point Pleasant, West Virginia, 1967

(*References:* Bennett and Mindlin 1973, Ross 1984, Petroski 1985, Lichtenstein 1993)

The most tragic highway bridge collapse ever to occur in the United States was the December 1967 failure of the 40-year-old suspension bridge over the Ohio River between Point Pleasant, West Virginia, and Gallipolis, Ohio. Called the "Silver Bridge" because of its silver paint, this bridge was not suspended from the wire cables that support modern suspension bridges, but from two chains made up of 15-m (50-ft)-long "eyebar" links. The Silver Bridge was the first eyebar suspension bridge in the country in which the eyebars performed double duty as the top chords of the stiffening trusses in both the main and anchor spans. In addition, this bridge was the first to use high-strength, heat-treated carbon steel eyebars.

The Silver Bridge was two traffic lanes wide and had a total length of about 530 m (1750 ft), including a main suspension span of 213 m (700 ft) and two approach anchor spans of 116 m (380 ft) each (Figure 4.9). Completed in 1929, the original bridge had a wood plank deck topped with rock asphalt. The deck was replaced by a concrete-filled steel grid in a 1941 general renovation.

The collapse occurred Friday, December 15, 1967, when the bridge was fully loaded with Christmas shopping traffic. All three of the spans and the two towers collapsed into the river. When the bridge fell, the main span

Figure 4.9 Silver Bridge, Point Pleasant, West Virginia, 1967. (Courtesy of the National Transportation Safety Board.)

overturned, dropping the vehicles into 20 m (70 ft) of water. The span then fell on the vehicles, and both towers collapsed into the wreckage. Eyewitnesses described the collapse as sudden, taking from a few seconds to half a minute.

The rescue mission and the investigation were greatly hampered by the tangled condition of the bridge debris and vehicles and by the murky condition of the water, flowing under a current of 3 m/s (6 knots). Most of the structural members in the wreckage were cut to get to the submerged vehicles, destroying evidence that might have facilitated the investigation. In addition to the concern for trapped victims, there was a desire to clear the busy shipping channel as quickly as possible. Of the 64 persons in 31 vehicles that fell with the bridge, 44 died and two were never found. There were 18 survivors, nine of whom were injured.

Early in the investigation, authorities focused on several potential contributing factors: overloading, age, fatigue, and corrosion. The unique design of the bridge was also studied, as was the record of inspections performed by West Virginia's highway department.

First, it was determined that the bridge was not "overloaded" at the time of failure, after the Corps of Engineers had pulled the vehicles out of the river that had been on the bridge at the time of collapse. This had been an initial concern since the bridge was designed for a much lower live load than is used today for interstate highway bridges. Live loads on bridges had increased considerably during the 40-year life of the Silver Bridge. It should be noted, however, that the ratio between live- and dead-load stresses was greater near the end of the life of the bridge than at the beginning. Under cyclic loading, this condition can contribute to fatigue failure.

The "nonredundant design" of the Silver Bridge structure was of concern to investigators. Most modern suspension bridges use cables that are comprised of a large number of cold-drawn steel wires. Failure of a single wire in such a cable will not bring about a sudden collapse. However, the Silver Bridge, with its chain of eyebars, was susceptible to total catastrophic failure due to fatigue fracture of a single eyebar link in the chain.

The eyebars were each about 50 mm (2 in.) thick and 300 mm (12 in.) wide. They were placed in pairs of varying length to maintain equal panel lengths in the stiffening trusses. Each eyebar of the sidespan suspension chain passed through a vertical strut. The eyebars and struts were connected by pins. The total collapse could have been initiated by the failure of a single eyebar, a single strut, or a single connecting pin. The wreckage was so damaged and tangled that investigators first thought they would have little physical evidence to determine which member or joint was responsible.

Eventually, however, the cause was determined to be eyebar 330, the second upstream eyebar west of the Ohio tower. The companion eyebar to number 330 had been found intact, but with the edge of its eye rolled back, suggesting that this eyebar was carrying the entire load just prior to the collapse, its companion having fractured earlier.

The final report was issued in April 1971 by the National Transportation Safety Board, three years after the failure. The report concluded that the collapse was initiated by an eyebar failure. The eyebar fractured through development of a critical-size flaw, as a result of the combined effects of stress corrosion and corrosion fatigue. Initial flaws had no doubt been present in the eyebar. The combined conditions of aging, cyclical stresses, and temperature fluctuations eventually enlarged the crack to a critical state.

The connection detail made regular inspection difficult and also made corrosion more likely than if all the steel had been exposed. Thus progressive deterioration went undetected. Small flaws, originating from the holes in the steel links, propagated through the now well-known process of fatigue-crack growth accelerated by corrosion.

Other than the obvious need for designers to provide greater redundancy, the principal lesson from the Silver Bridge collapse was the importance of regular, diligent inspection for aging facilities. The bridge had last been inspected by West Virginia Highway Department engineers in April 1965, at which time nothing of concern was reported. A state bridge inspector admitted that the 1965 inspection was somewhat incomplete and that the last full inspection had been in 1951. Even in 1951, some of the prescribed procedures had been circumvented. Inspectors had worked with binoculars rather than climbing the towers to inspect the upper members.

The Silver Bridge failure immediately triggered a number of states into promptly investigating the safety of their older bridges and led directly to federal regulations delineating bridge inspection requirements. The National Bridge Inspection Standards, passed by the U.S. Congress in 1968, for the first time set uniform standards throughout the country. Were it not for the tragedy of Silver Bridge, inspection guidelines might still be subject to the political differences of individual states.

It is ironic that the flurry of regulations increasing bridge inspection activity was stimulated by a failure that could not have been predicted by the visual inspection methods that were available at the time. The critical detail in the Silver Bridge would have been impossible to inspect for corrosion without dismantling each joint. Thus it is clear that maintenance is not exclusively the responsibility of owners and operators of facilities; there are lessons here for designers as well. For long-term performance, the structure must be designed with accessibility and inspectability in mind.

In a letter to the editor of *Scientific American* (July 1993), J. A. Bennett discussed the detail of the eyebar-to-strut connection:

> As a metallurgical study by the National Bureau of Standards showed, the eyebar had fractured suddenly because of a stress corrosion crack less than one-eighth of an inch deep that had started on the surface of the hole in the eye. The hole was almost completely filled by the pin that coupled successive links in the eyebar chain. The end of the pin and the hole in the eye were also covered by a plate that prevented visual inspection. At the time of the collapse of the Point

Pleasant bridge, an identical bridge was in service a few miles upstream [at St. Marys, West Virginia]. Naturally, there was great interest in determining whether its eyebars could be inspected. The best brains in the nondestructive inspection field concluded unanimously that it could not be done. Consequently, the bridge was removed. . . . (Bennett 1993)

Dunker and Rabbat, authors of an earlier article on bridge inspection (1993a), replied to Bennett's letter:

"Ironically, lax inspection noted at the time of the Silver Bridge collapse helped to trigger a massive federal bridge inspection program, and yet state-of-the-art nondestructive testing would not have detected the hidden defect. . . ." (Dunker and Rabbat 1993b)

To replace the Silver Bridge, a cantilever through-truss was built in 1969. A routine inspection in 1977 found a crack in a top-chord weld of a steel girder loaded in tension. The bridge was closed immediately for repairs on 16 welds. The residents of Ohio and West Virginia may have been spared a second tragedy as a result of important lessons learned from the first.

West Virginia is not the only state to experience a catastrophic failure due to deficient inspection of aging bridges. The last two examples in this section, one in Connecticut and one in New York, illustrate that no state can afford to be complacent about bridge maintenance.

4.2.1.f *Mianus River Bridge, Connecticut, 1983*

(*References:* Browne 1983, Levy and Salvadori 1992)

The sudden collapse of a 30-m (100-ft)-long three-lane section of the Connecticut Turnpike killed three people and injured three others on June 28, 1983. The collapse occurred at 1:28 A.M., causing four vehicles to plunge 20 m (70 ft) onto the tidal flats and bank of the Mianus River (Figure 4.10). The adjacent three-lane section remained intact. Timing of the incident was very fortunate; there could have been many more deaths had the accident occurred later in the day. This section of the Turnpike normally carries 100,000 vehicles per day.

The fallen span was hung between two cantilever sections, suspended from the ends of the supporting cantilevers by pins and hangers. The section that fell was comprised of two 2.7-m (9-ft)-deep plate girders and four stringers connected by cross-beams. It weighed 450 Mg (500 tons), including the 190-mm (7.5-in.)-thick concrete deck and 50-mm (2-in.)-thick asphalt topping. Ends of the span were skewed at 53 degrees to the direction of travel (ENR 1983).

Five days after the accident, divers found a fragment from the hanger connection—a missing section of a pin that was fractured. The rest of the

Figure 4.10   Mianus River Bridge, Connecticut, 1983. (© Hank Morgan, Rainbow, Housatonic, Massachusetts.)

broken 180-mm (7-in.) pin was still connected to the cantilevered end of the bridge. These pins had been used to pass through the hangers in the connection and through the face of the steel girders. Washers were used to separate the hangers and girders, and a cap was used to cover each end of the pin and hold the connection tightly together. The cap was secured by passing a threaded rod through the center of the pin, tightening it with nuts at each end.

The end caps, intended to keep the hangers from slipping off the pins, were clearly inadequate to perform this function. They were only 7.5 mm (0.3 in.) thick. Several of the caps had been deformed. There was a heavy concentration of rust found behind the cap plate and the 350-mm (14-in.)-diameter washers. The corrosion would not have been directly visible to inspectors, but it was substantial enough to exert sufficient pressure to deform the cap plate visibly.

The significant amount of corrosion found in the hanger connection region caused investigators to look for the source of water responsible for it. They found that 10 years prior to the collapse, roadway drains had been paved over. This permitted water, deicing salts, and other residue to flow through the joint between the cantilevered and suspended spans, dropping directly onto the hanger assembly.

One investigator implicated the skewed angle of the bridge design, which tended to cause a lateral force, pushing the straps off the pins. This angle would have exaggerated the horizontal forces applied to the bridge by braking and accelerating vehicles. However, test data showed that forces caused by corrosion alone were sufficient to overcome friction between the pin and the hangers and to move the hangers off the pin. The National Transportation Safety Board report agreed with this conclusion, rejecting the "skewed angle" argument (ENR 1985a).

Designers of the 26-year-old bridge were exonerated by a superior court jury in a decision that was upheld by the Connecticut Supreme Court. The court held that deficient inspection by the state of Connecticut Department of Transportation was responsible for the collapse.

Failure of the Mianus River Bridge again stimulated concerns for inspection and maintenance. In an emergency program, the Connecticut Department of Transportation repaired hundreds of bridges at a cost of $1.2 billion, with a budget five times greater than that available before the Mianus River Bridge failure. Cracks in hangers were discovered and repaired on the Yankee Doodle Bridge over the Norwalk River in Connecticut in 1984, a more redundant multiple-girder bridge. In Boston, two broken hangers were found in the Harvard Bridge over the Charles River as a result of the intense inspection activity immediately following the Mianus River Bridge collapse.

In particular, the Mianus River incident pointed out the vulnerability of pin-connected hung spans for bridges that lack redundancy. There are many bridges throughout the country that were built using similar pin-hanger connections. Such bridges were quickly identified and prioritized for retrofit projects to reduce their failure potential. Bridges having multiple girders are not as critical, since loss of one pin connection will not bring about a total collapse. In the case of the Mianus River Bridge, calculations easily established that the failure of only one pin or hanger was sufficient to bring down the entire section that collapsed.

The Silver Bridge collapse had earlier indicated the desirability for redundancy. Of course, the Mianus River Bridge was designed in 1955, with construction complete in 1958. This was a decade prior to the collapse of the Silver Bridge, so the lessons of that collapse were not available to the Mianus Bridge designers.

However, the lessons of nonredundancy are now part of standard bridge design procedure. Concerns for structural integrity attempt to eliminate the potential for catastrophic collapse due to failure of a single member or connector. Creative details are being applied to retrofit existing bridges that are deficient in this respect (Christie and Kulicki 1991). Retrofits to increase redundancy include adding beams and placing "slings" under the existing pin-hanger connections.

4.2.1.g Schoharie Creek Bridge, New York, 1987

(*References:* Thornton, Tomasetti, and Joseph 1988, Levy and Salvadori 1992)

On April 5, 1987, a bridge on Interstate 90 in New York State collapsed into the storm-swollen Schoharie Creek. Three automobiles and a tractor-trailer fell 24 m (80 ft) into the creek, along with 90 m (300 ft) of the 165-m (540-ft)-long bridge. Ten lives were lost in the failure.

The Schoharie Creek Bridge was 35 m (115 ft) wide and carried four lanes of traffic. It was comprised of five simple spans of 30 to 37 m (100 to 120 ft) in length, made of built-up riveted plate girders supported by reinforced concrete piers on spread footings. The bridge was 31 years old at the time of its collapse. It was originally built in 1956 and rehabilitated in 1981.

The cause of the collapse of the Schoharie Creek bridge was scour of the footings at the base of the supporting piers. The failure initiated with collapse of a single concrete pier into the fast-moving creek. Because of the simply supported condition, two spans fell simultaneously with the pier, along with a car and truck occupying the spans. Three more cars drove off the bridge into the raging creek, and within 90 minutes, a second pier and span collapsed.

Floodwaters had scoured the bridge footings and piers. Scour eroded away the bearing surface under the footings, and there were no piles under the footings. Insufficient riprap was provided to protect from scour, even though such provision was originally called for in a 1981 bridge rehabilitation project.

It was noted earlier that National Bridge Inspection Standards mandating biennial inspection of all bridges were established following the 1967 Silver Bridge collapse. The New York Thruway Authority followed a more stringent maintenance schedule than the federal guidelines, with annual inspections of all Thruway bridges and underwater inspections every five years. An underwater inspection had not yet been conducted at Schoharie Creek, although one was scheduled for the year of the collapse. The bridge had most recently been inspected in April 1986 and was reported to be in "good condition." Footings were not inspected at that time; inspectors reported that the water was too high to conduct a proper underwater investigation.

The reports submitted by New York Thruway Authority inspectors indicated that they assumed incorrectly that the pier footings were supported on piles. The annual inspections never uncovered structural problems at the bridge, although a 1977 inspection noted that protective riprap around the base of the piers was visible only on the downstream side of the piers. That report recommended placing additional riprap around the upstream sides, but a nonengineer state employee deleted that portion in the 1980 maintenance contract. After observing the bridge from the river bank, he decided that stone riprap was not necessary.

When the Schoharie Creek Bridge was originally built, the piers were surrounded by steel plate cofferdams. These cofferdams were used to keep the pier construction areas free from water. However, the construction documents also implied that the cofferdams were to remain and be filled with stones as a protection against erosion of the piers.

The investigators determined that the depth of scour under the pier that first failed was about 3 m (9 ft). This caused the pier to rotate into the scoured depression and removed support from the girders supporting the bridge deck. It was noted that had the girders been continuous, the failure of only one pier might not have caused total collapse of the system. At the least, there would have been some large deformations, giving warning of impending collapse. This redundancy might have been just enough to save the lives of the 10 victims of the Schoharie Creek collapse.

The dramatic Schoharie Creek failure, along with that of the Chickasawbogue Bridge near Mobile, Alabama, in April 1985, immediately strengthened public support for bridge inspection programs, especially underwater inspections (Thornton, Tomasetti, and Joseph 1988). While the Alabama bridge did not fail due to scour, the National Transportation Safety Board report blamed ". . . undetected deterioration of steel H-piles due to the inadequate inspection of the underwater bridge elements." Many new nondestructive geophysical methods are now available to aid investigation for scour conditions.

Bridges are vulnerable to many hazards, but scour has been responsible for more bridge failures than all other natural hazards combined. There were 17 other bridge failures due to scour in the northeastern states during the spring of 1987, and there have been over 500 such failures recorded since 1950 (Murillo 1987, Huber 1991). Scour was blamed for the collapse of the Hatchie River Bridge in Tennessee on April 1, 1989. Eight people died in that failure when their vehicles fell into the flooded river. Three 8-m (27-ft)-long bridge sections collapsed along with two piers. The 55-year-old bridge had passed an inspection in 1987, although scouring was noted by the inspectors. One important contribution to the Tennessee failure was that the river channel had meandered, migrating 25 m (83 ft) northward. The pier that failed was not protected by riprap since it was not originally in the river channel. Two prior inspections, however, had identified the need for such protection (ENR 1990a).

4.2.2 Bridge Failures during Construction

Most bridge failures come during the erection stage. Some of these failures are discussed in Chapter 11; a few others are presented here. The most common bridge failures during construction are the result of insufficient temporary support or bracing, problems with connection details, improper sequencing of construction, or inadequate consideration of construction loads.

On May 22, 1952, a 29-m (96-ft)-long highway overpass span collapsed in Boston, Massachusetts. The span moved off one of its supports and the ends of 17 beam stringers fell to the ground. Although not all the engineering reports agreed as to whether the cause stemmed from design or erection errors, the facts indicated an unwise detail at the support. The structure consisted of parallel stringers supported on three-legged plate girder bents. The ends of the spans that dropped were mounted on expansion rockers that were designed to be vertical under full dead load at 21°C (70°F) temperature. The same bent carried an adjacent 9-m (30-ft) span that had an expansion rocker at this end, so that the bent was without lateral support. The concrete on the 9-m (30-ft) span had been placed but not on the 29-m (96-ft) span.

Failure came only a short time after completion of the concrete placing. After failure, the freestanding bent was found to be inclined away from the collapsed bent, with two anchor bolts sheared off and two others bent at the base of the columns (Figure 4.11). It was agreed that any nonvertical position of the rockers would induce a thrust into the bent, with bending of the columns. Such bending of the bent would then increase the inclination of the rockers, with successive steps repeated until the bent supports at the base gave way and the beams were free to fall. The improper placement of the rocker arm supports created an unanticipated thrust on an already unstable assembly.

Disagreement in the investigation reports was whether the initial out-of-vertical position of the rockers came from erection in such a position, or from a thermal and loading change as the adjacent span was concreted. Certainly, it is wiser not to design such a loose link in a chain of supports. Small details alone may not be sufficient to cause a collapse, but they may be an important contributor to an already deficient condition. In the Boston failure case, the inherent instability of the system was revealed by a particular connection detail.

Demolition of a concrete bridge is sometimes more critical than erection, especially when the load of the demolition equipment is on the bridge. In 1958, a 46-m (150-ft) span concrete arch bridge at Topeka, Kansas, was being demolished by an iron ball slung from a crawler crane on the bridge deck near one abutment. The operator was quite surprised when the bridge and

Figure 4.11 Sullivan Square highway overpass, Boston, 1952.

the crane fell into the river together as the two concrete arches sheared near the abutment.

4.2.2.a Quebec Bridge, Canada, 1907 and 1916

(*References:* Ross 1984, Brown 1993)

The great Quebec Bridge over the St. Lawrence River was started in 1902, but it was three years before any steel of the main span was erected. The cantilever bridge, with a clear center span of 550 m (1800 ft) and 150-m (500-ft) anchor spans, went far beyond any precedent for such structures. Modeled on the Firth of Forth Bridge, the main span of the Quebec Bridge was 30 m (100 ft) longer, and the structure was much more economical, appearing even fragile in comparison with the massive Forth Bridge.

The bridge is very well known for two tragic failures associated with its construction. The first was a collapse of the main structure in August 1907, and the second was a fall of the center span as it was being lifted into place in September 1916. Together, these accidents killed 85 construction workers.

At the time of the first collapse, the south anchor and cantilever arms were complete and about one-third of the suspended 206-m (675-ft) truss span had been cantilevered out from the balanced arms. The steelwork projected 244 m (800 ft) over the river from the main pier when some movements of the bottom compression chord of the south anchor arm were observed. Field representatives were sent to New York and to Phoenixville to seek advice from the consulting engineer and from the steel fabricator. A telegram sent to the site to suspend work arrived just after the collapse, which started at the bottom compression chord in the second panel shoreward of the south river pier.

At 5:15 P.M. on August 29, 1907, the entire structure suddenly collapsed. Eighty-five workers, all due off-site in only 15 minutes, fell into the St. Lawrence River along with 18,000 Mg (20,000 tons) of steelwork. Only 11 workers survived the incident. The collapse claimed 74 lives, more fatalities than any other accident in the history of North American construction.

The subsequent investigation uncovered numerous design deficiencies. The Quebec Bridge was designed by Theodore Cooper, James Eads' chief engineer at St. Louis, Missouri. The bridge was designed to be the world's longest span bridge. However, at the time of construction, the aging design engineer stayed in New York, leaving construction supervision to others. Had he been closer to the site, his orders to halt construction may have been received in time to prevent the catastrophe.

The Royal Commission of Inquiry blamed the collapse on failure of the lower chords in the anchor arm near the main pier. The report stated:

> The failure of these chords was due to their defective design . . . We do not consider that the specifications for the work were satisfactory or sufficient, the

unit stresses in particular being higher than any established by past practice. The specifications were accepted without protest by all interested. . . . A grave error was made in assuming the dead load of the calculations at too low a value and not afterward revising this assumption. This error was of sufficient magnitude to have required the condemnation of the bridge even if the details of the lower chords had been of sufficient strength because, if the bridge had been completed as designed, the actual stresses would have been considered greater than those permitted by the specifications. (Royal Commission 1908)

Design deficiencies were noted in both the member design and connection detailing. Theodore Cooper's career was finished; he died a few years later.

By 1914, the debris from the 1907 collapse had been removed, the design had been improved significantly, and reconstruction had begun using a revised method. This time, two cantilevers were to be built in place and the center span, constructed separately, would be lifted into position to close the gap. The new design used two and a half times as much steel and included massive straight chords in place of the slender curved members in the original design.

The anchor and cantilever arms were completed and a completely assembled suspended span had been floated into position for lifting to the deck level on September 11, 1916. The suspended span weighed 4700 Mg (5200 tons). Jacks began to raise it into place between the cantilever arms 46 m (150 ft) above the water. When the structure had only risen 3.7 m (12 ft), a sudden collapse occurred. The first five lifts of 600 mm (24 in.) each had been entirely successful. Barges had been removed to leave the span suspended by cables from the main framework after the third lift. While retracting the jacks for the next lift, the lower shoe at the southwest corner failed and the span slid sideways and fell into the river. Eleven more workers were killed in this accident.

The cause of the second collapse was the failure of one casting on which the span was suspended during the lifting. This same casting had served as a support during the span assembly, and a greater reaction had actually been previously imposed on it; it had earlier supported the weight of erection equipment on the completed span before it was barged to the site.

Lessons learned from the Quebec Bridge incidents included awareness that major collapse can result from disregard of the safety of small details and that both compression and tension supporting members require close attention. The suspended span was rebuilt and lifted into position successfully. The Quebec Bridge was opened to traffic in August 1918.

4.2.2.b Murrow Floating Bridge, Seattle, Washington, 1990

(*References:* ENR 1990b, 1991; Dusenberry 1993; Firth 1993; Dusenberry et al. 1995)

Each bridge, like each building, is a unique site-specific project. Lessons learned from the performance of one type of structure may not be directly applicable to dissimilar structures. However, there is nearly always some value in studying cases where problems were encountered. This last set of examples is from a unique set of bridges used in Washington State. Because of certain conditions in the Puget Sound area, the Washington State Department of Transportation has chosen to build several long floating bridges. While the overall performance of these bridges has been quite satisfactory, several interesting failures have occurred, both during construction and in use.

While under construction, the precast sectional floating bridge across Washington State's Hood Canal experienced a set of mishaps between December 1958 and January 1960. This highway bridge, 1972 m (6470 ft) long, consisted of 23 concrete pontoons supporting an elevated roadway structure. Individual pontoons were 15 m (50 ft) wide and 110 m (360 ft) long. Pontoons were joined together with a steel-transition hinged span at each shore to adjust for a 5.5-m (18-ft) tidal range. Except for the two center pontoons, which could be retracted to provide a ship passageway, each unit was anchored with two 500-Mg (550-ton) concrete blocks.

The pontoons with their sections of roadway were fabricated at a casting yard 60 km (35 miles) from the bridge site. In December 1958, two pontoons sank in the channel leading from the precasting yard. One unit containing a completed deck section became completely submerged. The other pontoon had only the columns constructed and sank on a 22-degree slope, with the columns projecting above water. The sunken pontoons, each representing an investment of $500,000, were salvaged by building walls from the steel panel forms attached to the sides into a floating cofferdam on the totally submerged unit and by winch lifting from barges for the other unit. The sunken pontoons had to be salvaged or removed since they blocked the entrance to the dock where the fabrication yard was located. The cost of salvage was about 75 percent of the amounts already spent on the two units.

Cause of the sinking was discovered a year later when a third pontoon started to sink. It was found that wood blocks plugging holes at one end of the pontoon had worked loose and allowed water to flow into the internal compartments. The holes were near the water line and were used to hold anchor lines that strung the pontoons together into the floating bridge.

In December 1959, two pontoons collided and the end wall of one was punctured by the projecting shear keys of the other. Repairs were necessary before the pontoons could be joined together. A storm the following month caught the bridge while the grout in the joints was not fully hardened, and deformations of the prestressed concrete bottom slabs and ordinary reinforced concrete top slabs became cracks. Strengthening of the design was then agreed upon, chiefly by adding 24 post-tensioned cables through the pontoons and by replacing the portland cement with a fast-setting epoxy in the joint fillers. Each joint required 2.8 m³ (100 ft³) of grout, making this project one of the largest applications of epoxy in the construction industry to that date. The completed Hood Canal Bridge performed without significant problems until the winter of 1979, when half of the bridge was lost in a severe wind and rain-storm.

In one of the most interesting and costly recent bridge construction acci-dents, the landmark Lacey V. Murrow Floating Bridge over Lake Washington, sank during a rehabilitation project, on November 25, 1990. This bridge, a 2616-m (8583-ft)-long structure, opened to traffic in 1940. It was the world's first floating concrete highway bridge. The bridge was named for a former director of the Washington State Department of Transportation. Lacey Mur-row presided over that department at the time of the Tacoma Narrows Bridge failure; he was a brother of the famous broadcast journalist, Edward R. Murrow.

Pontoon sections for the 50-year-old bridge were hollow watertight steel compartments encased in concrete and anchored by steel cables in 60 m (200 ft) of water. Pontoon sections were 18 by 107 m (59 by 350 ft) and 4.5 m (14.5 ft) deep. Each pontoon weighed 4400 Mg (4900 tons). They were designed for a 2.1-m (7-ft) freeboard, but a solid concrete railing gave them an effective height of 3 m (10 ft) from the water's surface to the top of the rail. Adjacent pontoons were bolted together to form a continuous, rigidly connected structure (ENR 1990b).

In the accident, eight of the bridge's 25 pontoons and half of another one sank, eventually taking down 850 m (2800 ft) of the roadway with them. As the structure sank, it sheared 13 of 52 steel cables anchoring the midsection of a new parallel floating bridge located 18 m (60 ft) away. The new $110 million bridge was closed in anticipation of further damage. Other pontoon sections of the Murrow bridge carrying some 215 m (700 ft) of roadway that had been set adrift were towed to shore so that they would not crash into the neighboring structure.

The pontoons had flooded during a severe storm. The bridge was undergoing a $35 million rehabilitation project at the time. Hydroblasting was being used for concrete removal. Environmental protection requirements prevented the contractor from discharging the hydrodemolition water back into Lake Washing-ton. It was also required that all rainwater falling on the bridge be disposed of away from the site to avoid contamination of the lake. In response to these requirements, the contractor was using the hollow pontoons for temporary water storage. The water level was monitored, and water was regularly pumped into trucks for hauling away. The storing of this hydroblasting water in the pontoons may have increased the cracking in the pontoons.

During the Thanksgiving vacation weekend, in addition to the hydrodemoli-tion water, there was a severe rainstorm. In the pontoons, a gradual accumulation of rain and lake splash water occurred. Allegedly, access holes cut in the pontoons were not covered properly by the contractor to prevent water entry. The contrac-tor had cut 67 holes into the north sides of a number of the pontoons, 300 to 380 mm (12 to 15 in.) above the water line to facilitate worker access. The holes were 1.2 to 1.5 m (4 to 5 ft) high, and only the lower 600 mm (24 in.) had been covered by plywood to keep water out (ENR 1991).

Investigators for the state found evidence that the cutting of access openings into the pontoons had weakened the concrete and increased cracking. The severe, turbulent Thanksgiving weekend storm caused water to gush into the pontoons through the cracks and access openings (Dusenberry 1993, Dusenberry et al. 1995).

In an opposing opinion, an investigative consultant for the contractor concluded that the widening cracks in the pontoons were the result of progressive bond slip at lap splices due to fatigue in bond. This interpretation blamed design deficiencies and implied that cycles of wave-induced stress reversals over the life of the bridge had contributed to the failure (Firth 1993).

Closing these bridges had a great impact on traffic in the Seattle area for a number of months. This section of the transportation network in the Seattle area is a major commuting link. In addition to the direct cost of repair and replacement, the state and citizens absorbed other indirect costs.

A panel set up by the governor concluded that both the contractor and the Washington State Department of Transportation underestimated the risk of sinking during the rehabilitation project. The panel recommended a third-party review for future projects involving floating bridges. Panel members noted that Washington Department of Transportation officials could easily consider themselves experts, since four out of the five major floating bridges in the world are in Washington State. However, the report suggested that marine specialists, as well as bridge designers, should be consulted, since ". . . considerations to prevent sinking are not normally a typical part of bridge design."

Certainly, one of the lessons learned through experience in Washington State is that concrete pontoons do not float when they are full of water. A floating bridge has much in common with a marine vessel. During construction and rehabilitation, such a facility is quite vulnerable to failure. Unfortunately, providing a "drydock" environment for maintenance of a floating bridge is not as easily accomplished as for a marine vessel.

Prior to reconstruction, the bridge pontoon system was redesigned. The revised design called for 20 prestressed concrete pontoons averaging 110 m (360 ft) long. Prestressing was used to improve resistance to cracking and for increased strength. Silica fume concrete was specified to reduce shrinkage and cracking during casting and to lower the permeability of the pontoons. To avoid the progressive type of failure that occurred with the old pontoons, the new design provides 48 closed cells in each pontoon. The intention is to confine flooding to small compartments. Each of the cells contains a water warning sensor and a piping system to pump it out (ENR 1992). The Lacey V. Murrow Bridge was completed and reopened to traffic in the summer of 1994.

A mediated settlement in August 1992 called for the contractor to pay the state $20 million. The state had brought a suit against the contractor for $69 million.

In 1994, when the floating Evergreen Point Bridge, also over Lake Washington, was scheduled for rehabilitation, the Washington Department of Transportation selected planing equipment rather than hydrodemolition for removal of the old pavement.

4.3 SUMMARY

In this chapter, several examples of large-scale catastrophic bridge and dam failures have been presented. Some of these events, while tragic, have contributed much to advances in engineering theory. Others have emphasized the critical need for maintenance of the infrastructure.

At the present time, there is general recognition of the need to commit effort and funds for repair and maintenance of existing dams and bridges. This recognition has come as the result of several of the costly failures discussed in Sections 4.1 and 4.2. The work under way includes structural repair and retrofit projects—for both bridges and dams—as well as routine maintenance for nonstructural components, such as bridge decks. There is a renewed emphasis on regular inspection, and new tools have been developed to aid in nondestructive evaluations, including acoustic emission monitoring techniques and sophisticated photogrammetric methods. These new tools were created primarily to respond to the agenda for infrastructure repair. They have been used successfully to identify many problems, some needing immediate attention (Fisher and Mertz 1985).

New materials and methods for upgrade and repair have also been developed to respond to the critical needs of the existing infrastructure. New composite materials are being used to upgrade existing highway structures to incorporate information learned from the 1971 San Fernando, California, earthquake (and relearned in the 1989 Loma Prieta and 1994 Northridge earthquakes). Roller-compacted concrete methods have been refined for dam repair projects; especially to raise dams, to provide additional spillway capacity, and to reduce the potential for overtopping (McLean and Hansen 1993). New grouting materials and methods, advances in prestressing technology, and the use of improved rock anchors for increased stability are all finding applications in this repair and seismic remediation effort.

Still, there is a long way to go before the vast inventory of deficient bridges and dams will be addressed adequately. Funding appropriations have not been sufficient to keep pace with the priority projects that have already been identified.

In 1980, the Federal Emergency Management Agency (FEMA) estimated that $7.5 billion would be needed to repair some 14,500 unsafe dams in the country. The *majority* of nonfederal and privately owned dams inspected were found to be unsafe (ENR 1980b). This study was initiated as the result of the 1972 Buffalo Creek tailings collapse in West Virginia, the 1976 Teton Dam failure, and the Toccoa Dam collapse in Georgia in 1977, all discussed in Section 4.1.1.e. The FEMA report's credibility was strengthened by news in July 1985 of the collapse of a privately owned dam at a mine in Italy that killed 269 persons in the village of Stava. The Italian dam was found to have been ". . . poorly engineered, badly constructed and improperly maintained," the same words that were used to describe hundreds of privately owned dams in the United States (Robison 1985).

A similar study of the nation's bridges was released in 1981 by the General Accounting Office. This alarming report concluded that four out of every 10 U.S. bridges—at least 200,000 bridges—were deficient. A national bridge inventory conducted by the Federal Highway Administration identified 98,000 bridges that were structurally weak or unsound, requiring closure, replacement, or rehabilitation. The other 102,000 deficient bridges were deemed functionally obsolete. At that time (1981), the cost of remedial measures was estimated at $41 billion (ENR 1981) Since then, despite accelerated efforts, little progress has been made in reducing the inventory of deficient bridges. The inventory of U.S. bridges requiring remedial work remains at slightly over 40 percent (Dunker and Rabbat 1993).

One of the more notable problems, of course, is that live loads on bridges, and the volume of traffic, have increased beyond that anticipated by designers. This problem is not unique to the United States. Many important bridges throughout the world are being rebuilt to accommodate heavier traffic. One such project was a $54 million upgrade of the 25-year-old Severn and Wye bridges in England, initiated in 1988.

The failure of an important bridge or a large dam—whether from design error, construction deficiency, or maintenance and operational error—is a frightening event, with the potential for significant loss of life and irreparable damage to the environment. Such facilities require extraordinary care because the consequences of failure are so severe.

The late Lev Zetlin made the following comments when discussing the June 1983 Mianus River Bridge collapse in Connecticut:

> Even a fatal accident like the Mianus bridge collapse can have a positive side. When engineers first designed metal airplanes, the designs worked perfectly on paper, but the planes crashed. Eventually, they discovered that even a minor nick, one-thousandth of an inch deep, could decrease structural strength by 70 percent. They began polishing metal to eliminate nicks, and their planes stayed up. I hope and believe that the things we find on the Mianus River bridge will keep other bridges up. . . .
>
> What is needed is preventative engineering. I look at everything and try to imagine disaster. I am always scared. Imagination and fear are among the best engineering tools for preventing tragedy. (L. Zetlin, as quoted in Browne 1983)

4.4 REFERENCES

Ammann, O., T. von Kármán, and G. Woodruff, 1941. *The Failure of the Tacoma Narrows Bridge,* Federal Works Agency, Washington, DC.

Bennett, J., and H. Mindlin, 1973. "Metallurgical Aspects of the Failure of the Point Pleasant Bridge," *Journal of Testing and Evaluation,* American Society for Testing and Materials, Philadelphia, PA (March).

Bennett, J., 1993. Letter to the editor, *Scientific American,* Scientific American, Inc., New York (July).

Brown, D., 1993. *Bridges,* Macmillan Publishing Co., Inc., New York.

Browne, M., 1983. "Disaster on I-95," *Discover,* Disney Magazines, New York (September).

Christie, S., and J. Kulicki, 1991. "New Support for Pin-Hanger Bridges," *Civil Engineering,* American Society of Civil Engineers, New York (February).

Daniels, G., ed., 1982. *Flood,* Planet Earth Series, Time, Inc., Alexandria, VA.

Dunker, K., and B. Rabbat, 1993a. "Why America's Bridges Are Crumbling," *Scientific American,* Scientific American, Inc., New York (March).

Dunker, K., and B. Rabbat, 1993b. Letter to the editor, *Scientific American,* Scientific American, Inc., New York (July).

Dusenberry, D., 1993. "What Sank the Lacey Murrow? The State's Case," *Civil Engineering,* American Society of Civil Engineers, New York (November).

Dusenberry, D., M. Zarghamee, A. Liepens, R. Luft, and F. Kan, 1995. "Failure of the Lacey V. Murrow Floating Bridge, Seattle, Washington," *Journal of Performance of Constructed Facilities,* American Society of Civil Engineers, New York (February).

Engineering News-Record, 1928. McGraw-Hill, Inc., New York (March 22).

Engineering News-Record, 1929. McGraw-Hill, Inc., New York (June 6).

Engineering News-Record, 1930. McGraw-Hill, Inc., New York (May 22).

Engineering News-Record, 1963. McGraw-Hill, Inc., New York (October 24).

Engineering News-Record, 1964. McGraw-Hill, Inc., New York (January 16).

Engineering News-Record, 1980a. McGraw-Hill, Inc., New York (February 14).

Engineering News-Record, 1980b. McGraw-Hill, Inc., New York (May 8).

Engineering News-Record, 1981. McGraw-Hill, Inc., New York (August 27).

Engineering News-Record, 1983. McGraw-Hill, Inc., New York (July 7).

Engineering News-Record, 1985a. McGraw-Hill, Inc., New York (February 28).

Engineering News-Record, 1985b. McGraw-Hill, Inc., New York (August 22).

Engineering News-Record, 1989. McGraw-Hill, Inc., New York (January 5).

Engineering News-Record, 1990a. McGraw-Hill, Inc., New York (June 14).

Engineering News-Record, 1990b. McGraw-Hill, Inc., New York (December 3).

Engineering News-Record, 1991. McGraw-Hill, Inc., New York (May 6).

Engineering News-Record, 1992. McGraw-Hill, Inc., New York (January 20).

Firth, C., 1993. "What Sank the Lacey Murrow? The Contractor's Case," *Civil Engineering,* American Society of Civil Engineers, New York (November).

Fisher, J., and D. Mertz, 1985. "Hundreds of Bridges—Thousands of Cracks," *Civil Engineering,* American Society of Civil Engineers, New York (April).

Frank, W., 1988. "The Cause of the Johnstown Flood," *Civil Engineering,* American Society of Civil Engineers, New York (May).

Gasparini, D., and M. Fields, 1993. "Collapse of Ashtabula Bridge on December 29, 1876," *Journal of Performance of Constructed Facilities,* American Society of Civil Engineers, New York (May).

Hokenson, R., 1988. "Lightning Spurs Hydro Rehab," *Civil Engineering,* American Society of Civil Engineers, New York (May).

Huber, F., 1991. "Update: Bridge Scour," *Civil Engineering,* American Society of Civil Engineers, New York (September).

ICOLD, 1965. *Proceedings of the Eighth International Congress on Large Dams, Edinburgh, Great Britain,* May 4–8, 1964, Commission Internationale des Grands Barrages, Paris.

Johnson, W., 1980. "Inspecting Dam Construction," *Civil Engineering,* American Society of Civil Engineers, New York (November).

Jansen, R., 1983. *Dams and Public Safety,* U.S. Dept. of the Interior, Bureau of Reclamation, U.S. Government Printing Office, Denver, CO.

Levy, M., and M. Salvadori, 1992. *Why Buildings Fall Down,* W. W. Norton & Company, Inc., New York.

Lichtenstein, A., 1993. "The Silver Bridge Collapse Recounted," *Journal of Performance of Constructed Facilities,* American Society of Civil Engineers, New York (November).

MacDonald, C., 1877. "The Failure of the Ashtabula Bridge," *Transactions,* American Society of Civil Engineers, New York (February).

McCullough, D., 1968. *The Johnstown Flood,* Simon & Schuster, New York.

McLean, F., and K. Hansen, 1993. "RCC: Rehab Results," *Civil Engineering,* American Society of Civil Engineers, New York (August).

Murillo, J., 1987. "The Scourge of Scour, "*Civil Engineering,* American Society of Civil Engineers, New York (July).

Petroski, H., 1985. *To Engineer Is Human,* St. Martin's Press, Inc., New York.

Petroski, H., 1993. "Failure as Source of Engineering Judgment: Case of John Roebling," *Journal of Performance of Constructed Facilities,* American Society of Civil Engineers, New York (February).

Puri, S., 1994. "Discussion: Failure as Source of Engineering Judgment," *Journal of Performance of Constructed Facilities,* American Society of Civil Engineers, New York (May).

Robison, R., 1985. "New Safety for Old Dams," *Civil Engineering,* American Society of Civil Engineers, New York (September).

Robison, R., 1988. "Brighter Future for Stay Cables," *Civil Engineering,* American Society of Civil Engineers, New York (October).

Ross, S., 1984. *Construction Disasters: Design Failures, Causes, and Prevention,* McGraw-Hill, Inc., New York.

Royal Commission, 1908. "Quebec Bridge Inquiry Report," Sessional Paper No. 154, Edward VII, Ottawa, Ontario, Canada.

Terzaghi, K., 1944. "Discussion on P.C. Rutledge, 'Relation of undisturbed sampling to laboratory testing,' *Proceedings,* Vol. 69, ASCE," *Transactions,* American Society of Civil Engineers, Vol. 109, New York.

Thornton, C., R. Tomasetti, and L. Joseph, 1988. "Lessons from Schoharie Creek," *Civil Engineering,* American Society of Civil Engineers, New York (May).

USCOLD, 1975. U.S. Committee, International Commission on Large Dams, *Lessons from Dam Incidents, USA,* American Society of Civil Engineers, New York.

USCOLD, 1979. U.S. Committee, International Commission on Large Dams, *Lessons from Dam Incidents,* American Society of Civil Engineers, New York.

Watson, S., and D. Stafford, 1988. "Cables in Trouble," *Civil Engineering,* American Society of Civil Engineers, New York. (April).

5

TIMBER STRUCTURES

In the next five chapters, performance problems specific to particular construction materials are presented. A few selected case studies illustrate common recurring causes of failure when using each material. It will be seen that there is no "ideal" material, completely free from problems. Each material has unique characteristics. For satisfactory performance, designers and builders must make wise choices in the selection of appropriate materials. Then they must detail and construct facilities with an understanding of the unique qualities of these materials.

5.1 TIMBER AS A STRUCTURAL MATERIAL

Wood is one of the oldest and most commonly used building materials. It is used for a large variety of structural members, including beams, columns, girders, decking, arches, panels, trusses, and foundation piles. Temporary timber shoring and formwork is used for the construction of steel, masonry, and concrete structures. Wood is also used as a finish material and for numerous nonstructural applications, such as for furnishings and cabinetwork.

Since wood is a traditional building material, much knowledge has been gained about its properties and their effects on service behavior. Experience, including many failures in wood buildings, and laboratory research have established safe methods of construction, connection details, and design limitations. There exist a number of excellent texts on the use of wood in construction.

For example, Somayaji (1990) gives an extensive account of the properties of timber that influence its performance in constructed facilities.

Wood structures can be highly weather resistant when properly built and maintained. Historic structures throughout the world illustrate the capacity of timber to endure (Figures 5.1 and 5.2). In the United States, wood has proven its dependability in a wide range of structures, from the covered bridges of the nineteenth century to modern long-span athletic stadia. Two blimp hangers in California, recently designated national historic civil engineering landmarks, are a testament to the versatility of timber. The hangers, built in 1942, are thought to be the largest wood-framed structures in the world and may contain the largest clear-span space of any buildings in the world. Each facility is 18 stories high, 332 m (1,088 ft) long and 91 m (297 ft) wide, covering 28,000 m² (300,000 ft²). The structures and their pile foundations have survived a number of earthquakes; the buildings are still in use for military purposes (ASCE 1994).

Most of the low-rise structures built today in North America, including school buildings, residences, and commercial and apartment buildings, are of wood construction. Timber in the form of glue-laminated products is also used increasingly for heavy construction projects, such as in bridges and for industrial applications.

Some of the advantages of wood are good strength/weight ratio, excellent insulating properties, good fire resistance (in large dimension cross sections),

Figure 5.1 Telč Castle, Czech Republic, an example of a historic timber structure.

Figure 5.2   Historic temple in Kyoto, Japan.

and aesthetic appeal. It is also a renewable resource, and wood products are biodegradable. Timber structures are relatively easy to construct, and future modifications usually are inexpensive.

Because of their light weight and ductility, timber structures respond quite well to earthquakes, although earthquake performance is influenced by quality of connections. In windstorms, the light weight of typical wood construction can be a disadvantage. However, when connections have been constructed properly, light-frame timber structures have performed quite well in severe hurricanes. Numerous proprietary connectors are available for improving the performance of timber structures in windstorms and seismic events. In such extreme hazard events, the increased resistance of timber to short-term loading is also an advantage.

One unique quality of timber as a construction material for industrial applications is its resistance to certain chemical compounds that corrode steel and deteriorate concrete. Buckner, Huck, and Whittle (1987) describe a steel and aluminum bulk storage building for calcined coke that deteriorated after only six years of service. Coke dust combined with condensation moisture to form a slurry that was highly corrosive. A heavy timber structure was designed to replace the original structure, with special care given to protection of the metal connectors. The replacement facility has performed satisfactorily.

Of the problems related to wood use, perhaps the best known is the need to guard against decay. Durability is excellent when the structure is detailed and maintained properly to protect from the effects of weather and moisture. Untreated wood can provide long service if the moisture content is kept below 20 percent. With prolonged exposure to moisture, however, untreated wood will lose its strength and stiffness as decay progresses. The principal cause of decay is the presence of moisture from trapped rainwater or condensation. When wet conditions are expected, pressure-treated lumber must be used. Certain woods, such as redwood and cedar, have greater natural decay resistance than other species.

Weather and moisture protection are achieved only by consciously applying proper construction techniques. High-quality workmanship may be as important a factor in long-term performance as quality design. The owner must also provide adequate maintenance throughout the life of the facility to prevent deterioration of the details that protect wood from the elements. Particularly troublesome details are those that trap moisture in hidden locations, allowing unseen deterioration of critical structural members. The prevention of decay in timber structures is explored further in Section 5.3.

Since wood is a natural material there is a great deal of variability, even within a particular species. This variability is taken into account through proper safety factors that reflect the reliability of the material. However, once in awhile an unseen growth defect in a critical location has led to failure.

The properties of wood (stiffness and strength) are influenced by moisture content. Variable moisture content also leads to dimensional stability problems. The shrinking and swelling of wood, if restrained by connection details, causes secondary stresses in the structural members and splitting at the connections.

It is important to detail all the connections in timber structures so that the force transfer between members is adequately considered. It is also important to understand the interaction of timber with other materials it may contact: steel, concrete, or masonry. Metal fasteners must be protected from moisture to prevent their corrosion. The importance of proper detailing of connections is discussed in Section 5.2.

Wood is an orthotropic material. Its properties vary in three mutually perpendicular directions: longitudinal, radial, and tangential. This anisotropic nature causes problems with connections, especially when structures are subjected to unusual loads such as earthquakes and extreme winds. Direction of load in relation to direction of grain is important. Shear strength is weak parallel to grain, and many timber beams fail in shear parallel to grain. The compression strength of timber is weak perpendicular to grain, causing crushing of fibers of beams at the supports or concentrated loads unless bearing plates are provided to distribute the concentrated forces adequately.

Many new wood products have been developed to improve the material's efficiency, durability, fire resistance, and structural performance. These new products include glued-laminated members and panel products, such as plywood, particleboard, and oriented-strand board. The development of glued-laminated products has addressed some of the deficiencies of the natural material. Advantages include dispersion of growth defects, improved dimensional stability, and unlimited size and shape. There are fewer moisture-content problems, since each lamination can be dried prior to assembling the member. Large-dimension members are now available, reducing the number of required connections.

The widespread use of these new wood products has generally been beneficial, but some new modes of failure have been introduced. There have been some incidents of delamination of glued-laminated beams, although this has been quite infrequent, due to strict standards in the industry. The ability to create long-span structures means that each connection is more critical. When there is a connection problem, a larger portion of the structure is affected adversely. Such structures have less redundancy. This problem is not unique to timber structures; lack of redundancy is a common characteristic of many modern long-span structures constructed in steel and prestressed concrete. The ability to manufacture deep timber glued-laminated beams has increased the potential for certain stability problems, such as torsional buckling of inadequately braced members. These problems have traditionally been associated with steel structures rather than timber systems.

One final source of contemporary problems is the proliferation of numerous proprietary structural timber products and "preengineered" timber systems. While most of these products have proven to be satisfactory, an increasing number of failures involving proprietary systems are finding their way into the forensic engineering literature. The problem with such systems seems to be the tendency of designers, builders, and owners to view these products as "already engineered," therefore requiring no further engineering expertise in their application to specific projects. Nothing could be further from the truth. It is clear from the record that competent engineering advice is most definitely needed in the use of such proprietary products. Such expertise is required to interpret site-specific and project-specific details, such as concentrated loads, construction loads, and proper temporary bracing during erection. Examples of failures involving proprietary systems are given in Section 5.6.

5.2 IMPORTANCE OF CONNECTIONS

The structural integrity of light-frame and heavy-timber construction is highly dependent on the quality of connections. Shear failures at the supports are quite common when beams are notched to facilitate connections. Connections are often the cause of failure of older timber structures, where deterioration has occurred due to trapped moisture. Another common source of failure is splitting due to restraint at the connections that resists dimensional changes caused by moisture content variations. When wood is treated with waterborne salts to improve fire resistance, extra precautions must be taken to keep the wood free from moisture. Such treatments can be quite damaging in that they can contribute to accelerated corrosion of metal fasteners (Zollo and Heyer 1982).

Connection quality (or lack thereof) is especially evident when structures are exposed to extreme winds and seismic events. Structural members, shear walls, and floor and roof diaphragms must be adequately interconnected for the proper force transfers to be made. In addition, the structure must be securely fastened to its foundation (Figure 5.3). When the designer does not provide details for the connections, or when construction quality is not monitored, connections often prove to be the weak point in timber construction. Earthquakes, tornados, and hurricanes are particularly skillful in seeking out these weak points.

In August 1992, more than 60,000 homes were damaged or destroyed by Hurricane Andrew in Florida (Keith and Rose 1994). The great majority of these failures were due to improper connection details. Inadequate connections fell into several categories. One deficiency was poor quality connections of wall units to other walls, roofs, and floors. Much of this type of failure was associated with tract houses that used panelized construction. Top plates were not lapped at corners, and panel sections were attached to adjoining wall sections with as few as three 10d nails. One of the most common observations was loss of gable end walls. This was nearly always accompanied by loss of roof sheathing immediately adjacent to the gable end walls due to inadequate fastening of the sheathing. As soon as the sheathing system began to fail the roof truss assembly was left without lateral support, often leading to total progressive collapse (Figure 5.4a).

(*a*)

(*b*)

Figure 5.3 (a) This residence was moved off its foundation during a windstorm; (b) the absence of foundation anchor bolts for the residence shown in part (a) is evident in this photograph. (Courtesy of the Wind Engineering Research Center, Texas Tech University.)

(a)

(b)

Figure 5.4 (a) Total loss of structure, initiated by loss of roof sheathing, roof trusses, and gable end wall due to inadequate connections; (b) straps intended to prevent uplift were not connected to roof trusses. (Courtesy of APA—The Engineered Wood Association.)

When tie-down straps for roof and floor systems were provided and installed properly, site-built construction performed quite well during Hurricane Andrew. However, in several cases the straps were bent out of the way to permit ease of truss erection, then completely ignored after the trusses were in place (Figure 5.4b).

The general methods used in light-frame construction for the collection and transfer of static gravity loads have changed very little over the past century. Balloon frame and platform frame construction techniques have served very well for many years. However, major advances have been made in recent years that can improve the ability of light-frame construction to resist lateral dynamic loads. Considerable research and observation have established the essential qualities for shear wall, floor, and roof diaphragm performance, including proper connection details. Of special importance are nailing patterns and other connections near ridges of roofs and at corners of buildings, where stress concentrations exist in windstorms. Numerous proprietary connectors are now available that provide positive connections. These connectors are substantially more effective than simple nailing. The availability of these new techniques and products should have a beneficial impact on improved safety and reduction of property loss in future natural hazard events.

Failures regularly occur during construction of timber structures due to inadequate connections. On August 14, 1979, the $8 million Rosemont Horizon Arena sports complex under construction near Chicago's O'Hare Airport collapsed, killing five workers and injuring 16 others (Figure 5.5). The collapse brought down the entire roof of the 20,000-seat arena. The roof was framed with 16 glued-laminated arches spanning 88 m (288 ft) and bearing on 17-m (56-ft)-long precast concrete buttresses. The stadium is 116 m (382 ft) long. Wood was selected to diminish the noise from the intense local air traffic (ENR 1979a–e).

The 1.85-m (6.1-ft)-deep arches were tied together by 940-mm (3.1-ft)-deep girders interspersed with groups of three purlins. At the connections between arches and girders, an angle iron was connected to the arch with several bolts and to the girder with three. The last arches at the east end were being erected, and the stadium was 90 percent complete when the collapse occurred. Missing bolts were identified as a key factor early in the collapse investigation. The bolts in question were those that should have connected the laminated wood roof arches to the cross girders that ran perpendicular to them.

The arches were made up of three separate pieces, two 25-m (83-ft)-long end sections and a 37-m (122-ft)-long center section. The erection sequence for each arch involved three stages. First, a crane lifted one of the end pieces into place and workers bolted it to the top of a support column. While the crane continued to hold up the free end, another crane lifted girders and purlins into place and workers bolted them on, omitting two of three required bolts in the girder connections. The other end arch section then went up in the same manner. In the third stage of the erection, a crane lowered the

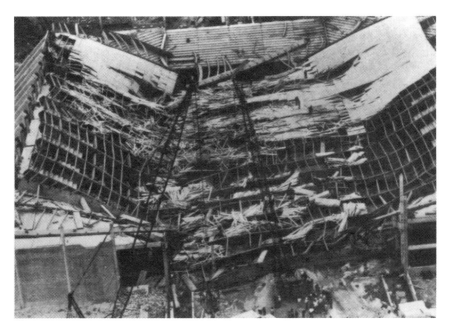

Figure 5.5 Rosemont Horizon Arena, Rosemont, Illinois, August 14, 1979. (This photograph was taken by permission from *Toward Safer Longspan Buildings,* published by the American Institute of Architects, 1981.)

37-m (122-ft)-long center section into place. Workers bolted it to the end sections and placed the final girders and purlins (ENR 1979b).

The contractor claimed that the supervising engineer approved the concept of temporarily leaving two of the three bolts out of the girder–arch connections. The intention was to allow the arches to deflect under the load of the purlins and deck prior to installing the other two bolts. To compensate for the missing bolts, a steel plate was designed to be placed across the tops of the 280-mm (11-in.)-wide arches. The plate was to be bolted with two lag bolts into the cross girder and one lag bolt into the arch. On the last arch, under erection at the time of the collapse, the steel tie plate was not yet in place.

According to the report by the Occupational Safety and Health Administration (OSHA), several factors contributed to this collapse. Missing bolts were cited as the triggering cause, but a number of additional improper construction practices were identified, including inadequate bracing and stockpiling of decking materials on the roof. The OSHA regional administrator noted: "The building was in such unstable condition that anything could have set off the collapse. You could have blown on it and knocked it down" (ENR 1979c).

The most alarming violation cited by OSHA was the omission of over 53 percent of the required connection bolts in the stadium roof. Of 966 girder bolts needed for the connections already installed, only 444 were in place. Of these 444 bolts, 338 had no nuts. Furthermore, some of the nuts that were in place were only fingertight. With regard to the contractor's claim, OSHA found only 27 percent of the "compensating tie plates" installed as required. Another 43 percent were installed, but three of five holes in them had been enlarged by a torch from 24 mm (0.938 in.) to about 65 mm (2.5 in.) in length.

Fines for a number of other violations were levied against the contractor, architect, and subcontractors. In an interesting side note, the forensic engineering firm hired by the city to study the collapse was cited by OSHA for unnecessarily exposing its employees to fall hazards while they were performing their field investigation.

Connection deficiencies continue to contribute to failures in timber structures. As was the case for the Rosemont stadium, they are often found along with other deficiencies, reflecting a general complacent attitude toward construction quality. In July 1990, the roof of a recital hall at the Long Beach campus of California State University collapsed. No one was in the building, and there were no injuries, but repairs were estimated to cost $1.3 million. The 6-m (20-ft)-high one-story building was built with 6 by 9 m (20 by 30 ft) tilt-up concrete panels 200 mm (8 in.) thick. The roof was a composite membrane over insulation, a 75-mm (3-in.)-thick lightweight concrete slab and 19-mm (0.75-in.)-thick plywood decking. The roof supporting structure was 24-m (78-ft)-long wood trusses 1 m (3.3 ft) deep at the supports and 1.6 m (5.3 ft) deep at midspan.

Heavier than anticipated dead loads contributed to the collapse. An investigation report noted that the actual dead load may have been as high as 158 percent of the dead load that designers used to select the trusses. It was also alleged that critical roof diaphragm connections were inadequate for stability and load distribution. Nailing of the sheathing was found to be inadequate and was entirely absent in some places (ENR 1991).

Moore (1991) describes a 1982 failure in Boston, Massachusetts, in which connection deficiencies may have contributed to sudden total collapse of a burning building, causing serious injuries to several firefighters. The building was an abandoned army training facility that had been moved in the late 1950s. The integrity of the roof structure was dependent on shear ring connections between the roof trusses and the columns. Moore postulated that shear rings (not visible in the reassembled timber structure) were not installed when the building was relocated. This could explain the collapse that occurred without warning, before the heavy timber members had suffered severe reduction of their cross sections from the fire. This investigation was reconstructed from photographs after all the debris had been cleared away, so it is impossible to verify Moore's conclusion without doubt. However, photographic evidence did support his theory, and the lesson is clear. Since timber connections are such an integral component of structural integrity, competent engineering advice is needed for design and supervision during erection at the site. Provision of members of the required cross-section dimensions is not sufficient.

5.3 PROTECTION FROM DETERIORATION

It has long been recognized that timber requires protection against moisture change, yet failures due to timber disintegration are still reported. Deterioration attacks the ends of beams and trusses embedded in masonry or so covered by roofing that air circulation is prevented. In 1909, a 20-year-old six-story factory collapsed in New York City because deterioration had weakened the oak columns set in cast-iron socket seats. The lobby of the Strand Theater in Pittsburgh, Pennsylvania, collapsed in 1922 when the 38 by 240 mm (2 by 10 in.) joists disintegrated in the crawl-space environment after only eight years of use.

In March 1986 a roof over a swimming pool in Mt. Laurel, New Jersey, collapsed, trapping several children (ENR 1986). Rescue workers had to cut up wood roof trusses with chain saws to free the children. The building was about 25 years old and measured 17 by 37 m (56 by 122 ft). The wood trusses were connected by metal gusset plates and were supported by steel columns and beams. At the beams, the trusses sat on wood sills that were found to be rotted completely through. The trusses were also saturated and stained a dark color. The plywood roof decking was saturated, contributing to the load on the roof. In addition, the owners had added new insulation and new roofing materials over the original materials, increasing the dead loading. Such collapses continue to occur in poorly ventilated interior environments despite centuries of experience with wood.

Wood can be an extremely durable structural material if proper construction procedures and good quality maintenance is provided (Somayaji 1990). Without this care, wood can be destroyed by the environment. Environmental attacks come from various types of fungi. Fungi are low forms of plant growth that produce thin branching tubes that spread through the wood, using the cell walls and lignin as food. Conditions necessary for fungal growth are proper temperature, moisture content over 19 percent, oxygen, food (wood fiber), and a favorable environment (Meyer and Kellogg 1982). Elimination of any of these will effectively control the growth of the organisms. Usually, the most effective measure is to remove moisture from the environment.

Termites and certain other insects can cause wood destruction. As Jagjit Singh, editor of an authoritative textbook on building mycology, notes: "The world is full of bugs and beetles—most of whom enjoy eating wood" (Singh 1994b). Wood-eating insects are more common in warmer climates and/or areas with high humidity. In these regions, untreated wood must not be close to the soil; details must prevent insect access to the wood.

Some of the best references regarding the adverse effects of moisture on building materials are by authors from Denmark, Norway, and the United Kingdom (Addleson 1987, Singh 1994a). Experience with the true dry rot fungus *Serpula lacrymans* and other molds is legendary in that region of the world. At a recent timber preservation conference, a researcher from Denmark credited William Shakespeare as the world's "first building mycologist." The evidence given was Hamlet's famous observation: "There is something rotten in the state of Denmark" (Bech-Andersen 1994). Numerous creative techniques have been devised by professionals in this part of the world to deal with the problems of moisture and wood-decaying fungi. Trained dogs are used to seek out hidden active spores. Biological remediation techniques are being introduced as alternatives to chemical treatment in some cases. Natural biological enemies to decay-producing fungi have been identified by mycologists. These are sometimes effective in eliminating the decay-producing fungi; then they die, due to elimination of their own food source.

The owners and operators of facilities play an important role in the preservation of timber structures through adequate maintenance. Frauenhoffer (1989) discusses the failure of a taxi company maintenance garage roof in Champaign, Illinois. The roof was supported by heavy timber trusses. The roof membrane had been allowed to deteriorate, causing rotting of the roof deck and the top chords of the trusses. If the roof membrane had been maintained properly and replaced when its useful life had expired, the collapse would not have occurred.

Many costly problems with structural wood members have occurred as the result of failures of protective membranes, even in new structures. The roof over the football stadium at the University of Idaho in Moscow, Idaho, is an excellent example (Gorman and Reese 1991). The roof is a cylindrical vault with plan dimensions 122 by 122 m (400 by 400 ft). The structure system is unique, using composite trussed arches made of tubular steel web members and lightweight top and bottom laminated veneer wood chord members (Figure 5.6a). The trussed arches, 2.3 m (7.5 ft) deep, span the entire 122 m (400 ft) and bear on steel support trusses connected by pins to supporting walls. Completed in 1975, the project received the Outstanding Structural Engineering Achievement Award from the American Society of Civil Engineers the following year. Initial construction cost for the facility was $4.1 million.

Almost immediately after completion, the roof membrane began to leak. The original membrane was a 0.38-mm (15-mil)-thick seamless elastomeric material covering the entire 16,000-m^2 (4-acre) roof. The membrane was applied directly over 38-mm (1.5-in.)-thick polyurethane foam sprayed on an asphaltic film primer applied to 10-mm (0.375-in.)-thick plywood sheathing. The sheathing was laid in a herringbone pattern to provide diaphragm support to the top chord of the trusses (Figure 5.6b). The working height, roof curvature, and occasional wind gusts made it difficult for workers to apply the foam uniformly to the thickness specified.

After several attempts to repair the roof membrane, the university concluded in 1980 that the original membrane was not protecting adequately the top chord of the trusses. It was also discovered that condensation was collecting in the foam and that freeze–thaw cycles were causing the foam to disintegrate. A new roof was constructed at a cost of $2.1 million. The new roof is composed of either built-up roofing or asphalt shingles (depending on slope) over a layer of 16-mm (0.625-in.)-thick plywood on 38 by 140 mm (2 by 6 in.) rafters to

(a)

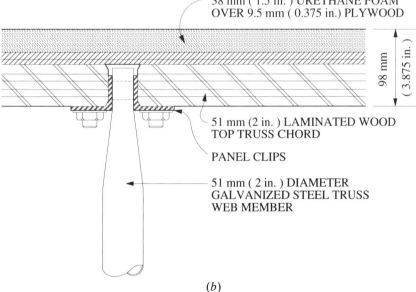

38 mm (1.5 in.) URETHANE FOAM
OVER 9.5 mm (0.375 in.) PLYWOOD

98 mm (3.875 in.)

51 mm (2 in.) LAMINATED WOOD
TOP TRUSS CHORD

PANEL CLIPS

51 mm (2 in.) DIAMETER
GALVANIZED STEEL TRUSS
WEB MEMBER

(b)

Figure 5.6 University of Idaho stadium, Moscow, Idaho: (a) trussed arches, shown during construction (Courtesy of T. Gorman); (b) original roof construction; (c) reconstruction detail; (d) stadium after roof reconstruction (courtesy of T. Gorman.)

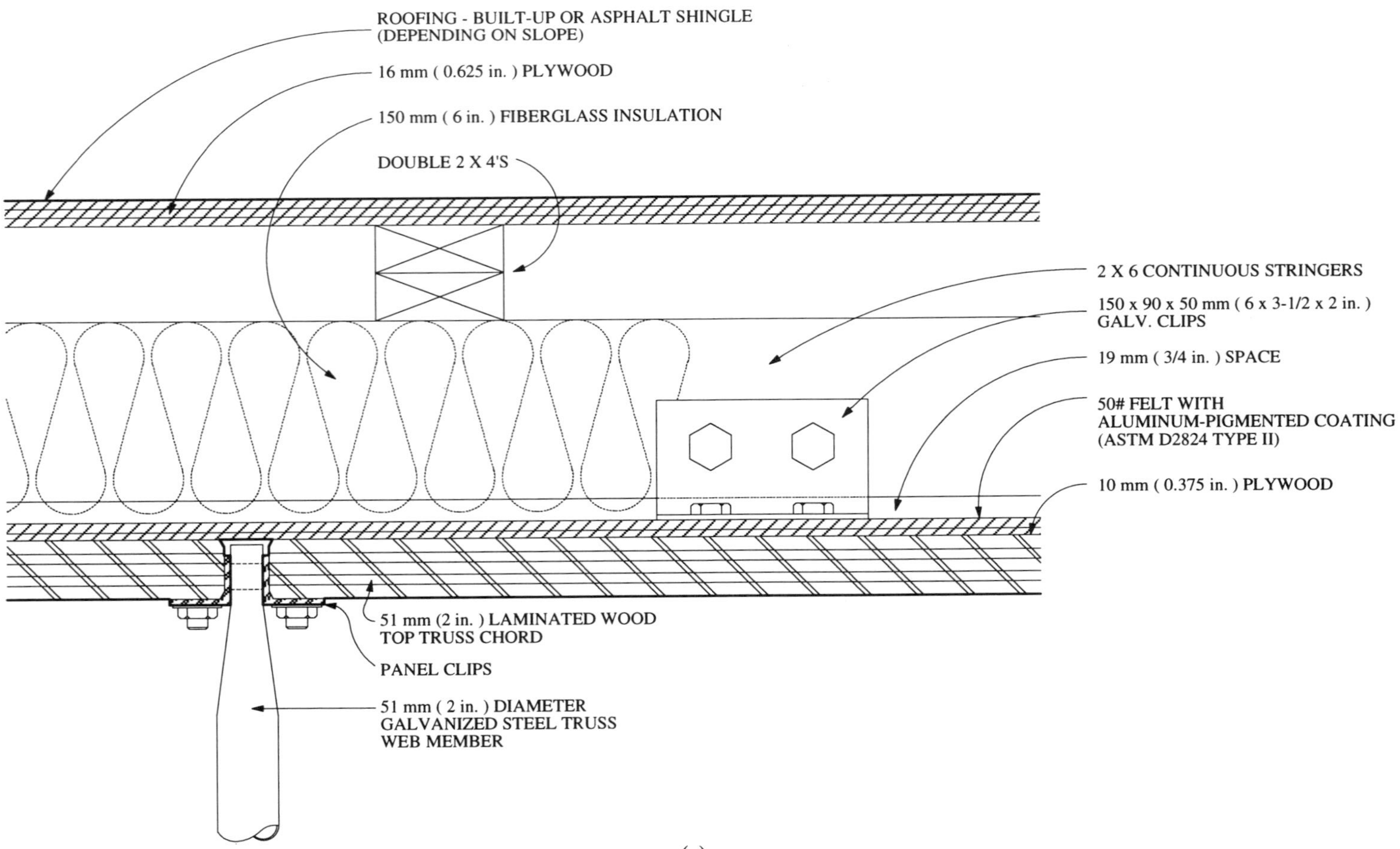

ROOFING - BUILT-UP OR ASPHALT SHINGLE (DEPENDING ON SLOPE)

16 mm (0.625 in.) PLYWOOD

150 mm (6 in.) FIBERGLASS INSULATION

DOUBLE 2 X 4'S

2 X 6 CONTINUOUS STRINGERS

150 x 90 x 50 mm (6 x 3-1/2 x 2 in.) GALV. CLIPS

19 mm (3/4 in.) SPACE

50# FELT WITH ALUMINUM-PIGMENTED COATING (ASTM D2824 TYPE II)

10 mm (0.375 in.) PLYWOOD

51 mm (2 in.) LAMINATED WOOD TOP TRUSS CHORD

PANEL CLIPS

51 mm (2 in.) DIAMETER GALVANIZED STEEL TRUSS WEB MEMBER

(c)

Figure 5.6 (Continued)

(*d*)

Figure 5.6   *(Continued)*

provide space for fiberglass insulation, and finally, another built-up roof over the original plywood diaphragm (Figure 5.6c). Much of the original decayed plywood had to be replaced, and some of the upper chords of the trusses had to be routed out and reinforced with fiberglass and steel plates. The new superstructure was completed in October 1982. It has provided satisfactory service since then (Figure 5.6d).

This case study demonstrates the need for an inspection or monitoring program for projects that use innovative structural concepts or untested materials. Maintenance personnel may not be able to diagnose or correct problems when new designs and new materials are involved. In this case, several years of attempts to address the moisture problem were undertaken without success. During this time, extensive deterioration of the structural diaphragm occurred, adding significantly to the roof replacement costs.

5.4   REPAIR AND REHABILITATION

A frequent problem in renovating older buildings is the need to repair or reinforce timber structural members to conform with current codes and standards. One very common element of these older structures is wood trusses that were not designed to carry today's required loadings at current allowable stresses. Also, members may have been subjected to conditions that have produced decay.

Much has been learned about wood since these truss roofs were constructed. Current standards for tension members are more conservative, and more is known about the long-term behavior of timber subjected to high temperatures. The effect of vibrations in producing fatigue problems in timber members is now a consideration, as are new standards for snowdrift, seismic, and wind loading. Both connections and members commonly need remedial attention, and lateral bracing is often deficient according to current guidelines. Ebeling (1990) provides an excellent guide for the analysis, repair, and strengthening of older timber trusses.

As might be expected, rehabilitation projects usually uncover problems that were unforeseen at the start of the project. Lavon (1994) discusses the case of a synagogue in Port Washington, New York. While undertaking a routine maintenance project in the building, a contractor discovered severe deterioration of the glued-laminated timber arches. The damage was at the base of each arch, hidden from view behind a vinyl strip. Since imminent collapse was a possibility, the structure was immediately shored and braced so that the building could be used safely while repairs were made.

The building, built about 30 years prior to the repair project, is 44.5 by 21 m (146 by 70 ft) in plan. The arches are 7.6 m (25 ft) high and span 21 m (70 ft). Investigations determined that a significant amount of damage had progressed in concealed areas. In some cases the damage extended 4 m (12 ft) up into the arches from the base, causing severe reduction of the effective cross section. Architectural details contributed to rainwater penetration of the building and the trapping of water in hidden locations. Repairs consisted of replacement of damaged sections of the glued-laminated members, new steel shoes that were welded to the original base plates in the concrete floor slabs, and the addition of new threaded steel reinforcing rods, installed in grooves cut into the sides of the glued-laminated members.

There are a number of destructive and nondestructive tests that can determine the degree of decay in timber structural members. Visual access to enclosed spaces, using modern borescopes, requires only very small holes to be drilled in the members or adjacent construction (Figure 5.7). Other techniques, such as stress-wave propagation, do no damage to the wood (ASCE 1991). However, from the preceding example, it is clear that life-threatening damage can occur in timber structures in locations completely invisible to the human eye. Certain architectural details that permit moisture to accumulate in hidden areas should be avoided, and owners should be particularly concerned about these conditions in older buildings.

5.5   PONDING FAILURES AND DRIFTING SNOW

Wood roofs have experienced failures due to snow and rain loads. As discussed in Section 2.1, considerable attention has been given to snow-drift patterns

Figure 5.7 At historic Telč Castle, Czech Republic, (shown in Figure 5.1), investigation of aging timber framing using borescope.

on roofs since the winters of 1977–1978 and 1978–1979. Those two winters were especially severe in the northern United States. In 1978–1979, more than 200 roofs collapsed in the Chicago area alone. Improved standards for snow loading were developed, including provisions for nonuniform or drift loads.

One interesting roof collapse due to heavy snow-drift loading was the failure of a new junior high school in Waterville, Maine (Zallen 1988). The collapse occurred on February 9, 1978, the first day of classes in the new school. The night before the collapse there was a heavy snowstorm. The primary cause of this collapse was failure to consider the load from drifting snow in the design (see Figure 2.1).

Fortunately, the classroom where the collapse occurred was evacuated prior to the failure, so there were no injuries or lives lost. The ceiling had deflected noticeably and dust was falling from the ceiling, giving warning of impending failure.

The structure is a one-story building. The plan is structurally divided into six sections separated by masonry walls. The roofs of adjacent sections are at various levels. The roof deck, comprised of tongue-and-groove cementitious wood fiber planks, is supported by bottom-chord-bearing open-web roof joists with wood chords and steel-tube web members. These joists are of a proprietary design and span between masonry bearing walls. Depths and spans of the open-web joists vary from section to section.

Selection of the roof joists was based on the assumption of a uniformly distributed snow load. No provision was made in the design for additional loads caused by drifting snow. The open-web joists were severely overloaded by the actual conditions from the snowstorm (Figure 5.8). In fact, the prevailing wind combined with the configuration of the roof at various levels to create large drifts. The wind cleared snow from a higher roof area and deposited downwind drifts over 2.1 m (7 ft) in depth on lower roofs. In the collapse area, joists were loaded to approximately 2.2 times their allowable load. This failure serves as an example of the need for competent engineering expertise in the application of proprietary designs to specific projects, a subject that is discussed further in Section 5.6.

Ponding failures in flat roofs during heavy rainstorms are briefly discussed in Section 2.1. Such failures are best prevented by designs that eliminate the potential for water accumulation. There is a need for some positive slope and for good drainage systems that are adequately maintained. Scupper details, or other methods to permit fail-safe drainage, should be provided in case drains are restricted by debris.

Estenssoro (1989) discusses an interesting collapse where a malfunctioning fire sprinkler system flooded an interior attic space, causing a ponding failure of a ceiling, accompanied by ultimate failure of a long-span roof. The ceiling consisted of fire-protecting gypsum-board panels nailed securely to the bottom of closely spaced wood joists. The joists, in turn, were securely nailed to the bottom chords of long-span bowstring timber trusses. The good structural integrity of the ceiling system actually proved to be detrimental to the ultimate performance of the roof structure. An ideal "bathtub" situation was created. The accumulation of water eventually caused collapse of the entire bowstring truss roof.

5.6 PROPRIETARY SYSTEMS

The past two or three decades have seen the development of a number of excellent proprietary wood products and preengineered timber systems. Most

Figure 5.8 Fractured open-web joists due to snowdrift loading, junior high school, Waterville, Maine (see also Figure 2.1). (Courtesy of R. Zallen.)

of these products have performed admirably in service. They have improved the efficiency of timber structures and addressed many of the inherent deficiencies of the material in its natural form. However, these products and systems have suffered occasional failures. Sometimes the cause has been deficiency in the product itself. More commonly, failures have been the result of lack of competent engineering expertise in the application of these products and systems to particular projects.

The principal problem with these systems seems to have been an attitude among designers, builders, and owners that presumes these products to be "preengineered," therefore needing very little or no further engineering expertise in their application. Such expertise is clearly required, however, to interpret site-specific and project-specific details, such as concentrated loads, construction loads, and proper temporary bracing during erection. A case study is presented in Section 5.5, for example, where proprietary preengineered trusses were selected from the manufacturer's literature for a uniform load condition, whereas the actual loading was not uniformly distributed.

In the state of Oregon there were as many as seven roof collapses in the 1980s involving the same type of manufactured wood truss. One was the sudden failure of the 10-year-old gymnasium roof of the Athena, Oregon, elementary school on August 21, 1987. No one was injured, and there was no live load whatsoever on the roof at the time (ENR 1987). The trusses were open-web wood trusses with glued finger joints. This type of truss was manufactured in Oregon between 1974 and 1980. Some of the failures involving these trusses may have been the result of installation errors, but investigators also noted the lack of redundancy in the joist design. Since joints were not reinforced with metal side plates, failure of the glue joint would prove to be catastrophic.

A very popular system for light-framed structures is metal-plate-connected wood roof trusses based on standard designs. The manufacturers of these products have made handling and bracing recommendations available to designers and users, but the fact is that many truss installers ignore the guidelines. Progressive collapse of unbraced roof trusses during the process of construction is all too common (Figure 5.9). The forensic engineering literature con-

Figure 5.9 Progressive collapse of unbraced roof trusses. (Courtesy of H. Kagan.)

tains much evidence to suggest that leaving the provision of temporary bracing during construction up to the contractor's judgment is a questionable practice. Damage to critical truss members and connections during transport and erection is also common (Kagan 1993). The installation of these trusses is usually performed by carpenters, most of whom seem to be unaware of the documents produced by the Truss Plate Institute (1991) regarding these matters.

In many failures of this type, the designer of record has had no field inspection responsibilities, at least by contract. This practice is an abrogation of professional responsibilities. Designers should be aware that in many cases, truss fabricators are not franchised, licensed, or trained by the truss-plate manufacturers. There are exceptions, but the general practice is for truss-plate manufacturers to sell the fabricator a book of standard designs and the truss plates. All "nonstandard designs" are supposed to be submitted to the plate manufacturer for review, but this may not happen, and the plate manufacturer bears no responsibility for making the decision as to whether or not the design is "standard." The truss-plate manufacturer and the truss-plate engineer rarely see the contract documents for specific projects (Turk 1994).

Other manufactured structural members also have experienced failure from the lack of participation of an architect or engineer of record. Padgett (1992) describes a floor system failure involving web buckling of manufactured wood I-joists. Contractual arrangements allowed the project to be built without a structural engineer of record. No engineer inspected the framing system during construction. The architect did not review the joist manufacturer's drawings nor ask to have them reviewed by an engineer. Beam and column calculations were not checked by either an architect or engineer; these were presumed by the architect to be the manufacturer's responsibility.

Failures of timber projects during construction due to inadequate temporary bracing are certainly not limited to preengineered products. Erection failures of site-built timber trusses and arches are quite common. Such erection failures occur even in well-conceived designs (Carper 1987). Several examples are given in Section 11.6, including the failure of a 73-m (240-ft)-span timber arch dome in Alexandria, Louisiana in 1964 (see Figure 11.7). The dome was framed with 36 glued-laminated timber arches that were all in place, ready for the installation of a wood-fiber concrete deck. However, two of the four erection cables failed before the deck was installed, causing total collapse of the structure.

5.7 REFERENCES

Addleson, L., 1987. *Building Failures: A Guide to Diagnosis, Remedy and Prevention,* Van Nostrand Reinhold, New York.

ASCE, 1991. *ASCE Standard: Guideline for Structural Condition Assessment of Existing Buildings,* American Society of Civil Engineers, New York.

ASCE, 1994. "California Blimp Hangers, Framed in Wood, Are Landmarks," *Civil Engineering,* American Society of Civil Engineers, New York (October).

Bech-Andersen, J., 1994. "The True Dry Rot Fungus: Origin and Destination," *Proceedings, Workshop on Conservation and Preservation of Timber in Buildings,* Prague, The Czech Republic, September 22–23.

Buckner, C., R. Huck, and C. Whittle, 1987. "Deterioration and Replacement of Roof System for Bulk Storage Building," *Journal of Performance of Constructed Facilities,* American Society of Civil Engineers, New York (February).

Carper, K., 1987. "Structural Failures During Construction," *Journal of Performance of Constructed Facilities,* American Society of Civil Engineers, New York (August).

Ebeling, D., 1990. "Repair and Rehabilitation of Heavy Timber Trusses," *Journal of Performance of Constructed Facilities,* American Society of Civil Engineers, New York (November).

Engineering News-Record, 1979a. McGraw-Hill, Inc., New York (August 16).

Engineering News-Record, 1979b. McGraw-Hill, Inc., New York (August 23).

Engineering News-Record, 1979c. McGraw-Hill, Inc., New York (November 22).

Engineering News-Record, 1979d. McGraw-Hill, Inc., New York (November 29).

Engineering News-Record, 1979e. McGraw-Hill, Inc., New York (December 20).

Engineering News-Record, 1986. McGraw-Hill, Inc., New York (April 3).

Engineering News-Record, 1987. McGraw-Hill, Inc., New York (September 10).

Engineering News-Record, 1991. McGraw-Hill, Inc., New York (January 7).

Estenssoro, L., 1989. "Two Roof Failures Due to Water Ponding and Related Code Requirements," *Journal of Performance of Constructed Facilities,* American Society of Civil Engineers, New York (August).

Frauenhoffer, J., 1989. "Roof Collapse, Taxi Company Garage, Champaign, IL," *Journal of Performance of Constructed Facilities,* American Society of Civil Engineers, New York (May).

Gorman, T., and D. Reese, 1991. "Roof-System Replacement on Sports Arena," *Journal of Performance of Constructed Facilities,* American Society of Civil Engineers, New York (May).

Kagan, H., 1993. "Common Causes of Collapse of Metal-Plate-Connected Wood Roof Trusses," *Journal of Performance of Constructed Facilities,* American Society of Civil Engineers, New York (November).

Keith, E., and J. Rose, 1994. "Hurricane Andrew: Structural Performance of Buildings in South Florida," *Journal of Performance of Constructed Facilities,* American Society of Civil Engineers, New York (August).

Lavon, B., 1994. "In Rehab We Trust," *Civil Engineering,* American Society of Civil Engineers, New York (September).

Meyer, R., and R. Kellogg, eds., 1982. *Structural Uses of Wood in Adverse Environments,* Van Nostrand Reinhold, New York.

Moore, R., 1991. "Fire-Related Roof Collapse: Case Study," *Journal of Performance of Constructed Facilities,* American Society of Civil Engineers, New York (February).

Padgett, T., 1992. "Manufactured Wood Joists: Noncollapse Failure," *Journal of Performance of Constructed Facilities,* American Society of Civil Engineers, New York (February).

Singh, J., ed., 1994a. *Building Mycology: Management of Decay and Health in Buildings,* E & FN Spon, London.

Singh, J., 1994b. "Causes of Decay and Damage to Heritage and Its Prevention," *Proceedings, 4th International Conference on Lessons from Structural Failures,* Prague, Czech Republic, September 19–20.

Somayaji, S., 1990. *Structural Wood Design,* West Publishing Company, St. Paul, MN.

Truss Plate Institute, 1991. *Commentary and Recommendations for Handling, Installing and Bracing Metal Plate Connected Wood Trusses,* HIB-91, Truss Plate Institute, Madison; WI.

Turk, A., 1994. "Discussion: Common Causes of Collapse of Metal-Plate-Connected Wood Roof Trusses (by H. Kagan)," *Journal of Performance of Constructed Facilities,* American Society of Civil Engineers, New York (November).

Zallen, R., 1988. "Roof Collapse Under Snowdrift Loading and Snowdrift Design Criteria," *Journal of Performance of Constructed Facilities,* American Society of Civil Engineers, New York (May).

Zollo, R., and E. Heyer, 1982. "R_x: Treated Lumber and Metal Corrosion Ailments," *Civil Engineering,* American Society of Civil Engineers, New York (August).

6

STEEL STRUCTURES

Several incidents in which steel was the structural material and in which failure was caused by improper design or detail are included in Chapter 1. Steel structural failures caused by natural or human-made disasters are discussed in Chapter 2. A great deal of our knowledge about iron and steel structures has come from the bridge failures reviewed in Chapter 4. Chapter 11 contains several examples of erection failures involving collapse of steel structural elements during construction. In this chapter the unique characteristics of steel structures are reviewed and additional illustrative examples are given, with the understanding that many of these incidents could have occurred were other structural materials used.

6.1 STEEL: THE "IDEAL" STRUCTURAL MATERIAL

Steel has a number of characteristics that contribute to outstanding structural performance. Among these are its high strength per unit weight and high strength per unit volume compared to reinforced concrete and reinforced masonry structural elements. The lightweight quality of a structural steel frame reduces the dead load on the foundations and supporting soil. Since the inertial mass of a structural steel frame is relatively small, the effects of seismic accelerations are reduced.

Unlike cast-in-place concrete elements, steel members maintain dimensional stability over time, and the predictable nature of the material in the elastic range makes for reliable calculation of initial and long-term elastic deformations. The ductile performance of mild structural steel when stressed beyond the elastic range is a great advantage in resisting abnormal loads such as those encountered in earthquakes or explosions. In addition, steel has similar strength in tension and compression. This permits the use of symmetrical cross sections that have inherent ability to resist stress reversals, assuming the presence of proper connection and bracing details. The nearly isotropic nature of structural steel aids in resistance to multidirectional loading in windstorms and seismic events.

When one considers all these desirable characteristics, it is not surprising that steel has been called the "ideal" structural material. In the engineering literature there are many impressive examples of steel structures that survived extraordinarily dramatic events. These include numerous examples from earthquakes as well as the well-documented performance of the World Trade Center towers in New York City during the February 26, 1993 terrorist bombing (Puri 1994, Ramabhushanam and Lynch 1994).

Yet there are also characteristics of steel structures that require special attention to prevent failure. Some of these stem from the nature of the material itself, once again proving that there is no "ideal" material free of the need for diligent care in design and construction. The problem of fatigue under cycles of stress reversal is quite well known, especially among bridge designers (Fisher 1984; Gasparini and Fields 1993; Lichtenstein 1993; Zwerneman, West, and Lim 1993). Experience with airplane performance first led to the understanding that metals weaken when they are subjected to stress reversals. The catastrophic failure of seven of the 21 Comet commercial airliners built between 1952 and 1954 taught designers to be especially cautious of fatigue in the vicinity of stress concentrations at openings or where flaws exist in the base materials. Brittle fracture can also occur at cold temperatures under impact loads.

Another undesirable characteristic is the behavior of unprotected structural steel elements in fire. Detailing for fire protection requires insulation of the steel to prevent distortion under high temperatures. The traditional method of providing such protection is encasement of the steel members in concrete. This increases the dead load and the dimensions of the otherwise lightweight and slender members. Corrosion protection is an additional design and maintenance challenge when steel is selected for the structural material.

Ironically, the most catastrophic failures of steel structures, both during construction and in service, result from the very efficiency presented as an advantage earlier. The slender structural members require special detailing to prevent stability failures. Unbraced structures frequently fail during erection, and there are repetitive examples in the forensic engineering literature of other stability problems in completed structures, such as lateral, torsional buckling of steel beams.

Dov Kaminetzky, a forensic engineer with many years of experience, has written that he could not recall a single failure in which the cause was a deficiency in strength of the base structural steel material (Kaminetzky 1991). Occasionally, there have been problems with the composition of the base material, leading to insufficient toughness or to flaws and surface defects in fabricated members. However, the preponderance of the literature on the subject confirms Kaminetzky's observation. Most of the catastrophic steel failures have been due either to connection problems or to insufficient provisions for stability. This is particularly noteworthy since we so often praise steel for its ductile behavior prior to fracture. The large deformations accompanying high stresses would seem to give adequate warning at overload, thereby minimizing injuries and loss of life. The fact is, however, that stability failures and connection failures can occur instantaneously; such failures in steel structures can clearly be life threatening.

The case studies presented in this chapter illustrate stability problems and the critical importance of connections in steel structures. These are followed by a brief discussion of concerns related to corrosion.

6.2   STABILITY PROBLEMS IN STEEL STRUCTURES

While stability problems contribute to failures involving all structural materials, steel structures are particularly susceptible to such occurrences (Galambos 1987). When strength alone is considered, the efficiency of steel as measured by strength per unit volume permits large loads on slender members. Erection failures are common when insufficient temporary bracing is provided, and stability failures also occur from time to time in completed structures when proper considerations relative to stability are not taken.

6.2.1   Erection Failures

As noted in Chapter 11, some of the more dramatic structural failures during construction are those that result from a temporary lack of stability during the construction phase. Many of the construction-related steel bridge failures discussed in Section 4.2.2 involved stability problems.

Section 11.6 reviews two steel erection failures due to missing or insufficient temporary bracing: the 1958 collapse of an 11-story welded steel frame in Toronto, Canada, and failure of the University of Washington football stadium in 1987. These are only two of the many repetitive costly failures of this type. The engineering literature contains countless examples, most involving well-engineered structures that would have been safe if construction had progressed to the point where final connections, shear walls, and floor and roof diaphragms had been present (Carper 1987). These include a number of wind-induced failures of preengineered steel buildings, where erection practices in the field did not conform to established procedures for temporary bracing (Sputo and Ellifritt 1991, Beri 1993).

A steel-frame tilt-up procedure combined with hoisted lift-slab floors using the erected frame as a support for a crane hoist has been used in Paris, France, for low-cost apartments. In 1964, one such frame assembly for a 10-story building collapsed into a mass of twisted steel as the last bent was being raised into vertical position but before any slab had been lifted to brace the columns.

In 1987, 28 construction workers were killed in the collapse of L'Ambiance Plaza in Bridgeport, Connecticut. The L'Ambiance Plaza towers were designed with post-tensioned prestressed concrete floor slabs constructed by the lift-slab method. During construction the stability of the towers relied on steel frame column sections that were not adequately braced. The shear walls that would have provided stability in the completed structure were lagging behind the column erection and slab-lifting operation. This tragic failure is described in detail in Chapter 8. There were many complex and interrelated factors, but nearly all investigators agreed that frame instability was a significant contributor to the magnitude and suddenness of the collapse (Moncarz et al. 1992). The L'Ambiance Plaza failure and the earlier collapse in Paris illustrate the need to provide lateral stability as each floor of a multistory building is completed.

All elements that may experience compression stress in any form require special considerations for stability. These elements include columns, compression members of trusses, and compression flanges of beams. The potential for stress reversals under unusual loads during construction should be explored. A review of the record bears out the importance of temporary bracing. Many of these failures go unreported, but there have been sufficient published reminders to emphasize the critical nature of temporary provisions for stability during construction.

Typical of roof-truss erection failures is the 1937 case in Los Angeles, California, where eighteen 30.5-m (100-ft) spans were erected for the roof of a 46 by 91.5 m (150 by 300 ft) warehouse. All trusses were held by guy wires on one side only. Complete collapse followed the loosening of one guy wire.

Tapered girders for a long-span roof in East Los Angeles, California, failed in 1955 from torsional buckling, even though most of the roof purlins were in position (Figure 6.1). As a result, rulings were issued to require all tapered steel girders to be braced at not more than 9-m (30-ft) intervals with braces capable of resisting a 4.5-kN (1000-lb) thrust at either the top or bottom flange. Again at Long Beach, California, in 1956, eleven 49-m (160-ft) welded rigid frames twisted over before final plumbing but with the roof purlins and some framing connected on one side of the spans. The other side was held by guy wires, which required change to clear added framing. Failure occurred while the guys were being changed during a gusty wind, and 450 Mg (500 tons) of steel fell.

In 1956, the Cleveland Auditorium roof experienced an erection failure after four 21-m (68-ft) girders resting on columns were assembled for the roof. Guy wires were on only one side of the 915-mm (3-ft)-deep girders. A nudge by a crane boom may account fully for the collapse of the four girders and eight columns.

A rigid frame of 30-m (98-ft) span covering a gymnasium in Bombay, New York, collapsed in 1957, with another gymnasium in Newcastle, Indiana,

Figure 6.1 Erection failure of unbraced long-span steel beams, East Los Angeles, California, 1955.

making similar news in 1958. A third gymnasium, in Beaver, Pennsylvania, became a sideway failure in 1959. The roof consisted of five two-hinged tied arches with a rise of 11.2 m (37 ft) and a span of 62 m (204 ft). The steel was erected on timber falsework bents. The underfloor horizontal ties had been installed, and final bolting of purlins and knee braces was completed. Yet four arches fell over when a guy wire connected to the top chord was adjusted. In 1962, a one-story school in Atlanta, Georgia, with three lines of columns supporting conventional roof beams and bar joists, fell sideways while the interior line of columns was being plumbed by the steel erectors.

In 1960, an almost completed steel frame for a school at Bessemer, Alabama, 98 by 26 m (320 by 85 ft) in plan and 22 m (72 ft) high collapsed by lateral drift. Together with failure of a school at Orlando, Florida, and a shopping center in Birmingham, Alabama, the three failures resulted in strict control of foreign steel importation into the area covered by the Southern Building Code Congress, since it was reported that some uncertified foreign steel may have been used in these frames. Following the collapse of the University of Washington football stadium in 1987 due to inadequate bracing (see Section 11.6), one labor union official also blamed the imported steel used in the construction. However, there was no deficiency found in the steel. Furthermore, it should be clear to any knowledgeable observer that the properties of the steel itself have very little to do with stability failures of this nature.

Steel is normally erected by heavy moving equipment, with dangerously high impact loading if the moving equipment contacts previously erected steel. Collision accidents are quite common, usually resulting in a slight dent or repairable distortion. Some accidents are not that fortunate, however. In 1906, at the Landsdowne Stock Show building in Ottawa, Ontario, six roof trusses had been erected and braced with guys. A load of planking fell out of a crane sling and fell on a guy wire, pulling all the trusses over. In 1930, a four-story concrete building with steel column cores was being erected for a bakery in Brooklyn, New York. The 20-m (65-ft)-high columns had been erected full height and were guyed and braced at both the third- and fourth-floor levels. A crane boom hit one column, causing a slow progressive total collapse of all columns. Similarly, in 1934, a school building in Bayside, New York, had all steel for the four-story frame erected and guyed. A column was struck by a bundle of reinforcing steel being hoisted to a lower deck, and all the framing collapsed.

In 1939, the fifth roof truss spanning 34.5 m (113 ft) for a school auditorium at Everett, Washington, was being lifted into place. It struck the preceding truss and all five fell over in slow succession. A similar failure occurred in Russia in 1961 when a crane swung a precast concrete wall panel into an eight-story steel frame, causing complete failure.

A 91-m (300-ft)-diameter steel dome under construction in 1966 in Kingston, Jamaica, had 15 of 24 trussed ribs in place. The center compression ring was supported on a temporary wood tower and the outer ends were framed to a peripheral tension ring supported on concrete columns. Sudden collapse was blamed on failure of the center tower. Inspection of photographs reporting the failure, however, showed that one of two cranes had been moved after collapse of the ribs and the compression ring. The rig operators subsequently admitted moving the crane after the boom had contacted one rib. This movement, with rigging entangled in the steel work, pulled all the ribs to the ground in a corkscrew shape (Figure 6.2).

The need for lateral bracing for compression members (columns and compression chords of trusses) to avoid buckling has been well known from the early days of construction. This lesson has been continually reinforced the hard way, with costly collapses. In 1911, the Buffalo, New York, Porter Avenue Pump Station roof, covering 30.5 by 110 m (100 by 360 ft) failed completely when the sag of the roof trusses pulled the wall columns inward. The 23 trusses were located 16 m (53 ft) above the highest floor level. Some blame was cast on the steel supplier when it was determined that all steel was 3 mm (0.125 in.) shy of required thickness. In New York City, the balcony of the Orpheum Theater, located 18 m (59 ft) above the main level, collapsed in 1913 because of insufficient bracing of the high columns.

In 1915, at Oswego, New York, the steel trusses covering a 20 by 30 m (66 by 100 ft) warehouse turned over when a guy wire was loosened. A concrete roof at the lower chord level was being placed at the time. The upper truss chords had no permanent bracing. Similarly, in 1917, 26-m (85-ft) roof trusses

Figure 6.2   Torsional failure of steel-framed dome, Kingston, Jamaica, 1966.

over the Buick showroom in Kansas City, Missouri, fell over during roof construction when a gusset plate in the unbraced compression chord bent. The complete collapse in 1921 of a theater in Brooklyn, New York, was caused by the failure of an unbraced column at the end of a cross truss. The roof consisted of eight 24-m (80-ft) trusses, wall-bearing at one end and sitting unconnected on the cross truss at the other end.

In 1920, the 15-m (48-ft)-diameter concrete dome roof of the Christian Church in Long Beach, California, was being built on a steel frame of trusses with no bracing provided to the top chords. A truss tipped over and buckled, causing collapse of the entire dome.

The collapse of the main entrance tower of the Fuller Brush Company building at Hartford, Connecticut, in 1923 resulted in 10 deaths. The tower was a massive masonry structure projecting from the main building face, 8.5 m (28 ft) square and 34 m (113 ft) above the ground. The top of the tower contained a 190-m³ (50,000-gallon) water tank supported by four inclined steel columns seated on grillage in the walls at about 20 m (65 ft) above the ground. The bottoms of the legs were connected by horizontal channels to resist the outward component of the loads on the legs. After the masonry walls were completed, someone ordered the horizontal ties and the diagonal sway bracing removed in the belief that they were no longer needed because the walls would resist all wind loads. (The steel bracing interfered with a new proposed use of the space below the tank.) Collapse gave little warning, as the tank twisted the four legs in a clockwise direction, shearing off the tank support beams.

Contemplation of these past failures, and reflection on the fact that similar failures continue to occur today in the construction and renovation of buildings, provides a clear illustration of the need to learn from experience.

6.2.2 Stability Failures in Completed Structures

While stability-related failures are much more common during the construction phase, examples can also be found in completed structures. Design codes and standards have long recognized the need for details that ensure stability of the completed structure. However, these standards do not always find their way into the details of a specific project. One recurring source of problems has been the torsional instability of steel beams, where lateral support for the compression flange of the beam has been insufficient or nonexistent.

This problem is particularly evident in the case of overhanging, cantilever, and continuous beams. For simple spans, the top flange of the beam is in compression. Typical details for the roof or floor deck construction may provide some limited (accidental) lateral support to the top flange even when the designer is not aware of the need for such support. However, when the bottom flange is in compression, as it is for portions of overhanging, cantilever, or continuous beams, the designer must consciously provide details for lateral support or there will be none at all. As with most cases where thought and overt action are required, things occasionally slip through the cracks. Two examples of this type of failure are given here. It is no coincidence that both involved conditions of negative bending moment, where the bottom flange of the beam was subjected to compression.

6.2.2.a Station Square Shopping Center Roof, Burnaby, British Columbia, Canada, 1988

On April 23, 1988, part of the rooftop parking deck of a supermarket at the Station Square development in Burnaby, British Columbia, Canada, collapsed into the store (Figure 6.3a–c). The failure occurred on opening day, during a grand opening ceremony attended by neighborhood senior citizens. The customers had been directed to park on the roof of the 69 by 122 m (225 by 400 ft) one-story building, which was part of a regional community shopping center. The shopping center included a hotel, apartments, retail space, theaters, offices, and the flagship building for the Save-on-Foods retail chain. After a welcoming program, the 600 senior citizens began to shop, aided by 370 employees of the Save-on-Foods store.

About 15 minutes later a loud bang was heard, and water began to flow from a broken overhead fire sprinkler pipe. A remarkable photograph was

(*a*)

Figure 6.3 (a)–(c) Rooftop parking deck collapse, Burnaby, British Columbia, Canada, 1988; (d) lateral torsional buckling of beam–column connection (Courtesy of the Province of British Columbia, Canada.)

taken of the broken pipe and a severely distorted beam-to-column connection (Figure 6.3d). The photographer had been hired by the store to document the grand opening celebration. He did not at first recognize his precarious position as he looked through the viewfinder of his camera. Note in the figure the horizontal position of the web of the beam. The top of the column has been displaced by nearly 600 mm (24 in.).

After this distortion and pipe rupture, the supermarket staff acted promptly and efficiently to clear people from the immediate area and then began evacuation of the entire store. Approximately $4\frac{1}{2}$ minutes later the roof in four bays collapsed into the shopping area, along with 20 automobiles. The collapsed area was 26.5 by 22.8 m (87 by 75 ft). Fortunately, no one was killed, but 21 people were injured.

A commissioner inquiry was appointed two weeks after the failure by the government of British Columbia. The commissioner's report detailed the probable technical cause of the collapse and also reviewed the many procedural deficiencies that led to this failure. The recommendations included in the report continue to influence important revisions to standards of practice in Canada.

(*b*)

Figure 6.3 *(Continued)*

The 8400 m² (90,000 ft²) supermarket building is a single-story steel column-and-beam structure that uses the cantilevered or Gerber beam system. Wide-flange steel beams pass over hollow steel-tube columns, extending as cantilevers at each end. The beams support open-web steel joists, which in turn support a composite concrete and corrugated metal deck. This form of construction is quite common to commercial facilities of this type. The most probable technical cause of the failure was insufficient stability of the beam-to-column connection. There was no provision in the design for lateral support to the bottom flange of the beam, at a condition of bending moment that placed this flange in compression. The tube columns were HSS305 × 305 × 13 mm (HSS12 × 12 × ½ in.) sections, but with a cantilevered height of 7.3 m (24 ft) the columns were not stiff enough to prevent torsional distortion of the W610 × 113 mm (W24 × 76 in.) beam. This type of failure has occurred in the past, and the need to investigate this mode of failure has been well-established.

Perhaps of greater interest than the technical cause of this failure were the many procedural deficiencies in the project delivery system that permitted the design deficiency to go unrecognized. The commissioner's report cited numerous contributing procedural problems, including competitive bidding for design services, unclear assignment of responsibilities, inadequate involve-

(c)

(d)

Figure 6.3 *(Continued)*

ment of designers during the construction phase, poorly monitored changes during construction, incomplete peer review, and inadequate professional liability insurance. These procedural problems are certainly not unique to Canada but are the same set of deficiencies that plague the construction industry in the United States. The commissioner's report contained 19 recommendations, including (1) independent project peer review funded by increased permit fees, (2) special examinations for structural engineers and mandatory professional liability insurance, (3) development by the provincial government of a manual to clarify the responsibilities of all parties in the construction process, (4) a minimum fee schedule for design services, and (5) strengthening of certain steel industry guidelines and design manuals, particularly with respect to beam–column connection support requirements (Jones and Nathan 1990).

6.2.2.b Magic Mart Store, Bolivar, Tennessee, 1983

A similar failure occurred at a regional shopping center in Bolivar, Tennessee, on July 2, 1983 (Bell and Parker 1987). The entire roof structure of a one-story Magic Mart retail store collapsed suddenly during a severe rainstorm. No one was killed, but the store was occupied and 52 people were injured, several of them seriously. The fact that none of the occupants was killed was attributed to the strength of the merchandise shelving, which proved to have more integrity than the building's structure. The display racks actually held the collapsed roof above the floor slab and protected the people in the building (Figure 6.4a; see also Figure 2.2).

Construction details were quite similar to those in the 1988 British Columbia failure discussed previously. Open-web steel joists were supported on continuous steel beams. Failure occurred due to ponding of rainwater at a beam-to-column connection. No lateral support was provided to the bottom (compression) flange of the beam as it passed over the column (Figure 6.4b). A detail in the structural drawings indicated that bottom chord extensions for the open-web joists were to be provided at the column lines (Figure 6.4c). This detail would have braced the top of the column and the bottom flange of the beam against torsional buckling. However, bottom chord extensions were not present in the as-built construction.

The postcollapse investigation found no evidence that either the architect or structural engineer had visited the site or provided any inspection services during construction. In fact, there was almost no agreement between the contract documents and the field conditions. Structural members were not the sizes specified in the drawings, and the structural framing was even placed at right angles to that indicated in the structural plan. This failure case study serves as yet another example of the critical importance of field inspection. The forces of nature do not read the contract documents; they simply find the weak points in the as-built structure.

(*a*)

Figure 6.4 (a) Magic Mart Store, Bolivar, Tennessee, 1983; (b) beam–column connection used in the Magic Mart Store; (c) proposed detail showing bottom truss chord extensions at column lines. (Courtesy of Simpson, Gumpertz & Heger, Inc.)

6.2.2.c Civic Center Coliseum, Hartford, Connecticut, 1978

As noted in Section 2.1, a number of roofs collapsed due to exceptionally heavy snow loads in the northeastern United States during the winters of 1977 and 1978. One of the most dramatic of these failures was the collapse of the 1300-Mg (1400-ton) steel space truss roof of the Hartford, Connecticut Civic Center Coliseum (Figure 6.5; see also Figure 1.2). Stability deficiencies in the design of the space truss are generally blamed for the collapse.

The 9700-m^2 (2.4-acre) coliseum, with a seating capacity of 12,000, was completed in 1973. It collapsed five years later on January 18, 1978, at 4:00 A.M. There were no deaths or injuries; no one was in the building at the time of failure. Six hours earlier, a basketball game was played in the coliseum, with 5000 spectators in attendance. The snow load on the roof was estimated at between 480 and 770 Pa (10 to 16 lbf/ft^2). This was the heaviest snow experienced during the life of the building, so it was the triggering cause of the collapse. However, this was not an overload; it was well within the required

(*b*)

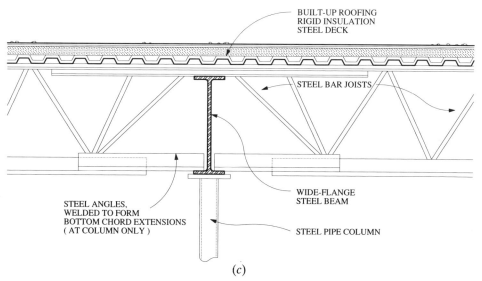

BUILT-UP ROOFING
RIGID INSULATION
STEEL DECK

STEEL BAR JOISTS

STEEL ANGLES,
WELDED TO FORM
BOTTOM CHORD EXTENSIONS
(AT COLUMN ONLY)

WIDE-FLANGE
STEEL BEAM

STEEL PIPE COLUMN

(*c*)

Figure 6.4 *(Continued)*

Figure 6.5 Hartford, Connecticut, Civic Center Coliseum roof collapse, 1978. (Courtesy of W. Turek.)

design load. The live loads assumed for design were 1440 Pa (30 lbf/ft^2) for snow and 960 Pa (20 lbf/ft^2) for wind.

The roof was designed by state-of-the-art computer methods. Comprised of 2300 members, the 112 by 93 m (366 by 306 ft) space truss was thought to be highly redundant structurally. It was supported on four pylons near the corners. The truss was 6.4 m (21 ft) deep. Top and bottom chords were arranged in a 9.14 by 9.14 m (30 by 30 ft) grid. Main diagonals were also 9.14 m (30 ft) long. Intermediate bracing 4.57 m (15 ft) in length connected the midpoints of the main members. Most of the main members were composed of four equal-leg steel angles shop-bolted back to back to form a built-up cruciform shape. The angles were separated by spacer plates for end connections at gusset plates. End connections were quite complex, with as many as eight members coming together, each with three components of slope. The truss was flat. A positive slope was achieved by supporting a sloping roof envelope on stub columns of varying heights above the top chords of the truss.

The coliseum project was built in a nontraditional fast-track manner, with the contract subdivided into five smaller contracts, all coordinated by a construction manager. The steel space truss assembly and erection was one of the five contracts. The truss was assembled on the ground and then lifted into place, 23 m (75 ft) above the arena floor level, on top of temporary supports. The four reinforced concrete pylons were then constructed. This procedure was used for reasons of safety and economy. The truss was on the ground for four months, followed by 13 days spent lifting it into place. The roof contract was completed on January 16, 1973.

The city of Hartford responded to the failure by forming a citizens' committee and by selecting a technical consultant. The technical consultant was asked to perform an independent analysis of the truss, while the citizens' panel collected information on the history of the project, including depositions of involved parties.

During the investigation it became quite evident that the building had consistently given warnings of impending failure throughout its entire life. Symptoms of structural deficiencies were even seen in construction photographs taken while the truss was being lifted into place. Excessive deflections (twice those predicted in the computer analysis) had been measured. The fascia subcontractor had documented numerous problems through change-order requests and extra costs associated with field coping and welding of fascia panels to conform to the deformed shape of truss members at the edge of the roof.

There was considerable interest in this collapse, since a number of similar structures are located throughout the United States. The widespread alarm was somewhat diminished by the technical consultant's report, which stated that the space-truss roof was ". . . based on a configuration which is inherently stable and widely used, with a long list of successful projects to its credit . . ." (LZA 1978). The most probable cause of the collapse was a design error that was specific to this particular building.

The technical report found substantial design deficiencies, in that the lateral support assumed in the design of certain compression members of the space truss was not provided in the construction details. The most critical deficiency was this lack of bracing, which caused some compression members to have an actual unbraced length of 9.14 m (30 ft) rather than the assumed 4.57 m (15 ft) used in the computer analysis. Also, some unanticipated eccentricity was found at the complex joints. (The centroids of the connected members did not always intersect at a common point, causing rotational moments at the ends of the members.) In fact, the combination of these two factors caused a computed overload of more than 800 percent on some of the main compression members. The true capacity of these members was only 9 percent of that assumed in design, according to one report, due to combined bending and axial stresses on the already buckling members.

There were some other contributing factors, but none so dramatic as the inaccurate assumption of unbraced length:

1. In some of the four-angle cruciform members, space plates were placed too far apart. This permitted individual angles to buckle independent of the built-up section.
2. The construction manager had changed the roofing materials during construction, increasing the dead load by 20 percent.

3. Some of the steel did not test according to specifications.

4. Some of the diagonal members were misplaced.

However, the inadequate bracing of the compression chords and the neglect of secondary bending stresses were in themselves sufficient to cause this failure.

The design assumed full bracing at midpoints of main members. All persons involved in construction should be aware that the capacity of a slender compression member depends on how and where it is braced. In the case of built-up members, interconnection of the components is also important. In this particular roof, the roof diaphragm did not provide any bracing to the top chords, since it was supported above the roof on stub columns to provide a sloping roof membrane surface.

As for the remarkable number of observed deficiencies in performance of the roof during construction, the technical report decried the "absence of a full-time registered structural engineer experienced with the design of long-span special structures" (LZA 1978). Neither the project architect nor the structural engineer were retained to provide regular inspections during construction.

The citizens' panel found evidence in the minutes of project meetings that the architect had called for the retention of an independent engineer to observe construction. On two occasions, the architect recommended that a qualified structural engineer should be hired to provide full, continuous, on-site inspection during the space truss assembly and erection. However, these recommendations were overruled by the construction manager as a "waste" of the client's money. The construction manager asserted that inspection was under his purview. By this assertion, he appeared to have assumed all inspection responsibilities.

Following the collapse, however, the construction manager disclaimed responsibility for monitoring and interpreting field performance. He pointed out that the technical report clearly blamed a design error. The construction manager's position was that it was not his responsibility to correct design deficiencies but simply to assure that the design *as submitted* was constructed in the field. In particular, he noted that he did not have any professional engineers on his staff and should not therefore be held accountable for technical interpretation of the performance of the project, including the excessive deformations encountered by the fascia subcontractor.

This interesting response should be considered relative to the long and successful history of field inspection performed by the design professional under more traditional contractual relationships. One of the most valuable aspects of retaining the designer for field inspection is that the designer is regularly present at the construction site. The designer can evaluate field performance and compare field conditions with design assumptions. Without doubt, there are many facilities that have been saved from tragic failures simply because designers caught their own errors while performing field inspections. In the case of the Hartford Coliseum, the construction manager's interpretation of field inspection circumvented this opportunity completely.

The Hartford Coliseum experience calls into question the use of traditional design factors of safety when projects are built using nontraditional approaches to project delivery. Accepted factors of safety have evolved over the years assuming traditional division of project responsibilities. A system of checks and balances helps mitigate errors made by a single person. However, in the case of the Hartford Coliseum, one party (the construction manager) had the authority to make decisions involving design, workmanship, and materials. Of course, an arbitrary reasonable increase in design factor of safety would not necessarily compensate for such gross errors as were present in this building.

The citizens' panel noted that ". . . inspection of his own work by the construction manager is an awkward arrangement. . . ." In a report on long-span buildings, the American Institute of Architects also referred to this case as an illustration of the ". . . need for adequate and well-delineated inspection assignments . . ." (AIA 1981).

The citizens' panel further characterized the Coliseum as a fast-track project with ". . . all review processes poorly managed. . . ." The panel was quite critical of the city of Hartford staff in the department of licenses and inspection, which approved the contract documents. The city had established a tradition of requiring an independent project peer review of the technical design of large or complex projects submitted by private developers. However, the staff did not require a peer review of the space-truss roof, since it was a project owned by the city. The intent of this decision was apparently to reduce design costs. Should there have been some injuries or deaths associated with this collapse, this may have become a significant liability factor for the city, particularly since the design error was so evident. In fact, the forensic consultant's technical report clearly states the following: "It is our opinion that a design and code review of the space truss documents by either the city or an outside consultant retained by the city would have detected the design deficiencies which caused this collapse" (LZA 1978).

There were several interesting studies of the Hartford Coliseum collapse made by other investigators. One iterative computer simulation explored the progressive collapse potential of this particular truss configuration (Smith and Epstein 1980). This study indicated that space trusses may not be nearly as redundant as formerly assumed. Following the initial failure of a few members, just a slight increase in load caused progressive failure in a roof of this particular geometry. A fold line developed completely across the roof, at which point overall stability of the roof was lost. Another analysis focused on the susceptibility of the cruciform shape of the truss members to torsional instability (Loomis et al., 1980). This theory was supported by the physical evidence of twisted compression members found at the collapse site. A third, unusual theory suggested that a defective weld on the scoreboard (attached at the center of the roof) failed, causing an explosive release of energy that brought down the entire truss.

On April 10, 1984, over six years after the collapse, an out-of-court settlement was reached in which all parties participated. The city of Hartford settled for about half the $25 million originally claimed for damages. Although the settlement was probably in the best interests of the involved parties, a number of important nontechnical questions were left unanswered: Did the construction manager indeed carry any liabilities or were they all "contracted out" to other parties? If so, did the other parties recognize that their liabilities were not reduced from those attendant to more traditional contract relationships? What is the function of field inspection, and who is best qualified to provide this service? This could have been a landmark case, helping to establish the role and responsibilities of the construction management professional.

Subdivision of this contract resulted in the general contractor eventually assuming responsibility for the space truss, which had already been completed by another contractor. Was the general contractor then responsible for inspection?

Until questions of this nature are answered clearly by precedent established through normal litigation, it appears to be quite risky to undertake unusual projects under nontraditional contractual relationships. Architects and engineers should be aware that the courts may continue to hold them responsible for performing traditional services, even when their contracts and fee agreements are based on reductions in the level of service. This is because they are the licensed professionals, and therefore the relevant jurisdiction must hold them accountable for public safety.

One clear lesson from the Hartford Civic Center roof collapse is that the computer is simply an analytical tool. It must be used carefully by competent and experienced designers. All design assumptions must be diligently checked with as-built conditions, and actual measured field performance should be compared with predictions. Precision of analytical computations is never a guarantee that one has achieved the correct solution.

6.2.2.d C.W. Post College Auditorium, Brookville, New York, 1978

On January 21, 1978, four days after the Hartford failure, a dome collapsed at C.W. Post College in Brookville, New York (Figure 6.6). No one was injured in the collapse, which occurred early on a Saturday morning during the Christmas vacation. Levy and Salvadori (1992) note that stability problems contributed to this failure. The 52-m (171-ft)-diameter 3500-seat auditorium was constructed in 1970. Forty steel pipe trusses served as meridians for the shallow dome, with steel parallels for hoops. Around the perimeter were steel columns and a canopy that functioned as a tension ring. Cross-bracing was provided to the pipe trusses, but only in alternate sectors. The cross-bracing was not sufficient for the effects of the nonuniform loading caused by drifting snow and ice buildup. The diagonal braces buckled and the system lost its stability.

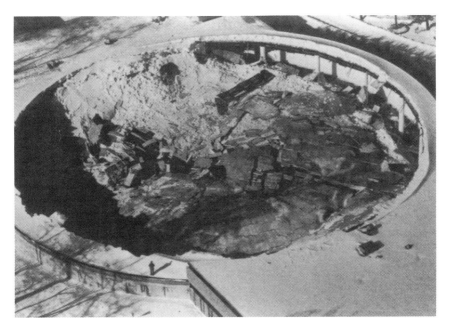

Figure 6.6 C.W. Post Auditorium, Brookville, New York, 1978. (This photograph was taken by permission from *Toward Safer Longspan Buildings,* published by the American Institute of Architects, 1981.)

6.3 STEEL CONNECTIONS

In Section 6.2 several failures caused by instability of steel structures were presented. The other common source of steel structural failures is deficient design and execution of connections. Problems with connections, like those of stability, plague projects both during construction and after completion. There are a number of ways of joining steel members, including rivets, welds, and bolts in bearing or friction-type connections. Each fastening method has merit, and the appropriate method is generally dictated by the specifics of the particular project, including concerns for economy. Failures have occurred using each type of connection.

6.3.1 Critical Details

Responsibility for the design of steel connections has been a subject of heated debate since several dramatic failures of the late 1970s and early 1980s. Confusion over the division of responsibility between the steel fabricator and the structural engineer of record still remains, despite many attempts at clarification (Dallaire and Robison 1983, Robison 1984, Newman 1994). Part of the problem appears to be a shortage of experienced detailers in the industry.

Some steel connections fail due to inadequate provision for movement. On Kyoshu Island, Japan, in 1979, a 115-m (376-ft)-long pedestrian suspension bridge over a river collapsed, killing seven and injuring 15 persons. Saddle connections at the top of the towers had become stuck in position, causing excessive cable wear on the 10-year-old bridge, which provided access to a shrine.

In some cases, problems with connections have caused substantial cost increases and construction delays. The New York City Exposition and Convention Center was plagued by defects in many of the 18,000 hub castings for its space-truss elements. The fast-track construction schedule was interrupted when hairline cracks and porosities were discovered by magnaflux and x-ray tests, most likely caused by the heat-treating process.

Steel connection deficiencies are frequent contributing factors to accidents during construction. The 1987 collapse of L'Ambiance Plaza is discussed in Chapter 8. The project was a post-tensioned prestressed concrete building constructed by the lift-slab method. This method relies on the transfer of forces through a steel connection during the lifting operation. Many investigators found a deficiency in this temporary construction load connection to be the triggering cause of the collapse.

A collapse at a large U.S. post office construction site in Chicago killed two ironworkers and injured five others in 1993. Beams that were 10 m (32 ft) long and weighing between 2.3 and 4 Mg (2.5 and 4.5 tons) were being erected at the time of the collapse. A temporary connection failed, causing a chain reaction that pulled down some 60 to 70 steel members, metal decking, and sheeting covering an area 20 × 20 m (70 × 70 ft). More than a dozen workers were on the beams at the time; some fell or rode the beams over 15 m (50 ft) to the ground (ENR 1993).

When properly designed and constructed, bolted connections are reliable and economical. Inspection is required and there have been a few isolated problems with mislabeled, counterfeit fasteners. For friction-type bolted connections, some method must be used to certify the proper degree of tightening of the nut.

The reliability of welding, by certified workers, that conforms to American Welding Society (AWS) specifications has also been established. However, once in awhile failures involving welded connections create new questions for the fracture mechanics discipline. Problems with welding are commonly encountered when thick plates having impurities or insufficient fracture toughness are welded in restrained configurations. The residual shrinkage stresses, when movement is not permitted, can be the cause of fracture.

The difficulty of ensuring competency of field-welded connections continues to be a problem, despite the availability of numerous nondestructive testing methods. Field conditions are not always conducive to reliable welding. Many designers choose to use a combination of shop welding and field bolting, so that the welding can be accomplished under ideal controlled conditions.

Although ultrasonic testing and radiography can easily establish the depth of weld penetration and can find discontinuities in the steel, the testing must be accomplished by competent and honest technicians. An example of a breakdown in quality control is the case of falsified weld test certifications discovered in 1975 during construction of the Trans-Alaska Oil Pipeline. Fraudulent practices were exposed when construction of the 1400-km (900-mile)-long pipeline was about 50 percent complete. Because of the sensitivity of the environment to pipeline leaks, the testing of field welds was strictly specified. Both visual inspection and radiography were to be employed. Metal tags were placed at each joint to certify the completion of x-ray inspection. The specified procedures were very detailed, perhaps more stringent than for any other pipeline in the world. Yet an employee reported that he was terminated for refusing to falsify inspection records. The testing equipment frequently failed to perform, and test personnel were lagging behind the welders.

The Trans-Alaska Pipeline case illustrated what can happen when construction workers receive conflicting messages from management. Those conducting the tests knew that management had high standards for quality, but the stronger signals and the most costly penalties were focused on failure to meet project completion schedules. Management was quite intolerant of delays, and the intended quality assurance program was affected adversely. The end result was a costly project delay and a loss of public relations credibility for the entire pipeline project. All welds had to be reinspected.

Welding techniques have improved considerably since the days of the blacksmith and forge, but all new techniques bring new problems. A review of the history of failures in welded connections illustrates how engineers learn as new techniques are applied in the field. Bridges require special attention because of the typically thick plates and sections. The investigation by P. P. Bijlaard of the 1937 collapse of a Vierendeel welded span at Hasselt, Belgium, noted that steel under certain high-stress concentration and low temperatures will lose its ductility; brittle cracks will form. Such a condition exists in welded thick plates where shrinkage in all directions is restrained. The Belgium bridge collapse occurred during cold weather, fracturing the heavy steel plates.

In 1958, during assembly of the second Carquinez Bridge in San Francisco Bay, California, welded sections were found to have minor fractures, sufficient to require corrective work. High-strength, 620-MPa (90,000-psi)-yield-point steel had been chosen to reduce the weight of the new bridge. Defects were discovered by magnaflux testing of the fillet welds in H-sections used as tension members. A total of 56 members required correction, removing the defective weld metal with a carbon arc torch and replacing the weld manually. Exposed flame-cut edges of plates had been prepared carefully for welding by removal of hardened metal, sharp corners, and abrupt surface irregularities. The detailed study of procedures and materials resulted in safer methods for welding such heavy members.

The failure in 1962 of the Kings Bridge in Melbourne, Australia, contributed further information on proper welding methods. The bridge over the Yarra

River and a railroad consisted of two parallel 700-m (2300-ft)-long one-way bridges with spans up to 49 m (160 ft). Like the Duplessis Bridge near Three Rivers, Quebec, failure came on a cold day. Cracks started in tension flanges at welds connecting different size plates. Both bridges used steel alloys not considered the best for welding. The failure at Melbourne came more than a year after completion, when a 34-Mg (38-ton) crane crossed the bridge on a day with below-freezing temperatures. The bridge sagged 460 mm (18 in.) and wide cracks opened in the deck for a distance of about 30 m (100 ft). Investigators reported poor control of fabrication, with traces of the red-lead shop coat found in the cracks of the welds, improper steel for welding with considerable variation in metallurgy of samples taken, and lack of awareness of the difficulty in proper production. The project was a design–construct contract with no outside supervision.

Brittle fracture of some steels at low temperature was correlated with notch effects and high local stress concentrations by the 1957 report of the National Research Council Committee on Ship Structural Design. This report was the result of 13 years of research instigated by brittle fractures in welded ships during World War II. Elimination of sharp corners at openings reduced failures in Liberty ships and tankers from 7 per 100 ship-years to $\frac{1}{2}$ per 100 ship-years. The improvements consisted of changes at hatch openings, riveted shell seams, and some crack arrestors. Some rather minor revisions in design and details, together with more rigid control of welding enforced in 1947, resulted in complete elimination of Liberty ship losses up to 1955. The cost of the research program was $3.5 million, about half the cost of one merchant ship without cargo, certainly a profitable investment.

The Wolftrap Performing Arts Center in Fairfax, Virginia, experienced a dramatic brittle fracture in one of its steel box roof girders in January 1985. The girder was 40 m (130 ft) long and 1.8 m (6 ft) deep. A crack 38 mm (1.5 in.) wide extended through the 76-mm (3-in.)-thick bottom flange of the girder and 1.5 m (5 ft) up one side and 0.9 m (3 ft) up the other side plate. Fortunately, there was sufficient redundancy to the roof structure so that the roof did not collapse, although the girder deflected 240 mm (9.5 in.) when the crack occurred. The subsequent investigation cited multiple causes for the crack: weld flaws, extremely cold temperatures, and poor metallurgy (steel with insufficient fracture toughness) (ENR 1985a, Kaminetzky 1991). The Wolftrap Center was repaired by cutting away the cracked plates, welding in new plates, and then post-tensioning the girder with cables.

Brittle fracture problems are more likely when high-strength steels are used. The higher-strength steels generally have lower fracture toughness. Brittle fractures have been experienced in steel storage tanks for industrial or agricultural products, frequently caused by defective welds (Gurfinkel 1988, 1989).

Rupture of a large storage tank can cause widespread environmental damage, as was the case in Pennsylvania in January 1988 when the Ashland Petroleum oil storage tank collapsed (Gross, Smith, and Wright 1989). The capacity of the tank was 15,000 m^3 (4 million gallons). Over 3800 m^3 (1 million gallons) of oil was spilled into the Monongahela and Ohio rivers, affecting the water supply of many downstream municipalities. Investigators determined that the steel had inadequate toughness at operating temperatures to prevent brittle fracture propagation. The crack initiated at a defect near a weld. In fact, the weld was not defective; the flaw was present prior to welding. However, welding near the defect contributed to embrittlement of the metal.

Problems are also common with welded jumbo sections, especially large cross-section members subjected to tension forces. Many designers prefer bolted connections for such members. When thick plates are welded, steel with good fracture toughness must be specified; otherwise, the residual thermal stresses will produce cracks. Some preheating may help. These problems have caused costly delays in a number of projects, including the Washington State Convention Center in Seattle. The structure spans over a freeway and contains many jumbo sections loaded in tension. Repairs involving splice plates were required for about 100 cracked connections.

Lamellar tearing associated with large welds in thick steel plates has been studied by fracture mechanics specialists for a number of years. Significant problems with lamellar tearing became more common in the 1970s as designers began to use larger built-up sections on long spans. Lamellar tearing is a failure of the base material in the through-thickness direction. Steel has been shown to be less than perfectly homogeneous, and this is the direction in which steel has the weakest strength. The problem is related to constrained details, where the heat generated in the welding process cannot dissipate and becomes locked in. As the weld cools, microscopic cracking occurs, sometimes below the surface, where it can only be detected by ultrasonic methods. Impurities in the steel contribute to the potential for lamellar tearing.

Lamellar tearing has affected several high-profile projects, including the 100-story Hancock Tower in Chicago, the El Paso Civic Center in Texas, and the twin 52-story Atlantic Richfield Towers in Los Angeles. Ross (1984) gives an informative chronology of the controversies generated by these cases, and the response of the research community. New steels with fewer impurities are now available because of this research. The extensive research activity generated by lamellar tearing resulted in recommendations that reduce localized internal strains. These recommendations have improved the quality of welded connections in general and enhanced our understanding of the properties of structural steel. This is a case where the desire for longer spans and more creative structures caused designers to go beyond the limits of established technology. The materials science research community was forced to catch up with the practicing architects and engineers.

6.3.2 Failure Case Studies Involving Connections

In this section, two building failures are discussed in detail, both coincidentally located in Kansas City, Missouri. One is the 1979 collapse of the long-span roof over the Kemper Memorial Arena; the other is the 1981 failure of the suspended pedestrian walkways in the Hyatt Regency Hotel. Both failures were attributed to connection deficiencies. These failures became the object of much discussion regarding responsibility for connection design in steel structures. Emerging from the discussion is a greater awareness of the critical nature of connections, the importance of shop drawing review, and a general improvement in the assignment of responsibilities for safety of the completed project. One important lesson has been the need for cooperation between the designers and fabricators of steel structures, so that the details are carefully considered.

The Kemper Memorial Arena roof incident was a dramatic collapse of a major portion of a sophisticated long-span roof. Failure of such a prominent structural element has the potential to cause a large number of injuries and deaths, but no one was killed or injured in this failure because the building was unoccupied. The Hyatt Hotel walkways, on the other hand, were minor architectural circulation elements, not even components of the primary building structure. The failure involved a connection at the end of a 3-m (10-ft)-long simple span beam and errors that were quite unsophisticated technically. Except for the fact that 114 people were killed in the collapse, the failure would have taken place with very little notice. However, the magnitude of human suffering and of the related litigation made this incident extremely significant. The National Bureau of Standards report on the collapse called it ". . . the single most devastating structural collapse in the history of the United States."

Following the review of these two failures are some comments regarding the disappointing performance of welded connections in moment-resisting steel frames in the January 1994 earthquake in Northridge, California. The failure of a large number of beam-to-column connections in new buildings was a great shock to the design and construction community. This was especially disconcerting in view of the moderate nature of the seismic accelerations recorded in the event, which were considerably less than the magnitude for which the structures were supposedly designed. The Northridge experience illustrates that the process of learning from failure continues today, despite the sophistication of modern design theories and computational tools. The physical world still has some surprises and some humbling lessons for designers of the built environment.

6.3.2.a   Crosby Kemper Memorial Arena Roof Collapse, Kansas City, Missouri, 1979

A large portion of the roof of the 17,600-seat Crosby Kemper Memorial Arena in Kansas City, Missouri, collapsed on June 4, 1979 (Figure 6.7a; see also Figure 1.3). The collapse occurred at approximately 7:10 P.M. during a severe wind and rain storm. The storm was not unusual for the area; such thunderstorms are common in Kansas City. Of the 12,000-m^2 (130,000-ft^2) roof, approximately 4000 m^2 (43,000 ft^2) had to be replaced. The falling roof caused internal

(*a*)

Figure 6.7 (a) Crosby Kemper Memorial Arena, Kansas City, Missouri; (b) connection detail.

air pressures that blew out some of the walls of the building, so the repair costs were quite extensive. There were no deaths or injuries; only three building maintenance personnel were present at the time of the failure.

Kemper Arena was first opened in 1974. It was widely acclaimed for its innovative design. The building was given the 1976 Honor Award by the American Institute of Architects (AIA). Ironically, the AIA was holding its annual national convention in another building in Kansas City during the week of the Kemper Arena failure and was presenting another Honor Award to the designer of Kemper on the very night of its collapse. The original construction cost was $23.2 million.

During the six years preceding its failure, a number of important events were held in the arena, including the 1976 national convention of the Republican Party. Two nights prior to its collapse, over 13,000 people had attended a trucker's show in the building.

The building enclosure, 95 by 129 m (310 by 424 ft), is formed by three triangulated space frames 16.5 m (54 ft) wide spaced 30 m (99 ft) apart. The three space trusses, formed of steel pipe sections up to 1.2 m (4 ft) in diameter, serve as a dramatic exoskeleton, exposed entirely outside the structure. The space frame's members are bolted together with high-strength A490 bolts loaded in shear. The failure occurred in the connections between the space frames and the secondary roof structure spanning between them. The space frames remained in place after the collapse, with virtually no damage.

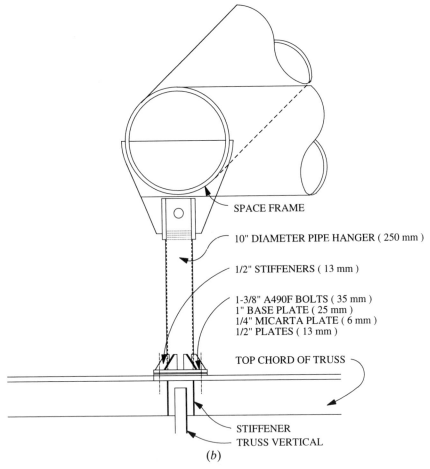

SPACE FRAME

10" DIAMETER PIPE HANGER (250 mm)

1/2" STIFFENERS (13 mm)

1-3/8" A490F BOLTS (35 mm)
1" BASE PLATE (25 mm)
1/4" MICARTA PLATE (6 mm)
1/2" PLATES (13 mm)

TOP CHORD OF TRUSS

STIFFENER
TRUSS VERTICAL

(b)

Figure 6.7 *(Continued)*

The secondary roof structure (steel plane trusses) is suspended from the three space trusses by steel pipe hangers at 42 different panel points. Each hanger was fastened to its supporting space frame by a welded gusset plate with a pin connection. A steel base plate at the bottom of the hanger was connected to a steel plate that was welded to the top chord of the secondary roof truss. For some reason that remains unclear, a flexible plastic resin layer (Micarta plate) 6.4 mm (0.25 in.) thick was placed between the two steel plates. Four high-strength A490 bolts, vertically loaded in tension, completed the connection, passing through the two steel plates and the Micarta plate (Figure 6.7b).

The collapse investigation for the city was conducted by James L. Stratta, a consulting engineer from Menlo Park, California. As is usually the case in a catastrophic structural failure, there were several interrelated contributing factors. James Stratta's report noted that the causes included ". . . a materials problem, an installation problem, and a design problem . . ." (Stratta 1979).

The investigation found that contrary to public rumors, none of the roof drains had been blocked, and any accumulated ponded-water weight would have been well within the calculated capacity of the roof structure. The principal cause of the failure was fatigue in the bolts making up the connection of the secondary trusses to the hangers.

Wind loading during the five years of occupancy of the building caused many cycles of loading and unloading of the bolts in the tension connection. Recurring movement (rocking) subjected the bolts to dynamic loading they were never intended to carry. In addition to tension loading and unloading, the hanger assemblies were subjected to bending motions from horizontal movement of the entire roof structure in variable winds. The connections finally gave way when subjected to the shock of the intense wind and rain storm.

High-strength steel bolts, which are relatively brittle, should not be used under dynamic load conditions. Steel design specifications warn against use of A490 bolts under variable loads and recommend their use only under static load conditions. A footnote containing this warning appeared in the *Manual for Steel Construction* published by the American Institute of Steel Construction (AISC). This failure illustrated the reasons for that recommendation and the desirability of ductile connections.

The design load requirement for each connection was 934 kN (210,000 lb). The capacity was well above this, at about 1400 kN (320,000 lb) initially. But the connections failed under much less loading, due to fatigue under the repeated cycles of motion. Stratta estimated that in the six years of exposure to the wind, the connections were subjected to at least 24,000 oscillations. Weakened by fatigue, the bolts exhibited a tension strength that was as little as 20 percent of their original strength. Under dynamic loading in tension, A490 bolts deteriorate dramatically. Testing showed that when tightened and loosened five times, the 22-mm (0.875-in.)-diameter A490 bolts failed at one-third of the tension load they carried initially.

A contributing factor was the way in which the connection was constructed, in that considerable movement through prying action was permitted. The nuts were not fully tightened, permitting more movement than would have been the case otherwise. The relatively flexible Micarta plate absorbed about 40 percent of each bolt's torque load, subjecting the bolts to increased dynamic loading.

Stratta also cited a general lack of redundancy in the basic design as a contributing factor to the magnitude of the collapse. This is a common problem in modern long-span structures, where there are few connections, so that a single connection failure becomes more catastrophic. In this case, fatigue failure in a single bolt was able to bring about a progressive collapse of the entire center section of the roof and major damage to the rest of the structure. Simple calculations showed that if a single hanger failed because of bolt fatigue, the adjoining hangers would rapidly fail in a chain reaction as a result of lack of redundancy.

Kemper Arena was rebuilt; the building was reopened 10 months after the collapse (*Civil Engineering* 1981). The four bolts, two steel plates, and the Micarta plate in the original design were replaced with a single A36 steel bar for each hanger, welded to the truss diagonals. The new connections are much more ductile. The roof slopes were also increased and additional roof drains were provided. Trusses were strengthened and deeper joists were used. Nearly $6 million was spent in the repairs, which included the installation of electronic monitoring devices to warn of motion problems in the roof. The instrumentation is activated when winds exceed 56 km/h (35 mi/h).

A campaign to restore public confidence in the building included distribution of brochures that clearly explained the failure cause in lay terms and outlined the repairs that had been made. This campaign was quite successful in that the building opened to a capacity crowd in March 1980. However, one year later, the Hyatt Regency Hotel pedestrian walkways collapsed downtown. It is surely understandable that some of the unfortunate residents of Kansas City continue to question the competency of the construction community.

Sorting out the liabilities for the Kemper Arena collapse was quite difficult because like the Hartford Civic Center Coliseum that failed one year previously, the arena had been built under a nontraditional design–build contract. The designer worked under a direct contract with the builder rather than under the traditional owner–architect agreement. Four years after the collapse, and two days into the trial, a settlement of the city's lawsuit against the designers and builders of the facility was reached. All parties, including the city, contributed to the nearly $6 million rebuild cost. The alternative would have been a lengthy, expensive litigation.

The Kemper Arena collapse was only one of several dramatic structural failures that occurred within a couple of years. These included the Hartford Civic Center Coliseum roof (1978) and the Hyatt Regency Hotel walkways (1981). Taken together, these failures had a sobering effect on the construction industry. Attention was focused on the need to provide greater structural integrity and redundancy in the design of structures. In particular, the importance of connection design was emphasized. Detailing and execution of connections, including provisions for their adequate review during the shop drawing phase, have been the subject of heated discussion since these failures occurred. In addition, this failure and others have contributed to the understanding of wind forces on buildings and the evolution of improved wind design standards that more rationally consider dynamic effects and pressure concentrations.

6.3.2.b *Hyatt Regency Hotel Pedestrian Walkways, Kansas City, Missouri, 1981*

On July 17, 1981, two suspended walkways in the atrium area of the Hyatt Regency Hotel in Kansas City, Missouri, collapsed suddenly. The failure caused 114 deaths and 185 injuries. Because of the large number of casualties, this failure had a greater impact on the design and construction industry than any other failure in recent times. The magnitude of litigation and discussion in the engineering literature also exceeded that for any other single structural failure in history.

The project was completed and dedicated on July 1, 1980, one full year prior to the collapse. The walkways, suspended by tension rods from the atrium roof structure, were arranged such that the second-floor walkway was suspended directly below the fourth-floor walkway. A separate walkway at the third-floor level was not involved in the collapse.

When the failure occurred, just after 7:00 P.M., between 1500 and 2000 people were crowded into the area on the floor of the lobby and all three walkways, listening to music and participating in a dance. Many in the crowd were local Kansas City residents. The rescue operation took over 12 hours with 58 Mg (64 tons) of debris removed. The third-floor walkway was also removed.

This was a building with a history of problems during construction. During the investigation a number of prior incidents came to light. On October 1, 1979 (one year prior to completion), 250 m² (2700 ft²) of atrium roof collapsed due to a connection failure—slotted holes for an expansion joint did not permit sufficient movement. The owner, concerned by this collapse, paid the engineers an additional fee to recheck every member and connection in the atrium.

In a formal report, the National Bureau of Standards (NBS) called this the "most devastating structural collapse ever to occur in the United States." This statement is arguable, considering events such as the 1889 Johnstown flood, in which over 2200 people died (see Section 4.1.1). However, the Hyatt case was undoubtedly the most heavily litigated failure in history. There were class-action suits, individual suits, civil suits, criminal suits, administrative reviews, and ethical conduct reviews. At one time, the claims under review totaled over $3 billion. The criminal investigation against the engineers was not dropped until December 1983. One class-action suit settled for $143 million. As many as 30 parties were named as defendants in one case. One victim was awarded $14 million in a jury trial.

In one of the more unusual claims (submitted five years after the collapse), 72 rescue workers sued for $150 million punitive damages against the owner of the hotel for negligence in not preventing the tragedy. The police, ambulance, and fire personnel claimed that the emotional trauma of the experience caused long-term psychological effects. The suit noted that the rescuers had been ". . . overwhelmed by the gruesome work, which included severing limbs from bodies and collecting body parts for official identification." This unique claim was settled out of court in January 1988 for $500,000.

Although some of these claims are arguably excessive, the magnitude of litigation was clearly related to the loss of life and injuries, not to property damage. The entire original hotel and convention center complex cost $43 million; the reconstruction cost was only $5 million.

The atrium space in which the failure occurred was a four-story open lobby between a wing of meeting rooms and restaurants and a 40-story, 733-room hotel tower. Each of the three 37-m (120-ft)-long walkways was supported by six 32-mm (1.25-in.)-diameter tension rods, dramatically suspended from the roof structure (Figure 6.8a). The walkways were 2.7 m (8.7 ft) wide. The fourth-floor walkway was suspended from the ceiling by one set of steel rods; the second floor walkway was suspended from the fourth-floor box beams by a second set of rods. A separate walkway located at the third-floor level was not involved in the collapse but was later also found to be structurally deficient.

The floor of each walkway was a concrete deck 83 mm (3.25 in.) thick over a 38-mm (1.5-in.) ribbed metal deck. The deck was supported by four 9-m (30-ft)-long, 400-mm (16 in.)-deep steel wide-flange beams connected end to end along each outside edge of each walkway. These edge beams were welded to steel box beams running transverse to the direction of travel at three evenly spaced points. The box beams were comprised of steel channels welded toe to toe. The tension rods passed through holes drilled through the welded seams of the box beams. The channel box beams were made of MC 8 × 8.5 sections, with a web thickness of only 4.5 mm (0.179 in.) and a flange thickness of only 7.9 mm (0.311 in.). This is the lightest-weight channel of that depth. (A channel, C 8 × 22.8, is commonly available and would not have changed the appearance since all the steel was concealed by finish materials.) The failure of the connection was accompanied by crippling of the web and bending of the flange in each channel of the box beam (Figure 6.8b and c).

The most probable technical cause of the failure was easily established. A deficient connection, where the steel suspension rods were connected to the box beams that supported the fourth-floor walkway, failed. This caused the fourth- and second-floor walkways to fall to the floor of the atrium area.

Soon after the investigation was initiated, it was discovered that a change had been made to the failed connection detail during construction. A drawing in the contract documents showed a single continuous rod-and-nut assembly at the fourth-floor box beam, but it was not constructed that way. The change, to a second set of rods, doubled the stress in the vicinity of the nut under the fourth-floor beam. In the original design, that nut would have supported only the fourth-floor load, but after the change, it supported both the second- and fourth-floor loads. These nuts pulled through the box beams, causing the failure.

The National Bureau of Standards (NBS) was asked by Kansas City officials to investigate the collapse. The scope of the NBS investigation was to determine the most probable technical cause of the failure, not to establish responsibility. Tests conducted at the NBS laboratory in Gaithersburg, Maryland, demonstrated conclusively that the connections as originally shown on the design drawings were incapable of supporting the gravity load required by the relevant building code. This deficiency was further compounded by the change made to the detail during construction, which doubled the load on the connection, making its failure inevitable (Figure 6.8d).

(*a*)

Figure 6.8 (a) Atrium space in Hyatt Regency Hotel, Kansas City, Missouri, showing suspended pedestrian walkways; (b) tension rod to box-beam connection at fourth-floor walkway (concrete slab on metal deck is not shown for clarity; (c) fourth-floor box-beam connection failure (concrete slab on metal deck removed for clarity); (d) detail shown on the project documents indicated one continuous tension rod through the fourth-floor box-beam (left); during construction, a change was made to use two tension rods (right); (e) test assembly at the National Bureau of Standards, Gaithersburg, Maryland (now the National Institute of Standards and Technology); (f) tests at NBS were able to reproduce the box beam-to-tension rod connection failure under static gravity loads less than the relevant codes required; (g) Laboratory analysis of box-beam test specimen, National Bureau of Standards, Gaithersburg, Maryland. [(a)–(d), (f) Redrawn from National Bureau of Standards report.]

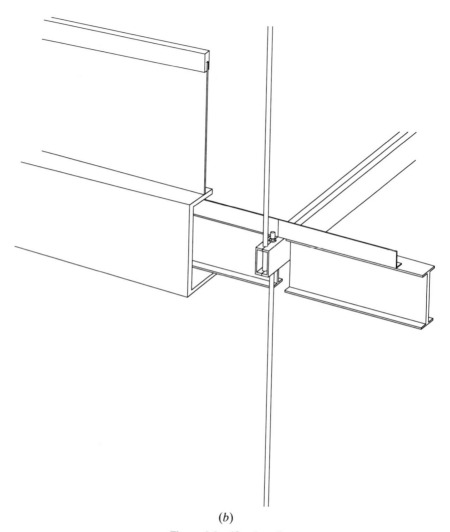

(*b*)

Figure 6.8 *(Continued)*

A number of theories had been advanced to explain the collapse. Some of these blamed the dynamic vibrations caused by people dancing on the walkways. At the NBS laboratory, numerous full-scale mockups (67 in all) were tested to failure (Figure 6.8e–g). The conclusion was that dynamic effects were insignificant and that neither the original detail nor the as-built detail was adequate for the design gravity static live and dead loads.

The original detail proved to be capable of developing only 60 percent of the design load, and after the change the capacity was reduced to about 30 percent of the required load. Evidence collected from the separate third-floor walkway confirmed that the original design was deficient, since that walkway

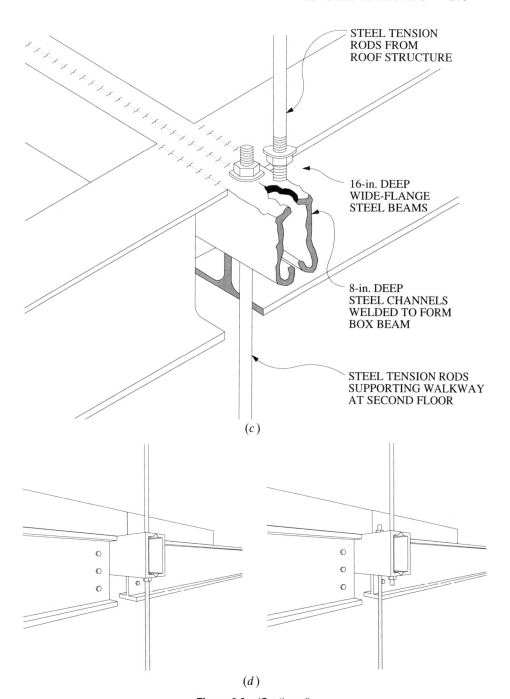

STEEL TENSION
RODS FROM
ROOF STRUCTURE

16-in. DEEP
WIDE-FLANGE
STEEL BEAMS

8-in. DEEP
STEEL CHANNELS
WELDED TO FORM
BOX BEAM

STEEL TENSION RODS
SUPPORTING WALKWAY
AT SECOND FLOOR

(*c*)

(*d*)

Figure 6.8   (*Continued*)

(*e*)

Figure 6.8 *(Continued)*

(*f*)

Figure 6.8 *(Continued)*

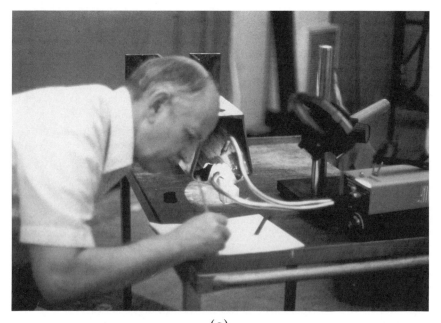

(*g*)

Figure 6.8   *(Continued)*

connection was also visibly distressed. Failure had been progressing for some time prior to the collapse. With the change to the detail during construction, the ultimate capacity was so reduced that the connection had only minimal capacity to resist the dead load and virtually no live-load capacity. Even with the weight of only one walkway, the separate third-floor walkway connection was at a critical load state (NBS 1982, Pfrang and Marshall 1982).

The NBS report noted that the primary cause of the failure was insufficient load capacity of box beam-to-hanger rod connections. Contributing factors included inadequacy of the original connection design, the change during construction that doubled the load, and an actual dead load 8 percent higher than anticipated. The load on the walkways at the time of collapse was probably only about 31 percent of the ultimate capacity required by the code in effect in Kansas City.

NBS found no problem with workmanship or quality of materials, but rather with the connection detail itself: "The box-beam hanger rod connection detail used is not among the connection details suggested in the AISC manual. Without a properly validated analytical model or supporting test data, there can be no reasonable certainty as to the load capacity of such a connection" (NBS 1982).

The report further recommended that concentrated loads should never be applied to flanges of steel sections. The use of a load distributing plate is necessary. A better detail would have applied the load in the plane of the web of the channel (placing the channels back to back rather than toe to toe). Also cited was the obvious lack of redundancy in the overall design of the walkway support system. Failure of all connections was imminent. Failure of any one would lead to progressive failure and total collapse; there was no reserve capacity at other connections.

Nearly all investigators agreed with the technical explanation of this failure given in the NBS report. However, argument about the chain of events that permitted this collapse to occur was extremely heated. Questions regarding deficiencies in the project delivery system were the focus of landmark litigation and administrative hearings. These questions were quite basic: Who designed the original connection? Who is responsible for connection design—the fabricator, or the structural engineer? Was the original connection buildable? Who initiated the change to the connection during the construction phase? Why? Who approved it? What is the meaning of "shop drawing review"? This failure led directly to ongoing activities aimed at improving quality assurance and quality control in the design and construction process.

In terms of human suffering, the Hyatt Regency Hotel walkway collapse was one of the most compelling tragedies ever to confront the construction industry. The technical cause of the failure was elementary and easy to understand. However, there were many lessons for design and construction professionals. The important lessons involved procedural issues. Clearly, there is a need for all parties to understand their responsibilities and to perform their assignments competently. The structural engineer's responsibility for overall structural integrity, including the performance of connections, was firmly established in this case. This failure also lends credibility to the practices of project peer review and constructability review.

A landmark negligence suit involved the state of Missouri versus the structural engineers of record. The administrative law judge's decision in this suit had a major impact on defining the standard of practice for structural engineers. The state's position was that ". . . total responsibility for structural safety rests with the structural engineer of record no matter what. . . ." The structural engineer claimed that the steel fabricator is responsible for connection design, and that ". . . when the engineer reviews the connection shop drawings, he does so only to see that the fabricator has correctly copied the size and arrangement of members from the engineering drawings . . ." (ENR 1984).

In the midwestern states, it had been common practice for fabricators to design the connections. This generally led to more economical construction, since the fabricator was free to use familiar details. (One alternative now recommended is that the structural engineer should design the connections but permit the fabricator to submit alternative details for review and approval by the structural engineer.)

The steel fabricator claimed that he and his subcontractors did design every connection except this one, which must have been designed by the structural engineer. Although many detailers that work for fabricators are not licensed engineers, this particular detailer was a licensed engineer with 28 years of experience.

Following eight weeks of testimony, the administrative hearing concluded with a report on November 20, 1985 (ENR 1985b). The 442-page decision agreed with the state's position. The judge ruled that state registration of structural engineers implies their ultimate responsibility for the safety of the design.

Commentary accompanying the decision noted that no original calculations for the connection design were ever found—in fact, it may never have been designed. The engineers were made aware of the change to the detail during construction and were asked to check it, both by the detailer and by the project architect. The engineer's own technician called his employer's attention to the change during shop drawing review. The engineer's response was that he "looked at it" but didn't "review it."

The judge's ruling came as a great surprise to many structural engineering consultants in the region, who felt that it was quite unfair (ENR 1985c). As a result of the hearing, the licenses of the engineers were suspended on January 22, 1986. One of the project engineers made the following comments:

> The tragedy of the skywalk collapse with its devastation and loss of life will haunt me the rest of my life. Because of these hearings my company is in liquidation, and the stockholders, which include myself, have lost in excess of a million dollars, with over $200,000 in unpaid legal bills still remaining. All of the fixed assets have been sold, and the employees are now with another company. . . . I have resigned as a director of the company and, as an employee, have no engineering responsibility. I have been publicly denounced and chastised by the press, but, most of all, my reputation . . . has been destroyed. (ENR 1986)

Subsequently, the project engineers were suspended from membership in the American Society of Civil Engineers for violating the ASCE's Code of Ethics: ". . . engineers shall hold paramount the safety, health and welfare of the public in performance of their professional duties. . . ."

A case such as the Hyatt Regency walkway collapse contains many lessons. Some of these are related to technical aspects of engineering design: the nature of connections and fracture mechanics. Others are related to procedural issues: the way in which decisions are made and communicated to others and the importance of shop drawing review (Rubin and Ressler 1985, Boehmig 1990).

Some practicing engineers claim that the engineers involved with the Hyatt Regency case were simply the unfortunate victims of a complex construction environment in which such events are bound to occur. Others see important lessons for all design professionals. Rubin and Banick (1987) write that these engineers were ". . . *not* hapless victims of a failed system." The right people did ask the right questions and there were several opportunities to correct the error. According to Rubin and Banick, the failure was the result of an attitude of cavalier negligence on the part of the project engineers:

> How can their conduct be explained? An understanding of their conduct is perhaps the most important lesson that can be drawn from the Hyatt collapse

because it represents, more than anything else, a human failure to which all professionals are subject. Some succumb, some do not; most are just plain lucky in that they do not get caught. Our errors are picked up by others, or although our errors go undetected, no tragedy ensues. Complacency is a human failure. It creeps into a professional's approach to practice as the newness, excitement, and other early rewards of the profession fade. The professional becomes indifferent and stops worrying and agonizing. He takes shortcuts and gets away with it, and then takes more shortcuts. It becomes a way of life. This is human. The shock of an occasional failure brings him to his senses and forces him to reevaluate his conduct. . . . "There but for the grace of God . . ." is an appropriate response to the Hyatt failure and the engineers' license revocation. But at the same time, it is inappropriate and counterproductive to blame the system, the administrative law judge, or others who also ought to have caught the error. (Rubin and Banick 1987)

The Hyatt Regency hotel was repaired and was back in service within five months of the tragedy. The reconstructed building was reopened in October 1981. The three walkways were replaced by a single 5-m (17-ft)-wide walkway at the second-floor level. This walkway is not suspended dramatically from the ceiling; it is supported on 10 large reinforced concrete columns spaced 9 m (30 ft) apart. The new design will support a 9.6-kPa (200-lbf/ft^2) live load, twice that required of the lobby floor and all other corridors in the building (ENR 1981).

6.3.2.c Moment-Resisting Frame Connection Failures in Northridge, California, Earthquake, 1994

Initial damage reports following the January 17, 1994 earthquake in Northridge, California, focused on the collapse of poorly detailed and nonductile reinforced concrete structures and the dramatic failures of portions of precast concrete parking garages. Comparisons were quickly made to the seemingly invulnerable steel frame structures. However, subsequent investigation found a surprising amount of damage to welded connections in moment-resisting steel frame buildings (ENR 1994a).

Steel moment-resisting frames have generally been regarded as the most earthquake-resistant type of structure available. Performance in the Northridge earthquake caused confidence in such structures to be reduced considerably. Postearthquake inspection of buildings currently under construction, where connections were still visible, first revealed the problem. Some fractures went unnoticed until owners removed insulation, fire protection, and other finish materials. The indicators of damage were very subtle; with frame connections usually concealed behind finish materials, damage was not readily apparent. Inspection alone was extremely expensive.

Beam flanges and the interface of web and column flanges were subject to cracks (Figure 6.9). All the failures were brittle fractures, occurring at a relatively low level of stress, well within the nominal elastic ranges of the beams and columns. There were no reports of the formation of plastic hinges

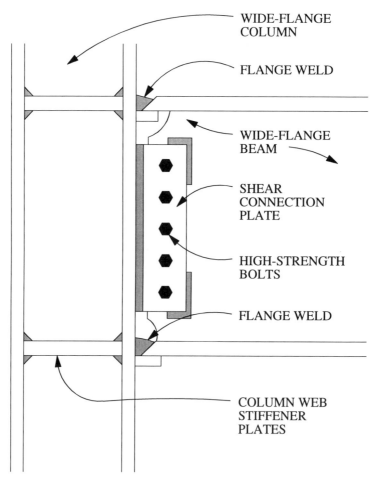

Figure 6.9 A commonly used moment-resisting steel frame connection experienced surprising brittle fractures during the Northridge, California, earthquake of January 1994.

in the members. Fractures were discovered in so many buildings that serious questions were raised about codes, design standards, and welding practices. In some cases the quality of welding was suspect, but in most the workmanship conformed to accepted standards. Weld cracking was not prominent. Engineers were astounded by the many cracks that appeared in well-executed connections with no evidence of porosity.

Up to 90 percent of these connections cracked in some buildings, and most of the buildings affected were less than 10 years old. It should be noted that there were no structural collapses related to these connection failures despite fairly large accelerations and numerous aftershocks. The fractured connections continued to perform as semirigid joints. However, the Northridge earthquake registered only 6.8 on the Richter magnitude scale and lasted only 10 seconds. The recorded accelerations were about half the design earthquake level.

This connection had previously been assumed to be almost seismically invulnerable. Over the years, the moment frame has evolved from its early configuration. In the 1960s and 1970s, most of the connections in a frame were typically moment resistant. More recently, only a few bents of some frames were made moment resistant. In southern California and other seismically active regions throughout the world, designers created hundreds of low and midrise buildings in the 1980s featuring few bays of long spans supported by very large columns and deep beams, requiring huge moment connections. Some investigators believe that there has been too much reliance on too few welds.

Larger connections seemed to have fared most poorly in the Northridge earthquake (ENR 1994d). Tests in the 1970s were based on beam and weld sizes much smaller than those commonly used in systems today.

Tests since the earthquake have shown that large deformations and joint rotations coupled with large member sizes result in high strains around the beam flanges of moment connections. The rotations also result in high strain rates when the building is subjected to an earthquake shock (Zarghamee and Ojdrovic 1995).

Although all the special moment-resisting frames (SMRFs) were still standing after the earthquake, their resistance to future shocks was compromised by the connection damage. The potential for catastrophic failure of damaged buildings during future earthquakes prompted the profession to reevaluate the recommended connection details for such systems. Numerous professional societies and working groups are currently working to develop improved details (AISC 1994, SAC 1994). The improvements are aimed at reducing the concentration of stress at the beam-to-column joint and reducing the amount of field welding required. Tests have demonstrated that repair procedures that merely strengthen the weld but continue to produce high local stresses at the connection may not substantially improve performance under dynamic loads.

At the time of this writing (16 months after the earthquake), repair methods and guidelines for new construction are still under study. The lack of consensus among designers and researchers on the appropriate repair procedures is adding to the anxiety of building owners, who are facing the prospect of costly repairs with no guarantee of improved performance.

The Northridge earthquake is an excellent example of the continuing evolution of design practice as a result of trial-and-error experience. Reflecting on the earthquake, structural engineer W. Gene Corley noted: "We realize there are still a lot of things we don't know; the problem is staggering" (ENR 1994c). Another researcher and structural engineering professor at the University of California at San Diego, M. J. Nigel Priestley, commented: "Steel is a wonderfully ductile material—as long as you don't try to connect it" (ENR 1994b).

6.4 CORROSION

Steel exposed to moisture and oxygen will turn to rust (iron oxide) and be reduced continuously in section, causing a loss in structural integrity. Corrosion affects not only structural steel members, but all metal elements that are exposed to moisture, including the connectors used in timber structures (see Section 5.2). Metal connectors are especially at risk when in contact with chemically treated wood, as for fire-treated wood using waterborne salts. It is important that these connections are cautiously detailed to avoid moisture.

Corrosion-induced failures are common in masonry curtain walls, where metal anchorages and ties are used. Corrosion is also a serious problem in reinforced concrete and reinforced masonry structures. The reinforcing steel is an integral component in such structures and it must be protected from moisture. Prestressed concrete relies on the integrity of high-strength steel tendons. Corrosion of these tendons can produce sudden explosive failures. Corrosion problems involving steel used in concrete and masonry structures are reviewed in Chapters 7 to 9. Diagnosing and correcting corrosion problems is often very difficult; corrosion specialists may be necessary. Ill-conceived "corrective" work may actually exacerbate the condition.

Certain chemicals accelerate the corrosion process. Buckner, Huck, and Whittle (1987) discuss the interesting case of a large industrial bulk storage facility that deteriorated to the point of replacement in only six years of service. There are even some rare cases of corrosion from biological attack, known as microbiologically induced corrosion (MIC). Most of the cases in which MIC has been documented are in the oil, gas, and chemical processing industries, but there are examples involving encased structural columns in buildings (Scott and Davies 1992). The diagnosis and treatment of such conditions is similar to that used by the mycologists, who more typically address decay problems in timber structures.

In 1987, the Statue of Liberty restoration project was named that year's Outstanding Civil Engineering Achievement. This was recognition of the heroic efforts sometimes required to defend valuable metal structures against the deteriorating influence of the elements. Corrosion is an extremely expensive problem; mitigation techniques are costly. But the consequences of deficient maintenance are even more costly. Many steel failures occur annually due to the weakening effects of corrosion in aging facilities—failures that might have been prevented by diligent maintenance. Such failures can cause environmental damage (in the case of pipelines), can result in substantial replacement costs for bridges and other facilities, and can result in injuries and deaths to the users of bridges and the occupants of buildings.

Two of the fatal bridge failures discussed in Section 4.2.1 were at least partially caused by corrosion: the 1983 collapse of the Mianus River Bridge in Connecticut and the 1967 Silver Bridge failure in Point Pleasant, West Virginia. These two bridge incidents had a significant impact on enhancing awareness among public officials of the need for rigorous inspection and

maintenance of steel structures. Numerous bridges need emergency maintenance or replacement. This activity is expected to require large public expenditures on an ongoing basis.

The traditional method used to protect steel bridges from corrosion is to
encase the steel with an impervious coating. Approximately 80 percent of the
nation's existing steel bridges are coated with lead-based paint. Lead-based
coatings have been identified as environmental hazards. Containment and
disposal of lead-based paints is costly when they must be removed prior to
recoating. Some agencies are now specifying overcoating rather than removal,
to avoid the high costs associated with removal. The lead paint is not a hazard
if left on the bridge. Many new high-performance coatings are available for
protecting steel, and galvanizing continues to be an alternative method.

Corrosion is a serious concern for pipeline operators as well. Especially in
environmentally sensitive regions, a great deal of attention is focused on the
diagnosis and mitigation of corroded pipelines. One example is the ongoing
effort to maintain the 1400-km (900-mi) Trans-Alaska oil pipeline. A major leak
in this pipeline could cause irreparable damage to the fragile tundra ecosystem.

Since corrosion results in reduction of the cross section, those components
that have small cross sections to begin with are most susceptible to failure.
Such components include bolts and rivets used in connections, and cables
carrying high stress. These are the items that require the most careful attention
and maintenance.

When corroded materials are visible, the owners and operators of facilities
usually recognize the need for corrective action. Corrosion in hidden locations,
however, can cause progressive damage with disastrous consequences. In 1985,
corroded stainless steel hangers supporting the ceiling over a swimming pool
near Zurich, Switzerland, cracked and dropped the ceiling, causing 12 deaths
and four injuries. Corrosion was caused by the chlorine vapor from the disin-
fectant used in the water of the 13-year-old pool. The ceiling was a 90-mm
(3.5-in.)-thick cast-in-place reinforced concrete slab, weighing 145 Mg (160
tons). It fell suddenly, in one piece, covering the entire pool area. Rescuers
had to use jackhammers to get to the injured. The stainless steel hangers,
which suspended the ceiling below the structural framing, were 1-m (3.3-ft)-
long Z-shaped hangers 8 mm (0.32 in.) in diameter, placed on 2.1-m (7-ft)
centers. Hangers were overdesigned for the load. The designers' error was in
assuming that stainless steel was immune to corrosion. Two engineers and
the architect were found guilty of negligence and involuntary manslaughter
in the subsequent litigation and were sentenced to prison (ENR 1987). In such
corrosive, humid environments, the need for competent ventilation systems is
clear and the expertise of specialized metallurgists is advisable.

Frauenhoffer (1988) describes a similar situation involving open-web steel
joists and steel beams concealed by a suspended ceiling over a swimming pool
in Illinois. The space in which the structural framing was located was poorly
ventilated, and the entire interior space suffered from severe humidity. Despite
long-standing evidence of corrosion in the form of discoloration of the ceiling

and other finishes, the owner took no action other than to paint the discolored surfaces regularly. Finally, a maintenance contractor, recognizing the seriousness of the situation, refused to proceed with cosmetic repairs. When the ceiling was removed, it was discovered that the corroded bottom flanges of some of the beams had completely separated from the webs and were lying on top of the ceiling. Thus a life-threatening collapse was narrowly averted.

In 1982, excessive vibrations were noticed in the seating stands at the 60-year-old University of Illinois Memorial Stadium in Champaign, Illinois. The investigation that followed uncovered numerous instances of severe corrosion in structural steel and reinforced concrete structural members—corrosion that was concealed behind finish materials (Wilkinson and Coombe 1991).

As noted in Section 1.4.7, not all corrosion failures should be blamed on the owners of facilities. Although regular maintenance is the owner's responsibility, it is also necessary for designers to consider the maintainability of the structures they create and to communicate with their clients regarding maintenance requirements. This communication should include discussion of the consequences of neglect. Durability should be one of the principal design criteria, along with strength, stiffness, stability, economy, and aesthetics.

Certain types of structure and specific details require special consideration during the design phase. For example, much recent attention has been focused on the potential for premature deterioration of the cables in cable-stayed bridges. When appropriate precautions are taken, cable-stayed bridges perform very well in service. But the designer must be especially diligent when selecting and detailing materials and connections. Unlike the suspender cables in typical suspension bridges, stay cables are primary load-carrying elements. Failure of a single cable can affect the stability of the bridge, as was the case when a corroded cable in the 10-year-old Lake Maracaibo, Venezuela, bridge broke in 1979. The design should include provisions for replacing individual stay cables, if necessary, while the bridge is in service.

Designers must beware of new materials and systems that promise to eliminate the traditional problems inherent in construction. It is quite dangerous to assume that a new "miracle material" or protective coating is going to render unnecessary all the traditional details that resist corrosion. We have already noted the Zurich swimming pool ceiling collapse, where the assumption that stainless steel is immune to corrosion was a fatal error. Another example of designer idealism is the misuse of "corrosion-resistant steels," heavily promoted and widely used in the 1960s and 1970s by architects and engineers in buildings and bridges (Robison 1988).

"Weathering steel" was promoted as a maintenance-free material. It was formulated to rust just enough to protect itself from further rust penetration. No coatings were thought to be necessary, so the extra cost of this steel could be quickly recovered in reduced maintenance costs. Although the material has performed as expected in some bridges built in certain states where the climate is ideal, there have been a number of disappointing failures in other bridges and in nearly all architectural applications. It has been learned that to develop the protective oxide coating, sufficient wet–dry cycles are required. If there is too much water, or conversely, if there is too little moisture, the unprotected steel will just continue to corrode; in such environments weathering steel must be coated in the same manner as standard steel. In particular, problems have arisen at unprotected connections where ice-melting salts can collect in hidden locations that are not rinsed periodically by rainfall.

Because of the many failures associated with thin sheet applications of weathering steel in building facade panels and roofing, it is no longer recommended that this material be used in sheet thicknesses. This has virtually eliminated its use in building construction. A number of costly facade replacements in multistory buildings were necessary (Frauenhoffer 1987). At the present time, the use of weathering steel is limited to bridges where the climate is appropriate. Outdoor sculptures and other landscape elements also continue to be constructed in weathering steel, although caution must be exercised even in these applications. In 1991, a 24-m (80-ft)-tall weathering steel lamppost in a park in Fort Worth, Texas, fell and killed two people (ENR 1991). The base of the 860-mm (34-in.)-diameter post had been exposed to water leaking from a city water main. It was completely corroded just above the base plate, but the corroded section was concealed by plantings.

The New Haven, Connecticut, Coliseum parking structure shows what happens when creative designers place too much faith in a new "miracle material." The project was widely published and proclaimed a model for innovative design when it was completed in 1972. Eighteen years later, a \$20 million repair project was needed to completely replace 74,000 m^2 (800,000 ft^2) of composite concrete deck in the four-level parking structure, including installation of cathodic protection (ENR 1990). The structure was designed using weathering steel, and all details assumed that there would never be problems related to corrosion. The only tension reinforcement provided for the composite slabs was exposed weathering steel decking, which suffered from extreme corrosion. The floors were designed to be perfectly level and no floor drainage system was designed into the structure.

Parking garages are among the most vulnerable facilities for material degradation (Robison 1986). Open to the weather, they experience extreme temperature variations. In northern U.S. winters, vehicles constantly deposit ice-melting salts on the concrete decks. The decks are not periodically rinsed by rainfall as is the case for bridges, a provision that is now known to be critical for successful performance of this material. The lesson to be learned from the New Haven parking garage is that new materials should be used with care, especially in applications that subject them to extreme environmental conditions, until their properties are fully understood.

6.5 REFERENCES

AIA, 1981. *Toward Safer Long-Span Buildings,* Longspan Building Panel, American Institute of Architects, Washington, DC.

AISC, 1994. "Northridge Steel Update No. 1," American Institute of Steel Construction, Chicago.

Bell, G., and J. Parker, 1987. "Roof Collapse, Magic Mart Store, Bolivar, Tennessee," *Journal of Performance of Constructed Facilities,* American Society of Civil Engineers, New York (May).

Beri, P., 1993. "Cyclonic Damage to Structural Steel Skeleton During Erection," *Journal of Performance of Constructed Facilities,* American Society of Civil Engineers, New York (May).

Boehmig, R., 1990. "Shop Drawings: In Need of Respect," *Civil Engineering,* American Society of Civil Engineers, New York (March).

Buckner, C., R. Huck, and C. Whittle, 1987. "Deterioration and Replacement of Roof System for Bulk Storage Building," *Journal of Performance of Constructed Facilities,* American Society of Civil Engineers, New York (February).

Carper, K., 1987. "Structural Failures During Construction," *Journal of Performance of Constructed Facilities,* American Society of Civil Engineers, New York (August).

Civil Engineering, 1981. "Kemper Arena Roof Collapse and Repair," American Society of Civil Engineers, New York (March).

Dallaire, G., and R. Robison, 1983. "Structural Steel Details: Is Responsibility the Problem?" *Civil Engineering,* American Society of Civil Engineers, New York (October).

Engineering News-Record, 1981. McGraw-Hill, Inc., New York (August 20).

Engineering News-Record, 1984. McGraw-Hill, Inc., New York (July 26).

Engineering News-Record, 1985a. McGraw-Hill, Inc., New York (June 13).

Engineering News-Record, 1985b. McGraw-Hill, Inc., New York (November 21).

Engineering News-Record, 1985c. McGraw-Hill, Inc., New York (November 28).

Engineering News-Record, 1986. McGraw-Hill, Inc., New York (January 30).

Engineering News-Record, 1987. McGraw-Hill, Inc., New York (February 26).

Engineering News-Record, 1990. McGraw-Hill, Inc., New York (June 7).

Engineering News-Record, 1991. McGraw-Hill, Inc., New York (April 8).

Engineering News-Record, 1993. McGraw-Hill, Inc., New York (November 15).

Engineering News-Record, 1994a. McGraw-Hill, Inc., New York (March 14).

Engineering News-Record, 1994b. McGraw-Hill, Inc., New York (April 4).

Engineering News-Record, 1994c. McGraw-Hill, Inc., New York (July 25).

Engineering News-Record, 1994d. McGraw-Hill, Inc., New York (September 19).

Fisher, J., 1984. *Fatigue and Fracture in Steel Bridges,* John Wiley & Sons, Inc., New York.

Frauenhoffer, J., 1987. "Weathering Steel Cladding Failure," *Journal of Performance of Constructed Facilities,* American Society of Civil Engineers, New York (May).

Frauenhoffer, J., 1988. "Natatorium Reconstruction," *Journal of Performance of Constructed Facilities,* American Society of Civil Engineers, New York (August).

Galambos, T., ed., 1987. *Guide to Stability Design Criteria for Metal Structures,* 4th edition, John Wiley & Sons, Inc., New York.

Gasparini D., and M. Fields, 1993. "Collapse of Ashtabula Bridge on December 29, 1876," *Journal of Performance of Constructed Facilities,* American Society of Civil Engineers, New York (May).

Gross, J., J. Smith, and R. Wright, 1989. "Ashland Tank-Collapse Investigation," *Journal of Performance of Constructed Facilities,* American Society of Civil Engineers, New York (August). (See also Discussion by G. Derbalian in May 1991 issue.)

Gurfinkel, G., 1988. "Large Steel Tanks: Brittle Fracture and Repair," *Journal of Performance of Constructed Facilities,* American Society of Civil Engineers, New York (February).

Gurfinkel, G., 1989. "Brittle Fracture and Collapse of Large Grain-Storage Tank," *Journal of Performance of Constructed Facilities,* American Society of Civil Engineers, New York (August).

Jones, C., and N. Nathan, 1990. "Supermarket Roof Collapse in Burnaby, British Columbia, Canada," *Journal of Performance of Constructed Facilities,* American Society of Civil Engineers, New York (August). (See also Discussion by B. Baer in February 1992 issue.)

Kaminetzky, D., 1991. *Design and Construction Failures,* McGraw-Hill, Inc., New York.

Levy, M., and M. Salvadori, 1992. *Why Buildings Fall Down,* W.W. Norton & Company, Inc., New York.

Lichtenstein, A., 1993. "The Silver Bridge Collapse Recounted," *Journal of Performance of Constructed Facilities,* American Society of Civil Engineers, New York (November).

Loomis, R. S., R. H. Loomis, R. W. Loomis, and R. W. Loomis, 1980. "Torsional Buckling Study of Hartford Coliseum," *Journal of the Structural Division,* American Society of Civil Engineers, New York (January).

LZA, 1978. *Report of the Engineering Investigation Concerning the Causes of the Collapse of the Hartford Coliseum Space Truss Roof of January 18, 1978,* Lev Zetlin Associates, Inc., New York.

Moncarz, P., R. Hooley, J. Osteraas, and B. Lahnert, 1992. "Analysis of Stability of L'Ambiance Plaza Lift-Slab Towers," *Journal of Performance of Constructed Facilities,* American Society of Civil Engineers, New York (November).

NBS, 1982. *Investigation of the Kansas City Hyatt Regency Walkways Collapse,* NBSIR 82-2465, National Bureau of Standards (National Institute of Standards and Technology), U.S. Department of Commerce, Washington, DC.

Newman, A., 1994. "Debating Steel-Connection Design," *Civil Engineering,* American Society of Civil Engineers, New York (February).

Pfrang, E., and R. Marshall, 1982. "Collapse of the Kansas City Hyatt Regency Walkways," *Civil Engineering,* American Society of Civil Engineers, New York (July).

Puri, S., 1994. "Trapped in an Elevator During the World Trade Center Bombing: A Personal Account," *Journal of Performance of Constructed Facilities,* American Society of Civil Engineers, New York (November).

Ramabhushanam, E., and M. Lynch, 1994. "Structural Assessment of Bomb Damage for World Trade Center," *Journal of Performance of Constructed Facilities,* American Society of Civil Engineers, New York (November).

Robison, R., 1984. "Structural Steel Details: Comments on Divided Responsibility," *Civil Engineering,* American Society of Civil Engineers, New York (March).

Robison, R., 1986. "The Parking Problem," *Civil Engineering,* American Society of Civil Engineers, New York (March).

Robison, R., 1988. "Weathering Steel: Industry's Stepchild," *Civil Engineering,* American Society of Civil Engineers, New York (October).

Ross, S., 1984. *Construction Disasters: Design Failures, Causes, and Prevention,* Engineering News-Record, McGraw-Hill, Inc., New York.

Rubin, R., and L. Banick, 1987. "The Hyatt Regency Decision: One View," *Journal of Performance of Constructed Facilities,* American Society of Civil Engineers, New York (August).

Rubin, R., and M. Ressler, 1985. "Shop Drawing Review: Minimizing the Risks," *Civil Engineering,* American Society of Civil Engineers, New York (March).

SAC, 1994. "Program to Reduce Earthquake Hazards in Steel Moment Frame Structures," The SAC Joint Venture Partnership, Structural Engineers Association of California, Sacramento, CA.

Scott, P., and M. Davies, 1992. "Microbiologically Induced Corrosion," *Civil Engineering,* American Society of Civil Engineers, New York (May).

Smith, E., and H. Epstein, 1980. "Hartford Coliseum Roof Collapse: Structural Collapse Sequence and Lessons Learned," *Civil Engineering,* American Society of Civil Engineers, New York (April).

Sputo, T., and D. Ellifritt, 1991. "Collapse of Metal Building System During Erection," *Journal of Performance of Constructed Facilities,* American Society of Civil Engineers, New York (November).

Stratta, J., 1979. *Report of the Kemper Arena Roof Collapse of June 4, 1979, Kansas City, Missouri,* J. Stratta, Consulting Engineer, Menlo Park, CA.

Wilkinson, E., and J. Coombe, 1991. "University of Illinois Memorial Stadium: Investigation and Rehabilitation," *Journal of Performance of Constructed Facilities,* American Society of Civil Engineers, New York (February).

Zarghamee, M., and R. Ojdrovic, 1995. "Northridge Postscript: Lessons on Steel Connections," *Civil Engineering,* American Society of Civil Engineers, New York (April).

Zwerneman, F., A. West, and K. Lim, 1993. "Fatigue Damage to Steel Bridge Diaphragms," *Journal of Performance of Constructed Facilities,* American Society of Civil Engineers, New York (November).

7

REINFORCED CONCRETE STRUCTURES (CAST-IN-PLACE)

A number of examples of failures involving reinforced concrete structures are presented throughout this book. Chapter 1 includes examples of design and construction errors, and Chapter 2 contains several cases related to natural disasters and aging facilities. There are many examples of concrete foundation failures in Chapter 3 resulting from the widespread use of reinforced concrete in below-grade applications. Chapter 4 includes failures of concrete bridges and dams. In Chapter 11, numerous examples of construction accidents involving cast-in-place concrete structures are detailed, most of these related to premature removal of formwork or collapse of systems providing temporary support to the incomplete structure. In this chapter we provide additional examples and discussion unique to cast-in-place reinforced concrete structures.

7.1 REINFORCED CONCRETE AS A STRUCTURAL MATERIAL

Concrete is an artificial stone containing sand and stone aggregate bonded together through a chemical reaction, using a paste of cement and water. Also present in the mix may be air-entraining agents and other admixtures intended to improve workability and durability or to control curing time.

Natural stones are weak in tension and shear, and they tend to be brittle. These are also the underlying deficiencies in unreinforced concrete. The principal purposes of introducing reinforcing steel into the concrete are to give the artificial stone some capacity to resist tensile and shear stresses and to provide the needed ductility in unusual load conditions such as large seismic events. In some cases, as in reinforced concrete columns and doubly reinforced beams, the reinforcement also increases the compressive resistance of the structural member.

Because of the many ingredients that go into the making of this material and the procedures used in forming it at the construction site, the factors that can lead to failure or deficient performance are complex and interrelated. In the early years, concrete was hand-mixed at the construction site. This is still done in many developing countries and on smaller projects where equipment access is difficult. The reliability and uniformity of the material has greatly improved with the increasing availability of quality-controlled concrete supplied and delivered to the site by mixing companies. Most of the problems now involve human error at the construction site, and most of these errors are related to formwork and temporary supporting structures. These particular problems are discussed in detail in Section 1.4.5 and in Chapter 11.

7.1.1 Advantages of Reinforced Concrete

As with all construction materials, reinforced concrete has some properties that are advantageous as well as some undesirable characteristics. Among the desirable qualities are:

a. *High Compression Strength.* The compression capacity of concrete is generally reliable and predictable. Considerable advances in compression strength of concretes have occurred in recent years, using various admixtures that increase the plasticity of the mix without adding water.

b. *Good Resistance to Fire and Water.* The fire resistance of well-detailed concrete structures and the durability of such structures in marine environments are well known. Reinforced concrete structures are not completely immune from the effects of fire and water, however. The reinforcing steel is an integral component of the material. In Chapter 6 the problems of corrosion and fire were detailed for steel structural elements. It is particularly important to provide adequate and durable concrete cover for the reinforcing steel if the structure is to survive exposure to fire or water.

c. *Rigidity.* Since reinforced concrete does not have as high a strength/volume ratio as steel, reinforced concrete structural members tend to be more bulky than steel members. Thus there are fewer problems related to stability, unwanted deformations, and vibrations than is the case for typical steel-framed structures.

d. *Low Maintenance Requirements.* Properly detailed and competently constructed reinforced concrete structures generally require less maintenance than do facilities built with other materials. There are exceptions, of course, and one recurring source of problems with reinforced concrete structures is the tendency of owners and operators of such facilities to treat them as permanent, passive objects requiring no maintenance. However, if properly cared for, concrete structures can provide a very long and useful life. In fact, many of the structural properties of the material actually improve with time.

e. *Economical Material for Below-Grade Structures.* Because of its good resistance to the effects of moisture, reinforced concrete is nearly always used for retaining walls and for the foundations of structures, no matter what the choice of structural materials for the superstructure. Thus reinforced concrete elements are common on nearly every construction project.

f. *Architectural Potential.* Cast-in-place concrete can be used for an extraordinary variety of structural elements of nearly any imaginable size or shape. These range from linear elements, such as beams, columns, and arches of any dimension, to large surface structural elements, such as flat slabs, folded plates, and thin shells of single and double curvature. The versatility and potential for architectural expression of this "plastic" material is unlimited.

g. *Use of Inexpensive Locally Available Materials.* Except for the reinforcing steel and the cement, all the materials used in concrete construction are locally available throughout the world. Concrete may therefore be the only material available for construction of large structures in some of the economically disadvantaged developing nations. The quality of locally available materials may be a problem, however, as discussed in the following section.

h. *Inexpensive Labor.* Many of the tasks associated with the mixing and placement of concrete can be accomplished by labor having less skill and training than that typically required for steel structures. There is always the need for supervision and direction of such labor by persons who are trained and skilled in use of the material, but the construction of concrete structures may employ large numbers of minimally educated workers. In the developing countries there is a large workforce of persons who are minimally skilled and in desperate need of employment. Concrete is an ideal material for this workforce.

7.1.2 Undesirable Characteristics of Reinforced Concrete

The unique disadvantages and difficulties of working with reinforced concrete are illustrated by many of the case studies in this book. Understanding the undesirable characteristics of materials is often the key to designing and constructing successful and durable facilities. Some of the unique problems associated with ordinary reinforced concrete are:

a. *Composite Material Complexities.* Since the structural member is composed of a variety of materials, the interaction of these materials is difficult to control and to predict accurately. In particular, the sharing of stresses between steel and concrete, and the differential response of the various materials to temperature variations, are difficult to quantify. For designers to have confidence in their theories, much research has been required. The bonding together of all materials is integral to the satisfactory performance of the composite material. When the bond between the reinforcing bars and the concrete is lost, reinforced concrete ceases to exist.

b. *Low Tension and Shear Strength Without Reinforcement.* The reinforcing steel is intended to assist in overcoming this fundamental problem, but even properly reinforced concrete structures will experience some cracking due to the inability of concrete to resist the tension stresses present in structural members. Cracking in reinforced concrete structures introduces moisture problems and affects adversely the durability of facilities, even when structural performance is not immediately affected.

c. *Forms, Falsework, and Shoring Are Required.* Since concrete does not develop its design strength for some time and is incapable of supporting its own weight for a time, the construction of cast-in-place structures must rely on temporary supporting structures and on formwork to contain and mold the concrete into the desired shape. The cost of formwork and temporary support is often the greatest cost of a concrete structure, even though it has no place in the completed structure. Also, the vast majority of construction accidents associated with concrete construction are related to premature removal of forms and collapse of inadequate temporary support structures.

d. *Low Strength per Unit Weight.* When compared with steel, the lower strength/weight ratio means that a concrete structure will weigh more than a typical steel frame. This will impose greater loads on the foundations and increase the inertial forces in seismic events. Heavier structural elements also result in greater life-threatening loads when they do fail, as was tragically illustrated in the 1985 Mexico City earthquake.

e. *Low Strength per Unit Volume.* Although this characteristic results in more rigid structural elements than are typical for steel structures, and fewer problems with stability and vibrations, the larger, bulkier structural elements in concrete buildings will occupy greater volume. This means that floor-to-floor heights may be greater and columns will occupy more floor space than is the case for a typical steel-frame building.

f. *Wide Variability.* The properties of reinforced concrete are strongly influenced by a variety of interrelated factors that are quite difficult to control and predict. The characteristics of local sands and aggregates used in the mix have a significant impact on the structural properties and durability of the product. The quality of workmanship is also immensely important, and the common use of moderately skilled labor can result in wide variability of results even when the constituent ingredients are carefully controlled. The need for quality supervision and field inspection is more evident with this material than for many others. Minimally trained workers generally do not understand even such basic concepts as how the water/cement ratio influences the eventual quality of the concrete. Nor do they understand the necessity for proper vibration so that voids are not left in the forms. The literature is filled with examples of ignorantly constructed shoring and formwork, incapable of resisting the horizontal dynamic loads associated with the construction process—unbraced and out-of-plumb shoring that collapses progressively and totally with the application of a small horizontal impact load. There are examples of wall and column formwork exploding from unanticipated hydrostatic pressure. Thus the use of unskilled labor, although economically advantageous, often affects not only the quality of the finished product but even the safety of the workers. The properties of the product vary from place to place and even from one portion of a project to another on the same construction site. Temperature and humidity during the curing process also influence the characteristics of the concrete. Weather is impossible to control, but its effects can be anticipated and provisions can be made to minimize its detrimental effects. However, the variabilities of weather are often the source of unpredicted deficiencies in cast-in-place concrete structures.

g. *Dimensional Stability Problems.* The shrinkage and creep that take place in cast-in-place structures can contribute to long-term problems, particularly in long-span structures or structures comprising large uninterrupted surfaces. Accommodating the movements associated with these changes is a difficult design challenge. Because of dimension changes in the material over the life of a structure, it is difficult for the designer to predict long-term deformations in structural members with the same degree of accuracy as is present in the design of steel structures.

h. *Lack of Ductility.* Without reinforcement, or when the reinforcement is not carefully detailed, concrete structural elements are brittle. They have very little capacity to absorb energy prior to fracture. Unusual loads, particularly earthquakes, have time and time again illustrated the explosive life-threatening hazards associated with nonductile concrete buildings. Much research has been devoted to solving this problem, and the resulting details have performed very well in recent earthquakes, but this deficiency has been corrected in building codes only recently. Concrete buildings and highway structures built before the mid-1970s are expected to be among the most hazardous facilities in future earthquakes.

7.1.3 Evolution of Concrete Design and Construction Standards: Early Failure Examples

Concrete as a structural material differs from all others because it does not come entirely ready-made to the site and is quite prone to misbehaving if any one of the many controls is either disregarded or improperly implemented. Compared to masonry and timber, reinforced concrete is a relatively new construction material, and, like steel, has had to go through a period of empirical development to prove its feasibility. Despite intensive research in the field of concrete structures, many of the succession of new developments must depend on actual performance, and such trial-and-error procedures naturally have their share of failures. Regulation of concrete design procedures came from a series of spectacular collapses. The designer's full freedom of action, which as a "master builder" was considered a right and privilege, was diminished by regulations.

One such case was the failure of the reinforced concrete viaduct built by the Celestial Globe Company in 1900 over the Avenue Suffren in Paris so that visitors could visit the attraction without leaving the enclosure of the Exposition grounds. The deck was 114 m (375 ft) long, 5 m (16.5 ft) wide, and some 5 m (15 ft) above the street. It consisted of 9-m (30-ft) spans with concrete girders on 300-mm (12-in.)-square concrete columns. The two-way reinforced concrete slab was 150 mm (6 in.) thick. It was supported on light steel beams resting on the girders. Longitudinal stressed cables bounded the walkway as hand railings. Originally designed as a straight line, two axial deviations were introduced to avoid removal of some trees. The resulting oblique forces practically kicked the deck off the piers, and when the bridge fell, a number of people were trapped beneath it. The Board of Experts and the Court pronounced the consulting engineer and the contractor guilty of "the most culpable negligence."

Hennebique designed a five-story hotel for Basel, Switzerland, that failed on August 21, 1901. The city building department had given the designer full control of the work. The commission of investigation found that the columns had been designed for full allowable axial compression stress in both steel and concrete, so that in view of the possible bending and eccentric loadings, there was only a small factor of safety. Also, the steel in the columns was not continuous but ran only from story to story. Although this was a serious deficiency, the actual failure cause was attributed to the cellar masonry piers, which were designed but had not yet been built. Temporary timber bents were provided while the concrete superstructure was being built, and these were apparently partially removed to permit construction of the masonry piers, leaving the columns above unsupported. The experts listed this as the chief cause but also as additional causes: (1) insufficient dimensions of columns in the first story, together with lack of control of the engineer-contractors over the construction; (2) use of improper material (unwashed sand and gravel); (3) careless construction of the concrete work (bottoms of the columns were less dense than the rest); (4) lack of tests of the cement and concrete; (5) lack of organization among the various subcontractors; and (6) haste in construction. This report could be used without a single change to explain similar, more recent incidents.

Deficiency in the basic design, such as amount of reinforcement at points of maximum bending moment or dimensions of concrete section to provide compressive and shear resistances under normal loadings, is an extremely rare cause of failure. One case that was caught just before placing concrete in roof beams spanning 19.5 m (64 ft) over a school auditorium in Yonkers, New York, in 1925 was described in Section 1.4.4.

With growing experience in all types of reinforced concrete designs becoming common in all the controls through which a job passes—design, estimating, detailing, field supervision, and construction—the possibility of a gross error in design actually getting into the construction stage today is indeed remote. Yet some examples cited elsewhere in the book indicate that the possibility still exists. The complex organizational schemes under which some contemporary construction projects are delivered may lead to compromises in the cross-checking of critical design decisions. Such was the case in the highly publicized Stamford, Connecticut, train station project, which was halted after 85 percent completion in 1984 due to substantial deficiencies in the design. Subsequent redesign added over $11 million to the original $15 million construction budget.

In 1993, cracks occurred in vertical members of the Vierendeel trusses for the Miami, Florida, Metropolitan Detention Center. The cause was too few shear stirrups shown in the design drawings, despite proper calculations. The project was still under construction when the cracks appeared.

These examples show that reinforced concrete designs need more than a numerical check; an independent examination of the design and connection details by an experienced engineer or constructor may expose gross design errors. It is often the small items of underdesign in the category of secondary stresses from volume or shape change that are the troublesome difficulties, as was the case for structural steel design. An arbitrary assumption of the location of points of inflection in continuous beams and frames, a procedure quite popular for many years, often resulted in cracks. The structures refused to accept the assumption when it was not consistent with actual relative stiffnesses of the contiguous members. The moment-distribution analytical method introduced by Hardy Cross in 1930, with its simplified mathematical computations, reduced the number of such failures.

Although reinforced concrete design theories have been refined, human errors in the design and construction process persist. Errors show up from time to time in the interpretation and communication of design calculations. For simplicity in detailing and reinforcing bar installation, it is often expedient to use the same size and spacing of bars for a roof slab as for the typical floor of a multistory building, even though floor loads are usually greater. This tendency may explain why in a New York school building in 1953, the main reinforcement over an interior support of the floors was incorrectly copied from the design notes, and the much lighter reinforcement of the roof design was shown and used. Although this was only one of several noncompliances discovered in the investigation, it may have had a considerable influence on the resulting large slab deflections found after the forms were removed. As a correction, a 100-mm (4-in.)-thick cap slab was introduced after shoring the deflected slab and wedging it upward. A high-strength steel-wire mesh welded to studs was shot into the concrete over part of the deflected span and the complete (shorter) adjacent span. A considerable portion of the unwanted deflection was eliminated and a full-scale load test showed good composite action of the unusual poststressing procedure.

Errors can occur as a result of the large number of construction trades that work on a cast-in-place concrete building, even when construction supervision is present. Omission of designed reinforcement is not a common cause of failure, but it can exhibit itself as an otherwise unexplainable deflection. In a three-story concrete waffle slab classroom building at a military installation in Virginia, the ceilings showed large deflections in every room. The first explanation was that students were marching in step to class, but when this policy was discontinued, the deflections still increased and the building was vacated.

Electromagnetic nondestructive probing surveys indicated a lack of reinforcing steel bars in the center ribs in each ceiling area, and when the concrete was cut away, no bars were found. Since the construction had been performed under close supervision and every form had been inspected and approved, the omission of the steel could not be explained. To make the building usable, a complete frame of steel beams and columns was incorporated into the structure as a support for the concrete slabs.

Further inquiry finally resulted in the true explanation. The reinforcing bars had been delivered to the site and placed, two in the bottom of each rib form. Inspection was made and recorded. The electrician then came on the deck to install lighting fixture outlet boxes and conduit. The boxes could be placed in the bottom of the ribs only by removing the bars, which he did. Not being permitted to place steel, as that was the task of the iron workers, he left the bars on top of the pans used to form the waffles. The electrical work was then inspected and the concreting done under the guidance of another inspector. The "surplus" bars were thrown off the deck before placing the concrete on the pans. The same thing happened at every panel on each of the three floors and the roof—perhaps a case of too much inspection by the book without any knowledge of what was actually being built.

7.2 FORMWORK PROBLEMS AND FAILURE OF TEMPORARY STRUCTURES

The vast majority of failures of cast-in-place concrete construction occur during construction and are related to collapse of temporary supporting structures or premature removal of formwork. Such failures often involve serious injuries to workers or even deaths. The dramatic 1970 failure of the 17-story Commonwealth Avenue building in Boston, Massachusetts, killed four workers. Fifty-six lost their lives in the Willow Island, West Virginia, cooling tower collapse in 1978. In 1982, 13 construction workers were killed in the collapse of the temporary structure supporting the Riley Road Interchange Freeway Ramp in East Chicago, Indiana. All of these are discussed in Section 11.5. The loading of concrete before it has reached adequate strength usually results from the desire to reuse forms quickly, but sometimes the concrete set has been delayed by uncontrolled factors, such as cold temperatures.

One additional case study involving premature removal of shoring is presented here: the dramatic progressive collapse in March 1973 of the Skyline Plaza apartment building at Bailey's Crossroads in Fairfax County, Virginia

(Figure 7.1). The 26-story building was part of a $200 million residential–commercial complex and was one of several identical apartment buildings, two of which had already been completed. The towers were 118 by 23 m (386 by 76 ft) in plan and used 203-mm (8-in.)-thick cast-in-place flat plate floor slabs (NBS 1977).

Construction of the building was proceeding at a rapid rate; the schedule was one floor slab per week. Concrete was just being placed on the twenty-fourth floor while shoring was being removed from recently placed concrete at the twenty-second-floor level. The weight of debris from the failed concrete of the twenty-second through twenty-fourth floors and the weight of one of two climbing cranes on the twenty-fourth floor fell all the way to the ground, in a progressive failure that tore a 20-m (60-ft)-wide gap through the building from top to bottom. After the failure, the building was left standing, but in two separate sections. The failure also collapsed the completed portion of a post-tensioned parking garage under construction beside the building.

The collapse killed 14 workers and injured 35 others. All investigators concluded that the concrete simply had not attained sufficient strength to carry the construction loads imposed on it. The most probable triggering cause was punching shear in several columns at the twenty-second-floor level. The design was not deficient; the shear stresses would have been acceptable had the concrete attained the prescribed strength prior to removal of shoring.

This failure became a very important case in the ongoing debate about the responsibilities of structural engineers and architects on the construction site. A federal jury found the architect and the consulting engineer guilty of negligence, and large damages were awarded to the plaintiffs (ENR 1975). The general contractor and building owner were not named in the suit because the plaintiffs had already received awards under the Virginia Workers' Compensation Act, which shielded them from further legal action. Thus, while the failure was clearly the result of mismanaged construction procedures, the court held the designers responsible. The jury agreed with the plaintiffs that the relevant building code required the designers to visit the job site, inspect the work, and warn the contractors regarding any unsafe conditions observed. The designers were held responsible even though the architect's and engineer's specifications for required shoring were not being complied with at the site.

The Skyline Plaza collapse is an important example of the need to improve the structural integrity and redundancy of structures so that progressive collapse is less likely. Construction loads are often in excess of the design loads anticipated for the completed structure. This implies that progressive collapse mitigation strategies must be implemented during the construction phase, especially with regard to the design of structures providing temporary support.

7.3 QUALITY ASSURANCE OR QUALITY CONTROL PROBLEMS

The properties of reinforced concrete are highly dependent on the degree of quality control exercised during construction. Control over the characteristics

(a)

Figure 7.1 Skyline Plaza collapse, Fairfax County, Virginia, 1973. (Courtesy of the National Bureau of Standards.)

(b)

Figure 7.1 *(Continued)*

of the constituent elements in the mix (aggregates, water/cement ratio, and admixtures) and in the proper placement and consolidation of the concrete in the forms is important. The mix must be of good quality, and proper curing time and conditions must be provided for successful results. Contaminants in the mix will not only affect strength but will also cause durability problems. Proper consolidation may require vibration. The use of mechanical vibrators to produce dense concrete without segregation requires an experienced equipment operator.

Special problems with concrete construction in the developing countries have been noted, where lack of skill among workers and poor quality control during the mixing and placing operations are common (Hadipriono and Sierakowski 1988). However, these problems are not unique to the developing countries; many examples of lapses in quality control can be found in the technically advanced countries as well.

One recurring source of problems is the tendency among designers to specify only the expected strength of the cured concrete. Although strength is an important characteristic of the finished product, the long-term performance of the concrete is also affected by a number of other factors. Durability cannot be predicted on the basis of strength specifications alone. Durability is defined by the American Concrete Institute as the ability to resist weathering, chemical attack, abrasion, or any other process of deterioration (ACI 1982). Durability is affected by strength, but also by density, permeability, air entrainment, dimensional stability, characteristics and proportions of constituent materials, and construction quality. Durability is harmed by freezing and thawing, sulfate attack, corrosion of reinforcing steel, and reactions between the various constituents of the cements and aggregates. Many of the costly problems related to concrete durability can be reduced by careful selection of concrete ingredients, based on the particular exposure conditions for the completed facility.

The past few years have seen the introduction of a considerable array of new admixtures for improving the density, corrosion resistance, and durability of reinforced concrete. Thus proper considerations for the mix recipe have become quite complex. At the same time, significant advances in concrete compression strength have been achieved with new superplasticizing admixtures. These new high-strength concretes tend to be more brittle. Proper selection of aggregates and admixtures is critical to improving toughness and durability (Berry and Malhotra 1980). The extremely sensitive effect of very small percentages of some admixtures must be considered carefully before permitting their use. Such analysis must keep in mind that accurate control of all other ingredients in a concrete mix is desirable but seldom available.

Improper aggregates in a concrete mix can be the cause of much trouble in appearance and even use of a facility. The use of improper aggregates can accelerate the deterioration of pavements, causing costly repairs to airport runways. Aggregates used in a concrete mix must be compatible with the other ingredients. Proper and consistent grading of aggregate size is another important factor. Consistency in aggregate size and distribution helps with finishing and improves durability.

Alkali-reactive silica contained in some aggregates can cause expansion, cracking, and deterioration. An alkali aggregate reaction occurs when a combination of certain aggregates and cements react to form an expanding gel in the hardened concrete. Eventually, the concrete bursts from the internal pressure.

In 1947–1948 an investigation was made of abnormal cracking in concrete highway structures in Georgia and Alabama. An analysis of the concrete records, compared with actual performance and checked by extensive laboratory tests of aggregates and cements, showed that the defects seemed limited to a concrete mix in which natural siliceous aggregates from Montgomery, Alabama, were used with three cements having alkali content of more than 0.6 percent. Investigation covered the history of 294 bridge structures, all at least five years old. Cracking from the alkali–silica reaction seldom starts at a younger age. Defects were observed in piers, pier caps, wing walls, railings, and curbs. The work was corrected by adding concrete covering after removal of the cracked and loose concrete. On less-affected surfaces, paint sealers were used. All future work was performed under a specification restricting alkalies to 0.6 percent in cements, which has been effective in reducing most failures of this type. However, low-alkali cements may not completely eliminate the problem because alkali-bearing aggregates are sometimes found in the mix.

Steel blast-furnace slags used as aggregate can also cause a damaging chemical reaction. This effect was responsible for runway pavement deterioration after three years of use at the Tampa, Florida, International Airport in 1983. Strict quality-control procedures have been established in the United States to prevent alkali aggregate reaction and similar chemical reactions. Aggregate coming from rocks that have expansive qualities, usually classed as dolomites, can be responsible for damaging effects. Examples have been found in the southern Appalachian area and in the vicinity of Kingston, Ontario, Canada. The American Society for Testing and Materials (ASTM) publishes standards for aggregate suppliers, but the possibility of contaminated aggregate still exists.

In 1980, over 130 buildings in the San Francisco Bay area were found to have serious structural defects caused by poor-quality aggregate. Structural members experienced severe spalling, which was traced to expansive brick aggregate. Several tons of the brick had been accidentally dumped onto an aggregate pile at a cement products plant.

What is to be done when a building has been designed and constructed on the assumption that the concrete will test at a certain strength in 28 days and the required specification is not met? In the writers' experience, such events have led to embarrassing and expensive delays. Often, the "conservative" designer's conservatism is valued highly at this point, when tests and reexamination of the design shows that the lower-strength concrete will still be satisfactory. However, a little more integrity in the furnishing and curing of concrete with the proper aggregates, water/cement ratios, and other ingredients is a better way to avoid such embarrassing and costly problems.

Improper aggregates have caused orders to demolish concrete already placed because of low indicated strengths. In 1963, 28-day cylinder tests of the concrete for Public School 90 in Brooklyn, New York, indicated 40 percent below the 24 MPa (3500 psi) specified design strength. When the city ordered all of the concrete removed (approximately one-third of the total required for the building) the contractor asked for determination of the strength on the basis of the New York City Building Code. This entailed core tests to check the cylinder tests, analysis of the design to see if the indicated concrete strengths were sufficient for the intended loadings, and static proof load tests using 150 percent of the design live load. Some 1300 m^2 (14,000 ft^2) of floor area was involved and the concrete was a lightweight mix, the first use of such aggregate in New York City schools.

Much of the concreting had been performed in cold weather. An inspection at the mixing plant, correlated with analysis of the concrete cores, indicated that the aggregate had been piled without frost protection, contained much ice, and was completely uncontrolled with regard to separation of aggregate size. The core tests confirmed a wide variation in concrete strength. Interestingly, low results were not found in some of the suspected areas and were found in areas where the earlier cylinder tests had indicated sufficient strength. A program was undertaken of 26 proof load tests in the areas where low-strength concrete was suspected. Tests were carefully monitored, and every test passed the requirements of the code. Technically, this should have been sufficient for approval of the completed work, but the Board of Education refused to accept the results as conclusive. Instead, the board insisted on the removal of about half of the originally rejected areas. Reconstruction was with a modified concrete mix design. Aggregate gradation and temperature were more carefully controlled, and the project was completed after a nine-month delay.

The rate of concrete strength gain depends on the temperatures of the mix as well as of the outer air. Frost effects are very important, as are conditions in extremely hot climates, especially with low humidity. Cold weather, even if not freezing, can delay the chemical action that sets the concrete. Temperature is an important factor in determining how long the forms and temporary support must remain in place.

Concrete can be safely installed in freezing weather if precautions and protection equipment are prepared and available before work is started. Witness the great number of successful winter operations in Canada, where contractors have learned to live with the cold weather and protect against it rather than hope for the thaw that does not come when needed.

There have been some serious concrete failures caused by lack of proper protection from cold temperatures. The collapse of a concrete bath house at Atlantic City, New Jersey, in March 1906 was attributed almost entirely to frost penetration into the wet concrete, resulting in separations between beams and columns and between slabs and girders.

A most spectacular frost-damage collapse occurred when the seven stories of a proposed eight-story concrete hotel building at Benton Harbor, Michigan, collapsed on January 28, 1924. The building simply melted away over a 30-hour period. Most of the concrete in the debris could be hand-shoveled away; only a few of the lower story columns and first floor beams remained intact. Concreting was completed in December and January at a rapid speed, with very few concessions to the cold weather. Although no one was injured in this collapse, every detail of specification compliance was checked. As far as could be determined, the ingredients and the mix were correct, the aggregates were thawed before mixing, and the water, in which some calcium chloride was dissolved, was heated. Protective tarpaulins were available for only one story at a time. Temperature records showed some below-freezing days in December and almost no day without freezing in January. Compressive strengths of five samples taken from the debris ran from 1.7 to 4.95 MPa (247 to 718 psi), not enough to carry the dead weight.

In 1925 the four-story concrete John Evans Hotel, Evanston, Illinois, was shored completely to the roof and collapsed when the cellar shores were removed. The concrete mix was of poor quality (dirty sand from the excavation, containing clay and silt, had been used) but failure was caused primarily by the lack of expected strength in the frozen concrete. In 1927, in Buffalo, New York, a 12-story steel frame apartment with concrete floor slabs suffered a partial collapse when the shores were removed. Several spans had frozen. All the concrete below the ninth floor fell to the basement. The ninth-floor slab was properly set and remained in place.

One recurring type of frost damage results from the practice of placing wooden mud-sills on frozen ground. The ground softens from the drip and the frost protection heaters, causing unwanted sags in the supported structure. An example of a less common type of frost damage occurred in a substructure contract where completed piers with embedded anchor bolts were left exposed during the winter. The bolts had been placed in oversized sleeves to permit some adjustments for fitting with the steel base plates. Water filled the sleeves, froze, and cracked off the edges of many piers. Such open sleeves, even if only for future handrailings, should be covered or filled with nonfreezing materials.

The formation of growing ice crystals, exerting extremely large pressures, is a cause of many other problems. The continually expanding cracks in exposures to water during below-freezing temperatures result in progressive surface cracking and spalling of roads and walls. Frost penetration of the subgrade on which concrete footings and slabs are placed will heave the footings and slabs if moisture is available to form ice crystals. Spring thaws will then bring settlement of the structure.

An interesting situation of this nature occurred in Watertown, New York, following a cold winter. In the fall, footings and piers for an apartment house were completed up to grade and the work was backfilled and left idle over the winter. With the next construction season the timber sills and framing were set on the piers and work was well advanced when considerable distortion was observed. A study of conditions showed that frost action in the fills around the piers had lifted the undoweled piers off the footings, some by as much as 300 mm (12 in.), and had held them up during the construction work until the ground thawed. Repair work required the exposure of all piers. They were each grouted solid to the footings and the entire floor construction was releveled.

The advisability of providing an impervious surface on concrete having exposures to water is often seen in dams and bridges as well as buildings. A typical example of the havoc raised by a series of freezing and thawing cycles during a long winter was experienced in 1959 by the Indian Lake dam near Brussels Point, Ohio. The entire spillway surface required replacement after chipping away all the loose concrete.

Finishes on slabs placed in cool weather where frost occurs before the concrete is set are vulnerable to rapid disintegration. In 1957, at a school in Platteville, Colorado, such conditions existed and the tile floor finish was very irregular. Exposure and examination of concrete cores showed that the top 12 to 30 mm (0.5 to 1.25 in.) had been frozen before hardening, was severely fractured, and disintegrated continually under traffic. After setting, the internal cracks permitted formation of ice crystals that did further damage to the top layer of concrete and formed the wavy surface.

The failure to clean snow and ice accumulations from the surface of concrete forms is a common error. Workers sometimes expect that the concrete mix will melt the snow and clear the space. Since heated liquids rise, the warm concrete mix water will not thaw out packed snow and ice near the bottom of the forms. Such a condition in the cellar walls and column piers of a New York Housing Authority project resulted in a honeycomb condition that was beyond correction and an entire section of walls and piers was ordered demolished and rebuilt (Figure 7.2). In the same project, pile caps poured on frozen ground did thaw some of the soil, and the concrete below the reinforcement mat settled. The result was a horizontal cleavage in the concrete caps at the level of the bars. Some of these were removed and replaced; minor separations were grouted to protect the steel.

The dangers of concrete work in freezing weather are well documented and the necessary frost protection procedures are known, but this knowledge is of no use if the necessary equipment is not assembled and made available before frost strikes. After concrete is frozen, no frost protection methods will undo the damage; the concrete must not be permitted to thaw partly and refreeze. Any concrete program in the winter where frost is a possibility must be preceded by a complete assembly and distribution of all wind breaks and heating devices to protect against the coldest day for the largest expected area of work. Only such insurance will properly mitigate against weather damage to the work.

Figure 7.2 Concrete replacement required because of frost and debris in forms.

7.4 SHRINKAGE, EXPANSION, AND PLASTIC DIMENSIONAL CHANGES

Providing connections in a reinforced concrete structure that accommodate movements during its service life is essential to successful performance. The designer must understand the character and source of these movements, whether from drying shrinkage, temperature effects, creep, or varying live loads. Details at expansion joints need careful attention. Sliding supports must have the freedom to slide; otherwise, the brackets or supports will spall and crack. There are very few expansion joints that work properly and show no distress. Antifriction bearing plates are never perfect. Many attempts to provide sliding expansion joints by placing concrete beams on piers with a troweled surface or even a paper-separation layer have proven unsuccessful.

A number of materials and details are available for addressing movement problems, including the use of neoprene pads that permit rotation and vertical movement, and sliding Teflon pads that allow horizontal movement. However, the best intentions of the designer are sometimes thwarted by friction or by debris that negates the function of a movement joint.

In 1988, a section of a 12-m (40-ft)-high viaduct in New York City dropped 100 mm (4 in.), only one month after it was inspected. The damaged 15-m (50-ft)-long span failed because one of the steel bearings had frozen with corrosion, refusing to permit the intended movement. Movement joints in pavements will fill with debris if they are not properly maintained and kept filled with compressible joint material. Florida's Department of Transportation has experienced the unusual problem of replacing expansion joint filler material eaten away by fire ants. One suggestion was to include an additive in the silicone material that is distasteful to the ants.

Kaminetzky (1991) gives numerous repetitive examples of the damaging effects of distress caused by restrained volume changes in reinforced concrete. Cracks, spalls, disintegration at connections, deflections, and deformations appear, apparently without cause. The problems are exacerbated with longer spans and larger uninterrupted surfaces.

Immediate elastic response to loading, and delayed creep changes in dimension and in shape of concrete structures, cause deformations that may necessitate expensive maintenance and repairs even though they may not seriously affect the strength of the structure. Connections between structural members introduce secondary stresses not normally computed. Stress computations from normal loading may indicate no reason for cracking, but the strains resulting from secondary stresses may give trouble. Such conditions always exist, but the intensity in usual designs is not high enough to exceed the available (but usually disregarded) tensile strength of the concrete.

Structural elements such as flat slabs transmit loadings to supports in the most efficient way, involving the least work, which may not be the pattern assumed by the designer. Cracks then result as the frame forms hinges and tries to conform to the assumed pattern. Slab loads are assumed to travel to the columns by bending of the allocated bands. Actually, because of the unbalanced bending moments transmitted to them, the spandrels rotate and carry the reactions to the columns partly by torsion. Diagonal torsion cracks then appear in the face of the spandrel section, especially if a deep beam is used. Such rotation has been known to push masonry veneer out of position and form horizontal cracks in the mortar, with subsequent rain infiltration into the building. Spandrels must be designed to resist such torsional strains. Similar torsional action causes corners of roof slabs to curl, distorting roof flashing and cracking the exterior faces of exterior columns.

Concrete floor slabs with monolithic finish, especially in long-span designs, are seldom constructed to satisfactorily level tolerances. The customary procedure of screeding the concrete from midspan to column lines pulls the denser materials toward the columns, and on hardening and drying, the surface is dished. Built-in camber in the formwork is lost. Actually, the only resistance to future deformation is the compression of the form supports from the dead weight of the concrete. To neutralize deflection of the slab after the form supports are removed, the top surface should be finished as a slight dome. Even then, the later plastic flow of the finished concrete plus any creep due to loadings is not compensated, and a dished floor results. Where moisture accumulation can be expected, as in a parking garage structure, it is always a wise precaution to provide a slope of about 2 percent for satisfactory drainage over the life of the structure.

All structures change dimension with varying temperature. In buildings, uniform control of internal temperature increases the range of differences at the roof level. Change in size of the roof slab where rigid attachment is made to the walls will result in cracked walls, usually at about the head of the top-story window, since the resisting horizontal wall section is suddenly reduced at this level. Thermal expansion of the roof indicates itself in many buildings by such cracking, with accompanying leakage, wall staining, and heat losses in the top story. Most older designs called for a projecting brick course at this level to shade the crack and protect the interior against water infiltration.

Lateral expansion of structures often causes brick covering to push and crack. The cracks never close completely with colder weather; some dust always finds its way into the crevice, and subsequent expansions increase the crack size. To control this cracking, all brick walls covering concrete frames should be built with vertical continuous caulked joints at the first window edge on each end of a wall surface. This does not eliminate the expansion, but at least stops the cracking of the brick since the tendency to open follows a continuous expansion joint. This and other problems related to brick veneer on concrete frame buildings are discussed further in Chapter 9.

Roof expansion troubles in concrete structures were analyzed very early in the history of the art. In 1885, a rectangular concrete reservoir for 38,000 m^3 (10 million gallons) was built in Colombo, Ceylon. Walls were vertical face inside and 5:12 batter outside, the floor was 300 mm (12 in.) thick, and a concrete slab covered the 9-m (30-ft)-deep box. When this was filled in 1885 to a depth of 7.5 m (24 ft), cracks were noted near the corners. After repairs and adding another 300 mm (12 in.) of floor concrete, refilling in 1886 to a depth of 8.5 m (28 ft) again resulted in cracks near the top of the walls, which traveled to the ground level in 15 minutes. Sir John Fowler, a past president of the British Institution of Civil Engineers, was consulted and determined the cause to be expansion due to the high local temperatures. He recommended cutting strips out of the walls and inserting brick piers covered with asphalt. This was done and the reservoir was filled again in 1889. Cracks again appeared at the corners. Measurements were then taken of the roof dimensions and daily variations of 0.25 mm (0.01 in.) were found. The roof was then covered with earth, and no further cracking occurred. This five-year history of repair exemplifies British determination to make a design work.

Normal temperature ranges permit large structures and pavements to ab-sorb enough heat so that surfaces spall, plane areas become curved, and even structural defects occur. Failures of pavements can result from curling caused by temperature and shrinkage effects (Rollings 1993).

Athletic stadia are a good example of the necessity for providing free space for expansion. This is true for all stadia, whether steel or concrete frame. During a detailed investigation of the Polo Grounds in New York in 1940, the intensity of movements was evidenced by loud and continuous creaking of the structure in the early mornings as the sun started to warm it. It is not necessary to list the stadia that are under continuous repair programs; an inspection of almost any of them shows a shortage of open joints to permit expansion. Often, such freedom must be three-dimensional if excessive and expensive maintenance is to be avoided.

Thermal growth of concrete exposed to the sun accumulates over the years and can be the cause of ultimate failure. The curtain walls of the Texas State Fair exhibition building in Dallas were constructed in 1948 by shotcreting mortar on a pressed fiberboard backup to a 50-mm (2-in.) thickness. The layer was reinforced with 100 by 100 mm (4 by 4 in.) welded wire mesh and braced by 50-mm (2-in.) double-channel studs at 1.2-m (4-ft) centers. Panels were 7.3 m (24 ft) wide by 4.4 m (14.5 ft) high. Three such panels on the south wall collapsed in September 1966 with little warning.

Although few examples of catastrophic structural collapse can be blamed entirely on temperature and shrinkage loadings, lack of freedom for dimensional changes results in considerable cracking and spalling in concrete structures. Unfortunately, due to the thermal insulation afforded by the concrete to the inner volumes and to the nonuniform rate of shrinkages in elements of different thicknesses and shape, the dimension changes are not entirely linear. Rotational displacements result and seriously affect brittle masonry surfacing, windows, and door frames. The differential shrinkage between thick and thin reinforced concrete members in intimate contact can cause an eccentric pull on the members; this is the usual reason for the cracks forming in slabs connecting to heavy girders.

In 1983, shrinkage cracks appeared in the ribs of the top-floor two-way slab of a nine-level parking garage in Portland, Oregon. These cracks, which required immediate repairs to prevent structural failure, were explained by temperature differences between the top and bottom of the unshaded slab structure.

Normally, increased compression stress caused by thermal expansion can easily be taken by the concrete. However, if no reinforcement is provided to resist shrinkage, only a small amount of cooling may result in cracks when the low tensile strength of concrete is overcome. The usual amount of prescribed temperature reinforcement required by code merely extends the range of temperature that will not cause cracking. Experience in temperate zones indicates that 0.65 percent reinforcement is necessary to resist a 38°C (100°F.) temperature drop. This is considerably more than the nominal code requirements; it is therefore not surprising to find so few uncracked concrete exposures.

With the accepted architectural use of exposed concrete columns in multistory structures, a new problem has arisen. Interior temperatures are fairly constant, but the exterior columns change in length seasonally. As a result, a differential level develops in the upper floors between the exterior and interior supports. Floors are found to change in level, partitions crack because they cannot absorb the vertical warping, and slabs crack near the interior supports. Exposed concrete shear walls are equally vulnerable to these effects, and higher differentials are found in the sun-exposure positions. As would be expected, the effects are much greater in northern climates, where such distress has been observed in the upper 10 levels of 25-story buildings.

In four 32-story buildings with exposed concrete columns in the end wings, plaster cracking was found above the twentieth floor in the wings only and with greater incidence in the higher floors. To control the high maintenance costs from such action, the Federal Housing Administration (FHA), beginning in 1964, refused to accept any design with exposed concrete columns taller than 21 m (70 ft). Upon proof that the restriction was too drastic and unwarranted, especially that it disregarded variation in climatic exposure, the regulation was revised to permit local FHA offices to accept designs for taller buildings when the owner assured that the designs took into consideration the stresses from any changes in length or height resulting from temperature variations.

Drying shrinkage is defined as the reduction in volume resulting from a loss of water from the concrete after hardening. Some of this happens at an early age, with consequent development of plastic shrinkage cracking, forming internal weaknesses that later become open cracks. If volume changes are restrained by the connections, stresses produced by later shrinkages may cause distress and failure. For this reason, the construction sequence may call for the joints connecting beams and columns to be cast after some initial shrinkage in the structural members has taken place.

Transfer of loading, when columns in multistory frames are affected by delayed shrinkage, usually after the building is completed and in use, has resulted in cracked and spalled rigid brick and stone coverings. Use of flexible joints just below the continuous steel lintels supporting the masonry at each floor level can help to mitigate this problem, as discussed in Chapters 9 and 10.

The amount of long-term shrinkage of an actual concrete structure is dependent on the many factors of type of mix, temperatures during placing, and climate exposure conditions. The design must accommodate both external and internal loadings, including effects due to aging. These considerations exist for any material but seem to cause an inordinate number of problems in concrete. Whether such change is brittleness in dried-out timber, opening up of microcracks of metals from vibration or temperature change, or shrinkage and plastic flow in concrete, materials that change with age introduce internal stress conditions that must be considered in the design. Disregard of such changes will bring failure, either as collapse or as unsightly distress.

The great magnitude of thermal and shrinkage stresses compared to live-load stresses in some buildings is illustrated by the much-studied failures of the warehouse roof rigid frames at Wilkins Air Force Base, Shelby, Ohio, on August 17, 1955, and at Robins Air Force Base, Macon, Georgia, on September 5, 1956. Experimental and theoretical studies made in the attempt to evaluate the forces that caused these failures brought about a healthy discussion in the technical literature. It became apparent that the real causes were not known quantitatively at the time.

Control conditions were provided by the fact that similar warehouses were constructed in many other locations, all being 122 m (400 ft) wide and of various lengths but always in 61-m (200-ft) units with complete separation between units. The rigid frame bents were on 10-m (33-ft) spacing and consisted of six 20-m (67-ft) continuous spans, the third one from one end having an expansion joint.

The standard design used for preparation of contract drawings at various sites fixed the main girders but left open the choice of subframing in the 10-m (33-ft) spans and of the actual roof construction. In the various locations a great variety of roof types was used, from precast deep channel sections to span the 10 m (33 ft), to prestressed joists set into the forms of the girders and anchored to them, with cast-in-place gypsum roof slabs. Also, a great variety of cements and aggregates was used in the several locations. Yet some girders in almost every structure showed signs of distress, more so with the early designs, which did not require continuous reinforcement in top and bottom of the girders, and more so with designs having fewer stirrups.

None of the computational investigations explained the causes of these failures. After the failures, some modifications of the standard code were suggested and, in part, approved. A considerable test program was carried out in which model beams reproduced the various designs used. The tests to failure proved that under the design loadings, no failure or even cracking was expected. Yet failures did occur at two locations and cracks appeared at many others, with practically no live loads present.

Observations at several installations in geographically different parts of the country indicated the following pertinent data:

1. The expansion joints did not operate, and the 122-m (400-ft) length performed as a single unit as far as thermal and shrinkage stresses were concerned.

2. Both collapses came at about the date of maximum net heat absorption gain, the night loss being less than the day gain. The 1955 failure was after a most unusual hot spell.

3. Identical designs built at the same location and approximately over the same time period with the same aggregates did not show the same amount of cracking; the smaller building area units, built under slower construction completion schedules, showed much less distress.

4. Rotational strains of the columns about a vertical axis and of the girders at cracked sections were found. Except where caulking was used in place of cement mortar in the joint between columns and exterior block walls, spalling and horizontal cracking of the concrete masonry units was common. The vertical faces of the girders across one serious crack were not in a continuous plane.

5. Failure of the models did not occur under loadings below the total load (dead plus live load) unless an axial force was applied simultaneously to the girder model. With an axial stress of 1.4 MPa (200 psi) on the

minimum cross section, the ultimate load capacity of the model was substantially the equivalent of vertical dead load only.

6. Although some cracks appeared during the construction period, most of the distress was not observed until the frames were almost two years old.

The final reports of all tests, observations, and conclusions are available. Certainly, some lessons can be learned:

1. Just as it takes time to make good bread, it takes time to make good concrete; a rush job is vulnerable.
2. Expansion joints serve their purpose only when they operate; frozen joint details are worse than no joints at all.
3. Lateral restraint of all members to prevent torsional buckling is typically assumed in most designs; such restraint must be provided, especially at the compression faces of continuous structures, and that comes at the bottom of a girder, where it sits on a column. (Note that several stability failures in steel structures have also emphasized the need for designed lateral support; see Section 6.2.2.)
4. If torsional strain can absorb expansion with less internal work than axial expansion of the girder plus bending of the supporting columns, the girders will deflect laterally in a sine-curve pattern and rotate the columns; cracks will open if the girder is not braced or reinforced to compensate for such torsional strains. (In 1956, a bridge buckling in Queens, New York, was reported with a similar lateral distortion of the fascia wall and parapet on a rigid frame overpass. The distortion provided the extra length required by thermal expansion with less internal work than an axial elongation of the unbraced parapet.)

After the first indications of distress, an evaluation of the standard air force warehouse design showed that negative-moment steel had been stopped too close to the support, when the condition of unbalanced live load was considered. The immediate revision to the standard design was to provide continuous top steel and to extend the area in which stirrups were required in the design. These changes reduced the number of cracks but did not eliminate them. The Wilkins Air Force Base structure did not have continuous bars, and the diagonal shear failure left the ends of the bars exposed after pulling them out of the suspended part of the span. No stirrups were crossed by the failure surface. This design was in full conformity with the 1951 American Concrete Institute Building Code. About 450 m² (4800 ft²) of roof fell. Here the roof joists had been made up of 200 by 216 by 440 mm deep (8 by 8.5 by 17.3 in.) machine-made concrete blocks tied together with tensioned bridge strands into 10-m (33-ft)-long beams. Beams were set into the girder forms on 2.5-m (8.25-ft) centers, and negative reinforcement ran over the beams within troughs cast in the beams. The troughs were filled with the girder concreting. Some

20 similar projects were under construction at the time of the failure, and work on all of these was suspended until restudy of the design indicated the advisability of providing more top steel, addition of web reinforcement, staggering of bar splices, and an improved expansion joint detail. New designs were made available to all projects within two months of the first failure.

An immediate recommendation was to modify the ACI code requirements on the portion of shear to be resisted by steel, either stirrups or bent bars. The earliest (1907) standards allotted shear resistance to the concrete for the full carrying capacity and called for steel only for any excess shear stress. In 1916, European codes required the total shear, where in excess of the concrete allowable value, to be resisted by steel, with an abrupt stoppage at the point where the shear stress was balanced by the concrete value. Actual practice then was to provide bent bars and stirrups over the entire span, and this was the European code as of 1925. Yet in 1924, the Joint Committee permitted the 1907 standards, and with the almost universal elimination of bent bars because of high labor costs, stirrups were provided only over the range of excess shear stress and only for that excess.

Especially in rectangular continuous beams, most designers recognized the code as insufficient and called for at least two continuous bars in the top for the full length of the beam, with stirrups usually spaced 300 mm (12 in.) apart to make up the cage of steel. This precaution was not followed in the standard air force warehouse designs. Almost immediate action resulted in revisions to the ACI code; the new requirements called for shear reinforcement to carry at least two-thirds of the total shear or all of the excess wherever the shear stress exceeded the allowable value for the concrete alone.

Similar freestanding multispan frames had given trouble in previous years, usually at low-shear areas and near the normal points of contraflexure. Temperature and shrinkage loadings can shift the point of contraflexure a considerable distance. A rigid frame bridge model tested at the University of Illinois in 1938 showed such a failure. A two-span bakery roof of this design failed in 1954. A hanger built in Coronada, California, in 1943, consisting of nine rectangular frames each of seven 40-m (130-ft) spans, showed similar diagonal cracking and failed in 1954. Following this failure, warnings were given regarding the heavy concentration of longitudinal bars in narrow sections, another item that weakened the air force warehouse girders, exhibited by spalling and cracking in the bottom surfaces. The exterior spans had seven No. 9 bars in a width of only 500 mm (20 in.)

A program of analytical and experimental investigation was instigated by the several interested Department of Defense agencies to explore the warehouse failures further. The most extensive investigation was started in October 1955 at the PCA Laboratory in Skokie, Illinois. This investigation consisted of testing to failure 13 one-third scale models of the second span in the typical frame. The first seven beams were tested to failure under flexural loading only, and the last six were subjected to combined flexure and axial tension. Only the latter model beams failed under flexural loadings in a manner at all similar to that experienced in the actual frames. The tests also proved the value of a suggested retrofit design: adding exterior steel bands on the completed frames to provide the necessary shear resistance.

Inspections indicated that those frames designed for lower roof loadings in warmer climates showed more cracking, and the accumulation of temperature absorption seemed to have some effect. The almost daily observations at Robins Air Force Base near Macon, Georgia, showed widening of cracks in August, with one crack becoming 12.5 mm (0.5 in.) wide by September 4. Finally, some (560 m²) 6000 ft² of the roof, including two adjacent girders of the longer arm at an expansion joint, failed at 3:05 A.M. on September 5, 1956. This was in a warehouse 490 by 122 m (1600 by 400 ft), which had been in use for 18 months. A similar design in a building 122 by 122 m (400 by 400 ft) at the same base, constructed as a separate contract under a very slow project completion schedule, showed hardly a single crack. It seems that the extent of shrinkage with resulting axial tensions was somewhat related to the speed of concrete placement or to the extent of each separate pour.

The Robins design had the same shape as in the failed Wilkins frames in Ohio, but the top steel was continuous and nominal stirrups had been provided in the Georgia project. In both instances, however, the expansion joints were inoperative. All warehouse frames that showed cracks of consistent shape and larger than hairline width were reinforced by adding exterior stirrups of band steel stressed into shape by mechanical means. It was found necessary to tap the straps manually during the tightening operation so that the unit stress in all four legs of the band would be fairly equal. The cost of the repair procedure, plus the cost of temporary shoring of cracked girders, labor of moving and replacing stored goods, and final cleanup, was only a few percent of the value of the warehouses but ran into several millions of dollars.

The concrete construction industry owes much to the research accomplished on the actual performance of cast-in-place concrete structures such as the air force warehouses. Design theories need verification by measured and observed performance in the field before confidence can be placed in them.

All dimension changes can cause cracks in concrete. Of course, there are many new ways to repair cracks in concrete structures, using modern adhesives, epoxies, and grouts, and many new sealants that can reduce moisture-penetration problems. Those undertaking repairs, however, may find that they have only addressed the symptoms rather than the source of the cracking. The American Concrete Institute classifies the causes of cracking under these headings (ACI 1984):

1. Cracks occurring during hardening of the concrete
   a. Plastic shrinkage
   b. Plastic settlement
2. Cracks occurring after hardening of the concrete
   a. Drying shrinkage

b. Thermal stresses

c. Chemical reactions

d. Weathering

e. Corrosion of reinforcement

f. Poor construction practices

g. Construction overloads

h. Errors in design and detailing

i. Externally applied loads

The subject of cracking in concrete has been included in various levels of detail throughout the standards and recommendations of the ACI and PCA, but still many cracks appear in concrete structures that are preventable (Mullick and Rajkumar 1993). An interesting editorial on the subject appeared in February 1953 in the *Indian Concrete Journal.* These comments continue to be relevant despite the many advances in concrete construction:

Cracking is one of those problems in concrete construction which evokes frequent discussion among engineers. One attitude to this problem that seems to be largely prevalent is that cracking is a natural occurrence about which little can be done. And yet it is possible to investigate into the relationship and interconnection of different types of cracks, to detect their causes, and to apply remedial measures for their prevention.

Cracks can be broadly classified into those which occur before hardening and those which occur after hardening. The period before hardening is considered to extend from the time of placing the concrete up to the time when it has attained sufficient strength to resist alteration in form, in practice 2 to 8 hours after placing under normal conditions.

Subgrade movement due to moisture changes and movement of forms due to inadequate design or construction are two frequent causes of cracking in the pre-hardening period which can be prevented by adequate compaction and control of the subgrade and by careful attention to formwork. Settlement shrinkage is another cause of cracking in this group. When concrete is placed in position and before it sets the solid particles of aggregate settle and stabilize themselves. Continued settlement within the skeleton of these large particles can result in mortar settlement and ultimately in the accumulation of water at the upper surface or water gain. Now if the conditions are such that the upper surface of the concrete becomes partially set whilst the interior is still setting, and if the interior contains some rigid obstacles such as reinforcing bars, then the interior will settle around these bars and cause cracks in the partially set concrete above the bars. The remedy lies in the use of dense plastic mixes with well-graded aggregate, low water content, and adequate compaction.

Then there are those cracks which occur very soon after the placing and finishing, sometimes within a matter of minutes, even under a film of water. These are ascribed to plastic shrinkage caused by the expulsion of free water from the freshly formed silica gel or by a false set. The remedy here appears to be delayed finishing—pressing the plastic concrete to close the cracks. It is found that when this is done the cracks do not reappear.

Most engineers are familiar with the type of cracking caused by drying shrinkage. This can be particularly observed when freshly poured concrete pavements are exposed to a high wind, low humidity, or hot sun with temperature differential between the mass and the surrounding air. Prevention lies in covering the concrete with damp sand, hessian, jute bags, or straw as soon as there is no danger of their leaving unsightly marks on the concrete. Drying shrinkage, of course, continues after hardening. Reduction in the cement and water content of the mix with adequate curing to permit complete hydration minimizes the extent of this shrinkage.

The ramifications of the other group of cracks—those which occur after hardening—are considerable. Moisture movement here again plays a part as subsequent wetting of concrete after drying presents a reversible action which is comparable in magnitude with the original shrinkage. Therefore, structures which are subject to alternate wetting and drying must have provision for such movement or be able in some way to withstand the stresses which are set up, otherwise cracking may be expected.

The chemical causes of cracking include not only reactions within the constituents of concrete, but also from the presence of foreign bodies. Incomplete combination of the lime with the other raw constituents of cement, and excess of magnesium oxide and gypsum can cause cracking. Reactive aggregates present another source of cracking to which we have made reference in the past. Modern Portland cement specifications are framed in such a way as to prevent these types of cracking. An example of cracking due to the presence of foreign bodies is afforded by the oxidation or rusting of mild steel reinforcing bars. This failure can be prevented by the use of dense concrete and adequate cover to the bars.

Much work has been done to prevent cracks resulting from the differential temperature caused through the heat of hydration of cement, particularly in large mass concrete construction works such as dams. The precautions now employed include limitations on the heights of pours and periods between successive lifts, the use of low-heat cement, the addition of cracked ice to the mix, and elaborate cooling water systems including the use of refrigeration. External temperature variations also cause cracking in structures which can be prevented by the provision of adequate expansion joints.

Stress concentrations are another source of cracking. Stress concentrations can be caused by stress transfer in reinforcement or by structural form. An example of the former is when a bent bar is placed at the junction of the stairway and landing slabs with insufficient depth of concrete to withstand the stress concentration at the change in direction. Square openings or re-entrant angles such as door and window openings or openings for manholes in pavements cause considerable concentrations of tensile stress which must be resisted by suitably placed reinforcement if cracks are to be avoided. Foundation settlement and accidental happenings such as overloading, excess vibration, fire, storm, and earthquake are other sources of cracking in concrete structures.

It is apparent from this survey of the numerous causes of cracking in concrete that the subject is of considerable importance in concrete construction; also that cracking is by no means inevitable and much can be done in design and construction to prevent it.

7.5   ABRASION AND SURFACE DETERIORATION

The hardening of cement in concrete or mortar involves a chemical reaction, but the finished product continues to experience both chemical and physical changes during its useful life. The resulting crystallization of the amorphous cement is subject to alteration when in contact with many chemicals, corrosive materials, and abrasive exposures. Disregard of this possibility has resulted in much surface damage and some deeply formed disintegration of concrete, causing unsightly surfaces and expensive corrective work.

In foundations and slabs on grade, concrete can deteriorate due to sulfate attack by groundwater with excessive carbon dioxide or other undesirable chemical content. These problems occur when moisture is permitted to migrate into the concrete (Rzonca, Pride, and Colin 1990; Day 1995a).

The action of lactic acid on cement is so rapid and complete that use of concrete floors in process plants where lactic acid can form is most inadvisable. These include milk pasteurizing plants, cheese manufacturing plants, breweries, and slaughterhouses. Even the joints between the packinghouse brick used for floor surfacing, unless made of special cements or sulfur compounds, will deteriorate within a year. When complete sanitary control is required, even a surface defect in the joints permits the accumulation of milk or other wastes that soon alter to lactic acid and cause further chemical reaction with the cement.

Another source of concrete disintegration is the process of carbonation. Exposure to carbon dioxide in the atmosphere reduces the alkalinity of surface concrete over time. This decomposition is limited to a depth of only a few millimeters in the surface layer in dense, good-quality concrete. However, cracks may permit carbonation to penetrate farther into the material. If the eventual loss of alkalinity reaches the depth of the reinforcing steel, the natural resistance to corrosion provided by good concrete is reduced considerably. Combined with the presence of moisture, the corrosion will produce severe cracking and spalling. Problems related to carbonation increase with the use of lightweight aggregates and certain cements and can be partially mitigated by proper use of air-entraining or plasticizing admixtures (Roberts 1981).

Recognition of the danger of chemical and abrasion damage came as early as 1880, when Professor Prazier of the University of Aberdeen, Scotland, investigated the causes of concrete deterioration in the graving dock at Aberdeen Harbor. He warned that overcalcined cement used in concrete exposed to seawater will absorb magnesia, an ion-exchange phenomenon, resulting in expansive surface disintegration, which, in turn, exposes more surface to the same action. Similar trouble was encountered at many harbor works in England, and M. Vicat came to the same conclusion in France, in the study of disintegration of concrete blocks made of lime and artificial pozzuolana at Marseilles, Rochefort, Algiers, and Cherbourg, after only a few years of exposure to the sea.

Normal chemical action on concrete from salt spray, ice abrasion, tide, and weather is found in many waterfront structures. After about 21 years of use, the concrete piles and caps of the trestle bents of the James River Bridge at Newport News, Virginia, required a $1.4 million repair and replacement job in 1955 (Vaccaro 1956). Some 70 percent of the 2500 piles were found in need of repair. All the piles were therefore jacketed over the tidal range and the steel sleeves form left as a covering (Figure 7.3).

Concrete dams holding reservoirs with widely fluctuating water surfaces that are frozen most of the winter months are prone to serious surface disintegration in the zone on the upstream face exposed to many cycles of freezing and thawing. Protection against the natural aging influences of tide and wind has resulted in recognition that especially dense concretes are necessary and that moisture-tight surfacing is economically advisable. Disintegration by chemical action is slow unless there is frost action or continued mechanical abrasion, removing the decomposed cement products as they are formed.

A frequent item of observed concrete distress is the surface disintegration of pavements, curbs, and sidewalks. This can be the result of traffic action on a poorly surfaced concrete, or the chemical and abrasive additives to control snow and ice. Nevertheless, there are many good concrete surfaces of many years of life that hold up well in the most adverse exposure conditions. Details of design and construction can influence longevity. Although deicing chemicals are usually blamed for the extent and rapidity of surface damage, they are seldom the cause of the beginning of the disintegration. As a retardant of the action, higher air contents in the concrete, up to 8 percent, lower slump, and reduced finishing operations are beneficial.

At a plant built in 1947 in Syracuse, New York, excessive use of calcium chloride for snow removal, together with the leaching of tannic acid from the soil adjacent to new concrete curbs, caused complete disintegration of the curbs in only one winter. In some sections there was no evidence of a curb ever existing above the asphalt pavement when the snows melted in the spring. The remainder of the concrete was saturated with a tannic acid compound largely activated by the calcium chloride.

At the Yonkers Raceway in New York State, the top deck of the garage built in 1955 was covered with a 20-mm (0.75-in.)-thick application of tar macadam as a wearing surface and waterproofing protection for the lower levels. The deck sloped uniformly at a 1.5 percent rate to trough drains spaced approximately 75 m (250 ft) apart. After eight years use in summer weather only, the racing season was continued into the winter months. Some rapid method of snow removal was necessary to provide for the peak-load parking.

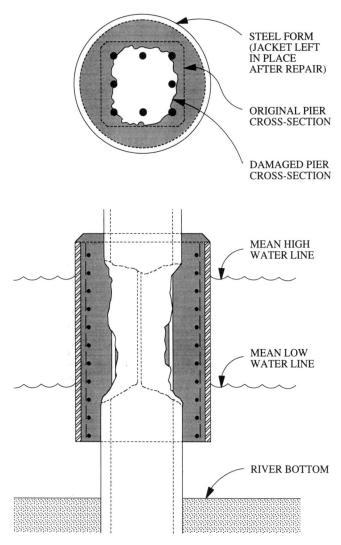

Figure 7.3 Repair of concrete piles.

Calcium chloride could not be used because it might be carried over to the stable area and is injurious to horses' hooves. Common salt was used in copious quantities. By the third winter the salt had filtered through construction joints of the tar surface, near the low area, penetrated through the 75-mm (3-in.)-thick concrete of the waffle slab, and covered the entire ceiling area with a continuous coat of white salt powder. It had also completely softened the concrete, so that a man stepped right through a pan area and was kept from falling only by the mesh layer that had been placed in the top of the slab.

Every dome area was sounded, and all hollow or soft indications were noted to be removed. All loose surface topping was cut out, soft concrete replaced, and the exposed concrete sealed with asphalt and recovered. Granular salt had been found piled in mounds on top of the deck when the extent of damage was being surveyed. Snow blowers were recommended as the safer method of keeping the deck available for winter parking (Figure 7.4).

The deicing problem for pavements has not yet been solved completely. Rock salt and calcium chloride remove snow and ice effectively and if used in moderation usually will not harm good normal concrete, and good air-entrained concrete even less. New agents have come on the market, but any that contain either ammonium sulfate or ammonium nitrate will actively attack any kind of concrete.

The detrimental effects of trapped water in causing premature deterioration of poorly drained pavements and exposed concrete structural elements is well documented (Day 1995b). Large open athletic stadia probably have more concrete exposure surface than any other structure. Disintegration is very common, starting as crazing and continuing as spalling when the reinforcement is attacked. A number of the very large stadia built during the 1920s, such as those at the University of Pittsburgh, Illinois, Ohio, Chattanooga, Chicago, and Northwestern, have been laboriously covered with double applications of surface sealants.

White (1925) was then the recognized authority on the subject and his explanation was often quoted:

> Water is the cause of both the life and death of concrete; Reaction of the cement with water gives it strength; Removal of the water (drying) causes it to shrink and crack; Restoration of the water causes it to swell; Alternate expansion and contraction, due to changes in moisture, is the greatest underlying cause of the destruction of concrete structures; If concrete is to be permanent, its moisture content must remain relatively constant. (White 1925)

Disregard of these warnings in many stadia designs caused the start of disintegration, with high annual maintenance costs. Some have been lifetime projects for the waterproofing industry and for companies specializing in repair using latex-modified concrete or other materials.

7.6 SHEAR FAILURES: FLAT-PLATE STRUCTURE

The strongest factor in concrete is its compression strength. If the concrete mix can provide the desired compression strength, there is little reason to suspect the columns as a cause of failure. Actually, very few failures have been attributed to overstressed columns, although distortion from eccentric loading and vertical change in length from temperature, shrinkage, and creep are largely connected with column action.

(*a*)

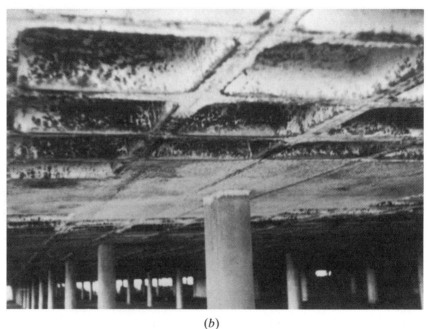

(*b*)

Figure 7.4 (a) Concrete deterioration due to salt used for deicing of concrete deck; (b) salt penetration to ceiling below.

In the early days of reinforced concrete, columns were designed using a lower factor of safety than that used in current practice. Seldom was any bending stress considered. In Basel, Switzerland, a five-story hotel collapsed in 1901 because of column failure. About 10 years later there were several column-induced collapses in the United States. In 1909, during a load test of the completed reservoir roof at Annapolis, Maryland, the 230 by 280 mm (9 by 11 in.) columns failed. In 1910, the Hencke Building in Cleveland, Ohio, collapsed with little warning when almost complete. The roof of the four-story building was then 12 days old and some timber reshores were still in place on the third floor. Failure started in the third-story columns, but the investigating committee also found the architect negligent for his lack of supervision and technical control, and the contractor negligent for careless construction, for a 20 percent deficiency in cement provided for the concrete used, and for lack of coordination with the architect.

Two foreign cases of complete building collapse caused by overload of columns are some eight-story apartment buildings on the Italian Riviera in 1965 and a 15-story structure in Piracicaba, Brazil, in 1964. The latter was designed as a 10-story apartment and commercial building, and the owner, over the protests of the structural engineer, insisted on adding five more stories. The concrete was completed to the roof level when one wing collapsed, killing 40 workers.

While failures in direct compression are rare, shear failures in reinforced concrete structures are quite common. Nature has a habit of finding the path of least resistance for propagating cracks, and in the case of low shear resistance, crack propagation nearly always means sudden failure. The designer and detailer must guard against the possibility of such weak paths. The reason that shear failures in concrete structures are so catastrophic is that there usually is no warning. Shear failures are preceded by few, if any, deflections and little cracking.

Probably more has been argued about shear resistance, and there is probably greater variation in the codes of various countries on shear resistance design, than about any other facet of reinforced concrete design. The basic unknown is whether and to what extent reinforcement acts together and simultaneously with the concrete in resisting shear, and how the torsional stress induced by nonsymmetrical loading influences the maximum shearing stress to be resisted.

Some of the most catastrophic concrete failures have occurred in flat-plate buildings. Flat-plate construction employs a concrete slab system that is of uniform thickness. There are no projections, such as column capitals or thickened drop panels, for increased shear resistance near the columns. This form of construction is very popular for multistory residential buildings because compared with alternatives it requires a minimal structural thickness and hence a reduction in overall building height. Many successful flat-plate buildings provide evidence that following well-known design procedures will prevent failure due to punching shear. But dramatic failures involving punching shear in flat-plate construction continue to occur when the rules are not followed. This type of failure nearly always causes loss of life, whether it occurs during or after construction. Punching shear was a contributing mode of failure in the 1970 collapse of the Commonwealth Avenue Building in Boston, Massachusetts (Section 11.5), and the 1973 Skyline Plaza failure at Bailey's Crossroads in Fairfax, Virginia (Section 7.2). Several other examples of this common failure mode are presented here.

On October 3, 1956, a four-story building under construction in Jackson, Michigan, collapsed, killing six construction workers and injuring 15 others. The building was cross-shaped in plan. While the fourth-floor concrete was being placed—an area of approximately 22 by 44 m (72 by 144 ft)—the 250-mm (10-in.)-thick flat plate fell vertically. The impact carried all the completed floors down into the cellar. The first and second floors were several weeks old. The third floor was 20 days old and had been reshored to the second floor. It was carrying the formwork for the fourth-floor concrete.

After the collapse, almost all the columns remained standing full height. The top-story forms were still attached, but there was hardly a single piece of reinforcing steel projecting from the freestanding columns at any floor level. Column sizes were 635 mm (25 in.) square below the first floor and reduced to 580 and 510 mm (23 and 20 in.) square above.

Some of the columns had 250 by 360 mm (10 by 14 in.) duct openings along two adjacent faces. These openings prevented running any slab reinforcing steel through the columns. The design called for a square spider around each interior column within the slab thickness, but how these assemblies could be placed within the zone of high shear and still permit the duct openings was not clear. The lack of load-transfer resistance from slab to column in this project underscores the critical need to investigate punching shear as a principal step in the design of any plate. In the modern air-conditioned building, the concrete frame becomes merely an enclosure of mechanical equipment. The many openings associated with sophisticated environmental control systems necessitate close coordination between structural and mechanical design. Otherwise, the holes, sleeves, and loads required by the mechanical design may prove disastrous to the structural system. Flat-plate floors are safe and economical only if uninterrupted continuity at columns is provided. No one would agree to a design for a plate footing with openings at the face of the columns. The structural requirements for a floor slab or flat plate are the same as for an independent column footing.

A similar partial failure occurred in New Jersey in 1961, when a shortage of proper reshoring material during the construction of a high-rise concrete apartment building caused overloading of the ninth floor. The slab sheared completely around three columns. Failure probably started at one column where a mechanical duct opening occupied the full width of the column face (as in the Jackson, Michigan, case) and no slab reinforcement could be carried through the column. With one support missing, the transfer of the total load to the adjacent columns exceeded their capacity, and the slab sheared there as well.

The cost of this failure included the demolition of both the ninth- and tenth-floor slabs and reconstruction, plus the cost of the considerable delay in project completion. The cause was the desire to save money by reducing the shoring under the seven-day-old slab, which was asked to carry a full deck of wet concrete. The severely reduced shear resistance at the one column made the failure inevitable (Figure 7.5).

Two Canadian school buildings with 150-mm (6-in.)-thick flat-plate floors and 300-mm (12-in.) circular columns experienced abnormal floor deflections. A typical crack pattern on the top of the slabs at each column emerged as a circle slightly larger in diameter than the column and four radial cracks along the diagonals of the slab panels. These incipient shear failures were quickly restrained by timber shoring until permanent concrete column heads 710 mm (28 in.) in diameter and 290 mm (11.5 in.) deep were added to each column. A 20-mm (0.75-in.) bite was cut out of the column perimeter for the depth of each cap, No. 4 spiral loops were placed, and the caps were gunited in place. A total of 140 column caps were added at each school to increase the shear resistance, which was inadequate because of low-placed steel and heavier interior partitions than assumed in design.

A concrete plaza deck, acting as roof of a garage in continuous use for almost three years, collapsed in New York City on December 23, 1962. There was almost no warning. Two garage employees, parking cars for the night, heard some cracking and left the area just before the 400-mm (16-in.)-deep two-way joist slab with approximately 1.2 m (4 ft) of earth cover crashed on top of the parked cars. About half of a symmetrical entrance plaza, an area approximately 14 by 15 m (45 by 50 ft) failed. The other half remained in place, apparently in sound condition. The structural bays were 7.3 by 9.8 m (24 by 32 ft) and supported on 350 by 760 mm (14 by 30 in.) columns.

Failure was a clean punching shear, with an almost perfectly vertical fall of the trees and little effect beyond the shear cut. Despite the extensive experience of the builders, contractors, and supervisors in similar construction, concrete caps designed to extend 300 mm (12 in.) beyond the column faces and 250 mm (10 in.) deep had been omitted at all nine columns in the failure area and at all nine columns in the symmetrical area that did not fail (Figure 7.6). One problem was that the column caps were not shown clearly on the drawings. Such a critical design element should be communicated so clearly that confusion is not possible, and field supervision should be focused on such details so that they cannot be constructed inaccurately.

The only difference in conditions on the two sides of the entrance was a plugged underground drainage system in the failed area, resulting in a frozen saturated earth cover. The other side was well drained. The difference in the dead weight between the dry and saturated soils was the difference between failure and stability. Computations of shear stress at the columns showed a factor of safety of 1.05 for the portion of the slab that did not fail.

(a)

(b)

Figure 7.5 Shear failure in flat plate: (a) top view; (b), bottom view.

Figure 7.6 Deck failure in shear, due to omission of column capitals.

Lightweight aggregate concrete was used on this project, and in the failed area a remarkable lack of bond adhesion was noted between the reinforcement and the concrete. The exposed bars had pulled cleanly out of the concrete, and the impressions in the concrete seemed dusty and only roughly formed. The slab was reconstructed with column heads (consistent with the original design) and new columns; the other half was also considerably strengthened by new girders and column jackets.

One final example will show the sudden and progressive collapse potential of flat-plate construction when punching shear stresses exceed the capacity provided by the design: the failure of the Harbour Cay condominium building in Cocoa Beach, Florida, on March 27, 1981 (Figure 7.7).

The five-story waterfront residential condominium project was a cast-in-place flat-plate reinforced concrete building. It was under construction at the time of the collapse, which occurred shortly after 3:00 P.M. during placement of the concrete for the roof slab. The entire structure fell vertically all the way to the ground, with each floor slab shearing through the columns. Eleven workers located at various levels of the building died and another 23 were injured. Some of the bodies recovered were buried under 6 to 7 m (20 ft) of debris. Descriptions of the collapse were given by 28 witnesses, 13 of whom were located in the building. Several were on the roof and rode the collapse

Figure 7.7 Harbour Cay Condominium construction collapse, Cocoa Beach, Florida, 1981. (Courtesy of the National Bureau of Standards.)

all the way to the ground. Some described seeing the slabs "just sliding right down around and over the columns" (NBS 1981).

Initial reports blamed the construction schedule, suggesting that premature loading of partially cured concrete was responsible. However, subsequent investigations found design errors to be the cause of failure.

The structure, comprised of nine structural bays, was 18 m (58 ft) in width with an overall length of 74 m (242 ft). The slabs were placed on flying forms made of plywood decks supported by aluminum trusses. Except for the roof, which was planned to be cast in one day, the floor slabs were cast in two separate operations, each covering one-half of the total floor area. At the time of the collapse, concrete had been placed on seven of the nine flying forms for the roof area.

Typical columns were 250 by 460 mm (10 by 18 in.), although some were only 250 by 300 mm (10 by 12 in.) and 200 by 300 mm (8 by 12 in.) The floor slab thickness was 203 mm (8 in.), with spans up to 8.4 m (27.7 ft). The first-story slab was cast on grade. Story heights were 2.64 m (8.7 ft). Columns were supported by pile caps at the foundation level, containing two to nine piles each. Continuous wall footings connected the exterior pile caps.

A number of critical omissions were noted by investigators who reviewed the design calculations (NBS 1981):

1. There were no calculations to indicate whether the deflection or minimum thickness provisions of the applicable design code were checked.
2. There were no calculations to indicate whether the punching shear and beam shear provisions of the code were met.
3. There were no checks to determine if the column reinforcement would satisfy the necessary spacing requirements specified by the code.
4. The steel area calculations for the flexural reinforcement in the slabs were based on ASTM grade 40 steel, which did not agree with the grade 60 steel specified on the structural drawings.
5. The effective depth of the slab in flexural reinforcement calculations was not defined explicitly but appeared as a constant multiplier of computed moments. Back calculations showed that the designer used a value of "around" 160 mm (6.3 in.) for the effective depth in steel calculations. For the 203-mm (8-in.) slab, the effective depth used is consistent with the 19-mm (0.75-in.) minimum cover specified by the ACI code.

Besides the obvious design deficiencies, there were some construction errors that played a minor contributing role in the failure. The bar chairs used (supports that hold the bars in place) were not high enough to put the top bars in the correct location. The chairs used to support the reinforcing bars placed them such that the cover was greater than the 19 mm (0.75 in.) specified. This meant that the actual effective depth measured between 126 and 136 mm (4.97 to 5.35 in.). This is about 25 mm (1 in.) less than if the 19-mm (0.75-in.) cover had been provided. This variance was important to the investigation because the effective depth of the slab is one of the critical parameters governing punching shear and bending moment capacity. (The other principal factors involved in punching shear capacity are the column plan dimensions and the strength of the concrete.)

Column steel was quite congested, so much so that the steel was squeezed together to permit slab steel to pass through the columns. Especially at splices, the bars were so congested that flow of concrete around the bars was prevented, and the bond between concrete and reinforcement was deficient. It should be noted that if the column size had been increased to address the steel congestion problem, this would also have improved the condition relative to punching shear, since the perimeter of the column would have been increased.

At the time of the collapse, shoring had been removed from the first story, meaning that the forces in the total shoring system no longer had a direct path independent of the columns to the foundation slab on grade. There were conflicting reports as to the existence of full shoring on the other floors. Based on worker accounts, the failure began in the center of the fifth-floor slab and propagated straight down to the ground. While all the slabs fell, many of the columns, three and four stories high, were still standing, indicating that the second- to fourth-floor slabs had been sheared at the columns on their way down.

Cracks and noticeable deflections had occurred in all the slabs after the flying forms were removed, according to workers. The cracks were located around the columns and at midspans. These conditions were pointed out to the structural engineer, who had responsibility for construction inspection, but he apparently did not recognize their significance (ASCE 1982.)

At the time of the collapse, gravity loads were the only significant loads on the structure. According to National Bureau of Standards investigators, the "most probable cause of the failure was lack of adequate punching shear capacity in the fifth-floor slab at several of the columns to resist the applied construction loads" (NBS 1981). This deficiency was brought about by design inadequacy, which resulted in a slab thickness significantly less than required by accepted practice and by placement of the top reinforcement at the column strips with greater cover than that which would have met the cover requirements specified in the structural drawings. The shear stresses at many column locations on the fifth floor exceeded the nominal shear strength. Once punching shear had initiated at one column, it propagated throughout the slab, causing total collapse of the fifth floor, which, in turn, caused the successive collapse of the lower floor slabs.

Most alarming was the fact that punching shear, the most common mode of failure for flat-plate structures, was not even checked by the design engineer for this project. A cursory review of the code requirements for resisting punching shear predicts a required slab thickness of 280 mm (11 in.) for the design loads and the column plan dimensions rather than the 203 mm (8 in.) shown in the design. Some construction deficiencies were noted, but the design error related to punching shear alone was clearly sufficient to bring about the collapse. Why was this critical design check not made by the structural engineer or by the regulatory agency that issued the building permit?

The structural engineer was a retired National Aeronautics and Space Administration (NASA) engineer, who, in turn, hired another retired NASA engineer to do the calculations. The project was built at a time when a considerable amount of construction was under way in Florida, straining the ability of local building department staff to keep pace with the developers.

Five of the parties involved in the Harbour Cay project were charged with negligence by the Florida Department of Professional Regulation. Both structural engineers surrendered their licenses and will never again practice in the state of Florida. The architect who designed the project was suspended from practicing in Florida for a period of 10 years, and two contractors were disciplined. This failure serves as a reminder that such tragic occurrences are still possible in uncomplicated low-rise projects, despite the availability of sufficient knowledge to prevent them.

7.7 DETAILING FOR DUCTILITY: LESSONS FROM EARTHQUAKES

Two of the deficiencies of concrete—its poor tension capacity and its lack of ductility—make concrete structures especially vulnerable to damage from earthquakes unless details are carefully considered. The stress reversals that will occur in seismic events require thoughtful placement of steel reinforcing on both sides of all columns in the frame, and in the top and bottom of all beams, so that positive and negative bending moments can be accommodated.

A ductile structure provides the ability to absorb energy through deformations. Under unusually large loads, such as those experienced in an earthquake, this capacity may prevent total collapse. For quite some time, the engineering community has focused on the desirability of ductility, and code provisions have favored ductile designs. It simply is not economically possible to design structures to accommodate the energy demands of a large earthquake with a purely elastic response. We must therefore rely on inelastic response to absorb or dissipate much of the energy produced by such an earthquake.

Without reinforcement, or when the reinforcement is not carefully detailed, concrete structural elements are brittle. They have very little capacity to absorb energy prior to fracture. The welded steel frame structure has traditionally been the preferred choice for major building structures in regions of high seismicity because of its ductile response. The reinforced concrete industry was forced to conduct a significant amount of research to improve the ductility of reinforced concrete, so that the material can continue to be acceptable and competitive in seismically active areas. In fact, modern ductile frames made of reinforced concrete have performed very well in recent earthquakes, illustrating that the quality of design and workmanship may be more important than the specific choice of structural material. However, the provisions of the codes that ensure ductility are quite recent. Concrete buildings and highway structures built before the mid-1970s are expected to be among the most hazardous facilities in future earthquakes.

A review of code development relative to seismic effects is given in Section 2.2. Observers have noted the advantages of ductility at least as early as the 1906 San Francisco earthquake. Codes began to focus on the problem of nonductile masonry and concrete structures following the 1925 Santa Barbara and 1933 Long Beach, California, events. In 1964, the Anchorage, Alaska, earthquake provided many lessons for the designers of concrete structures, including the need for details that improve confinement of the primary reinforcement. These lessons were relearned in the 1985 Mexico City earthquake. The 1971 San Fernando, California, earthquake brought about the failure of a number of important highway structures, proving that detailing for ductility is as important for bridges as it is for buildings. New code provisions for enhancing the ductility and overall stability of structures emerged from this earthquake.

Perhaps the most dramatic failure in the October 17, 1989, Loma Prieta, California, earthquake was the collapse of a portion of the elevated Interstate 880 Nimitz Freeway (see Figure 2.7). Again illustrating the problems that were observed in the 1971 San Fernando event, this collapse made evident the painful reality of economic limitations: There simply is not enough public money to replace the inventory of existing structures that are known to be deficient. Building codes have been directed primarily at new construction. It is now recognized, however, that there is a substantial seismic risk associated with existing buildings, bridges, and other facilities that do not conform with current design standards. Currently, much work is directed at the rehabilitation of unreinforced masonry buildings and nonductile concrete frame structures. The magnitude of the problem is quite impressive, since many of the nonductile reinforced concrete structures were constructed less than 20 years ago.

In the January 17, 1994, Northridge, California, earthquake, six bridges on major freeways collapsed, some due to problems at joints, some due to nonductile concrete details, and some due to deformation incompatibility, where long (flexible) and short (stiff) supports were incorporated into the same structure. Some of the most critically damaged highway structures had already been scheduled for seismic upgrade work within the year. This ongoing retrofit work was responsible for previous improvements to 114 bridges in the region, none of which were damaged in the earthquake. Prior to 1971, transverse reinforcement in freeway supports was woefully inadequate. Ties were often spaced 300 mm (12 in.) apart; now they are provided at a spacing of 50 to 100 mm (2 to 4 in.). Several dramatic collapses of precast and nonductile reinforced concrete buildings also occurred in Northridge (Figure 7.8).

In the Great Hanshin, Japan, earthquake of January 17, 1995, a large number of nonductile concrete buildings failed in a similar fashion to those in Northridge (Figure 7.9). The elevated Hanshin Expressway in Kobe was rendered useless by the complete collapse of a 500-m (1650-ft)-long section. All four lanes collapsed when 15 reinforced concrete supports crumbled due to lack of confinement (insufficient transverse reinforcement), brittle failures, and lack of redundancy (Figure 7.10). Built during 1964–1969, the inadequate details were similar to those causing failure of the Cypress Viaduct in the 1989 Loma Prieta, California, earthquake (see Figure 2.7). The Hanshin Expressway had not yet been retrofitted to current standards. Its performance was not really a surprise to engineers, just as the Cypress Viaduct failure in the Loma Prieta earthquake was not a surprise. Current retrofit projects are moving forward only as fast as public funding permits. In Kobe, older reinforced concrete buildings were severely damaged and nonductile concrete buildings were among the most common collapse hazards. High-rise buildings designed to current code requirements—both steel and reinforced concrete—performed relatively well.

The 1976 changes in the *Uniform Building Code* in the United States, and the 1981 *Building Standard Law* revisions in Japan have improved the ductility of modern concrete structures. Structures built before this in Japan and the United States are quite vulnerable to collapse in seismic events. Unfortunately, both countries have a large inventory of such structures. Code revisions, for the most part, have not been retroactive. Good performance of the newer structures in recent earthquakes shows that the code revisions have been in the right direction.

(a)

Figure 7.8 Northridge, California, January 17, 1994. Nonductile concrete and precast concrete failures. (Courtesy of EERC, University of California, Berkeley.)

(*b*)

Figure 7.8 *(Continued)*

(*a*)

(*b*)

Figure 7.9   Nonductile concrete failures in Kobe, Japan, January 17, 1995.

Figure 7.10   Hanshin expressway, Kobe, Japan, 1995. (Courtesy of Charles Kircher.)

The desperate need to retrofit nonductile reinforced concrete buildings and bridges was clearly illustrated in the Northridge and Kobe earthquakes. These facilities may present a greater risk to life safety than that posed by the older unreinforced masonry buildings. Some of the buildings are quite new, since the ductility requirements in the *Uniform Building Code* were not introduced until 1976. They have a long service life expectancy and really should be upgraded. But this work is particularly expensive. It has been estimated that a retrofit requirement for nonductile concrete buildings in Los Angeles alone would cost $35 billion to $50 billion.

In California, the current strategy for highway structure retrofit work is to prevent collapse of the existing elevated highway supports but allow controlled distortions. Designers concentrate on ductility, confinement details, and over-all stability. The goal is to accommodate potential movement with repairable damage. A thorough design requires incremental study of deformations and rotations without failure. Incremental displacements are investigated, considering changes in cross sections and stress–strain relationships as members yield and crack. Computer software is now available to help with the complex three-dimensional incremental analysis.

As discussed in Section 6.3.2.c, faith in the performance of the special moment-resistant steel frame (SMRF) was challenged by the 1994 Northridge, California, earthquake. In that earthquake, there were a number of frames that experienced brittle fracture in the welded beam-to-column connections.

Thus it became evident that a ductile response is highly dependent on the performance of the connections, even in a steel frame structure.

The same is also true for reinforced concrete structures. A great deal of contemporary seismic research is appropriately focused on the detailing of connections, especially to improve their inelastic performance and to increase the damping potential of energy-dissipating connections. Even a cursory look at the figures contained in this section reveals that the details are important. For ductile performance, closely spaced ties or spirals must confine the reinforcing bars in columns and beams and contain the concrete rubble within the core of the structural members. Integrity of the connections requires continuous steel both directions at beam–column connections, and adequate overlap (development length) of reinforcing bars where they are discontinuous. This continuity of steel reinforcement in the modern ductile concrete frame can create severe steel congestion in the forms. For a ductile frame design, the structural members may need to be larger than actually required for structural calculations simply to provide adequate spacing between reinforcing bars.

Detailing of the reinforcement patterns must be confirmed at the construction site with diligent inspection to be sure that construction conforms to the design intent. Adequacy of connections often relies upon inspection during the construction phase. Of course, quality design and workmanship is of utmost importance to survival in a large earthquake. The earthquake does not review the structural design drawings or the shop fabrication drawings; nor does it read the project specifications to interpret the designer's intentions.

7.8 CORROSION IN REINFORCED CONCRETE STRUCTURES

The deleterious effect of rust formation resulting from the expansive forces generated by the oxidation of iron has been covered in many examples of failure described in previous chapters. Corrosion is a complex electrical/chemical phenomenon; a thorough review of the process is beyond the scope of this book. The increasing cost of repair to the aging infrastructure has brought about a great deal of attention to the subject in recent years. Some excellent texts on corrosion control are now available (Uhlig and Revie 1985) and case-study articles involving repair to corrosion-damaged facilities are regularly published in the technical journals (Gurfinkel 1989, Wilkinson and Coombe 1991).

Although the processes of corrosion are technically complex, there are some common errors, easy to understand, that can increase the effect, both in extent and rapidity of action. Rust formation is a chemical change of iron requiring the presence of water and oxygen. It is accelerated by electrical phenomena such as stray currents and internally generated electric potentials. Galvanic corrosion is caused when dissimilar metals are in contact or connected by an electrolytic solution, such as where aluminum conduit or copper fasteners are used in conjunction with steel reinforcement.

The bibliography on corrosion of reinforcing steel in concrete is mainly a listing of cases where the mix contained free electrolytes, especially chlorine. Whether introduced by an admixture or in the water, chlorine content must be avoided. If it cannot be avoided, special precautions to produce a dense concrete must be taken.

The corrosion of iron embedded in concrete was studied in depth by the Zurich Institute for Testing Materials in the early 1900s. In 1923, a report written by Bruno Zschokke lists three precautions, which are still valid:

1. The concrete should not be too meager but should contain the best percentage of cement, so as to make it impermeable to air. Thus the proportion of calcium hydrate, which prevents oxidation, is increased, and the concrete is also more impermeable to carbonic acid, which neutralizes the lime.
2. There must be no substances in the concrete such as locomotive cinders, often containing sulfur, which exert a chemical action on the iron.
3. The layer of concrete covering the iron should be of sufficient thickness to cover it properly and should not crack under pressure or through shrinkage. (Zschokke 1923)

Seawater may be an economical necessity for concrete mixing in locations where fresh water is not available. It has historically served well in the making of "shell concrete" in many tropical islands, where crushed shells and beach sands were blended and packed into forms to harden slowly into a natural concrete without the addition of cement. When salt water is used in *reinforced* concrete, however, especially in tropical areas where high temperatures and humidity are prevalent, expansion and rusting of the reinforcement will rapidly destroy the structure. This warning was given as early as 1916 in a U.S. Bureau of Public Works Bulletin written to all engineers in the Philippine Islands. They were advised that use of salt water in concrete structures is dangerous and that the use of beach sands and gravels should be permitted only after thorough washing with fresh water.

An interesting comment on the durability of reinforced concrete in buildings is found in Special Report 25 of the Building Research Station of Great Britain, written in 1956: "The common defects that tend to shorten the life of a reinforced concrete building all stem from a simple cause. The enemy of durability is nearly always water, the vulnerable element is the steel reinforcement, and trouble is certain if the concrete fails to keep them apart" (BRS 1956). Design details must provide such protection and construction must not reduce the thickness of cover nor the imperviousness of the concrete.

In Chapter 6 it was noted that corrosion reduces the effective cross-sectional area of an affected structural steel member or component, thereby increasing the stress. Where corrosion causes pits and cracks in the metal, stress concentrations occur. These effects also exist in reinforced concrete, resulting from corrosion of steel reinforcement; the stress in the bar will increase with loss of effective cross section. This is a particularly serious problem with the high-strength steel tendons used in prestressed concrete, since a small reduction in area can cause a catastrophic tendon failure.

A more common problem related to corrosion of reinforcing steel in ordinary reinforced concrete is the spalling and cracking of structural members. The corrosion process replaces the steel with iron oxide that can occupy up to three times the volume of the displaced steel. This increase in volume generates large internal stresses in the concrete that may exceed 28 MPa (4000 psi). Long before unsightly rust stains appear on the surface, cracks parallel to the reinforcing bars may signal that corrosion is taking place below the surface. The cracking and subsequent spalling, especially where freeze–thaw cycles exist, will permit further penetration of moisture and lead to accelerated corrosion.

Certain facilities are more vulnerable to corrosion than are others, requiring particular attention to durability-conscious design and maintenance. These include industrial occupancies where concrete may be subjected to chemical attack and poorly ventilated facilities with high humidity and chlorine vapor content, such as swimming pool enclosures.

The indiscriminate use of roadway deicing salts in the 1960s caused widespread corrosion failures in bridges and parking garage structures (Figure 7.11). These problems, which captured the attention of the entire construction industry, began to surface in the 1970s (Robison 1986). The special vulnerability of parking garages to premature deterioration has already been discussed in Sections 2.7 and 6.4.

Corrodible materials must be protected from the elements: water, air, and damaging chemicals. To prevent corrosion damage, regular maintenance is necessary. This places some responsibility on the owners and operators of facilities. But there are many decisions made by the designers of facilities that also influence the durability and maintainability of structures. Several case studies of corrosion-related failures are given in Sections 1.4.7 and 2.7, some involving design decisions that made inspection and maintenance difficult.

Design strategies to prevent corrosion in new facilities include provisions for moisture protection and drainage of exposed structural members and concrete deck surfaces. Good-quality dense concrete with adequate cover over the reinforcement is the principal key to durability. Insufficient cover has led to many corrosion problems in modern structures, where idealistic designers have pushed the limits, given the strength capabilities of high-strength and prestressed concretes (see Section 8.6). On the other hand, the prestressing of concrete can be very useful in controlling cracks, thereby reducing the penetration of corrosion-causing moisture into the concrete. Often, even nonstructural components, such as curtain wall panels, are prestressed simply for the purpose of crack control.

Figure 7.11 Parking garage failure caused by corrosion, concealed beneath a layer of asphaltic paving. (Courtesy of Raths, Raths & Johnson, Inc.)

The corrosion resistance provided by the concrete mix can be enhanced by air entrainment and by corrosion-inhibiting additives that increase the density and alkalinity of the finished product. The natural alkalinity of concrete traditionally provides good resistance to corrosion, but this alkalinity can be reduced by carbonation in an aging facility (see Section 7.5). Dense concrete, with sufficient cover and supplemented by corrosion-resisting admixtures, is valuable insurance against premature deterioration.

Increased moisture protection can be provided by impervious coatings applied to exposed concrete surfaces in new construction. These include various sealers, liquid plastics, latex-modified concrete overlays, and polyurethane membranes with epoxy coatings.

The reinforcing bars themselves can be galvanized or epoxy coated. Use of galvanized bars or the more recent epoxy coatings as a protection against rusting should not be taken as a miraculous cure-all. Galvanizing does not guard against electrolytic action, especially in a damp environment; the process must be considered as a temporary safeguard, good only until the zinc coating is dissolved. Both galvanizing and epoxy coatings are susceptible to pinhole damage, defects, and fabrication and placement damage. Sometimes the ends of cut bars are not coated. If the surface coating is breached, the resulting trapped moisture may actually cause accelerated corrosion. If the bars are coated after fabrication and carefully placed, however, both methods can yield reliable service.

Avoidance of use of dissimilar metals and rejection of aggregates with undesirable chemical properties can also mitigate the potential for corrosion (see Section 7.3). Chlorides in the mix should be carefully limited. Maximum chloride content is controlled by ACI specifications. A small amount of chloride in the mix may be beneficial; it increases the early strength of the concrete and speeds up the set in cold weather to guard against freezing of the wet concrete.

At present, the "ultimate" and most expensive procedure used to combat corrosion in steel and ordinary reinforced concrete structures is cathodic protection (Tighe and Van Volinburg 1989). Cathodic protection is acclaimed as the only method that will stop corrosion in reinforced concrete structures, regardless of the chloride content of the concrete. It is important to note, however, that a cathodic protection system will not repair damaged materials; it will simply prevent further corrosion.

Cathodic protection systems introduce a low voltage and current that immunizes the steel against corrosion, halting the flow of electrons from anodic to cathodic areas of the structure. The method has long been used in pipelines, tunnels, and offshore structures. Only recently has it been applied to bridges, and even more recently to buildings. Cathodic protection systems were used successfully in pipelines for 40 years before the technology was applied to bridges. Various conducting systems are available, including mesh overlays, conductive polymer mounds, slotted systems, and conductive carbon-impregnated paints.

Cathodic protection has been installed in many parking garages. Most parking garages behave as efficient giant batteries, with the attendant severe corrosion problems. Cathodic protection can eliminate the structure's ability to maintain the corrosion cycle.

For all of their benefits, cathodic protection systems do have some drawbacks. They require continuous monitoring and introduce additional maintenance requirements for the systems themselves. Anodic materials must be replaced at regular intervals and the electrical circuits must be checked monthly. Although these systems seem to be performing satisfactorily at present, it remains to be seen how they will perform when the newness wears off; they may not receive the attention they require from owners, who typically think of their facilities as passive objects. These are not passive systems, and active systems require owner/operators who are committed to maintenance.

It should be noted that the possible "overprotection" provided by imbalanced cathodic protection systems may produce hydrogen as a by-product of electrolysis. For this reason, cathodic protection may not be an acceptable solution for prestressed concrete structures. The high-strength-steel tendons are susceptible to failures caused by hydrogen embrittlement (Pullar-Strecker 1987).

Research on the deteriorating infrastructure in the United States has led to numerous new techniques for the diagnosis and repair of corrosion-damaged concrete facilities. The available techniques have become so numerous and technically complex that corrosion-repair activity has become a specialized professional discipline. In 1989, a new International Association of Concrete Repair Specialists was formed. The exchange of information on an international level is quite helpful. A number of skilled concrete repair specialists exist in the developing world, where design deficiencies, construction errors, and high-chloride mix content have been compounded by severe weather exposures to produce accelerated corrosion (Al-Mandil and Ziraba 1990; Al-Mandil et al. 1990; O'Connor 1994; Raikar 1994a,b). In countries such as India, where housing is so desperately needed, it is disconcerting to see so much of the available capital being expended on the repair of avoidable deterioration (Figure 7.12).

Several useful textbooks and trade publications have been written on the subject of concrete repair (Sabnis 1985, Pullar-Strecker 1987). Solving corro-

Figure 7.12 Corrosion-damaged structure in India.

sion problems is a particularly difficult technical challenge. The diagnosis must be sound and specialized consultants may be necessary. It is entirely possible to apply an ill-conceived repair, based on an inaccurate diagnosis, that actually accelerates the deterioration process. Applying an impervious coating over contaminated concrete is not a valid solution. The National Association of Corrosion Engineers (NACE) provides many useful publications on corrosion topics, including practical repair procedures.

Nondestructive methods for diagnosing the source and extent of corrosion damage include the use of mechanical hammers, chain dragging over concrete decks, and acoustic emission techniques. Infrared thermography can find corrosion before evidence appears on the surface, since corrosion oxides emit higher temperatures than the surrounding materials. Active corrosion can be located with electrical half-cell methods. Voltmeters are used to map the electrical potential variations over the surface of the deck.

The damage done by corrosion is usually irreversible; the damaged concrete and corroded steel must be removed by hydrodemolition or other mechanical methods. The remaining surface is then prepared and new materials are applied. The new materials may include ordinary concrete, dense concretes with superplasticizers or silica fume additives to provide greater impermeability or electrical resistivity, corrosion-inhibiting admixtures, dry pack materials, shotcrete, preplaced concrete aggregates, latex-modified concretes, polymer concretes, epoxy injections or overlays, and/or protective coatings. Cathodic protection systems may be introduced to halt the spread of corrosion.

A new approach, similar to cathodic protection but using much higher electrical currents, has been successful on several projects in actually removing chloride ions from contaminated concrete. The chloride ions are drawn out by the electrical current and pulled to a temporary anode located outside the structure. First used experimentally on bridges, this approach, if it continues to be successful, offers a distinct advantage over continuous cathodic protection systems. The ongoing maintenance associated with an active system is not required (Green 1988).

7.9   REFERENCES

ACI, 1984. *Causes, Evaluation, and Repair of Cracks in Concrete Structures,* ACI Committee 224, American Concrete Institute, Detroit, MI.

ACI, 1982. *Guide to Durable Concrete,* ACI Committee 201, American Concrete Institute, Detroit, MI.

Al-Mandil, M., and Y. Ziraba, 1990. "Assessment of Damage to Concrete Girder-Slab Bridges in Saudi Arabia," *Journal of Performance of Constructed Facilities,* American Society of Civil Engineers, New York (August).

Al-Mandil, M., M. Baluch, A. Azad, A. Sharif, and D. Pearson-Kirk, 1990. "Categorization of Damage to Concrete Bridge Decks in Saudi Arabia," *Journal of Performance of Constructed Facilities,* American Society of Civil Engineers, New York (May).

ASCE, 1982. "Engineer's License Revoked After Condo Collapse," *Civil Engineering,* American Society of Civil Engineers, New York (February).

Berry, E., and V. Malhotra, 1980. "Fly Ash for Use in Concrete: A Critical Review," *ACI Journal,* American Concrete Institute, Detroit, MI (March–April).

BRS, 1956. "Durability of reinforced concrete," Special Report No. 25, Building Research Station, HMSO, London, UK.

Day, R., 1995a. "Damage to Concrete Flatwork from Sulfate Attack," *Journal of Performance of Constructed Facilities,* American Society of Civil Engineers, New York (November).

Day, R., 1995b. "Pavement Deterioration: A Case Study," *Journal of Performance of Constructed Facilities,* American Society of Civil Engineers, New York (August).

Engineering News-Record, 1975. McGraw-Hill, Inc., New York (January 16).

Green, P., 1988. "Structures Need a Low-Sodium Diet," *Engineering News-Record,* McGraw-Hill, Inc., New York (March 24).

Gurfinkel, G., 1989. "Restoring an Impaired Concrete Silo," *Journal of Performance of Constructed Facilities,* American Society of Civil Engineers, New York (May).

Hadipriono, F., and R. Sierakowski, 1988. "Concrete Construction Problems in Far East and Southeast Asian Countries," *Journal of Performance of Constructed Facilities,* American Society of Civil Engineers, New York (August).

Kaminetzky, D., 1991. *Design and Construction Failures,* McGraw-Hill, Inc., New York.

Mullick, A., and C. Rajkumar, 1993. "Assessment of Concrete Construction Projects in India: Multidisciplinary Approach," *Journal of Performance of Constructed Facilities,* American Society of Civil Engineers, New York (May).

NBS, 1977. *Investigation of the Skyline Plaza Collapse in Fairfax County, Virginia,* NBSIR 94, National Bureau of Standards (National Institute of Standards and Technology), U.S. Department of Commerce, Washington, DC.

NBS, 1981. *Investigation of Construction Failure of Harbour Cay Condominium in Cocoa Beach, Florida,* NBSIR 81-2374, National Bureau of Standards (National Institute of Standards and Technology), U.S. Department of Commerce, Washington, DC.

O'Connor, J., 1994. "Middle Eastern Concrete Deterioration: Unusual Case History," *Journal of Performance of Constructed Facilities,* American Society of Civil Engineers, New York (August).

Pullar-Strecker, P., 1987. *Corrosion Damaged Concrete: Assessment and Repair,* Butterworth & Co. (Publishers) Ltd., London.

Raikar, R. N., 1994a. *Durable Structures Through Planning for Preventive Maintenance,* R&D Centre, SDCPL, Raikar Bhavan, New Bombay, India.

Raikar, R. N., 1994b. *Diagnosis and Treatment of Structures in Distress,* R&D Centre, SDCPL, Raikar Bhavan, New Bombay, India.

Roberts, M., 1981. "Carbonation of Concrete Made with Dense Natural Aggregates," Building Research Establishment Information Paper, Building Research Establishment, Garston, Waterford, England.

Robison, R., 1986. "The Parking Problem," *Civil Engineering,* American Society of Civil Engineers, New York (March).

Rollings, R., 1993. "Curling Failures of Steel-Fiber-Reinforced Concrete Slabs," *Journal of Performance of Constructed Facilities,* American Society of Civil Engineers, New York (February).

Rzonca, G., R. Pride, and D. Colin, 1990. "Concrete Deterioration, East Los Angeles County Area: Case Study," *Journal of Performance of Constructed Facilities,* American Society of Civil Engineers, New York (February). (See also Discussion by B. Mather in February 1992 issue.)

Sabnis, G., ed., 1985. *Rehabilitation, Renovation and Preservation of Concrete and Masonry Structures,* American Concrete Institute, Detroit, MI.

Tighe, M. R., and D. Van Volinburg, 1989. "Parking Garage Crisis," *Civil Engineering,* American Society of Civil Engineers, New York (September).

Uhlig, H., and R. W. Revie, 1985. *Corrosion and Corrosion Control,* John Wiley & Sons, Inc., New York.

Vaccaro, G., 1956. "Salt, Tide Damage to Bridges Erased by Novel Facelifting," *Engineering News-Record,* McGraw-Hill, Inc., New York (May 3).

White A., 1925. "The Fundamental Cause of the Disintegration of Concrete," *Concrete,* American Concrete Inst., Detroit, MI, (May).

Wilkinson, E., and J. Coombe, 1991. "University of Illinois Memorial Stadium: Investigation and Rehabilitation," *Journal of Performance of Constructed Facilities,* American Society of Civil Engineers, New York (February).

Zschokke, B., 1923. "Corrosion of iron embedded in concrete," Zurich Institute for Testing Materials, Zurich, Switzerland.

8

PRECAST AND PRESTRESSED CONCRETE STRUCTURES

Precast concrete and prestressed concrete are relative newcomers to the construction industry. Despite the limited history of such structures, there are many examples of successful projects. There also exist some examples of deficiencies that can be corrected with improved design and construction practices. Precast and prestressed concrete offer some important advantages over other structural materials for certain projects. These include the potential for better quality control over the materials and workmanship, and enhanced economy through production line techniques where the design is based on elements of repetitive dimensions. Yet, as with all other materials, there have been some lessons involving design, construction, and operational errors. The industry is quite diligent in publishing case studies of experience with prestressed concrete, and there has been a steady decline in the number of performance deficiencies as a result.

The most common problems are associated with field connections between elements, bearing provisions at the supports, and erection accidents. With prestressed concrete, whether pretensioned or post-tensioned, there are some specific problems related to the prestressing equipment, the anchorage hardware, and sequencing of the prestressing operation. A prestressed concrete building, bridge, or tank is not a passive object. The intentional stresses locked into the system require consideration throughout the life of the facility, presenting unique challenges for owners and operators. During renovations the prestressing forces must be maintained. Alterations, maintenance activities, and even the eventual demolition of a prestressed concrete facility must be considered carefully lest the structure react in embarrassing or catastrophic ways.

290

In any project involving precast or prestressed concrete, the designer should be aware of the limitations and capabilities of local fabricators and specialty contractors. For economy and practicality the dimensions of elements and structural/architectural details should conform to local capabilities. Ideally, the designer will consult with representatives of local fabricators early in the design process to ensure that the design is constructable.

8.1 PRESTRESSED CONCRETE: THE MODERN STRUCTURAL MATERIAL

Although the concept of prestressed concrete has been in existence nearly as long as the existence of ordinary reinforced concrete, its implementation was made possible only by the development of high-strength steels, a product of modern material science. The initial stress in the prestressing steel must be sufficiently high so that its beneficial effect is not negated by shrinkage and long-term creep in the concrete. Thus prestressed concrete is a modern material, one in which the potentials (and some of the problems) have yet to be discovered.

Prestressed concrete uses the two materials—concrete and high-strength steel—in an extremely efficient manner, exacting the best innate characteristics from each. Unlike ordinary reinforced concrete, the entire cross section is designed to remain in compression. Because of this, all of the concrete contributes to the structural capacity of the member. Ordinary reinforced concrete is unable to take advantage of the new high-strength-steel strands as tension reinforcement; the cracks occurring in the tension zone of a structural member would be unacceptably large. But high-strength steels and concretes can be put to work efficiently in prestressed concrete.

The introduction of prestressing technology made possible some dramatic and innovative lightweight concrete structures of extremely long span. For buildings, long-span floor systems of minimal construction thickness are available, and thin prestressed concrete shell roofs of single and double curvature achieve statistics that are impossible in nearly any other material. The evolution of prestressing techniques brought about novel concepts for highway structures, combining modern materials and methods to create beautiful and economical bridges.

Yet these remarkable advances have not come entirely without disappointments. In some of the early projects, idealistic designers reduced concrete thickness to the point where there was insufficient cover to protect the steel from corrosion. These designs caused costly maintenance, and in some cases, catastrophic failure. Other poorly detailed facilities failed due to restrained movement stresses over the long term or even immediately following erection. The assumption that cracking would be eliminated by the compression prestress has not proven true. Cracking does occur in prestressed concrete members, for many of the same reasons that it occurs in ordinary reinforced concrete.

The prestressed concrete industry matured quickly in a wide variety of engineering applications and must guard against a repetition of the mistakes made in childhood. It is therefore quite proper to describe what did not work in the early days while the techniques of pretensioning and post-tensioning were rapidly evolving. Fortunately, several of the pioneers in the field of prestressed concrete, sometimes despite considerable resistance, were diligent in publicly presenting summaries of their investigations of troubles, nonsuccesses, and costly maintenance details. Contemporary prestressed concrete structures owe their success to the broad dissemination of information learned from the performance of early experiments with the material.

The history of prestressed concrete in the United States is younger than in Europe. Some very useful information about long-term performance of the material is published in European journals. There are records in both the United States and Europe of successful performance and of sufficient difficulties to warrant a summary of what not to do. It makes little sense to apply prestressing technology to designs that will not perform properly, where future maintenance costs will more than overcome predicted initial economies.

8.2   PREFABRICATION: RELIABLE MEMBER PRODUCTION TECHNIQUES

Precast concrete components are usually manufactured in a plant environment using assembly line techniques. Quality control procedures are rigorously applied to ensure the uniformity of the product. Thus precast concrete is not as susceptible as cast-in-place concrete to the many environmental and workmanship variables discussed in Chapter 7. The concrete quality is almost always superior to that achieved with site-cast concrete. The strength and density of the concrete is more uniform and predictable.

Problems in precast and prestressed concrete construction generally do not stem from concrete quality deficiencies, but rather, from the details. There are many examples of cracking and connection failure due to restrained movements, such as where weld plates have been used at beam-end connections. Sliding pads that permit joint movement due to shrinkage, creep, and load-related rotations must be provided to avoid these problems. Bearing length must be sufficient to allow for the long-term effects of shrinkage and creep, especially in long-span systems. Concentration of stress caused by notches or reduced sections at connections is another typical source of failure. The prestressing operation introduces failure modes involving high bearing stresses where jacking forces are applied and problems with the anchorages and other prestressing hardware. Adding to these technical problems are some procedural problems in the industry. Because several design specialists contribute to the creation of a prestressed concrete facility, the project delivery system may become rather complicated. Confusion of professional responsibilities and miscommunications of design intent have contributed to some prominent failures.

Useful documents and design guidelines for both member and connection design are published by the Portland Cement Association (PCA) and the Precast/Prestressed Concrete Institute (PCI). PCI certifies precast/prestressed concrete plants, helping to maintain quality control in the industry. The PCI Journal is a trade journal, and as such emphasizes the many "success" stories achieved in precast and prestressed concrete. However, the journal also contains articles on recurring problems in the industry, so that the product can be improved. PCI publishes design manuals that include successful connection and fabrication details (PCI 1971).

The designer of a prestressed concrete member must carefully consider the stress conditions at each stage of the life of the member, any one of which may prove to be fatal. These stages include (1) the stresses due to the initial prestressing force (and the sequence in which it is applied); (2) prestress plus dead load (the condition under which the member exists most of its life); (3) prestress plus dead load plus live loads; (4) future conditions, when prestress has been reduced by the long-term effects of shrinkage and creep; and (5) stresses under unusual load conditions, such as extreme winds or seismic events.

Prestressing concrete can help to control cracking. For this reason, many parking structures are constructed with prestressed concrete. Reduced cracking and the lower water/cement ratios generally found in precast concrete help to mitigate the corrosion problems that are so prevalent in such structures. Even nonstructural precast panels are often prestressed to reduce cracking. However, early claims that prestressed concrete will never crack because the concrete is always in compression were simply not accurate. There continue to be many incidents involving cracking in prestressed concrete, whether from design errors or from misuse of facilities (Dias et al. 1994).

Cracking can occur from the same sources that were outlined in Chapter 7. One recent example involves over $200 million in claims brought by a number of railroads in the northeastern United States against a supplier of precast concrete railroad cross-ties. The ties, approximately 500,000 in all, began cracking shortly after installation in the early 1980s. One explanation given was that the cracking resulted from a reaction between the alkali in the cement and the silica content of the aggregate, although high curing temperatures and sulfate levels may have also contributed to the expansive reaction in the concrete. The resultant cracking led to a loss of prestress and loosening of the rail-fastening hardware (ENR 1992).

Although rare, design and fabrication errors have occurred in the members of precast and prestressed concrete structures and continue to the present day. In 1957, at Omaha, Nebraska, five 30-m (100-ft) post-tensioned beams had been erected with partial prestress and were fully tensioned in position.

Before grouting could be accomplished, one of the beams failed when the center buckled upward 300 mm (12 in.). Bars used in the beams were 24 m (80 ft) long and were spliced, with seven of the 14 splices located at the point where failure occurred. To make room for the splices, 75-mm (3-in.)-diameter sheaths were used, which eliminated a large part of the concrete section. Such a large void in the cross section would not be permitted in ordinary reinforced concrete and must be avoided as well in prestressed members.

In August 1990, one worker was killed and another severely injured by a beam collapse at a $592 million airport terminal expansion project near Pittsburgh, Pennsylvania. A precast concrete beam failed near one end, shortly after a roof plank was placed on it. All investigators and parties involved in the design and construction agreed that the failure was due to insufficient reinforcement. The beam was fabricated such that a 500-mm (20-in.) section of the beam had no tensile reinforcement. The simply supported beam was 10.4 m (34 ft) long. At the supported end it was notched, narrowing the cross section to a 600-mm (24-in.) depth. The failure occurred near the notch. The engineering firms were cited for violations. It was alleged that design errors went unchecked and that critical dimensions were not shown clearly in the drawings. The precast fabricators and contractors were also fined for improper placement of the reinforcement. Hundreds of other beams and precast members already cast and erected were tested, and some were strengthened with steel plates. Those not yet fabricated were revised, adding additional reinforcement (ENR 1991).

A portion of a 28,000-m² (301,000-ft²) shopping mall parking structure in Raleigh, North Carolina, collapsed in 1993. The corbels at the top of the 460 × 460-mm (18 × 18-in.) precast concrete columns that supported the deck were insufficiently reinforced. The steel did not extend fully into the corbel. Cold temperatures caused the precast deck to contract so that the deck was bearing on the edge of the corbel that was not reinforced. Fifteen flawed columns were repaired by the precast concrete supplier (ENR 1993).

Gurfinkel (1988) describes a large precast concrete warehouse structure built in Champaign, Illinois, in 1974. Four years after completion, concrete fragments weighing up to 5 kg (10 lb) or more began to fall to the floor, posing a hazard to workers. An investigation uncovered improper reinforcement in the flanges of precast concrete I-girders that supported the double-tee roof deck. The falling debris was composed of fragments of the girder flanges. The potential for catastrophic failure of the roof deck members due to loss of support was mitigated by retrofit work in the structure.

The one-story warehouse was 80.5 m (264 ft) wide by 117 m (385 ft) long. Precast concrete columns 460 mm (18 in.) square and 7.6 m (25 ft) in height supported the precast girders that spanned the 8.9 m (29 ft) between columns. The girders were 910 mm (36 in.) deep, with top flanges 300 mm (12 in.) wide by 140 mm (5.5 in.) thick and bottom flanges 460 mm (18 in.) wide by 230 mm (9 in.) thick. The webs of the girders were 150 mm (6 in.) thick. Spanning the 19.5 m (64 ft) between girders were precast and prestressed double-tees 2.4 m (8 ft) wide with 560 mm (22 in.) deep stems and a 50-mm (2-in.)-thick slab.

Top flanges of the girders were severely cracked, with fragments missing in the vicinity of the stems of the double-tees. Less than 38 mm (1.5 in.) of support remained for bearing at a number of the connections. The cracks occurred because of insufficient attention to the character of loading on the top flanges of the girders; very high stresses concentrated at the extreme edge of the top girder flange because of the slope of the roof deck. The condition was exacerbated by restrained movements, since no elastomeric bearing pads were provided. Had the girders been designed with additional transverse reinforcement and with a chamfered edge, they might not have experienced the cracking problem. The provision of elastomeric bearing pads would also have reduced the horizontal forces that played a contributing role in the problem.

In this case the repair involved the introduction of shop-fabricated structural steel supports for each double-tee stem, cantilevering both directions from the existing precast girders. The innovative repair provided each stem with sufficient support, using a repetitive detail that proved to be economical to fabricate and easy to install.

Prestressed work relies on concrete of high and uniform strength. Obtaining predictable and reliable strength should take precedence over the desire to reduce the weight of the concrete. Lightweight aggregates save so small a part of the overall weight that they should be used only where the certainty of obtaining proper strength can be guaranteed. Weight reduction was the reason for changing to lightweight concrete in manufacturing the girders for the Kenai River Bridge in Alaska in 1955. The girders were cast in Oregon and transported to Alaska. The design specified 31 MPa (4500 psi) concrete, but about one-third of the end blocks cracked when the stressing was accomplished under the 150 by 150 mm (6 by 6 in.) plates holding the Freyssinet anchors. When the end diaphragms were tensioned, some cracks and spalls appeared at the bottom of two girder ends. By 1958, the cracks had increased in length to 1 to 1.5 m (3 to 5 ft) and some beam corners were loose. In 1962, complete replacement was ordered, even though samples of concrete then tested 33 MPa (4780 psi), a 5 percent increase in seven years.

Problems can also occur due to the casting methods used. A large number of both pretensioned and post-tensioned factory-produced I-beams were manufactured during the winter of 1957–1958 in a plant specifically built for a major highway project in Virginia. A few of the beams were rejected before shipment due to the appearance of cracks, but inspection in the field uncovered many more beams with serious problems. There were then 148 post-tensioned beams of 18 to 30 m (60 to 100 ft) span and 204 pretensioned beams of 9 to 15 m (30 to 50 ft) span intended for use in 18 highway bridges. The total order was for 572 beams. Some of the bridges were already complete with cast-in-place concrete deck already placed, others were in the process of erection, and many of the beams were in storage, both at the bridge sites and in the plant.

Inspectors found cracks of intermittent length outlining the curved metal sheaths in the web and along straight lines in bottom flanges. Of the 148 long beams, 101 had severe cracks, some at more than one location. Cracking was noted in 33 webs, 89 bottom flanges, and 4 top flanges. Of the 208 short beams, seven showed cracks in the webs, and in 16 beams, the bottom reinforcement was exposed, mostly in poor concrete covering. The latter beams were corrected by patching and epoxy–cement filling of cracks and honeycomb pockets. Most of the longer beams were too badly cracked for repair.

The first report placed the blame on freezing of the grout. Many of the beams were rejected and new ones were ordered. When similar difficulty was found in beams made in early summer, however, a more extensive investigation was initiated. The cause was then determined to be the unequal shrinkage of the different concrete thicknesses when the steam curing was shut off. The typical cycle of steam application consisted of 5 hours of normal setting at about 15°C (60°F). During the next $1\frac{1}{2}$ to 2 hours, steam was added slowly to raise the temperature to 74°C (165°F). The temperature was then held at 74°C (165°F) for 22 hours. After this, the steam was turned off and covers removed.

The metal conduit acted as a radiator to cool the thin concrete cover more rapidly and some fine separation cracks developed. The grouting under pressure after the post-tensioning was completed opened up the cracks. The combination of metal sheath inside the beam and the steam curing process generated uneven shrinkage. A search of the literature indicated that similar difficulties had been encountered elsewhere, and in some European countries the use of metal sheaths had already been prohibited when steam curing was to be used.

Shrinkage of the concrete during setting and curing of items cast in steel forms has been known to pull the flanges and webs apart, since the forms do not give. Precast concrete joists often exhibit fine cracks at the web junction with the flanges, but when reinforced with diagonal steel wires or mesh, full-scale load tests conducted in 1945 indicated no reduction in strength from such cracking.

Special attention must be given to stress concentrations under the bearing plate of a post-tensioned anchorage. Here the highest concrete strength is required, with some hoop or spiral reinforcement to resist lateral expansion under the high stress. The combination of bearing plate, often with attached anchors, cable or rod sheath, stirrups, and hoops, is not conducive to getting a dense, strong concrete cross section. Usually, there is no room for a vibrator, and densification of the concrete mix depends on manual rodding and tamping "as far as possible." Sometimes the result is not sufficient.

In one very large prestressed concrete viaduct project in New York, the bearing plates, set in the web face of the outer girders for cross-tensioning the roadway width, crushed the web when the jacking was applied. The concrete in back of the plate was found to be quite porous; there was too much steel hardware in back of the plate to permit proper filling of the web with concrete.

In a fine monumental building in Philadelphia, the main floor was comprised of circular sectors with field post-tensioning required to join the pretensioned castings. The narrow end of the sector consisted of two separate bearing areas, 180 by 360 mm (7 by 14 in.) each, with a total of four stressed rods. The bearing plates were located at the top of the casting with little clear distance along three edges. With each rod stressed up to 578 kN (130 kips), the average unit pressure under the bearing plates was 27.6 MPa (4000 psi) for a specified concrete mix of 34.5 MPa (5000 psi). When only the upper rod was stressed to full value the edge pressure was higher, depending on the closeness of contact. Some localized shear failures occurred in back of the bearing plates. The result was a complete stoppage of work, with a tentative order to demolish part or all of the structure built at that time. This order was later rescinded, but only after extensive tests, inspection, and arguments. In the broken end bearings it was found that two small steel stirrups had been omitted by the fabricators in the plant, "to provide space for the concrete."

Additional concrete strength was attained by filling the gap between the two separate bearing areas, a gap that had been left as an opening for vertical mechanical services. After the spalled and cracked ends were rebuilt, all subsequent post-tensioning was performed in a two-step procedure. Each rod was stressed to 50 percent value, then the other 50 percent was added in sequence. Since the rods were too closely located to provide clearance for jacks, this required four operations, but it reduced the high local eccentric stress concentrations that would have occurred if each rod had been fully stressed in one operation. The completed project was a credit to both the architect and the engineer. However, a great deal of trouble and unfavorable publicity could have been avoided by providing larger bearing areas and more easily accomplished details from the beginning.

Camber induced by eccentric prestressing can create problems if not considered carefully in the member design. Several long precast beams, up to 34.4 m (113 ft) in length, for a skewed bridge crossing were detailed with steel bearing plates to be anchored to the abutment seats. The post-tensioning induced camber in the beams, according to design, with straight profiles expected under full dead load of the stringers, deck slab, and paving. But the beams were erected in the abutments under their self-weight only, and this condition alone was insufficient to reduce the camber. Great care had been taken to set the bearing plates on the abutment seats in exact position, with the anchor bolts perfectly aligned to fit the templates of the concrete beam ends. Of course, the ends did not fit. Not only were the beams apparently short, but the shoe plates were slightly rotated. Repairs had to be made with 73-Mg (80-ton) beams held in the air by two cranes, an expensive and time-consuming operation. Shoe and bearing details must be designed to provide for the shape and length changes as a prestressed beam is loaded.

Details that work satisfactorily in structural steel or normal cast-in-place concrete design may not work at all well with precast prestressed members.

The end of a post-tensioned, cast-in-place girder carrying a building column over a garage door opening was designed to rest on an exterior concrete wall for the full thickness of the wall. The girder was cast short, however, to leave room for masonry veneer at the face of the wall. After the girder was tensioned and two stories of column load were applied, the bearing proved to be insufficient. The girder tore away from the slab and out of the wall. Fortunately, there were no workers in the immediate area, and only replacement cost was lost by the accident.

Prefabricated single- and double-tees must be laid out on the plans with sufficient tolerance for the usual deviations in line and camber of the component units making up the roof or floor area. In 1961, in the first use in New York City of precast pretensioned tees for a roof over a lecture hall, the deviations from straight lines at the contacts of the flanges required wider joints than expected and the total length of the roof had to be corrected by cutting several inches from the last unit.

8.3   CONNECTIONS: CRITICAL DETAILS

Although the quality of the materials and structural members encountered in precast and prestressed concrete is generally very high, the literature contains many examples of connection deficiencies (Birkeland and Birkeland 1966, Hanson 1969, Kaminetzky 1991). The stresses at connections between elements and at anchorages can be quite complex, involving many interrelated conditions. The design of the members themselves must take these conditions into account, as noted by several examples in Section 8.2.

Of course, the structural performance of some connections in transferring forces from member to member is critical to the safety of any system. For certain structures, such as multibeam prestressed precast concrete box girder bridges, the joints are integral to the load-sharing mechanism. Grouted shear keys tie adjacent girders together and guarantee the integrity of the deck waterproofing system (Huckelbridge, El-Esnawi, and Moses 1995). However, in other cases, the preferred details permit unrestrained relative movement among components. Failure to provide sufficient flexibility for movement is responsible for the vast majority of serious problems in precast and prestressed concrete facilities.

The largest failures can come from improper small details. Joints must allow for thermal expansion and contraction, creep, and shrinkage, all of which are time-dependent changes. Bearing stresses may be increased by axial forces, with significant bracket stresses and, sometimes, splitting of the concrete seat on the beam. Prestressed members continue to change camber with time; some provision for rotation at the support must be provided.

In a large industrial building on the west coast, the main roof was used for automobile parking. The roof was framed with precast post-tensioned beams and girders that were supported by precast columns. The bays were 12 by 13.5 m (40 by 44 ft); columns were 600 by 600 mm (24 by 24 in.) in section. Girders were 13.5 m (44 ft) long, pinned to the columns, and made continuous by adding heavy negative-moment steel over the supports in the 115-mm (4.5-in.)-thick cast-in-place concrete roof slab. Beams were located at the column lines and at the third points of the girder spans, sitting on brackets cast with the girder webs. The beam support was intended to permit horizontal movement, with an asbestos packing seat layer and concrete filling in the gap between beam ends and girder web.

To reduce total construction thickness, the end blocks of the beams were notched 300 mm (12 in.) above the bottom. Theoretical bearing areas were 300 by 125 mm (12 by 5 in.), but somewhat smaller areas were actually found, where shrinkage of the precast concrete resulted in beams of shorter than designed length. Such shrinkages reduced the bearing area by as much as 20 percent in some locations.

The concrete design mix was 27.6 MPa (4000 psi) in three days and 34.5 MPa (5000 psi) in 28 days. The concrete contained satisfactory glacial sand and gravel, a water/cement ratio of 0.435, and type III cement with $1\frac{1}{2}$ to 2 percent calcium chloride. After steam curing, test cylinders showed strengths of 41.4 MPa (6000 psi) in three days and 45.5 MPa (6600 psi) at 28 days for beams and 38.4 MPa (5565 psi) for girders. Beams were cast within steel forms, stripped after 3 hours, including 2 hours of steam. They were then steam cured for 20 to 24 hours at 38 to 60°C (100 to 140°F). Girders were stripped 5 hours after casting, including 4 hours of steam, and then steam cured for 30 hours at 32 to 38°C (90 to 100°F). Post-tensioning of the high-strength steel rods came at three days, when the concrete strength indicated by a Schmidt hammer test was at least 27.6 MPa (4000 psi). Of the 606 units manufactured, only four required rejection. Beams and girders had stirrup loops extended for composite action with the slab concrete.

Shortly after the completion of the roof slab, failures of the beam ends were noted at the expansion joints, and a more positive sliding seat detail was installed (Figure 8.1). After that repair, over a period of six years, progressive distress of end blocks developed in both the beams and girders. Typical cracks in the beams were vertical shear separations above the bearing area and diagonal tension cracks originating at the reentrant corner of the seat recess. It was evident that the post-tensioning did not provide a uniform compressive stress across the full section of the beam. An analysis of the beam just above the bearing indicated tensile stresses at the edge of the seat. A beam acting as a simple span, partly restrained by the roof slab, must have freedom to rotate when loaded. The sloping end modified the bearing reaction and caused the faces of the girder brackets to spall. Similarly, the shrinkage of the girder lengths caused a number of the columns to crack vertically.

Figure 8.1 Precast roof cracking from shrinkage.

Corrective work consisted of through-bolt anchored steel bridles added at each beam end to transfer the reaction without reliance on the girder seats and with pickup at the bottom of the full depth of the beam. Columns that showed distress were encased at the top with tensioned strap iron, installed as a bolted yoke, and gunited for fireproofing (Figure 8.2).

In 1984, two schools in Florida were declared structurally unsafe. Both schools, only nine years old, were closed. During a routine inspection, cracks were found in 22.4-m (73.5-ft)-long prestressed concrete double-tees. Where the double-tees were set on beams, they were welded at both the top and stem. The preferred detail recommended by PCI would have provided elastomeric bearing pads and no welds. The welds restrained the rotation that would have taken place under temperature stresses, so cracks inevitably appeared. The structural engineer of record alleged that the design of all the precast elements and their connections was not his responsibility but that of the precast manufacturer's engineer (ENR 1984).

In May 1994 it was announced that the Alabama Department of Transportation would begin repairs on some 74 cracked Interstate 565 viaduct girders. Cracks on the two-year-old 4-km (2.5-mi)-long structure were not considered life threatening. They were caused by upward camber in the 180-mm (7-in.) concrete deck that was created by sudden warming during hot summer mornings contrasted with cold night temperatures. Up to 28°C (50°F) changes were recorded in only a few hours. The deck curved upward, pulling the supporting

Figure 8.2 Saddle supports for beams to neutralize cracking.

30-m (100-ft)-long 1.4-m (54-in.)-deep precast, prestressed bulb-tee girders lengthwise and rotating the ends. The webs of the girders cracked, usually within 1.5 m (5 ft) of the supports. No expansion joints were provided at the piers. Instead, the girder ends were cast into concrete blocks extending 75 mm (3 in.) along the ends of the girders. Dowels extended into the blocks to provide continuity. Clearly, there was insufficient flexibility to accommodate the temperature-induced movements (ENR 1994).

The effect of shrinkage in changing the geometry of an assembly of members must be considered in the design and in the details of jointing. High-early-strength concrete, obtained by special cements, calcium chloride, and steam curing, is subject to greater shrinkages and over longer periods of time than are normal-mix concretes.

In all operations of hoop stressing, the diameter of the vessel is reduced and freedom for movement must be provided at contacts with the unstressed members. Such incompatibility of deformation caused serious diagonal shear cracks to develop in a concrete penstock pipe in Cuba in 1958. The joints were covered by a sleeve that was tightly grouted. When pressure was applied, the greater stiffness of the sleeve resisted expansion of the pipe ends and the shear cracks opened. The cure was to dig out the grout and caulk the joint with a plastic filler.

8.4   IMPORTANCE OF QUALITY CONTROL AT THE CONSTRUCTION SITE

As with all other construction materials, the best designs in precast and pre-stressed concrete can be ineffective unless the work done in the field is of high quality. If the design is marginal, construction deficiencies can compound the errors, increasing the potential for serious problems. In Section 8.3 the critical nature of details and connections was reviewed. Certainly, the execution of details that permit the kind of movements already discussed requires diligent field inspection and educated, capable workers.

Since the reliability of precast concrete members has been established by strict quality control and fabricator certification, the field work nearly always seems to be the weak point in the performance of precast facilities. One contributing factor to this problem is that precast systems are sometimes selected because they can be constructed quickly. An overriding emphasis on rapid project completion often comes at the expense of construction quality; less attention is likely to be focused on important details executed on the construction site.

Skilled supervisors who understand the design intent and can communicate it clearly to the field workers are needed full-time at the construction site while all prestressed concrete work is erected. M. D. Morris, founding editor of the ASCE *Journal of the Construction Division,* recounts an episode he witnessed in the former USSR. In 1961, Morris and several other journal editors were invited by the Soviet Institute of Construction and Architecture to observe some selected construction projects in Moscow. One of these was the site of the new Pravda Building. *Pravda* was the Communist Party's national daily newspaper. Its new showcase home was designed by the USSR's leading architects and engineers. The precast, prestressed concrete structure was manufactured of the finest materials under quality-controlled conditions and expert inspection. The workers who assembled the fabricated members as they were delivered to the job site were hand-picked skilled veteran craftsmen.

Yet halfway through the erection, the project experienced a total progressive collapse. Fortunately, this occurred during the night; no one was injured. In a candidly honest response to Morris's blunt question "Why?", his host told him that in those times, Soviet construction suffered from an abundance of great generals and super soldiers, but a shortage of sergeants to transliterate the architecture and engineering orders to the field force. Without proper site supervision, the best designs, using the best materials, may be doomed to fail.

In the case of post-tensioned structures, following the proper sequence of tensioning the strands is absolutely essential. The contractor is literally in a position to determine the character of stresses, both in magnitude and distribution, throughout the structure. Careful collaboration between the designer and contractor responsible for the post-tensioning will ensure that the history of loading corresponds to that anticipated in the design.

In 1966, a concrete foundation mat 900 mm (3 ft) thick for a multistory building was being constructed in accordance with a design using tendons draped downward under each column to conform to the bending moment diagrams (Figure 8.3). When post-tensioning was applied, the concrete in the column areas lifted and broke free of the mat. The design locating the tendons had been based on the assumption of loaded columns. However, when the mat was tensioned no columns had yet been constructed, and there was insufficient weight to resist the upward component of the curved tendons. The post-tensioning should have been accomplished in stages as dead loads were progressively applied to the foundation during construction.

To repair and complete the project, the prestressing forces were released and each column area was surrounded by heavy anchors drilled and grouted into the soil. In this manner, sufficient temporary loading was engaged to enable successful post-tensioning of the mat. This was a case where erection forces required temporary compensation, until the structure was completed to provide the necessary resistance. Closer coordination between the designer and contractor might have avoided this embarrassing situation.

Damage to grouted precast beams after post-tensioning from the freezing of the grout while sealing the cable sheaths has been reported at a number of locations. The expansive force of the freezing grout acting across unreinforced faces of concrete can completely shatter the highly compressed castings.

A tragic example of a large-scale structural failure, where poor design was compounded by shoddy workmanship, is the 1968 failure of the Ronan Point

Figure 8.3   Continuous foundation mat tendon pattern.

housing project in Canning Town, near London, England (Figure 8.4). Ronan Point was one of many prefabricated multistory projects built rapidly after World War II in the United Kingdom and throughout the European continent to respond to the severe postwar housing shortage.

The system used at Ronan Point involved factory-manufactured load-bearing panel walls. Each apartment level was stacked directly on the panels of the apartment below. Floor-to-wall and wall-to-floor joints were simply grouted, with little or no positive steel reinforcement to tie the structure together. As for lateral loads, the joints were held together by little more than friction. (Complete demolition of the tower in 1986 confirmed that there were also deficiencies in the quality of the grouted joints between the prefabricated components. The workmanship was substandard, with many ungrouted joints and a considerable amount of foreign debris in the voids.) The structure was essentially a "house of cards," 22 stories of stacked individual precast concrete panels above a cast-in-place concrete podium (Allen and Schriever 1973, Somes 1973, Levy and Salvadori 1992).

The collapse of Ronan Point on May 16, 1968, was triggered by an accidental explosion of gas that leaked from the connection of a kitchen range located in an apartment on the southeast corner of the eighteenth floor. The report of the inquiry into the collapse states:

> The explosion blew out the non-load-bearing face walls of the kitchen and living room, and also, unfortunately, the external load-bearing flank wall of the living

Figure 8.4   Ronan Point, near London, England, 1968. (Courtesy of the National Bureau of Standards.)

room and bedroom of the flat, thus removing the support for the floor slabs on that corner of the nineteenth floor, which collapsed. The flank walls and floors above this collapsed in turn, and the weight and impact of the wall and floor slabs, falling on the floors below, caused a progressive collapse of the floor and wall panels in this corner of the block right down to the level of the podium. (Report 1968)

The collapse affected both the living rooms and bedrooms above the six-teenth floor, while below this level, collapse was limited to the living rooms. Four people were killed and seventeen were injured. The report further stated: "The loss of life and injury might well have been very much worse. At 5:45 A.M., mercifully, most tenants were in their bedrooms. (Only one apart-ment of the four located directly above the explosion was occupied at the time of the explosion.)"

Apart from obvious lessons regarding continuity at connections, the Ronan Point case became quite influential in causing designers to consider the poten-tial for progressive collapse of nonredundant structure systems. There were no alternative load paths or structural integrity provisions at Ronan Point that would have enabled the structure to bridge over the local failure when the wall panels affected were blown out of the building.

As a result of the Ronan Point tower collapse, other projects in the United Kingdom deemed susceptible to progressive collapse were demolished. In 1970, England strengthened its building standards to provide an alternative means of support, even if a main structural member were to be removed or to fail. Steel bracing with floor-to-wall connectors was mandated, along with a minimum tensile strength of 44 kN/m (3000 lb/ft) across the length and width of floors (ENR 1970).

8.5 TRANSPORTATION AND ERECTION FAILURES

Transportation and handling of bulky precast prestressed elements can result in damage to the members. The transportation over some 500 km (300 mi) of lightweight precast tees 24 m (80 ft) long for a school gymnasium roof in Arizona may have released the bond of the pretensioned strands. Although lifted from the bed and placed on the trailer and erected on the walls without incident, the beams quickly developed a sag of 150 to 200 mm (6 to 8 in.). When a heavy rainfall caused further deflection, the roof was rejected. The combination of dynamic loading during transportation and the lower bond developed in lightweight concrete seems to be a reasonable explanation.

In a bridge realignment job in Yonkers, New York, 32 identical prestressed beams were to be shipped by rail on a schedule synchronized with the erection program. The last beam was dropped when it just barely failed to clear the side of the car. The replacement cost was not too serious, but the bridge construction schedule was upset. In retrospect, the completion time could have been bettered by using steel girders.

Transporting long precast members through congested urban areas requires special traffic permits and often must be scheduled during overtime work periods. Limited clearance in congested roadways has caused failures in transport, as illustrated by a 30-m (100-ft)-long girder in Providence, Rhode Island, which overturned while the truck tried to negotiate a sharp curve (Figure 8.5).

A more serious failure was reported in 1955 in New Zealand, where 80 beams of 32-m (105-ft) span for the Hutt Bridge near Wellington were post-tensioned and grouted before shipment. Each beam was prestressed with 96 wires of 6.78 mm (0.267 in.) diameter. A beam weighing 36 Mg (40 tons) tipped over in shipment and exploded so violently that the shock wave touched off failure of a second beam, still in storage, which also broke into fragments.

In 1957, a 30-m (100-ft)-long post-tensioned tee-beam at Omaha, Nebraska, buckled as it was unloaded by a somewhat off-center hitch. In 1962, a 30-m (100-ft)-long precast beam being transported in Seattle, Washington, on a low, two-unit trailer broke its moorings and slid down to the pavement, breaking into separate segments.

Erection failures involving unstable incomplete precast structures are common. They can be guarded against by proper care in lifting, guying, and stressing until all permanent bracing is in place and connected. Lack of such care can cause serious injuries to workers, or at the very least delay the job at considerable cost, eliminating the predicted economies of a prestressed design.

Figure 8.5 Transportation failure.

Precast elements have little reserve strength or stability until connected to the supporting frames. During erection of the roof over a one-story plant in Bermuda Hundred, Virginia, in 1963, some 1020 m^2 (11,000 ft^2) of 12-m (40-ft)-long double-tees fell to the ground.

The dynamic loads resulting from sudden problems with construction equipment or jacking operations can damage the members or precipitate failures in marginally stable incomplete structures. The jacking apparatus for post-tensioning must be in good condition and of sufficient stressing capacity. In 1960, in Montreal, Canada, a 32-Mg (35-ton) beam collapsed from the shock of sudden stress reversal when a jack failed during post-tensioning. At a viaduct construction project in Paris in 1963, a 62.8-m (206-ft) girder with 13 of its 14 cables stressed failed when a jack broke, as if the beam were eccentrically loaded.

In 1963, the British House of Commons ordered a technical inquiry into the collapse during construction of an officers' mess building at Aldershot. Owing to a severe shortage of construction labor, the Ministry of Public Buildings and Works had contracted for four similar buildings, chiefly of precast concrete, on a design-and-construct basis. The four-story buildings had precast columns, beams, floor, and exterior wall panels. The joints at columns and beams were made by cast-in-place concrete, and the center core area of the 18.3 by 18.3 m (60 by 60 ft) building was also cast in place. The framing was a 6.1 by 6.1 m (20 by 20 ft) grid.

Columns were cast in story height with bars projecting from the top and a hole at the bottom to receive the splice steel from the lower columns. Beams were cast full length, with recesses for the column steel, and formed part of the column when the recesses were concreted. Floor plates were ribbed sections of precast 38-mm (1.5-in.) slab and were doweled to the beams by bond splices between projecting rods of the beams and exposed rib bars in recesses of the casting. Bearing of the beams on the columns was noted as 38 mm (1.5 in.) but actually was found to be less, due to shrinkage in the castings. The ends of the beams were supposed to be rough to provide better bond to the filling concrete, but were actually manufactured smooth. Bent-up dowel bars were shown from the ends of the beams into the columns, but these were often omitted since the bar came close to the form face and could not be concreted easily.

Frames had been erected to full height for all four buildings when one building collapsed due to its inherent instability. The investigation recommended that the other three be demolished, but before action could be taken, another building collapsed. Collapse started by failure at a column joint. With no lateral stiffness, the entire assembly collapsed; there was no continuity between beams and no lateral bracing. The design had made no provision for wind loading or eccentric loading on columns.

The conclusions listed in the inquiry report are an important contribution to the literature explaining the failure of structures. The lessons are still being learned, sometimes the hard way:

a. Where a system of building using prefabricated structural components is extended by use in a new building type, a fundamental reexamination of the system design is necessary. This must include a reconsideration of all design assumptions and, if necessary, a recalculation of the structural design from first principles. This is especially true where, as in this case, a building frame which has normally been used with cross-wall stiffeners is used with full-length glazing and open plan. The removal of the stiffening effect of panels is particularly important if, as is likely in such circumstances, aesthetic considerations call for the columns to be as slender as possible.

b. When novel, or relatively novel building methods are used the thorough and systematic communication of the designer's intentions to the operative is more than ever essential. When traditional methods are used the familiarity of most operatives with the work may allow an occasional uncertainty or ambiguity in drawings and specifications to be corrected by intelligent interpretation. This is not so when structural techniques are used in which operatives have only a limited familiarity with the work. In particular, it is essential that fully dimensioned working drawings should be prepared for all joints. The designer is responsible for communicating his intentions, particularly with regard to those factors and dimensions which are of critical importance, in distinction from those where some tolerance can be allowed. It is also the designer's responsibility that he does not assume higher standards of accuracy and workmanship than can be attained.

c. In systems of construction depending on the assembly of prefabricated structural components the erection procedure is an essential part of the engineering design. It must be specified by the designer in sufficient detail to ensure that the structure is sound at all stages of construction. (BRS 1963)

A partial collapse of a seven-level parking garage during erection of the precast elements at the University of Michigan campus in 1966 was blamed on wind conditions. The collapsed section included nine columns 21.3 m (70 ft) high and fourteen 19.2-m (63-ft) single-tee beams. Four of the columns had seven tiers of beams connected and five were standing free except for four anchor bolts into the foundations. About one-third of the 61.9 by 75 m (203 by 246 ft) garage was erected full height at the time of failure, but only the items erected that day were involved in the collapse. Failure came after all work had stopped for the day and it completely wrecked the erection crane. One column was broken about three stories above the ground and one remained intact but tilted 10 degrees.

The domed roof over a Methodist chapel in Atlanta, Georgia, collapsed in 1966. The 12.2-m (40-ft)-diameter roof was covered by 12 precast radial sections, each supported by two precast wall columns and a center precast concrete compression ring. A weld in the outer steel tension ring gave way, and this warned all people out of the area. Failure came 10 hours later with the crash of 110 Mg (120 tons) of precast concrete.

To replace an old blimp hanger at the South Weymouth Naval Air Station in Massachusetts, a precast concrete design was being erected in 1967 when a 27.4-m (90-ft) roof unit lifted by four cranes crashed 21 m (70 ft) to the ground. The design of the 84 by 79 m (275 by 260 ft) hanger included 15 precast concrete arch ribs. The sequence called for half-arches to be lifted into position and pinned at the crown. Three arches were in place, and as the fourth was erected, some bits of concrete fell down and the arch members fell, crushing two cranes. The Navy investigation blamed the absence of tie-bars and inadequate vertical shoring during the erection phase.

The lifting of a prefabricated structural unit, whether a precast pile, the suspended span of a cantilever bridge, a precast or prestressed or ordinary reinforced beam, or a lift-slab section can be safely accomplished only if erection stresses do not go beyond the yield points and the lifting equipment provides sufficient continuous and uniform support. Failures of this type are nearly always caused by inadequacies in seemingly insignificant details or by nonuniform action of the lifting procedure. Small horizontal forces can easily upset the apparent equilibrium that seems to exist during an upward lifting of a large mass that is laterally unrestrained.

Wide acceptance of the technique known as *lift-slab construction* was delayed somewhat by several such erection failures when the technique was introduced. Much later, after many years of successful lift-slab projects and an excellent safety record, the industry was shocked again by the catastrophic 1987 L'Ambiance Plaza lift-slab failure in Bridgeport, Connecticut. A review of some of the early lift-slab failures will be helpful before discussing the L'Ambiance Plaza case.

There were three successive failures on three separate attempts to complete the erection of a 79-Mg (87.5-ton) lift-slab roof-slab section near Miami, Florida, in 1952. Only the last was attributed to a structural design deficiency. The first failure occurred when a 125-mm (5-in.) pipe stub supporting a jack on top of a permanent column buckled and dropped the jack while eccentricity in lifting was being adjusted. The second failure was triggered by stripping of defective threads on the inserted flange supporting the slab. The third failure was a shearing in the edge of the concrete cap above the column. The computed unit shear stress was 640 kPa (93 psi) over the circular concrete section. The final and successful design included some extensions of reinforcement and additional steel shear resistance.

In San Mateo, California, in 1954 a failure occurred during a demonstration of the lift-slab technique attended by about 60 spectators. While guys were being adjusted to counteract a 75-mm (3-in.) leaning of the columns, the 19.8 by 21.3 m (65 by 70 ft) slab unit drifted laterally 5 m (15 ft) and fell to the ground. In addition to six construction workers, four of the spectators were injured. They were subsequently awarded damages of over $100,000 as a result of litigation in a California court. The verdict was against the architect, general contractor, and lift-slab contractor. Fortunately, most of the visitors were cleared off the slab before it fell. They had been riding the slab as it was being lifted. The failure was partly explained by eccentric loading; the observers were concentrated on one side of the slab.

During a windstorm in 1956, with gusts of 55 to 80 km/h (35 to 50 mi/h)—normal late spring weather in Cleveland, Ohio—an eight-tier lift-slab parking garage tower drifted laterally about 2 to 2.5 m (7 to 8 ft) out of plumb. It was the west tower of two. The slabs were all raised into position, but none of the slab collars had been welded completely to the columns. The east tower had the second and third floors in place, with the rest of the slabs held temporarily within the next two stories, and it remained plumb. Emergency action stopped further movement. The *Cleveland Plain Dealer* newspaper reported the next morning (April 8, 1956): "As darkness fell, engineers apparently had won their frantic struggle to prevent their structure from crushing the two-story building next door. Cables attached to four winch trucks, telephone poles and other objects had stopped the building from listing more than the 20 degrees it had sagged earlier in the day."

Each floor slab weighed 82 Mg (90 tons). The accident occurred shortly after the top slab had been lifted into position. Within a month, design procedures were completed for righting the 27.7 by 6.4 m (91 by 21 ft) eight-story frame and its 10 columns. The complete success of the righting operation speaks well of the original design, but limber frames of such assembly clearly require special bracing precautions during erection.

The most dramatic and tragic accident related to lift-slab construction occurred on April 23, 1987, at approximately 1:30 P.M., when the structural frame and slabs of what was to be a 16-story apartment building in Bridgeport, Connecticut, collapsed during construction, killing 28 construction workers (see Figure 11.12). This was the largest loss of life in a construction accident in the United States since 51 workers were killed in the collapse of a reinforced concrete cooling tower under construction at Willow Island, West Virginia, in 1978.

The L'Ambiance Plaza project comprised two rectangular towers, each with 13 levels of apartments and three levels of parking. The plan of each tower was 34 by 19 m (112 by 62 ft) (Figure 8.6). The towers were being

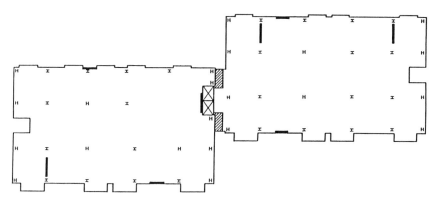

Figure 8.6 L'Ambiance Plaza, Bridgeport, Connecticut: floor plan. (Courtesy of the National Bureau of Standards.)

constructed using the lift-slab method. Floor and roof slabs were two-way, unbonded, post-tensioned flat plates. They were all cast at the ground level, post-tensioned, and then lifted into place. Steel columns (W and HP shapes) supported the floor slabs, which were lifted in stages by hydraulic jacks, threaded rods, and welded steel "shearhead" collars placed in the concrete floor slabs at each column (Figure 8.7).

The collapse investigation was complicated by a number of potential contributing factors and by the unavoidable damage to evidence that occurred during the frantic rescue operation. Several triggering causes were hypothesized, most of these centering on deficiencies in the design and construction of the shearhead collars. One theory proposed that a lifting angle in the shearhead deformed, allowing the nut at the end of the lifting rod to slip out

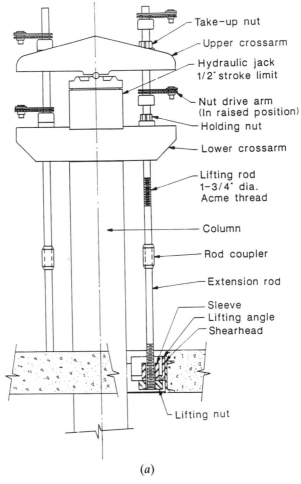

(a)

Figure 8.7   (a) L'Ambiance Plaza lift-slab jacking assembly diagram; (b) isometric of shearhead/lifting apparatus for L'Ambiance Plaza slabs. (Courtesy of the National Bureau of Standards.)

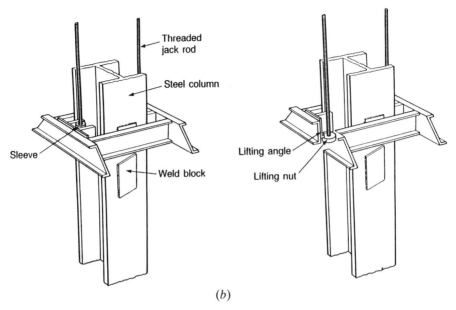

(*b*)

Figure 8.7 (*Continued*)

of the assembly (Figure 8.8) (NBS 1987, Scribner and Culver 1988). Another theory involved a rolling out of the wedges used to support slabs temporarily during the lifting operation (Cuoco, Peraza, and Scarangello 1992). Deficient welds were also discovered in the shearheads (McGuire 1992).

Whatever the triggering cause, the magnitude of the resulting catastrophe was the result of numerous contributing factors. Tests conducted at the National Institute of Standards and Technology (NIST) showed that failure of the lifting mechanism at only one point would not have brought down the entire project were it not for other deficiencies present in the design and in the construction sequence. These included improper placement of some of the prestressing tendons (Poston, Feldmann, and Suarez 1991) and general overall instability of the frame during construction (Moncarz et al. 1992). (Shear wall construction was lagging behind the lifting operation.) The instability of the overall construction on the day of the collapse was such that the apparent equilibrium was easily upset with a failure at only one location.

A remarkable and controversial mediated global settlement of claims resulting from this failure brought an end to forensic investigations in December 1988 (Felsen 1989). There may never be complete general agreement on the technical causes of the failure, but one result of the early settlement was that the various theories were discussed thoroughly in the engineering literature, much sooner than has been the case for other collapses of this magnitude. The Publications Committee of the ASCE Technical Council on Forensic Engineering invited all parties who had been working on technical forensic investigations to come together at a meeting in Arlington, Massachusetts, on

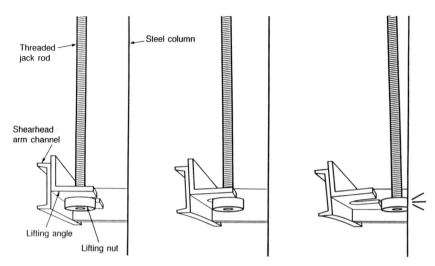

Figure 8.8 L'Ambiance Plaza: probable sequence for triggering cause of collapse, according to NBS investigators. (Courtesy of the National Bureau of Standards.)

March 14, 1989. The purpose of the meeting was to exchange information among the various investigators and to encourage dissemination of the incomplete results of their investigations in the engineering literature. As a result, numerous papers, technical notes, and discussions were eventually published in the ASCE *Journal of Performance of Constructed Facilities* and elsewhere.

The tragedy of the L'Ambiance Plaza collapse and subsequent discussion of the event gave impetus to efforts aimed at ensuring structural integrity during the construction phase of a project. The lift-slab method has an enviable safety record, better than that of cast-in-place construction in multistory buildings (Peraza 1995). But this failure emphasized the need to evaluate details and sequencing carefully for each project and each site. McGuire (1992) noted that lift-slab techniques and details have evolved over the years to the point that some of the fail-safe details and safeguards responsible for the good safety record may have been "optimized" out of the process.

Much discussion also centered on the convoluted and fragmented project delivery system for the L'Ambiance Plaza project. Responsibility for ultimate structural safety was confused by unclear relationships among the engineer of record, the lift-slab contractor, and the designer of the shearheads (Heger 1991).

Finally, it should be noted that neither the L'Ambiance Plaza collapse nor any of the other lift-slab examples cited were truly "prestressed concrete" failures. In fact, they were erection failures, and such failures may be associated with prefabricated elements of other materials as they are assembled on the site. Examples of erection failures involving timber and steel components have been discussed in Chapters 5 and 6. In the case of L'Ambiance Plaza, the most probable cause was a deficiency in a steel connection detail that was part of the lifting operation, a connection that would carry no load in the completed building. Components of the post-tensioning system (anchorages, strands, etc.) performed well at L'Ambiance Plaza, although the unbonded tendon design was blamed by some observers for the total disintegration of the slabs.

The complete breakup of the slabs at L'Ambiance Plaza was quite alarming; the entire building was reduced to rubble. However, this need not be the case for failures in future unbonded post-tensioned structures. The current American Concrete Institute code requires reinforcement detailing for redundancy and ductility so that a local failure will not precipitate a progressive collapse or loss of stability. For structures such as L'Ambiance Plaza, these new "structural integrity" provisions imply the need for auxiliary bonded mild steel reinforcement in the bottom of the slabs in the column strips. The presence of such bonded steel at L'Ambiance Plaza probably would have provided catenary action and reduced the magnitude of the catastrophe (Poston, Feldmann, and Suarez 1991).

8.6 PERFORMANCE IN EXTREME WINDS AND SEISMIC EVENTS

It has already been noted that the weak points in a precast concrete structure are generally found in the connections and that the proper design of such connections must be followed by proper execution in the field. Extreme and unusual loads (wind and earthquakes) are very good at discovering the vulnerable aspects of a structure. Thus it is not surprising that connection failures are the most noticeable features of precast and prestressed buildings that fail to perform satisfactorily in these events. Often, the individual members come apart so cleanly that they are hardly damaged. In some cases they appear to be in as good a condition as when they were originally assembled into the building.

The problem, of course, is how to detail a connection that will tie the building together under unusual and severe dynamic loading, yet permit the kinds of unrestrained movement discussed in Sections 8.2 and 8.3. This is a challenge, but the details that address the challenge are available and their proper use has been confirmed by outstanding performance in earthquakes and windstorms (Birkeland and Birkeland 1966, Hanson 1969, PCI 1971).

In August 1992, a number of lightweight prestressed precast concrete double-tee roof systems in Florida experienced failures during Hurricane Andrew (Suaris and Khan 1994). Some of these failures were caused by the negative bending moment created by uplift forces on the roof that was not anticipated by designers. Often, such roof members are subjected to combined internal pressure and external uplift forces when the integrity of the enclosing walls is penetrated. Some failures were due to inadequate anchorages at the supports, so that the members were simply lifted off their supports, becoming airborne. The quality of materials and detailing of these prefabricated members was admirable but hardly of value when they arrived suddenly and unexpectedly on the adjacent site. Failures also occurred at diaphragm shear connectors, where the anchorage or cover was insufficient. Recent hurricane experience has shown that there is not enough consideration given to uplift forces in the design of members and connections in low-rise buildings. This concern is not limited to precast and prestressed concrete buildings, but is also true for timber-frame structures.

Precast concrete facade panels on multistory buildings pose a particularly difficult connection challenge. If the connection to the structural frame is inadequate, the panels may detach from the frame in an earthquake, falling to the ground and threatening the lives of pedestrians. One famous example is the facade of the J.C. Penney building in the 1964 Anchorage, Alaska earthquake (Figure 8.9). On the other hand, a strong but improperly detailed connection may interfere with the beneficial flexibility of the structural frame, restricting frame deformations and leading to total collapse of the structure in a large earthquake. Considerable research has been focused on this problem, and ductile details are now available that satisfy these conflicting requirements (Lagorio 1990).

The structural implications of building facades in seismic events is a subject that warrants further study. The facade materials and details do have an influence on the behavior of the structural frame, whether or not this is considered in the design. They do not simply "go along for the ride." As our understanding of the complex interaction between the frame and facade advances, we may find ways to use the facade structurally to improve the resistance of the overall system to extreme loading (Cohen 1995).

The history of precast and prestressed concrete construction is quite short when measured in reference to geologic time, so the benefit of experience-based seismic performance is limited. In the Whittier, California, earthquake of 1987, some deficiencies involving the connections in precast tilt-up buildings were noted. Many of the deficiencies exposed by the earthquake were construction errors rather than design or fabrication problems. Earlier tilt-up buildings had performed well in some minor earthquakes, but these buildings had been of very regular and symmetric plan. Systems using tilt-up methods are now used for buildings with unique, irregular, and nonsymmetric configurations, and the stresses acting on the connections have become more complicated.

Of course, the tilt-up building relies on well-designed and executed connections between wall panels and the roof diaphragm for its stability and on connections between panels for lateral shear transfer. Unfortunately, this building system is often selected for its rapid completion time, and careful attention to the all-important field work is sometimes relaxed.

Of the buildings damaged in the January 17, 1994, earthquake in Northridge, California, there were several dramatic collapses involving precast concrete parking structures. Perhaps the most highly publicized collapse was the four-level parking structure at the Northridge campus of California State University

Figure 8.9 Facade failure, Anchorage, Alaska. 1964. (Courtesy of the Federal Emergency Management Agency.)

(Figure 8.10). The structure cost $15 million and was completed only two years prior to the earthquake. The failure was explained partly by deformation incompatibility (flexible and stiff structural elements) throughout the structure. It illustrated two important lessons: the need to consider displacements more thoroughly, and the need for ductile design of gravity-load columns. There were several tilt-up precast concrete buildings that failed due to inadequate placement of metal connectors. These failures cannot be blamed on the construction system, but rather, on a casual attitude at the construction site. The lesson is the need for diligent construction supervision and seismic education for the entire construction team. Improvements are needed in code implementation as well as in the code itself.

There were reports from Kobe, Japan, following the Great Hanshin earthquake of January 17, 1995, of good performance of precast prestressed concrete structures and well-designed precast concrete buildings with stiff shear walls. In Japan, quality control on the construction site is purported to be better, due to more centralized control of the construction project delivery system, which entails closer coordination among designers and constructors. There were many reports of excellent performance of individual precast concrete buildings in the 1994 Northridge, California, earthquake as well, where connections conformed to good practice and construction quality was carefully monitored. As already noted, execution of the details may be more important to

Figure 8.10 Parking structure failure, Northridge, California, January 17, 1994. (Courtesy of the National Geophysical Data Center.)

earthquake performance than selection of a particular structural material or construction method.

Existing deficient precast concrete structures can be strengthened against lateral loading, sometimes quite easily if members and connections are accessible. Mitchell et al. (1990) describe a project to improve lateral force resistance of a large 36.6 by 198 m (120 by 648 ft) precast concrete warehouse. Structural members were enlarged by encasing them in reinforced concrete. Such work requires careful investigation of the effects of load history, creep, and differential shrinkage.

8.7 CORROSION OF PRESTRESSING TENDONS

In Chapter 7 it was noted that the effect of corrosion on a steel component is a reduction of its effective cross section. Corrosion can be a serious problem indeed when the component is of small cross section to begin with and carrying high stresses. Thus it is quite evident that the designer must take every precaution to protect high-strength prestressing tendons from the damaging effects of corrosion.

There is a need to provide corrosion protection in the form of adequate cover for end anchorages and at end fittings of prestressing tendons, as well as for the tendons themselves. Some of the early designs with prestressed concrete experienced more problems with corrosion than is the case for contemporary facilities. In these early designs the designers may have been too idealistic about the capabilities of the new materials. Perhaps the misguided assumption that cracks would be entirely eliminated from the material by the compression prestress caused designers to relax considerations about moisture and chloride penetration. Also, the high strength of the new steels and concretes permitted dramatic reductions in cross section. Structural optimization then produced cross sections with insufficient cover to protect the steel adequately.

In fact, the low water/cement ratio generally found in precast concrete makes the material more impervious to water than is normal cast-in-place concrete. A 1985 study found a beneficial influence for low water/cement ratios, heat curing, and adequate concrete cover in reducing corrosion and chloride penetration (Pfeifer, Landgren, and Perenchio 1986). Tests were conducted on precast slabs showing that a low water/cement ratio was a principal factor governing resistance to chloride intrusion. The lower ratios used in precast concrete can therefore provide better corrosion resistance than is found in typical ordinary reinforced concrete. Recent precast and prestressed concrete designs, with improved connections and using galvanized or coated hardware, have proven to be very durable (Schutt 1996).

The phenomenon of hydrogen embrittlement of steel prestressing tendons has been documented by a number of investigators of bridge failures in the United Kingdom, but does not seem to be as prevalent in the United States.

Hydrogen embrittlement is caused by hydrogen atoms from atmospheric moisture that penetrate the steel and upset its molecular structure, causing disruptive internal stresses. It can be a significant problem in high-strength steels working at high stress levels and can be accompanied by accelerated corrosion. Ungrouted tendons are most susceptible to this phenomenon. The presence of contaminants, particularly sulfides, increases the potential for hydrogen embrittlement. Certain cathodic protection procedures can also introduce the opportunity for hydrogen embrittlement, and because of this, cathodic protection is not recommended for prestressed concrete structures (ENR 1981, Schupack 1989).

One of the more prominent early designs that experienced costly corrosion problems, including structural collapse, is the West Berlin Congress Hall. On May 21, 1980, one-third of the 3400-m^2 (37,000-ft^2) concrete shell roof, including one of two support arches, collapsed (Figure 8.11). Dynamic stress and corrosion in the prestressing tendons were the main causes. The condition of stress corrosion in the tendons was accelerated by inadequate protection against moisture (insufficient cover) and by the presence of corrosive agents. One person died and four others were injured in the collapse.

The West Berlin Congress Hall roof was a totally cast-in-place doubly curved surface structure 76 mm (3 in.) thick. The shell roof was supported by an edge arch on either side and a tension ring in the center. Steel tendons

Figure 8.11 West Berlin Congress Hall, 1980. (This photograph was taken by permission from *Toward Safer Longspan Buildings,* published by the American Institute of Architects, 1981.)

were anchored under tension in the arches and the central tension ring and embedded in thin reinforced concrete slabs for protection. Vertical movements of the edge arches due to concrete creep, temperature stresses, snow, wind, settlements, and other factors caused high fluctuating stresses in the tensioning steel. These stresses over the 23-year life of the structure cracked the roof. Humidity and carbon dioxide were able to penetrate through the cracks to the tensioning steel, causing severe corrosion and a gradual break of the wires.

The collapse was the result of details that permitted corrosion of the steel tensioning elements supporting the thin shell roof, according to an investigation report prepared by Jorg Schlaich of the Stuttgart Institute for Massive Construction (ENR 1980). The prestressing steel, embrittled by corrosion and hydrogen induction, finally broke. While the design was blamed for the collapse, there were also some construction deficiencies noted in the report. The roof joint concrete was of "insufficient, uneven quality, very porous and had, by today's standards, a too-high chloride content."

Precast concrete and prestressed concrete parking garage designs have improved greatly since the 1970s. Better joint details between tees and improved bearing pad details have reduced cracking. Denser concretes and corrosion-inhibiting additives are available as well as improved surface coatings. The industry is active in supporting research on topics that need attention and publishes application guidelines for designers and constructors that provide better corrosion protection for the prestressing strands, anchorages, and other hardware.

Morris Schupack is a structural engineer who has been involved in extensive surveys on behavior and repair of prestressed concrete structures and tendon grouting techniques. In an informative series of articles in the journal *Civil Engineering,* Schupack summarizes his observations regarding corrosion problems in prestressed concrete, both pretensioned and post-tensioned using bonded and unbonded tendons (Schupack 1982, 1989, 1993). He notes that corrosion is probably less frequent in prestressed concrete than in ordinary reinforced concrete, but that when it does occur, the results can be more dangerous and more costly. Just as is the case for ordinary reinforced concrete, corrosion is possible when an electrolyte is present, oxygen is available, the character of the portland cement has been damaged by contaminants, or when aggressive chemicals are permitted access to the steel. The causes are inadequate cover or poor-quality concrete, and deficient details. If ordinary steel reinforcement is also present, corrosion of that steel, and the resultant expansion and spalling of the concrete, will expose the prestressing steel to corrosion.

Schupack writes that the good performance of prestressed concrete indicates that prestressing steel, completely encapsulated in dense portland cement concrete or grout of adequate thickness and free of cracks and aggressive chemicals, does not suffer corrosion problems. Pretensioned concrete members, which are usually factory manufactured using established quality control procedures, have been relatively free of corrosion problems. Post-tensioned concrete structures (generally site fabricated) require more attention to construction details. The performance of bonded (grouted tendons) and unbonded tendons in post-tensioned concrete has been good, but long-term performance of unbonded construction relies upon the integrity and continuity of corrosion-inhibiting water-resistant grease contained in a protective sheath.

Those systems that use multiple strands have more redundancy to guard against corrosion failure. In bonded construction, problems generally manifest themselves by cracking and sagging of the structure if a significant number of tendons in a local area are affected. Few symptoms are visible at the anchorages unless the corrosion is located in the anchorage itself. Catastrophic failure is rare in bonded post-tensioned concrete facilities if the grouting is of good quality and continuous. Even if a bonded tendon does experience a local failure due to corrosion, that tendon still functions to provide prestress in other portions of its length.

Corrosion problems in bonded post-tensioned structures are usually associated with poor and incomplete grouting. In some cases, water has been able to penetrate at the anchorages and travel through the sheath through "bleed voids" caused by separation of the water and cement (sedimentation) before the pumped grout could reach its initial set. The resulting air voids can become a source of corrosion. It is possible for severe corrosion to take place with no observable effects, since the oxidation and expansion can be accommodated in such voids without causing the concrete to spall. Problems with grout bleeding have been reduced through improved grouting techniques using water-retentive admixtures in the grout, or fully encapsulated tendons. Fully encapsulated systems, with a plastic coating over all the hardware (strands and anchorages), provide more protection against all forms of corrosion if they maintain their integrity. If the cover is breached, corrosion damage can be detected using electrical monitoring techniques. Such systems are used in both bonded and unbonded post-tensioned concrete (Schupack 1993).

In unbonded construction, corrosion failures may be quite dramatic and may pose a hazard to occupants. The sudden energy release can cause unbonded tendons to explode suddenly out of their enclosures or to break through the concrete where the cover is minimal. Also, if an unbonded tendon fails at any point, the total prestressing effect of the tendon is obviously lost. These considerations clearly imply that corrosion control is a critical aspect of unbonded post-tensioned construction. Schupack notes that a small increase in care and materials during construction can be cost-effective over the long term. However, the unbonded tendon market is highly competitive. The continuity of protective grease in an undamaged sheath is essential, but these concerns may fall prey to misplaced economic priorities.

This is not to imply that unbonded post-tensioned concrete is plagued with many cases of strand corrosion. Only a very small number of problems have been reported, considering the many unbonded post-tensioned concrete projects that are performing satisfactorily. However, the tendons and anchorages must be protected against corrosion. Specific recommendations are available from the Post Tensioning Institute (PTI) and other sources.

Unbonded tendons were first used in buildings in the 1950s (Schupack 1989). The first unbonded tendons were typically wrapped with paper. These early structures have experienced most of the corrosion problems, especially where inadequate cover was provided. Since 1960, plastic-sheathed strands of various types have been used. The unbonded alternative has many advantages besides economy. The tendon, surrounded by grease, does not bond with the concrete. The grease protects against corrosion and reduces friction during tensioning. Since the sheaths are not grouted, there is no danger of freeze-induced cracks that can occur in bonded construction during winter. Unbonded tendons can be used in relatively shallow construction, and they provide economical deflection and crack control for longer spans. Unbonded tendons have performed successfully in earthquakes and unusual structural accidents. A further advantage over bonded construction is that the unbonded tendons can be monitored and easily replaced if necessary. For this reason, unbonded tendons have been the method of choice for post-tensioning containment vessels for nuclear power plants.

The debate over the relative merits of bonded and unbonded construction continues throughout the world, In 1992, the U.K. Department of Transportation imposed a ban on new construction of prestressed post-tensioned bridges using bonded tendons (Schupack 1993). This decision was brought about by concerns over potential corrosion problems with grouted tendons and the difficulty of detecting them. Corroded tendons caused the 1985 failure of the Yns-y-Gwas Bridge in Wales. Previously, U.K. officials preferred bonded construction over unbonded post-tensioning, but the new requirements state that the tendons must be inspectable and replaceable.

Some observers noted that the U.K. decision may be an overreaction to incidents that were, in fact, caused by poor design and construction practices. Methods are available to assure proper corrosion protection in grouted tendons. Procedures are also available to monitor the in-place corrosion protection. Properly detailed and constructed tendons have shown excellent durability and performance over many years. Inspection of demolished bridges has proven the effectiveness of these techniques.

Schupack (1993) describes extensive studies made on the demolished 35-year-old post-tensioned concrete Bissell Bridge over the Connecticut River at Windsor, Connecticut. Demolition of the bridge was necessitated by a highway widening construction project, thus providing an excellent opportunity to investigate the condition of the tendons. The bridge deck was carefully cut into 36.5-m (120-ft)-long, 109-Mg (120-ton) single-tee beams, that were transported to the laboratory for study.

The 512-m (1680-ft) simple-span Bissell Bridge was built in 1957. It was constructed of 14 spans, each 36.5 m (120 ft) long. Prestressing wires were used in a metal sheath with Freyssinet anchorages. (The currently used multistrand tendons are not as susceptible to blockage of air during the grouting operation as were the old bundled wires.)

A great variety of grout type and quality was discovered in the metal sheaths, varying in density, color, and additives. Where good-quality grout was used, there was no evidence of corrosion. Reasonably grouted tendons with chloride-free grout showed no corrosion problems, even where there were voids and the grout was not particularly dense. Pitting corrosion occurred only where water had been trapped and freezing action had spalled the concrete. Severe corrosion, of both the wires and the sheath, accompanied by longitudinal web cracking, was found in a few areas where the grout had been highly contaminated with chlorides. Some tendons had blockages, with resulting intermittent grouting, and others had never been grouted at all. Corrosion detection by selective probes would have been inconclusive and misleading, since the corrosion was intermittent.

Although none of these problems caused structural failure over the years, some cracking had been noted as early as 5 to 10 years after construction. This was in the vicinity of the grouts with high chloride content. External post-tensioning was added in 1985 to control the cracking. After 35 years of continuous use, corrosion in the Bissell Bridge tendons was not sufficient to diminish the ultimate strength of the tendons. Neither flexural deficiencies nor unacceptable deformations were noted.

8.8 REFERENCES

Allen, D., and W. Schriever, 1973. "Progressive Collapse, Abnormal Loads and Building Codes," in *Structural Failures: Modes, Causes, Responsibilities,* American Society of Civil Engineers, New York.

Birkeland, P., and H. Birkeland, 1966. "Connections in Precast Concrete Construction," *Journal of the American Concrete Institute,* Vol. 63, No. 3, ACI, Detroit, MI.

BRS, 1963. *The Collapse of a Precast Concrete Building Under Construction for the Ministry of Public Building and Works,* Building Research Station, HMSO, London, UK.

Cohen, J., 1995. "Seismic Performance of Cladding: Responsibility Revisited," *Journal of Performance of Constructed Facilities,* American Society of Civil Engineers, New York (November).

Cuoco, D., D. Peraza, and T. Scarangello, 1992. "Investigation of L'Ambiance Plaza Building Collapse," *Journal of Performance of Constructed Facilities,* American Society of Civil Engineers, New York (November). (See also Discussions by C. Culver, R. Marshall, and O. Rendon-Herrero in May 1994 issue.)

Dias, W., A. Jayanandana, M. Fonseka, and A. Perera, 1994. "Distress in Prestressed Concrete Roof Girders at Cement Plant," *Journal of Performance of Constructed Facilities,* American Society of Civil Engineers, New York (February).

Engineering News-Record, 1970 McGraw-Hill, Inc., New York (April 16).

Engineering News-Record, 1980 McGraw-Hill, Inc., New York (October 30).

Engineering News-Record, 1981 McGraw-Hill, Inc., New York (May 14).

Engineering News-Record, 1984 McGraw-Hill, Inc., New York (May 31).

Engineering News-Record, 1991 McGraw-Hill, Inc., New York (February 11).

Engineering News-Record, 1992. McGraw-Hill, Inc., New York (October 19).

Engineering News-Record, 1993. McGraw-Hill, Inc., New York (March 8).

Engineering News-Record, 1994. McGraw-Hill, Inc., New York (May 30).

Felsen, M., 1989. "Mediation That Worked: Role of OSHA in L'Ambiance Plaza Settlement," *Journal of Performance of Constructed Facilities,* American Society of Civil Engineers, New York (November).

Gurfinkel, G., 1988. "Precast Concrete Roof Structure: Failure and Repair," *Journal of Performance of Constructed Facilities,* American Society of Civil Engineers, New York (August).

Hanson, T., 1969. "The Structural Anchorage of Precast Prestressed Concrete," in *Mechanical Fasteners for Concrete,* Publication SP-22, Paper SP-22-9, American Concrete Institute, Detroit, MI.

Heger, F., 1991. "Public-Safety Issues in Collapse of L'Ambiance Plaza," *Journal of Performance of Constructed Facilities,* American Society of Civil Engineers, New York (May). (See also Discussions by C. Culver, R. Marshall, and C. Freyermuth in May 1992 issue.)

Huckelbridge, A., H. El-Esnawi, and F. Moses, 1995. "Shear Key Performance in Multibeam Box Girder Bridges," *Journal of Performance of Constructed Facilities,* American Society of Civil Engineers, New York (November).

Kaminetzky, D., 1991. *Design and Construction Failures,* McGraw-Hill, Inc., New York.

Lagorio, H., 1990. *Earthquakes: An Architect's Guide to Nonstructural Seismic Hazards,* John Wiley & Sons, Inc., New York.

Levy, M., and M. Salvadori, 1992. *Why Buildings Fall Down,* W. W. Norton & Company, Inc., New York.

McGuire, W., 1992. "Comments on L'Ambiance Plaza Lifting Collar/Shearheads," *Journal of Performance of Constructed Facilities,* American Society of Civil Engineers, New York (May).

Mitchell, D., W. Cook, D. Eyre, and G. Maurel, 1990. "Evaluation and Strengthening of Large Precast Concrete Warehouse Structure," *Journal of Performance of Constructed Facilities,* American Society of Civil Engineers, New York (May).

Moncarz, P., R. Hooley, J. Osteraas, and B. Lahnert, 1992. "Analysis of Stability of L'Ambiance Plaza Lift-Slab Towers," *Journal of Performance of Constructed Facilities,* American Society of Civil Engineers, New York (November).

NBS, 1987. *Investigation of L'Ambiance Plaza Building Collapse in Bridgeport, Connecticut,* NBSIR 87-3640, National Bureau of Standards (National Institute of Standards and Technology), U.S. Department of Commerce, Washington, DC.

PCI, 1971. *PCI Design Handbook: Precast and Prestressed Concrete,* Precast/Prestressed Concrete Institute, Chicago.

Peraza, D., 1995. "Collapse at L'Ambiance: What Went Wrong?" *Progressive Architecture,* Penton Publishing, Cleveland, OH (May).

Pfeifer, D., J. Landgren, and W. Perenchio, 1986. "Concrete, Chlorides, Cover and Corrosion," *PCI Journal,* Precast/Prestressed Concrete Institute, Chicago (July–August).

Poston, R., G. Feldmann, and M. Suarez, 1991. "Evaluation of L'Ambiance Plaza Post-tensioned Floor Slabs," *Journal of Performance of Constructed Facilities,* American Society of Civil Engineers, New York (May).

Report, 1968. *Report of the Inquiry into the Collapse of Flats at Ronan Point, Canning Town,* HMSO, London.

Schupack, M., 1982. "Protecting Post-tensioning Tendons in Concrete Structures," *Civil Engineering,* American Society of Civil Engineers, New York (December).

Schupack, M., 1989. "Unbonded Performance," *Civil Engineering,* American Society of Civil Engineers, New York (October).

Schupack, M., 1993. "Bonded Tendon Debate," *Civil Engineering,* American Society of Civil Engineers, New York (August).

Schutt, C., 1996. "Building Doctors," *Ascent,* Precast/Prestressed Concrete Institute, Chicago (Winter).

Scribner, C., and C. Culver, 1988. "Investigation of the Collapse of L'Ambiance Plaza," *Journal of Performance of Constructed Facilities,* American Society of Civil Engineers, New York (May). (See also Discussions by G. Gurfinkel, M. Brander, R. Clark, and J. Frauenhoffer in August 1989 issue.)

Somes, N., 1973. "Abnormal Loading on Buildings and Progressive Collapse," in *Building Practices for Disaster Mitigation,* NBS Building Science Series 46, National Bureau of Standards (National Institute of Standards and Technology), U.S. Department of Commerce, Washington, DC.

Suaris, W., And M. Khan, 1994. "Performance of Prestressed Concrete Roofs During Hurricane Andrew," *Journal of Performance of Constructed Facilities,* American Society of Civil Engineers, New York (February).

9

MASONRY STRUCTURES

One of the oldest construction systems is an assembly of stone or brick, set dry or with mortar, to enclose space or to support bridge decks. These materials are among the most time-honored structural materials in the designer's vocabulary. Masonry structures have survived for thousands of years. The Egyptian pyramids, the Great Wall of China, and the Roman aqueducts all testify to the impressive longevity of masonry. Yet this long experience with masonry has not totally eliminated failures, either during construction or after a relatively short life.

Many deficiencies in contemporary masonry buildings are related to new materials and combinations of masonry materials with other construction materials in nontraditional ways. In this chapter we review failures in traditional and contemporary facilities constructed of brick, block, and stone masonry. Discussed are both unreinforced and reinforced masonry used in structural and nonstructural applications for bearing walls, shear walls, curtain walls, and building facades.

9.1 CHARACTERISTICS OF MASONRY: TRADITIONAL AND CONTEMPORARY CONSTRUCTION

The compression strength of masonry has traditionally been its most valuable structural characteristic. There are few recorded masonry failures in modern times that can be attributed to excessive stress in pure compression. One masonry crushing failure was reported in 1911 in Boston, Massachusetts, where an exterior 178-mm (7-in.)-diameter cast-iron column with a 406-mm (16-in.) square base plate sank 1.2 m (4 ft) into a stone masonry pier (McKaig 1962). The column was supporting a corner of the building and approximately 36 Mg (40 tons) of load. Three men were killed in the collapse. A subsequent investigation uncovered poor-quality masonry in the interior of the pier, although the exposed masonry on the surface looked quite good even after the failure.

With unreinforced masonry, bearing failures are more common when the loads are eccentrically applied or when stress concentrations occur under nonuniform loads. The use of bearing pads to distribute loads can be helpful in preventing such conditions.

In addition to the inherent compression strength, there are several other characteristics of fired clay and stone masonry construction that are attractive to designers and builders and to human beings in general. We relate positively to masonry for its "timeless" quality—its aesthetic continuity throughout the history of the built environment. We appreciate the ease with which new construction can be integrated aesthetically with older construction. We like the play of light across a masonry surface—the varied textures and colors. Because of its mass, masonry construction has some unique thermal properties and fire-resistive qualities. The production of a masonry building is straightforward, usually progressing one floor at a time. The scale of a masonry building and its details is comfortable. The masonry units themselves are based on human dimensions, and their repetition gives the finished product a "human scale." Although masonry construction is labor intensive, there exists an uncommon opportunity for the expression of craftsmanship. We even admire the way a masonry building ages, typically decaying in a "nonpolluting" way.

Traditional construction of "permanent" works was based in a very limited number of natural materials, most of them masonry materials (Eppell 1981). The long tradition of construction in masonry would seem to suggest that all problems in masonry structures should have been eliminated by experience. However, failures do continue to occur. There have always been masonry failures, as has been the case for all other materials. Currently, there are failures due to new causes: the combination of masonry units with other materials in new applications and systems (Figure 9.1). Understanding the specific dynamic physical and chemical characteristics of the constituent materials in masonry buildings has become more important than in past eras.

The problems with traditional unreinforced masonry construction have been a lack of tension strength and ductility, as for unreinforced concrete. These deficiencies have caused numerous catastrophic failures in unreinforced masonry facilities during extreme windstorms and seismic events throughout the world. The idea of reinforcing masonry for large construction projects, to give it some tension capacity and ductility, is reported to have begun with the construction of brick caissons 760 mm (30 in.) thick, 15 m (50 ft) in diameter, and 21 m (70 ft) deep as part of the Thames River Tunnel in 1825.

Figure 9.1 Contemporary buildings use masonry in nontraditional applications. Note the spalling facade falling from this new hotel in New Delhi, India.

The aesthetic masonry tradition is based in forms that recognize the inherent compression strength of masonry and its weakness in tension: buttresses, columns, arches, and bearing walls. These are the forms we have come to associate with masonry, and they are still the most appropriate forms, although modern theory and materials permit some remarkable nontraditional applications. If contemporary designers would remember that a brick is a tiny, brittle, compression unit, their projects would be more expressive of the masonry tradition and would probably be more durable.

Historically, masonry was used in a structural capacity. It was often covered over by "finish materials" since it was felt to be utilitarian and therefore unworthy of being seen. Ironically, today we often use masonry veneer to cover the structural frame or backup walls, because we value its appearance as a finish material.

Older buildings built of unreinforced masonry relied on the application of large dead and live gravity loads to prestress the masonry components, forcing them to stay in compression even under extreme wind loads. For a tall building, this required massive wall thicknesses in the lower floors, resulting in very dense floor plans with little usable space.

With the introduction of reinforced masonry and related rational design procedures, the construction of tall buildings that could safely resist wind and seismic events with much less material was made possible. When detailed and constructed properly, the steel reinforcement gives the material the needed tension capacity and ductility. This new technology and other economic factors have brought about changing expectations for masonry walls. Traditional masonry details that were successful in low-rise buildings have not always worked well for tall buildings that are exposed to more severe weather, including wind-driven rain.

With the advent of the steel frame, buildings became taller and walls became thinner. Material modifications were required, including brick of higher strength and less porosity, stronger (and more brittle) portland cement mortars, cast concrete blocks, new window and flashing materials, and new shapes of masonry units that permitted quicker assembly. Economic considerations focused on erection speed; the mason's status began to change. "The craftsman became a mechanic—not a skilled artisan but a rapid and efficient assembly automaton" (Eppell 1981).

Mechanical ventilation and air-conditioning equipment, elevators, and other sources of vibration were attached to buildings having flexible structural frames. These dynamic vibrations have contributed to increased cracking in brittle masonry components. Controlling the interior building climate introduced new thermal and pressure stresses as well as condensation problems.

Along with these new sources of trouble, the modern user has higher performance expectations than in past generations. What used to be considered minor inconveniences are now considered unacceptable, such as waiting for an internal change in thermal conditions, or tolerance of cracks that are minor cosmetic defects.

A number of detailing problems have been introduced in modern masonry construction, which generally combines masonry products with other materials that have very different characteristics. Problems caused by differential deformations in such assemblages are not insurmountable, but the contemporary masonry building experiences conditions that simply did not exist in the traditional masonry structure. There are many constituent ingredients in modern masonry construction: (1) units of fired clay or cast concrete; (2) mortar to bond the units together, to level them, to compensate for dimensional variances between units, and to make the joints water resistant; (3) grout to bond the units and reinforcing steel together; and (4) the reinforcing steel. Failure can occur in any of these materials. It can also occur at the interface, or bond, between units and mortar, units and grout, or grout and reinforcing steel. To this may be added the complex condition of a backup wall constructed of dissimilar masonry units, or a structural frame of cast-in-place concrete or steel, to which the masonry is attached.

Understanding the composite action of such an assemblage of constituent components is a challenge for the contemporary designer, far beyond the complexity faced by the designers of masonry buildings in simpler times. Inspecting the condition of an existing masonry facility or determining the cause of a failure is equally challenging. Several helpful checklists and nondestructive investigation procedures are available to aid in this difficult task (ASCE 1991).

Mattar and Morstead (1987) note that up until the middle of this century, load-bearing exterior walls were heavy and massive. Designers compensated for a lack of technical understanding of the performance of masonry by using empirical methods and generous factors of safety. Today, however, the focus is on efficiency. Masonry is being used as a veneer and for wall panels in conjunction with structural steel or concrete frames. In these applications, where minimal thicknesses and reduced weight are usually desired, specific technical knowledge of material behavior is a prerequisite for successful design. Ironically, our apparent long familiarity with masonry materials and construction has caused designers and builders to assume that we know all we need to know about the materials. Thus the design has often been accomplished without rigorous attention to the specific properties of masonry assemblages. In fact, the education of architects, engineers, and construction technologists rarely includes any formal training in the design of masonry or its numerous contemporary applications.

At present, there is a great deal of litigation related to leaking brick-clad exterior walls and other more life-threatening facade failures. These problems are discussed in Section 9.6. Many costly moisture-penetration problems could be avoided by incorporating traditional masonry details into the design, details that have evolved over centuries of experience with buildings and the weather. Instead, contemporary designers have been all too willing to trust the integrity of the building envelope to new "miracle" sealants that seldom perform as promised.

The mechanisms of moisture penetration are fairly well documented. Details for constructing moisture-resistant walls are available from masonry industry sources. Consulting these publications prior to construction can reduce the large number of repetitive failures. Publications include standards for masonry materials and construction methods, information from laboratory research, and lessons from forensic investigations of deficient facilities. Some of the organizations providing practical masonry design and construction guidelines and details are the American Concrete Institute (ACI), the American Society for Testing and Materials (ASTM), the Brick Institute of America (BIA), the Concrete Reinforcing Steel Institute (CRSI), the Masonry Society (TMS), the Masonry Institute of America (MIA), the National Concrete Masonry Association (NCMA), and the Portland Cement Association (PCA).

9.2 FAILURES DUE TO AGING

The ability of masonry to sustain load over time is well known, but it must be recognized that aging will reduce the factor of safety against failure. Continued use is no guarantee of safety, as was shown in 1967 by the sudden fall of two arches in the 113-year-old Aricicia Bridge in the Alban Hills of Italy. The center part of this highway trestle consisted of three tiers of masonry arches between massive stone piers. A pier collapsed under the top tier, and two spans disintegrated in the top two tiers. The failure came at night without warning. At least two cars fell into the roadway gap.

The campanile in Venice, Italy, was constructed between the years 888 and 1517. (The 600-year project could hardly be classified as a rush job.) It was under continuous repair following completion. During some remedial work in 1902, workers discovered that the many successive additions of mortar in the grout space within the walls had caused the interior mortar to become the structural support rather than the masonry units. Before the condition could be corrected, the tower collapsed.

In the United States, an increasing volume of construction activity is concentrated on repair and retrofit work in older facilities, including masonry structures. Some of this work involves conservation of unique examples of high-quality architecture and engineering, or projects that have symbolic or cultural value. Repair of the aging monumental masonry buildings in Washington, D.C., is one example. Other building conservation projects are driven simply by economic concerns. Conservation, including structural retrofit work, may be less costly than replacement by new construction. Because of increasing activity in this area, articles on masonry repair are appearing more frequently in the technical literature (Beasley 1987).

Helpful publications for repair and renovation are also available in the European literature, where design and construction professionals have been working diligently for centuries to preserve stone and clay masonry structures. Among these are the many excellent publications by the Masonry Conservation Research Group of Historic Scotland (Historic Scotland 1994). Even the cleaning of aging stone masonry requires a high degree of knowledge and skill. Done improperly, cleaning can destroy the quality of the very detailing that one is trying to preserve (Figure 9.2). Experience in Europe has proven that simple housekeeping—regular cleaning of roof valleys and gutters and repairs to protective roofing and flashing—can be the most effective strategy against deterioration of masonry structures. Some unoccupied heritage buildings have fallen into irreparable condition from neglect of these minor details.

Older buildings have typically been changed from their original design. Walls have been removed, new openings in walls and floor diaphragms have been introduced, new loads have been applied, and the buildings have undergone changes in use. These revisions may have gone unrecorded and may have been accomplished with total disregard for the integrity of the original structure. While undertaking repairs in such structures, designers and construction workers must exercise extreme caution to be sure that existing load paths are identified, and that any new load-resisting systems include provision of a complete load path, all the way to the foundation. Examples of the catastrophic consequences of failure to provide such a load path are given in Section 11.7. In the case of seismic retrofit work, the connections between floor and roof diaphragms and the supporting walls are often identified as the weak links, and these are strengthened. Recent earthquakes have shown, however, that failure to follow the load all the way to the foundation merely moves the failure to another location. Again, a complete load-path study is necessary. Seismic retrofit work in masonry structures has generated considerable interest in methods of repairing cracks and improving the general integrity of older systems. This work is explored further in Section 9.8.

Old masonry walls weakened by the weathering of lime mortar joints fail regularly, sometimes with serious damage. In 1922, the Majestic Theater in Pittsburgh, Pennsylvania, originally built as a church, lost part of its roof when the end wall bowed and pulled over a 18-m (60-ft)-span timber truss. The brick wall was 330 mm (13 in.) thick and 15 m (50 ft) high. In 1947, a brick pier of the Empire Apartments in Washington, D.C., collapsed after 54 years of satisfactory support. The structure was seven stories high, with walls varying from 330 to 685 mm (13 to 27 in.) in thickness. The only change in the original wall was an opening cut 15 years before the failure.

Aging unreinforced masonry walls may lose part of their facing or may totally collapse from the accumulated effect of traffic vibration on a rough-surfaced street. A seven-story warehouse collapsed in New York in 1930 after 35 years of use with no observable distress when the wall moved progressively outward and allowed the unanchored floor beams to drop. Parts of walls, especially areas adjacent to chimneys, have fallen in a number of old buildings facing railroad trackage. In 1959, a fatal accident at a school in Houston, Texas, was caused by the toppling of a section of brick wall 4.3 m (14 ft) long and 2.4 m (8 ft) high adjoining a gate. Although the wall was reinforced and

Figure 9.2 Proper cleaning of older stone masonry requires skill and technical expertise. (Courtesy of Historic Scotland and the Robert Gordon University.)

tied to the footing, the rather small dynamic reaction of the gate action was apparently sufficient to cause overturning.

Temperature fluctuation, rain exposure, moisture absorption, chemical alteration of the masonry units and of the mortar, and elastic and plastic strains, all acting in many combinations, alter the appearance, weathertightness, and strength of masonry. Design details and construction procedures must expect and compensate for the changes that will occur during the reasonable life of the structure. Bricks with a high coefficient of saturation are vulnerable to premature surface deterioration (spalling), especially when exposed to freeze–thaw cycles.

Some time-related changes, but very few, can be beneficial. Ancient stone structures in which the lower contact face was carved out so that only a narrow rim of stone was in bearing became tighter with age as the plastic flow of the stone closed up the irregularities in the joint. Masonry units set in adobe mortar soon took on the appearance of a tight wall as rain exposure eroded the earthy filler. Older lime mortar jointing, although increasing in strength with age for several years, absorbed the expansion of the soft clay brick when it became saturated and the wall exposure became tighter. The newer cement or masonry mortars, when stronger than the brick, do not permit such compensation. When the brick expands, the mortar cracks or crumbles, and if the mortar is too strong, the brick spalls and cracks.

9.3 CONSTRUCTION ERRORS AND WORKMANSHIP DEFICIENCIES

Tall masonry walls such as those used for educational and industrial buildings are seldom really stable until tied by the roof and held by the floor. Unbraced wall lengths and heights exceeding normal practice are encouraged by economy, a false economy when the costs of failure cleanup and litigation are added to the estimate. Some examples of stability failures prior to completion of the supporting diaphragms and connections are given in Chapter 11.

Other construction-related masonry failures discussed in Chapter 11 include those associated with scaffolding, work platforms, and temporary structural support. One common error is the hanging of work platforms on one side of the wall being built, with through-ties and no exterior bracing to resist the overturning moment. Tying scaffolding or work platforms to the work being constructed is always a risky venture. Scaffolding should be supported by firm connections and a complete, reliable load path to the ground. Another typical construction failure occurs when edge-supported flexible steel joists are loaded by wet concrete. The slope of the loaded joists at the embedded wall support, usually not over 100 mm (4 in.), cripples the block or brick masonry wall. A lack of pilasters is usually given as the cause of failure. In 1956, the 194-mm (8-in.) concrete block walls of a school building in Orlando, Florida, buckled in a similar fashion when the 17-m (56-ft)-span double-tee precast concrete roof members deflected from rain loading. The wall was not as flexible as the embedded roof members. Such a failure at Waltham, Massachusetts, in 1959 involved roof members of hollow precast concrete slab section.

The high factors of safety typically used for design of structural masonry elements reflect uncertainties related to workmanship quality. While the quality of the masonry units can be carefully controlled during manufacture, the assembly of a masonry structure takes place on the construction site. The same uncertainties discussed previously for cast-in-place concrete and precast concrete construction are present in masonry construction, and to a greater degree. The importance of inspection to assure that field work is in accordance with design details and specifications cannot be overemphasized. Standards for the industry encourage the provision of independent inspection of masonry construction by permitting higher allowable stresses when such inspection is provided on a continuous basis.

For traditional masonry construction, designers relied on established styles and building methods that evolved both aesthetically and functionally. Craftsmen understood the details thoroughly, having only a few materials and tools with which to work. The masonry apprenticeship training had its own traditions and commitment to quality, with an understanding that quality of workmanship was reflected in permanence. Most masonry details evolved functionally as the means whereby water was kept out of buildings.

Some of the recurring problems in contemporary masonry construction result from the multiplicity of trades involved in the work. Poor workmanship by concrete construction workers, for example, may make it difficult or impossible for the masons to achieve the details and tolerances required by the contract documents. If the shelf angles are not in the proper place, if they are not continuous around the corners, or if, as is sometimes the case, they are missing altogether, one can hardly blame the masons for the ensuing difficulties. Proper masonry construction relies on the inclusion of metal ties that usually fit into tie slots and other miscellaneous anchorages embedded in the concrete frame or backup walls. There is an evident need for close coordination of the work accomplished by the various trades.

Improper execution of the stress-relieving joints at shelf angles at each floor level has been a continuing source of costly repair work in multistory masonry buildings. The intended effect of these "soft joints" is easily negated if they are of too small a dimension, or if mortar or other foreign matter is incorporated into the joint. In the case of stone masonry facing, the integrity of soft joints is critical. Failures in the stone panels have occurred from stress concentrations, sometimes caused by shims left in place in the joints near the edges of the panels.

Proper placement of flashing is integral to weather resistance of the building envelope and, in fact, to the durability of the masonry itself. The flashing must be protected from damage during construction. Torn flashing destroys the integrity of the envelope. Moisture can easily discover the location of discontinuities in the flashing—much more easily than can those who are trying to repair the subsequent damage. The time to inspect the flashing is, of course, while it is being placed. Masonry units are extremely durable. It would be well to consider the desirability of providing the same degree of durability for all the essential components that are hidden behind the masonry units. To repair or replace them later requires costly selective disassembly of the construction.

Performance and durability of masonry construction is dependent on the quality of the mortar. In many cases the deficient performance of a masonry facility can be traced to improper specifications for the mortar, despite the widespread availability of published industry guidelines (Isberner 1974). There is no one type of mortar that is ideal for all applications. A number of cements and admixtures are available for various exposure conditions. Since the mortar may contain many ingredients, its proper proportioning and mixing at the construction site require a high degree of skill.

Requirements for protection from the weather are similar to the precautions taken in cast-in-place concrete construction. The mortar must be allowed to set properly, without delay or freezing due to cold temperatures, or without proper curing time in hot, dry weather. A few decades ago, the practice of including calcium chloride in the mortar mix during cold weather construction was encouraged, but calcium chloride can be responsible for corrosion of metal components embedded in the assembly. Its use is discouraged in modern masonry construction.

The quality of the bond between the masonry units and the mortar is greatly affected by workmanship factors. Poor bond may result from either too much or too little water in either the mortar or the brick units. The units must be wet at the time of placement, especially if they have a high rate of absorption. Proper tooling and bedding of joints and maintaining uniformity in the size of mortar joints are additional characteristics of good workmanship that affect durability.

9.4 SETTLEMENT, EXPANSION, AND CONTRACTION

Like concrete structures, problems in masonry construction often result from restrained movements. The "secondary" internal stresses caused by elastic strains, creep, shrinkage, and expansion must be considered by the designer. The design must either compensate for the stresses induced by restrained movements, or provide isolation or separation joints so that movements are not restrained. Since masonry structures tend to be quite rigid, foundation movements involving uneven settlements or frost action in the soil can cause a complex distribution of stress and can easily lead to failure (Sputo 1993).

Movement in a masonry assembly can come from a variety of sources. Brick expands with moisture content from the air and from direct rain exposure. Water absorption always results in volume expansion, which unfortunately does not disappear completely when the water evaporates. At the same time, all masonry increases in volume with a rise of temperature. The dimensional changes are greater when the brick has not been baked sufficiently during the manufacturing process. Generally, the darker units have a lower coefficient of saturation, with greater durability and less dimensional instability. Cast concrete units, on the other hand, experience shrinkage in service (Mattar and Morstead 1987).

A common source of uneven expansion is the presence of waterborne chemicals. All masonry units, from spongelike porous cinder concrete block to almost impervious glazed brick, will absorb water in varying degrees. But the less porous the masonry unit, the larger the percentage of the moisture on the wall face that goes into the mortar joints. Saturation brings possible frost damage and, where conditions permit, sulfate reaction. Where masonry is likely to be wet over long periods, the bricks should be of low sulfate content and the mortar made of sulfate-resisting cement. Porous brick or block should not be exposed to frost action.

One interesting expansion phenomenon caused by moisture penetration is the almost universal condition of older brick masonry chimneys leaning into the prevailing winds (Figure 9.3). A dramatic example of sulfate action was the sudden mysterious bending of the top 6 m (20 ft) of one radial brick chimney 43 m (140 ft) high at a Westchester, New York, hospital. A similar companion chimney remained vertical. The explanation for the differing appearance was found after a study of use and exposure. Both chimneys had served well for about 25 years, one as the stack for the incinerator and the coal-fired hot water boiler, the other as the stack for three coal-fired heating boilers. Soon after the heating units were converted to fuel oil, the top of that chimney started to bend toward the northeast. The top was noticeably out of plumb, at least 300 mm (12 in.) when a careful inspection reported no cracking of the brick or the mortar. The southwest side was simply growing longer.

The oil being consumed had a high sulfur content, leaving some sulfur dioxide fumes congealed on the inner face of the chimney near the top. Prevailing wet winds were from the northeast, regularly saturating that side of the chimney and concentrating water on the opposite side at the inner face. There was an internal shoulder in the brickwork 6 m (20 ft) from the top and the sulfuric acid formed by the combination of the sulfur fumes and water accumulated up to that depth. The acid acted on the cement in the joints to form calcium sulfate, which takes up water readily and expands as it crystallizes. This expansion enlarged the joints in the southwest quadrant, pushing the chimney over into the wind without forming a single visible crack. The cure is to avoid use of oil with high sulfur content and to eliminate internal shoulders or other moisture-trapping details.

Thermal- and moisture-induced dimensional changes in masonry materials are not completely reversible. When expansion causes the formation of minute cracks or separation between brick and mortar, grit formed at the cracks and airborne dust partly fill the gap. Upon cooling, the wall cannot return to its original dimension. Each succeeding cycle adds to the deformation. Thus cyclic

Figure 9.3 Masonry chimney leaning into the wind.

moisture and thermal repetitions may lead to distress. It does not require many cycles of thermal change to form visible cracks, usually darker than the masonry exposure and subject to frost damage if moisture gets in during the winter (Figure 9.4). Walls in excess of 30 m (100 ft) in length have been known to grow, sliding on their foundations, until the ends of the building overhang the concrete foundation walls. The buried foundation walls are not affected by thermal changes to the same extent (Figure 9.5).

Masonry enclosures of roof areas that are exposed to the sun and high temperature, especially those surfaced with materials having high expansion coefficients, must be detailed to give the roof freedom of movement. Otherwise, the masonry develops cracks, bulges, and unstable corners. Parapets always present unique detailing problems (Figure 9.6). They are exposed to a wide range of temperature, differing from the thermal conditions in the wall below the roof. There is opportunity for moisture penetration from all sides and from poorly detailed roof intersections with the wall. The collection of moisture and freeze–thaw action creates significant movements. Careful detailing of movement joints, especially at the corners, is important to the satisfactory performance of a masonry parapet.

Commonly called expansion joints, isolation separations between rather limited lengths of masonry must be provided if cracks are not desired. The joint detail either works and protects the masonry from cracking, preventing infiltration of air and moisture into the structure, or it does not work. The great majority of joints do not work; this is the true explanation of why walls crack despite good intentions. In one such building, where mysterious cracking was followed by the usual "round robin" of placing the blame, the bottom joint in a two-story wall was of zero width, and the more visible top joint was a full 40 mm (1.5 in.) wide. In a large new multistory hotel with workmanlike jointing in the brickwork seemingly in accordance with construction details in the contract documents, a little probing proved that the expensive sealant, 19 mm (0.75 in.) deep, was backed with very strong mortar in the rest of the 300-mm (12-in.) wall thickness, explaining why the brick facing was cracked.

Masonry must be built in separate units, with isolation joints to give freedom of expansion so that cracking will not start, or else be properly reinforced with steel. In an apartment house in upper Manhattan, New York, a roof garden was added and a red ceramic tile floor installed. After one summer, the expansion of the tile pushed out the brick parapets that rotated about the top-story window lintels and cracked the brickwork at all returns in the wall. Cracks were sufficiently large, such that the top-story tenants complained of daylight coming through the walls. An expensive wall reconstruction included providing a wide expansion filler between the tile and the parapets. It should be noted, however, that exceptionally wide movement joints present problems, since most joint materials have limited tension strength. It is preferable to use a greater number of narrower joints.

Thermal changes depend on sun exposure more than on ambient temperature. Corners of buildings have to act as hinges when one face expands more

Figure 9.4 Masonry cracking from roof expansion.

Figure 9.5 Expanding clay masonry building moves off its foundation walls.

than the other, and vertical cracks typically form at the first vertical joint. The length of wall to be considered for expansion is the developed length, not the projected dimension. An exterior wall in a building with an L-shaped plan acts as if it were one long wall. This wall configuration is, in some respects, even more vulnerable to cracking than a straight wall, because of the hinge action and stress concentrations at the reentrant corner.

Disregard of the effect of restraint (i.e., omission of necessary expansion freedom) is the usual cause of the masonry distress seen in many buildings, more in recent structures than in older ones, which were built more slowly and with weaker components. The small amount of deformation in restrained condition that is required to cause failure may not be readily appreciated. Average values are for concrete: 0.10 and 0.01 percent, respectively, in compression and in tension; for brick: 0.20 and 0.016 percent, respectively, in compression and in tension; for stone: 0.25 and 0.007 percent, respectively, in compression and in tension; for steel: 0.13 percent; copper: 0.29 percent; and aluminum: 0.39 percent. Dimensional changes of this magnitude in masonry can easily occur due to thermal and moisture conditions. Where freedom for movement is not available, failure is inevitable.

For stone or brick masonry facade panels, the necessity of soft joints at regular intervals, filled with reliable compressible materials, has already been discussed. These joints can easily be negated by poor workmanship. Cracked and failing facades have sometimes been traced to mortar concealed behind a bead of sealant in the specified "soft joint."

Figure 9.6 Masonry parapets present unique detailing problems due to complex moisture and temperature effects. This parapet on a new university building is being held together by duct tape while a corrective solution is sought.

9.5 INCOMPATIBILITIES WITH OTHER MATERIALS

In modern buildings, dissimilar materials are used together in composite assemblies, even for load-sharing structural components. One common example is the juxtaposition of a reinforced concrete frame and masonry veneer. The concrete is subject to irreversible creep deformation under sustained dead loads, while the clay masonry units expand in service. Initial moisture expansion of the masonry is also irreversible. The resulting differential movement can cause bulging of walls, cracks at corners, and displacements and rotations of shelf angles. The differential deformations may cause structural components to tear themselves apart, given sufficient time, moisture, and temperature conditions. Prevention of such problems relies on careful detailing of movement joints at proper locations and provision of bond breaks and slip planes. Shrinkage of cast-in-place concrete slabs is another common problem if the slabs are indiscriminately anchored to the masonry walls. Floor slabs must be free to move or cracking may occur.

Incompatibility between brick and concrete masonry units making up a composite wall results in cracking along both horizontal and vertical lines in the brick facing. One of the earliest examples studied in detail was the Lincoln Park Homes development in Columbus, Ohio. Completed in 1942, the 26 two-story apartment dwellings had walls comprised of face brick backed with load-bearing structural clay tile. After a few months, the brick facing developed cracks, breaking up the faces generally 1 m (3 ft) apart vertically and 1.2 m (4 ft) apart along horizontal lines. Cracks were pointed, but they reopened and new cracks appeared. The interior plaster cracked consistent with the wall cracks. Tests of the masonry materials showed high moisture expansion of the red clay tile backing, about 0.2 percent. Repairs consisted of applying a thin cement paste with waterproofing admixture to the exterior face, followed by a burlap bag wiping. Cracks reappeared regularly and regrouting was required about every three years.

In Cincinnati, Ohio, a six-story wall-bearing apartment building was placed on a concrete-framed street-level garage. The wall consisted of a 100-mm (4-in.) cored face brick connected to concrete block by headers at every sixth course. Thickness of the wall varied from 508 mm (20 in.) at the second story to 305 mm (12 in.) at the top, but the brick layer was supported by a continuous steel lintel at the second floor above the garage, where there was no supporting wall. When the building was two years old, the lower three stories of facing brick peeled off several walls and the cored headers sheared. Since the incident occurred in July, neither failure could be explained by frost action or by omission of metal ties, the usual reason for such failures. The cause of this collapse was attributed to shrinkage of the concrete-block masonry backing.

Storage bins and concrete tanks with brick covering have similarly lost part of their ornamental veneer, resulting from the additional dimensional changes of the container under full and empty condition, together with thermal and moisture expansion. The highly variable effect of moisture expansion in clay masonry was not commonly recognized until the 1950s (Grimm 1975). Average expansion is 0.02 percent, but depending on the particular clay used and its manufacturing history, the moisture expansion can vary from near zero to over 0.05 percent.

Mixing brick and concrete or concrete block load-bearing walls in a building creates serious long-term deformation problems. Composite brick/block bearing walls carrying heavy compression loads in tall buildings are especially problematic. The cement products experience creep and shrinkage, while the clay masonry will expand. The mortar joints between the fired clay units are subject to creep, but the clay units are not. Since the mortar joints make up only 15 to 20 percent of the volume, differential deformations will arise in service. The result is unacceptable cracking of the load-bearing walls.

When the structural frame shrinks in service, load is transferred to masonry elements that may not be designed to resist the load. This is why soft joints are needed at every floor level to support stone or clay masonry veneer. Distressed veneer often requires repairs that entail cutting in new soft joints. Those doing the work must exercise great care as the locked-in stresses may be substantial. Highly stressed portions of the wall may literally explode when the saw cuts are made.

Elastic shortening of the columns in a structural steel frame from loadings applied after the wall enclosure is complete can seriously affect stone facing when no relief is provided in the horizontal joints. The common historic detail of lead joints at every second story for stone covering supported by steel framing seems to have been overlooked in the rush of construction starting about 1928. A monumental college building on Park Avenue in New York City, built in 1940, was covered with half-story-high limestone sheets, carefully set in thin high-strength mortar joints. Elastic compression deformations in the columns, along with thermal movements from the exposure of 60 m (200 ft) of wall to the west, warped the face. Under the almost horizontal morning sunlight the north and south stone facing appeared dangerously distorted. The shadows caused by small projections of each stone edge relative to its neighbor suggested much larger out-of-position displacement than had actually occurred. A number of the stones had to be removed and reset with lead joints.

The Veterans Hospital at Albany, New York, constructed in 1950, was a 10-story cruciform steel framed structure, with each of the four ends extended into an L-shaped unit. There were therefore eight end walls, about 12 m (40 ft) width, two at each of the four cardinal compass points. The walls were limestone covered, with a joint at the corners. Shortly after enclosure of the building, and following the first winter, cracks developed in the exterior corners, chiefly noticeable at the second-floor level, but some extended the full height of the building, with maximum width at about the third-floor level. On the east exposure, all four corners were cracked; the south and west had three corners cracked each, and at the north faces only the northwest corner opened up, showing the importance of thermal exposure. None of the 16 interior corners, all of which were at 135 degrees, showed any evidence of movement.

Masonry at roof lines, and especially at parapets, is often in a state of distress, as discussed in Section 9.4. The two-directional shrinkages at corners, together with lifting of the corners of the roof slab, tear the masonry apart. Long-term deformations in concrete or in timber cannot be resisted economically. Only properly designed and constructed isolation details, allowing room for movement, will protect against such cracking. Compatibility—indeed, almost equality—is needed in the moisture and thermal change of the various masonry units making up a wall. An ornamental limestone facing for a library in Brooklyn, New York, was backed with clay tile to make up the main walls. The limestone showed considerable distress and the wall had to be removed completely. It was rebuilt successfully with a brick backing. No older successful stone-covered building is ever found that used masonry backing with lower rigidity or dimensional stability than the facing material. Yet modern designers continue to apply stiff, brittle stone facade panels to flexible steel frames, with little consideration for the consequences.

Masonry covering on reinforced concrete frames is subject to all the troubles found in steel buildings, and some others as well. By 1948, the New York City Housing Authority had 10 years of experience with brick-covered reinforced concrete frames from 7 to 14 stories high, using stone or gravel concrete and solid walls of 100-mm (4-in.) brick backed with 150- to 200-mm (6- to 8-in.)

tile or concrete block, supported at each floor level by suspended steel lintel angles. An inspection of nine projects consisting of over 100 buildings found corner cracks in the horizontal joint at the bottom of roof spandrels in over 50 buildings. The cracks occurred in skeleton frame buildings and also in bearing wall buildings where a continuous concrete distributing wall beam was incorporated in the roof slab. Cracks were concentrated close to the corners, but occasionally they extended up to 3 m (10 ft) along the horizontal joints. Parapets were all 300-mm (12-in.)-thick solid brick, and in one project numerous generally vertical cracks appeared close to all corners and at any offset in the line of the parapet.

Subsequent project designs incorporated anchorage for the parapet wall to the roof slab with fairly close expansion joints or even complete omission of the masonry parapet to eliminate differential thermal and shrinkage movements between the concrete roof and the masonry. In the Mulford Houses Project, built in 1939 in Yonkers, New York, a 25-mm (1-in.) gap was provided between the face of the roof concrete and the masonry wall. All masonry cracking at the roof level was eliminated by this detail.

Brick facing on tall concrete frames requires provision for differential vertical dimension change in addition to lateral thermal length change. The delayed shrinkage and creep of the loaded columns, along with the elastic shortening from loads added after the walls have been built, especially where walls are started at lower floors before the framework has been topped out, impose compressive forces into the sheet of brickwork beyond the outer face of the concrete structure (Figure 9.7). Failure can easily occur in the veneer.

One such building is described by Grimm (1975). A concrete wall 33 stories tall exhibited a measured shortening due to shrinkage and creep of 81 mm (3.2 in.), or 0.069 percent. Total unrestrained expansion in the brick may typically have been 0.052, with a total differential strain of 0.12 percent. The ultimate compressive strength of masonry is reached when strain reaches 0.029 percent. But because the bearing of masonry on shelf angles is not exactly uniform, stress concentrations occur which can exceed the compressive strength of the masonry long before the ultimate differential movement occurs.

In the building analyzed by Grimm, the brick veneer was not well anchored to the concrete. The distress caused 10 m² (100 ft²) of masonry veneer to buckle, tear free of the building, and fall 20 stories through the roof of an adjacent building.

Where masonry headers connect the face brick with the backup, which is supported directly at each floor level, some of the induced compression is distributed into the floor construction. The shear value of such headers is the limit to such load relief. In some wall failures, the headers are found sheared at the lower floors. Where flexible metal ties are used in place of headers in solid walls (and almost always in cavity walls), no such relief is possible. The entire shrinkage effect of the columns is imposed on a 100-mm (4-in.) sheet of brickwork acting as a compression member. The steel angles, usually con-

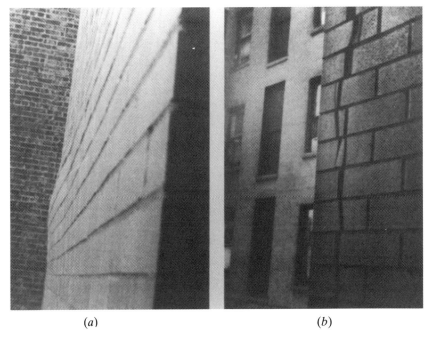

(a)   (b)

Figure 9.7 Masonry distress from load transfer as column shrinks.

nected to the floor construction at each story and rigidly mortared into the brickwork, become nothing more than metal fillers in the wall.

Use of cored brick with glazed face only worsens the problem. The brick is weaker and the face is more brittle. Use of lightweight concrete mixes, with higher shrinkage, more creep, and lower elastic modulus than those of normal concrete increases the amount of column shortening, and higher loads are then imposed on the brick. The result is a large inventory of buildings with vertical corner cracks, push-outs above the steel lintel levels at lower floors, brick face spalling, especially at the corners adjacent to angle supports, and cracked brick along the edges of wall openings, where concentration of imposed load is highest.

Problems such as these can be avoided simply by providing adequate horizontal expansion joints under the shelf angles that support the veneer. All of these observable defects, causing unsightly exteriors, leakage of walls, falls of brick faces on the streets below, and sometimes fall of complete wall sections, can be eliminated by proper introduction of horizontal relief caulked joints underneath the continuous steel lintels at each floor, together with vertical relief joints near each corner or offset in wall face. The horizontal expansion joints under the shelf angles must accommodate the combined effects of expansion of the masonry and shortening of the concrete frame due to creep, shrinkage, and elastic deformation.

9.6 MASONRY CLADDING, CURTAIN WALLS, AND FACADES

Ever since people began to build masonry walls of earth materials, sometimes waterproofed with the dung of their domesticated animals, the purpose has been to keep the weather out of the enclosure. Thousands of years of experience and experiment have not yet solved the problem for many modern buildings. Leaking walls are the second most common cause for building construction litigation; only roofing problems account for more litigation. While the general subject of leaking nonstructural building envelopes is covered in Section 10.1, masonry cladding problems are presented here. This is because masonry cladding problems are so integrally related to the general structural system, and their failure can be especially life-threatening, due to the weight of the materials involved.

The most common difficulties with any wall systems are related to moisture. A perfectly watertight facade is not possible. In modern masonry construction, fine cracks in the masonry and mortar permit moisture penetration, especially in tall buildings where wind-driven rain exposures are increased. The cracks stem from many sources, including the differential deformations of various dissimilar structural and finish materials discussed previously. Other moisture problems in contemporary "weathertight" buildings involve condensation. Proper location of an effective vapor barrier on the warm side of the exterior wall is necessary to prevent internal moisture from being drawn into the wall materials.

Old buildings sometimes leaked, but contemporary walls are more susceptible to moisture penetration. Traditional walls were massive and quite absorptive. They were able to store considerable volumes of moisture temporarily in storms, and they could breathe, eventually losing the water through evaporation. Modern thin, well-insulated walls do not have this capability. In addition to these factors, when high-speed production methods displaced craftsmanship in the mid-1920s, weather tightness became an unusual characteristic of masonry enclosures. Just as some very long span bridges have permanently assigned paint repair crews, there are tall buildings that require continuous joint and masonry caulking and repair.

The tendency toward denser masonry units and stronger mortars results in minute separations along all brick or stone edges. Every square meter of brick wall has about 21.5 m of joint. (6.5 lineal ft of joint per each square foot of wall). The walls of a building 30.5 by 61 m (100 by 200 ft) in plan, 6.1 m (20 ft) high, thus have over 24 km (15 miles) of joint. If only 15 percent of the joints are slightly open, water penetration starts to enter the masonry, although leakage starts with 30 percent of the joints open in the minute width of 0.125 mm (0.005 in.), which for each square meter of wall equals a slot 81 by 10 mm (for each square foot of wall, a slot 1 in. long by $\frac{1}{8}$ in. wide). The necessity for some absorption capability of the face brick and of the mortar is evident if the exterior surface of the wall is to act as a protection against the weather.

Sealants should be applied over masonry veneer materials to minimize moisture penetration, but it is not safe to rely on them entirely. Successful control of water movement depends on vapor barriers, flashing, and weep holes, all involving quality materials, detailing, and workmanship. Proper use of control joints will reduce the potential for cracking. Much can also be learned from traditional masonry details around door and window openings and the details at parapets and cornices. Although ornamental to some extent, these details evolved as an effective means to control the movement of water.

Water leakage in masonry-clad walls is always an expensive problem to correct (Beasley 1990). Usually, poor workmanship is responsible for water entry into the wall materials; poor design of the wall drainage system is at fault if the moisture penetrates further. A functioning flashing and weep system is critical to ridding the construction of moisture that penetrates the surface. Details and standards are readily available for constructing both barrier walls and cavity walls (where a continuous void is provided to guide the water out), but repairing a building in which these details were not followed is a difficult challenge indeed.

Frauenhoffer (1992) describes such a repair of a university building in Peoria, Illinois. The building was built under a fast-track construction contract in 1976. A two-story reinforced concrete frame provided the structure, which was infilled with masonry and bands of triple-pane windows. The building was plagued with numerous problems relating to the masonry wall and window system, including masonry cracking and water leakage. During repairs, as the construction was disassembled selectively, Frauenhoffer details the many design and construction defects uncovered that led to window seal damage, moisture penetration, trapped moisture, freeze–thaw damage, and masonry deterioration. In particular, the flashing details and materials were inadequate.

Sometimes an ill-considered use of new materials in an attempt to meet an unrealistic low budget results in serious failure. One such example was the use of newly available lightweight concrete blocks for 194-mm (8-in.) walls of two school buildings in Long Island, New York. Economy and "functional" design left the unfinished concrete block walls exposed in the classrooms, and only a pastel shade cement paint was applied to the exterior. Following occupancy, the blocks dried and shrank, to the extent that daylight could be seen through many joints and the students could not use the seats along the walls because of the draft. Litigation was brought by the school board against the designers and builders to collect the cost of repair and a waterproof stucco covering of the walls.

Masonry facades are placed on structural frames for weather protection of the enclosed space, for appearance, and as a protection for the structural elements against fire and against climatological change. Brick veneer covering on timber-frame houses has proven successful and of long life if proper waterproofing is placed between the timber and the brick and if the brick wall, usually 90 to 100 mm (4 in.) thick, is self-sustaining on a proper foundation.

Self-supporting masonry covering of steel-framed buildings became so massive as frames became taller that the panel wall, supported by the structure at each floor, became an economic necessity. The interaction of the frame and the more rigid masonry enclosure brought about new problems. Any change in relative position imposed shears in the more rigid and more brittle masonry. Normally, masonry cover was not installed until the steel frame was topped out and much of the floor slab weight in place. In a number of very tall steel-framed buildings, where vertical mortar joints at decorative corner bond or recess arrangements was not been provided, vertical cracks are practically continuous at all building corners about 200 mm (8 in.) from the edges. These result from thermal changes on the face having the greater sun exposure; the corner tends to act as a hinge. Where quoins or ornamental pilasters were built in more traditional masonry construction, the vertical mortar joint provided the hinge flexibility necessary for the very small angular motion.

Glazed bricks have been the source of many costly veneer problems. When the surface cracks, water is trapped behind the relatively impervious glazing. Freeze–thaw cycles in colder climates then can produce extensive spalling of the bricks. The Structural Clay Products Institute and Masonry Institute of America have issued many technical warnings since 1962 that successful use of ceramic glazed brick facing for exterior walls requires perfect control of moisture infiltration. That includes use of cavity walls with flexible anchorage to columns and floors, a vapor barrier on the warm side of the cavity, weep holes that are permanently open with ventilation holes near the top of free-standing walls such as parapets, adequate continuous flashing, and adequate expansion joints, not less than 20 mm (0.75 in.) wide per 30 m (100 ft) of wall on each side of each corner. Considerable information has appeared about the litigation resulting from masonry defects in concrete-framed buildings; examples are too numerous to enumerate (Figure 9.8).

Cavity-wall construction, in which two wythes of masonry units are provided with a gap and weep holes, has been promoted as a cure-all for masonry weather tightness. This form of construction relies on multiple components and several trades to construct the backup and facing materials and the system that connects them together. Cavity walls will not solve all the moisture problems unless proper care is taken to provide easy and continuous exit of any water that penetrates the cavity. This requires keeping the cavity and weep holes clean and provision of continuous flashing. Parapets are a common source of moisture penetration. It has already been noted that thermal movements of a parapet differ from those of the wall below. Flashing details at the wall–roof intersection are especially important.

A new exterior wall system, developed during the 1960s, involves a single exterior brick wythe, a cavity behind the brick, and a backup supporting structure of steel studs [spaced at 406 or 610 mm (16 or 24 in.) on center]. The steel stud backup wall is faced on both sides with gypsum board. This brick veneer/metal stud assembly was heavily promoted by the industry, and widely used, since it had numerous advantages over conventional masonry

Figure 9.8 Masonry distress at lintel from transfer of compression.

veneer walls, including economy, less thickness, improved thermal resistance, good appearance, lighter weight, rapid construction time, and good fire resistance.

Unfortunately, the system proved to cause problems in a large number of applications, leading to extensive litigation. This was a case where the technology of an excellent idea was not sufficiently developed to keep pace with its popularity (Gumpertz and Bell 1985).

Any curtain-wall system is intended to carry no vertical loads other than its own weight. The wall must, however, resist wind and moisture. When the brick veneer/metal stud assembly is loaded by wind, both the veneer and metal stud backup deflect, but composite action is not developed. The veneer and the backup move together, but because of the flexibility of the ties, restraint is not complete. The backup wall must provide the requisite stiffness to keep the veneer from cracking and resist the tendency for the masonry bed joints to open up when the assembly deflects.

When the system was introduced, industry publications recommended deflection limits in the stud backup that proved to be inadequate. These were on the order of $L/240$ to $L/360$, where L is the unsupported height of the panel. Current recommendations are for limits of $L/600$ to $L/720$, but research is ongoing. The system is a complex assembly that needs further research, including the collection of in-place performance data. The Masonry Research Foundation has identified this system as a high-priority research area since 1983. Structural problems result from the inherent incompatibility of the rigid (but weak and brittle) brick veneer and the flexible (but strong and ductile) metal stud backup. This incompatibility can lead to cracking of the bed joints of the brick veneer under wind loads. Cracking will cause loss of strength, masonry deterioration, and excessive water penetration. The water penetration affects durability to a marked degree, corroding steel studs, ties, and other fasteners. Walls less than 10 years old have been found with serious deterioration problems.

Some applications of the brick veneer/metal stud system have performed satisfactorily. But the system has not been entirely successful because of structural and water infiltration problems. As with all curtain walls, there are moisture problems caused by lack of proper attention to waterproofing, flashing, and insulation details. Condensation and external water infiltration have both led to serious structural and durability problems. Since the waterproofing and flashing surfaces are not accessible after project completion, all of the related materials should match the design life of the other components. This requires high-quality design, materials, and workmanship, just as for any curtain wall.

In the case of this system, however, moisture infiltration into the assembly can cause severe corrosion of the steel studs in a very short time. It should be recognized that a single wythe of brick veneer is not sufficient waterproofing for a wall system. Moisture will pass through the exterior veneer; brick masonry is permeable. Problems mentioned earlier associated with incompatible deformations between structural frame and masonry components can add to the cracking and further increase the moisture penetration. Functioning soft joints are needed, as with all masonry curtain walls, under shelf angles at the bottom of each panel. Vertical relief joints are also needed at close intervals to minimize cracking. Mortar quality is very important.

Redundancy must be provided in the water-resisting system design. Poor design means that water will penetrate the building interior, cause corrosion damage to anchors and steel stud components, and contribute to gradual deterioration of the water-soluble gypsum sheathing on both sides of the studs. A clean cavity, through-wall flashing, and weep holes are absolutely necessary to rid the assembly as quickly as possible from the moisture that gets past the veneer. Through-wall flashing and weep details are needed at each floor level. Flashing around door and window openings is critical. Ties between the masonry and the steel stud backup walls must be durable. Stainless steel anchors are needed, and the ties should include a drip in the cavity space.

Furthermore, the gypsum sheathing must be moisture protected. Full waterproofing is needed to prevent disintegration of the gypsum. Gypsum board sheathing braces the steel stud wall, provides fire protection, supports the waterproofing, protects against air passage, protects the wall insulation, and separates the cavity from the steel stud. Factory applied water-resistant coatings are not very effective because they are not continuous at cut edges. Condensation and airflow inside the stud wall must be prevented, suggesting the need for a vapor barrier on the warm side of the wall system, usually directly behind the interior gypsum wall board. All openings must be sealed.

The brick veneer/metal stud system will survive as a feasible curtain-wall alternative. But many of those who pioneered its use have found that the system is not as simple as originally thought. Reliable design guidelines are now available, and it is now generally known that the system relies on quality workmanship and good coordination of the various trades involved in the on-site assembly of its multiple components. Experience with the brick veneer/metal stud system indicates the need for research and experience prior to rapid use and promotion of new ideas. New, creative ideas deserve to be implemented, but they should be evaluated under the less-severe exposure conditions and then gradually applied in the more demanding situations, after their in-place behavior is well understood. In this case the system was widely used for facade applications on tall buildings exposed to severe wind and moisture effects before sufficient performance data were available. Designers should have been cautious given the inadequate and conflicting recommendations that were coming from the industry.

Failures of masonry facade materials, especially stone masonry panels, have been responsible for serious injuries and loss of life. Although not as common as the moisture-penetration problem, this problem has received considerable attention following serious accidents in several major cities.

Masonry decorative panels on the walls of shopping centers, schools, and industrial buildings are often eccentrically supported on steel spandrel beams which do not have any counteracting load to provide torsional restraint. The steel beams, usually connected to the columns by not-too-tightly bolted clip angles, rotate sufficiently to cause unsightly or dangerous bulging or even to tip the masonry off the lintel seat. Fortunately, these conditions are usually apparent before the building is occupied, so corrective measures can be taken.

Masonry cladding stress failures in older tall buildings, eight to 10 stories or more, stem from a gradual irreversible buildup of stress over many years. These failures appear as cracks, bulges, or spalls at masonry piers or corners, then create unstable wall segments and breach the building envelope to allow excessive water intrusion and subsequent damage to underlying wall elements (Beasley 1987). These effects may occur after many decades of good service, sometimes developing in walls and piers that were relatively free of distress only a few years previously. Proper diagnosis of the source of the problem is critical to making successful repairs. The conditions can be very hazardous (Figure 9.9).

In 1966, two incidents of large areas of brick veneer falling from concrete buildings caused some awareness of deficiency in code requirements for such construction. At a 10-story Tampa, Florida, bank building, a long diagonal crack developed in the brick covering in the unpierced lowest three stories. The masonry bulged out as much as 300 mm (12 in.) in some places. The wall consisted of 100-mm (4-in.) brick, with header bricks bonded in every seventh course to 194-mm (8-in.) concrete block. The cored bricks sheared at the

Figure 9.9   Bulging and spalling terra-cotta cladding panels. (Courtesy of K. Beasley.)

plane of contact with the block, which was also the plane of the concrete frame. One day after the crack developed, some 20-m (60-ft) length of the wall facing, three stories high, collapsed into the parking lot. Luckily, building personnel had anticipated the potential for the failure, and the area had been barricaded.

Several cities have adopted new building facade and cornice inspection ordinances as a result of specific failure events involving deaths and injuries to pedestrians (ENR 1980). Regulations in Columbus, Detroit, New York City, Chicago, Dallas, and Philadelphia, among other cities, now require periodic inspection of facade materials and anchorage details.

In 1980 a woman was killed in New York City by falling masonry facade materials. In 1984 a woman was seriously injured in Chicago when a 1.2-m (4-ft)-long piece of terra-cotta fell seven stories from a 69-year-old building. The woman's skull was fractured. A similar accident in 1974 killed a Chicago woman (ENR 1984a).

A 23-m (75-ft) section of limestone cornice fell from a building in Columbus, Ohio, in 1984, seriously injuring a city councilman and two other pedestrians. The councilman lost a foot and part of his leg in the accident. The cornice on the 60-year-old building was backed by brick and attached to the brick with iron anchor bars, many of which were found to be corroded (ENR 1984b).

In 1984–1985 there were three incidents involving brick facades falling from tall buildings in downtown Dallas, Texas. The cause was believed to be anchorage failures in high winds (ENR 1985). In 1987, granite pieces fell from a cornice on the twenty-fifth floor of the 73-year-old Municipal Building in New York City. The supporting steel had corroded and expanded (ENR 1987).

Repair projects often require stabilization or replacement of facades, veneers, parapets, and cornices. Relief angles are often overstressed or ineffective in reinforced concrete frame buildings having brick masonry curtain walls. The shrinkage of the concrete frame combined with expansion of fired clay masonry products can negate the effect of relief joints and produce severe unexpected compression stresses in the masonry. Sometimes replacement of the curtain wall is the only feasible option. In some cases new relief joints can be cut into the wall. This must be done with great care, considering the potential for explosive failure.

Many proprietary fasteners are commercially available for repair of curtain walls and for securing parapets and cornices. The properties of these connectors are detailed in the manufacturers' literature, but information regarding long-term performance under cyclic loading and potential material incompatibilities is often lacking.

A great deal has been learned in the past decade about the properties of marbles and other stones as the result of thin panel facade failures. Previously, not much was known about the irreversible expansion that comes with thermal cycles. Nor was there much reliable information regarding the decrease in strength and stiffness with freeze–thaw cycles and exposure time. Natural stones are neither as permanent nor as unchanging as assumed by conventional wisdom. It has been found that longevity, especially in thin panels, is highly dependent on joint and anchorage details (Widhalm, Tschegg, and Eppensteiner 1996).

Travertine has been used as an exterior cladding for over 30 years, but its physical properties are not well understood (Beasley 1988). The holes and veins that are beautiful characteristics of this limestone can cause severe problems under certain exposure conditions. Errors in supports and anchorage details are the most common source of problems. Tensile stresses existing along the weak bedding planes, inherent to the sedimentary process of limestone formation, must be avoided by proper anchorage detailing.

One example of a contractor's error having serious consequences is the Lyndon B. Johnson Library at the University of Texas at Austin. Seven months after the building opened in 1971, university officials noticed that some of the building's 2000 travertine panels were loosening. An investigation revealed that the masonry contractor had used the wrong kind of mortar, one that expanded when exposed to water. Every panel had to be removed, cleaned, and reinstalled with different mortar. Cost of the three-year repair project was $1.7 million (ENR 1980).

In 1984 it was reported that granite panels were slipping from the 13-year-old Pittsburgh National Bank Building. Stainless steel fasteners securing more than 4000 slabs to the outside of the 30-story building on one of the city's busiest corners had loosened. The stone slabs weighed 180 to 725 kg (400 to 1600 lb) each, posing a serious risk. Repairs were projected to take two years and cost $5 million (ENR 1984c). Later, as each panel was removed and reanchored, discrepancies between contract documents and as-built construction were found. The alleged source of the problem was confirmed: inadequacy of the anchorages, stemming from construction deficiencies and ". . . a lack of continuing supervision and inspection during granite cladding erection" (ENR 1984d).

It seems that thin marble panels have had more than their share of costly facade failures. Marble from Italy's Carrara Mountains has been treasured since well before the time of Michelangelo. Its use as a cladding stone on the exterior of buildings is quite recent, however. Such use has only been possible since the 1960s, when advances in stone-cutting technology permitted the fabrication of thin slices. The prestigious and costly material has proven to be a poor choice in certain climates and applications.

Current research is advancing as a result of the increasing number of deformation problems in marble cladding panels caused by atmospheric influences (Widhalm, Tschegg, and Eppensteiner 1996). Expensive Italian marbles (both Carrara and Laas) have failed due to lack of understanding of their thermal expansion characteristics. There are residual (permanent and irreversible) deformations that occur with thermal cycles, contributing to durability problems. Slender panels lose their dimensional stability in service, bending, cracking, and eventually detaching from the structure. With thermal cycles comes a loss in resistance to moisture penetration as well. This permits rapid deterioration in the polluted urban environment. Cohen and Monteiro (1991) provide a review of the durability studies on thin cladding panels for tall buildings. The record of thin marble panel facade failures is a clear example of what can happen when application advances beyond the research.

In 1982 it was announced that the 27-story Lincoln First Tower in Rochester, New York, would temporarily be sheathed in sheets of plywood painted white to match the 17,000 slabs of marble that had been removed from the facade. Eventually, the building was covered by aluminum panels. The 25-mm (1-in.)-thick Carrara marble panels had bowed in service, buckling under their own weight, and cracked due to weather effects and anchorage detail problems, causing a hazardous condition. The marble was dumped in a landfill. Repair costs were estimated at $18 million. The entire tower construction cost was $26 million when it was built in the early 1970s (ENR 1982).

Three marble panels fell from a 19-story, 23-year-old building in Kansas City, Missouri, in 1985 due to anchorage problems. In 1988, all 4400 panels were removed from the building and replaced with a white glass curtain wall, retaining the aesthetic appearance of the original marble color (ENR 1988b).

The most costly marble facade replacement project to date is that of the Amoco Tower in Chicago (ENR 1988a,d, 1989a; *Civil Engineering* 1989, Arndt 1989). The 82-story building was built for Standard Oil of Indiana, starting in 1971. Construction was completed in 1974 at a total cost of $120 million. The tower is 346 m (1136 ft) tall with a square floor plan 57 by 57 m (186 by 186 ft). Upon completion, the tower became the world's tallest marble-clad structure. The beautiful white Italian Carrara marble gave the building a unique visual presence in the skyline (Figure 9.10).

The original cladding was comprised of 43,000 panels 1.27 by 1.14 m (50 by 45 in.) in surface area and only 32 to 38 mm (1.25 to 1.5 in.) thick. Each panel weighed about 125 kg (275 lb). A new technology had been developed, permitting the cutting of marble sheets of this small thickness. The structure system to which the marble panels were bolted is a typical high-rise tube, using columns spaced closely around the perimeter and deep spandrel girders.

In 1988 it was reported that 30 percent of the marble panels had bowed out more than 13 mm (0.5 in.), some as much as 38 mm (1.5 in.). Plans were announced to restrain panels on the southeast corner with white stainless steel straps (painted to match the color of the marble). This corner was exposed to the most severe winds. The straps were bolted through to a heavy steel clip attached to the structural frame. Bending tests performed at the site with hydraulic jacks on 40 panels found that many of the panels had lost a substantial amount of their initial strength. These tests were supplemented with laboratory tests. Boundary-layer wind tunnel tests were also conducted at Ft. Collins, Colorado, and London, Ontario.

The tests confirmed that Chicago's famous winds had been a factor in the loss of strength, as was the cumulative irreversible expansion from thermal cycles. The corrosive urban environment also contributed to deterioration of

(a)

Figure 9.10 (a) Amoco building, Chicago. Built for Standard Oil of Indiana in 1974, the tower was originally clad in white Italian Carrara marble. (b) The dangerous bowed condition of the marble panels in the Amoco building facade can be seen in this photograph. The thin marble panels have been replaced with thicker sheets of granite. [(b) Courtesy of K. Beasley.]

(*b*)

Figure 9.10   *(Continued)*

the marble. Following the tests, it was announced that all 43,000 panels would be strapped with the detail originally planned for the southeast corner.

In 1989, the Amoco Corporation decided to replace all of the marble panels with 51-mm (2-in.)-thick granite slabs quarried in North Carolina. Granite is stiffer and more stable in the weather. It is not as easily corroded by carbon dioxide, acid rain, and soot in the atmosphere. (Marble is derived from limestone.) The *Chicago Tribune* (March 7, 1989) announced this cure for the "5900-ton headache": All panels would be stripped and replaced. Most of the panels were too warped to be of use in any salvage application; they would probably be hauled to a landfill or crushed into aggregate. The project was to occupy three years, at a rate of one floor per week, and was projected to cost over $75 million, more than half the original $120 million cost of the 17-year-old building.

The new granite panels weigh 180 kg (400 lb) each. Four work platforms were secured to towers at each corner and supported by window-washing tracks. The work progressed from the base of the building to the top. The building is still beautiful, and it now takes its place as one of the world's tallest granite-clad structures.

9.7   CORROSION

It should be clear from the previous discussion that corrosion can be a serious problem in modern masonry construction. The function and durability of a contemporary masonry wall depends on the integrity of numerous metal connectors and anchorages. In composite wall construction, masonry anchors tie the facing to backup, causing the two wythes to act as a structural unit. Loss of the tie strength due to corrosion can severely impact the structural integrity of the wall.

All metal components must be protected from the adverse effects of moisture. This requires diligent attention to details, as discussed previously, for control of the flow of moisture. Since moisture cannot be completely eliminated from the wall assembly, noncorrodible materials should be used for stone masonry anchorages and support, including stainless steel studs, nuts, and washers. Stainless steel shelf angles and anchors should be used to transfer the masonry panel loads to the structural frame. If stainless steel is not available, the shelf angles should at least be protected by a good-quality paint. Similar metals should be used in the assembly, since dissimilar metals, such as metal ties in close proximity to reinforcing steel bars, can introduce severe corrosion problems.

Even in more traditional masonry construction, corrosion problems are found wherever metal elements are used. Corroded pipes embedded in masonry walls can expand and cause failure of quite massive pieces of masonry. In one building, moisture saturation in the mortar joints of a high granite base course caused rusting of iron cramps inserted to hold the stones together.

After about 25 years, the rust accumulation and expansion were sufficient to spall the granite, shearing out sections 100 mm (4 in.) thick. In the same building, steel columns encased in the brickwork rusted to the extent that laminations of rust over 13 mm (0.5 in.) thick were found when the cause of the vertical brick cracking along edges of columns was investigated.

Vertical cracks at building corners are usually close to a steel column. Moisture running down the wall penetrates the masonry, enters at the crack, and soon saturates any mortar cover on the steel column. Rusting of the steel then forms an expansive iron oxide that pushes the masonry cover outward, aggravating the crack condition and speeding up the cycle of disintegration of the masonry cover. This problem is commonly encountered, especially where the masonry cover over the column is only 100 mm (4 in.) thick.

Calcium chloride should not be used in the mortar mix. It can cause severe tie corrosion. The first evidence of this problem may manifest itself in cosmetic discoloration of surface materials, as was the case for a church building in High Point, North Carolina, built in 1984 (ENR 1988c). If not corrected, the masonry veneer will eventually separate from the supporting backup material, with potentially life-threatening consequences. As late as the 1970s, many masonry contractors routinely added calcium chloride to mortar to accelerate the set in cold-weather construction. This practice is not recommended.

In an 11-story building constructed in 1972 in Sioux Falls, South Dakota, calcium chloride in the mortar mix was the alleged cause of corrosion and expansion of the facade's steel hangers, leading to severe cracking of the marble panels. The 19-year-old building's facade panels, 1.5 m by 3.7 m by 70 mm (5 ft by 12 ft by 2.75 in.) thick, were all replaced (ENR 1990, 1991).

In 1963 a new latex mortar additive was introduced to the construction industry which made possible a high-bond-strength mortar. The additive contained a copolymer of vinylidene chloride, which improved the structural qualities of portland cement mortar. The product was developed and marketed aggressively by a major U.S. chemical company. The additive was promoted on the basis of improved mortar strength (a bond strength about four times that of previously available mortars).

Using the latex additive permitted designers to detail prefabricated brick panels and single-wythe curtain walls, saving construction time and cost. But the additive was later found to contribute to accelerated corrosion of metal components in the wall assembly. Severe corrosion began showing up in numerous buildings around the country in the 1970s, apparently accelerated by the additive. The additive leaches out chloride ions when in contact with unprotected steel, causing accelerated corrosion of the steel.

Litigation involved hundreds of incidents of spalling, cracking, and staining where the additive was used. Some panels fell from buildings, and many buildings required extensive repairs, in some cases complete replacement of the facades. Facade replacement presented a challenging problem, particularly where exterior walls had been constructed with one wythe of brick glued to insulation and glued again to interior wallboard. On these buildings, the brick alone could not be removed (ENR 1986).

In two buildings in Minneapolis (21 stories and 16 stories), the original wall systems were retained behind new brick facades. The new walls are entirely independent structurally from the original walls. The new walls support the floors and are, in turn, supported by a newly constructed foundation system. In some cases, metal curtain walls cover the original masonry facades. These new metal curtain walls had to be designed to transfer wind loads to the floor diaphragms and to contain any spalling masonry from the deteriorating original wall left behind the new wall system.

A large number of lawsuits were filed against the chemical corporation. The list of affected buildings was very long and impressive, including many multistory applications. Hospitals, government buildings, office buildings, utility companies, university buildings, and housing projects were involved. Repairs were not only costly but entailed unacceptable inconveniences to tenants of the buildings.

In 1983, one jury verdict against the chemical company was announced in Cleveland, Ohio (ENR 1983). The case involved the brick facade of a 23-story building built in 1968–1969. Use of the high-strength mortar permitted a 100-mm (4-in.)-thick single course of brick the full height of the building without any backup block. One litigant noted that the mortar additive did everything the chemical company had claimed for it. "But it was what [the company] *didn't* say it was going to do that brought us into court." Repairs in 1979 and 1981 involved removing brick and mortar in contact with structural steel, sandblasting all the steel and painting it with epoxy, and replacing the brick with conventional mortar. In some places, corroded steel had to be replaced. The chemical corporation claimed the cracking was due to "faulty design and construction" and that all the metal in the assembly should have been protected by design with "cadmium or galvanizing." The jury award was for $26 million. In another Cleveland building, an out-of-court settlement of $12 million was reached. Numerous projects in New York State also experienced problems with the additive.

Even while numerous cases were pending in the courts, the additive continued to be sold and marketed. In fairness to the product, there were a number of successful applications. Most of these were in interior exposures, where moisture was not present.

In 1989 the chemical corporation won a seven-week jury trial in Denver, Colorado. The jury found that the mortar additive was indeed responsible for a somewhat accelerated rate of corrosion but that it also protects embedded steel somewhat by "plugging up pores in the mortar." This verdict was a great surprise to most observers, since the company had lost or settled previous claims out of court. The case involved two buildings, a 30-story hotel built in 1974 and a two-story hotel built in 1970 in Denver. The owners had claimed $33 million in damages. Prefabricated brick panels in the two buildings typically weighed as much as 1 Mg (more than a ton). They were transported to the site, hoisted into place, and bolted directly to structural members. Clearly, the strength of the mortar permitting such construction was remarkable. However, problems began to show up quite soon in service and were first explained to designers by the expert witnesses in court testimony.

Like any other chloride-based high-strength additive, the vinylidine chloride additive works by driving out water between cement particles, bringing them closer together. When this additive was developed in the 1950s, the addition of chlorides to mortar was routine. But organic chloride additives tend to break down as they weather, releasing salts that corrode the steel. Before bringing the additive to the market in the mid-1960s, the corporation claims to have tested it for roughly a decade. But company officials now concede that it was never tested in the presence of steel. The jury in the Denver, Colorado trial agreed that the company's premarketing testing was negligent, but also found that the company did not design or market the product deceptively. Expert witnesses and attorneys for the building owners termed the jury verdict "incredible and inconsistent" (ENR 1989b).

Surely, a product marketed for use in contemporary masonry assemblies must anticipate the inclusion of steel elements in the assembly. Once again, this experience teaches the need for caution when applying new "miracle materials." They should always be used on small projects first until the benefits of trial-and-error experience are available.

9.8 SEISMIC AND WIND PERFORMANCE

While massive, rigid masonry structures are quite wind resistant, older unreinforced masonry buildings have a record of disastrous performance in great earthquakes (Figure 9.11). The inertial mass generates large seismic forces, and the lack of tension capacity and ductility typically results in structural collapse. The performance of unreinforced or poorly reinforced masonry in seismic events is similar to that of older nonductile concrete construction. This problem has been recognized for many years, but there remain many older unreinforced masonry buildings throughout the United States and the world. These are known to be potential collapse hazards in future earthquakes.

Contemporary masonry construction must conform to codes that specify a minimum amount of reinforcement to provide the needed tension capacity and ductility. Attention is focused on the important floor–wall and roof–wall connections in masonry shear wall buildings. The floor and roof systems must be capable of functioning as diaphragms and must be able to distribute lateral forces to the walls. Reinforcement detailing requirements are similar to those specified for ductile concrete design (see Section 7.7). All of these current design provisions have been learned from studying the deficient performance of masonry construction in historic earthquakes.

Nonstructural masonry components can have a detrimental impact on the performance of the structural system in an earthquake. Rigid elements will

Figure 9.11 Nonductile unreinforced masonry buildings have typically experienced structural collapse in historic earthquakes. (Courtesy of the Federal Emergency Management Agency.)

resist lateral loads, whether designed to or not. A masonry partition can shift the center of rigidity of a building such that unplanned torsional effects are introduced. One very common problem has been the incompatible interaction of flexible structural frames with rigid masonry in-fill panels. Typically, in older buildings, the masonry panels fail when the frame begins to experience large deformations. The heavy masonry materials then fall into the street, causing injuries and death to people outside the building (Figure 9.12; see also Figure 10.6). There also exists the possibility that the panels will not fail, in which case the frame is not permitted to deform and total structural collapse becomes possible. If non-load bearing, partition or in-fill walls should be isolated from the structural load-carrying frame with a flexible joint separation material in order to minimize interference with the response of the frame to lateral loads. The details for providing such a connection are quite recent. Older masonry in-fill buildings are expected to continue to be hazardous in earthquakes.

Some of the more interesting repair projects currently under way throughout the world are seismic retrofit projects. These are intended to upgrade the inventory of existing bridges and buildings to conform to current design standards, or at least to improve the seismic performance of the facilities. Creative solutions and new strengthening techniques show promise not only for improved seismic performance but also for general applications where the intent is to upgrade structural integrity.

Figure 9.12 Masonry infill failure. (Courtesy of the Federal Emergency Management Agency.)

Some jurisdictions, such as Los Angeles, California, have mandated the retrofit of unreinforced masonry buildings, and the results are beginning to show benefits in recent seismic events. These upgrades have been costly and have not come without serious social implications. Many of the buildings identified for mandated retrofits are substandard, marginal buildings. Most are two-, three-, and four-story walk-up buildings that have little cultural value; they are not architecturally important buildings. Nevertheless, these buildings constitute a large portion of the affordable housing stock in the Los Angeles area and cannot easily be replaced. However, a modest increase in rent to pay for the upgrades is beyond the capabilities of many of the tenants (Comerio 1992).

Each seismic retrofit project is unique, requiring a site-specific solution. Older unreinforced concrete masonry unit walls typically have empty cavities. They may contain a nominal amount of horizontal reinforcement in the top of the wall. Retrofit of these walls usually requires the addition of both vertical and horizontal reinforcement. Vertical reinforcing bars can be grouted into the empty cavities. A clean-out at the bottom of the cell is necessary. High-lift grouting techniques are used after placement of the vertical bar in the empty cavity. The addition of horizontal reinforcement for the wall is a more difficult challenge. Sometimes, horizontal cores are drilled though the length of the wall. A reinforcing bar is placed in the new cavity and pressure grouted in place. An alternative procedure is to reinforce the wall from outside the existing construction. One or two continuous rods with threaded ends are placed at regular spacings from wall end to wall end. These rods are then stressed by bolting them to steel bearing plates at the ends of the walls.

All retrofit projects should be accompanied by a full load-path analysis, to follow the lateral loads all the way to the foundation. Some projects that merely upgraded the shear-wall connections at roofs and floors have not performed satisfactorily in subsequent earthquakes; the failures have simply been shifted to new locations in the structure.

Base-isolation technologies, used successfully in new buildings, are now being applied to historic masonry buildings such as the Salt Lake City Municipal Building in Utah. These methods reduce the seismic load on a structure. A greater degree of safety is provided without destroying the aesthetic integrity of the building with internal structural bracing. New lead-rubber bearings are installed under the existing foundation to increase the period of vibration of the structure and reduce the forces that are transmitted up into the structure.

One interesting method for strengthening historic masonry buildings is an in-wall reinforcement technique described by Sponseller (1987). Vertical holes 100 to 150 mm (4 to 6 in.) in diameter are carefully drilled at 1.2 m (48 in.) on center, from the parapet down through the entire height of the wall. A No. 6 or No. 7 reinforcing steel bar is then inserted in each hole and a polyester–sand grout is placed around the steel bar. The polyester grout migrates into minute cracks in the masonry, confining the masonry for some distance from the vertical hole. In addition, the method calls for roof–wall and floor–wall anchors. This approach is nondestructive, and the facility may continue to be used while the work is under way. The method does not alter the aesthetic appearance of the structure. The technique is appropriate for low-rise buildings constructed with masonry having good structural integrity. Alternative methods require the removal of bricks to place reinforcement in the wall, or the application of reinforcing ribs outside the structure, covered with shotcrete.

Although the past performance of traditional masonry facilities in large earthquakes has been disastrous, modern masonry construction can provide satisfactory resistance to earthquakes and wind effects. Properly designed and constructed, fully reinforced masonry walls have been quite successful in their response to recent earthquakes. Most earthquake damage to masonry construction continues to occur in older, unreinforced masonry buildings. In fact, contemporary masonry elements can be quite ductile, flexible, and lightweight. Wind effects, particularly deformations leading to cracks and the infiltration of wind-driven rain, are more prevalent problems in modern masonry construction.

9.9 INTERIOR PARTITIONS AND DECORATIVE SCREENS

Interior masonry partitions are subject to cracking if the tiles are not dimensionally stable. The serious distress in the partitions of the Virginia State Library, built in 1940 as a steel frame and concrete building, was explained when the plaster was removed in 1948. The fire-clay tiles making up the partitions were found to be crushed. Testing of the material showed the sample tiles to have undergone a moisture expansion, averaging 0.114 percent, with the mortar unchanged. Individual tests showed considerable variation in the amount of such expansion and in the length of time required for equilibrium to be reached.

In 1966, seven of eight elevators in the 30-year-old 24-story Suffolk County Courthouse in Boston, Massachusetts, were taken out of service when the clay-tile shaft walls fell apart. The damage was blamed on vibration effects from adjacent construction and from the dynamic elevator loads. Also in 1966, a freestanding 194-mm (8-in.) concrete block partition at a Columbus, Ohio, high school, used to carry bookshelves on each side, collapsed as a student was stacking books on one side. The unbalanced loading caused failure and a fatal injury.

Sculptured ornamental masonry screens are commonly used as decorative separations of public spaces. When used in exterior exposure conditions, they can become dangerous elements if not designed for wind loads. In 1964 at a shopping center in Norfolk, Virginia, a 2.4-m (8-ft)-high ornamental screen of 75-mm (3-in.)-thick perforated masonry elements on each side of an 200-mm (8-in.) air space was blown over by a gusty wind and hit a pedestrian.

Figure 9.13 Failure of decorative tile screen panel caused by shrinkage of concrete frame.

Wind exposure is not reduced by perforations on one side, and the suction of restricted orifices may even increase the wind effect over that of a solid surface.

In an eight-story concrete frame apartment building in New Brunswick, New Jersey, the outside entrance walkway at each floor level was made of 150-mm (6-in.)-long salt-glazed octagonal tile duct laid up as an ornamental screen between the concrete columns and floors. After a very hot day, it was noticed that a number of these screen units were loose, with gaps at the contact between mortar joints and the concrete. Without investigating the cause, which was the different thermal expansion of the two materials—not only in magnitude but in the time effect—all open gaps were filled with a rapidly setting epoxy mortar. In the autumn, when the concrete frame cooled, 22 of the tile panels crushed, some falling out completely. Although no exact record of panel location was made, the repair project records noted that 22 panels had been recaulked immediately following the hot day. Replacement was with a precast concrete block which had thermal characteristics similar to those of the concrete frame (Figure 9.13).

9.10 REFERENCES

Arndt, M., 1989. "Amoco Chucks All the Marble on Its Tower," *Chicago Tribune,* Chicago (March 7).

ASCE, 1991. *Guideline for Structural Condition Assessment of Existing Buildings,* American Society of Civil Engineers, New York.

Beasley, K., 1987. "Masonry Cladding Stress Failures in Older Buildings," *Journal of Performance of Constructed Facilities,* American Society of Civil Engineers, New York (November).

Beasley, K., 1988. "Use and Misuse of Exterior Travertine Cladding," *Journal of Performance of Constructed Facilities,* American Society of Civil Engineers, New York (November).

Beasley, K., 1990. "Leaking Brick-Clad Walls: Causes, Prevention and Repair," *Journal of Performance of Constructed Facilities,* American Society of Civil Engineers, New York (May). (See also Discussion by H. Schlick in the November 1991 issue.)

Civil Engineering, 1989. American Society of Civil Engineers, New York (June).

Cohen, J., and P. Monteiro, 1991. "Durability and Integrity of Marble Cladding: A State-of-the-Art Review," *Journal of Performance of Constructed Facilities,* American Society of Civil Engineers, New York (May).

Comerio, M., 1992. "Impacts of the Los Angeles Retrofit Ordinance on Residential Buildings," *Earthquake Spectra,* Earthquake Engineering Research Institute (EERI), Oakland, CA (February).

Engineering News-Record, 1980. McGraw-Hill, Inc., New York (January 24).

Engineering News-Record, 1982. McGraw-Hill, Inc., New York (October 21).

Engineering News-Record, 1983. McGraw-Hill, Inc., New York (March 3).

Engineering News-Record, 1984a. McGraw-Hill, Inc., New York (June 7).

Engineering News-Record, 1984b. McGraw-Hill, Inc., New York (July 5).

Engineering News-Record, 1984c. McGraw-Hill, Inc., New York (July 26).

Engineering News-Record, 1984d. McGraw-Hill, Inc., New York (August 30).

Engineering News-Record, 1985. McGraw-Hill, Inc., New York (March 14).

Engineering News-Record, 1986. McGraw-Hill, Inc., New York (March 6).

Engineering News-Record, 1987. McGraw-Hill, Inc., New York (April 16).

Engineering News-Record, 1988a. McGraw-Hill, Inc., New York (March 24).

Engineering News-Record, 1988b. McGraw-Hill, Inc., New York (March 31).

Engineering News-Record, 1988c. McGraw-Hill, Inc., New York (August 11).

Engineering News-Record, 1988d. McGraw-Hill, Inc., New York (September 15).

Engineering News-Record, 1989a. McGraw-Hill, Inc., New York (March 23).

Engineering News-Record, 1989b. McGraw-Hill, Inc., New York (May 18).

Engineering News-Record, 1990. McGraw-Hill, Inc., New York (September 13).

Engineering News-Record, 1991. McGraw-Hill, Inc., New York (March 11).

Eppell, F., 1981. *Rain Penetration of Masonry,* School of Architecture Report Series, Report 44, Nova Scotia Technical College, Halifax, Nova Scotia, Canada.

Frauenhoffer, J., 1992. "Masonry Wall and Window System Leakage Investigation for University Building," *Journal of Performance of Constructed Facilities,* American Society of Civil Engineers, New York (May).

Grimm, C., 1975. "Design for Differential Movement in Brick Walls," *Journal of the Structural Division,* American Society of Civil Engineers, New York (November).

Gumpertz, W., and G. Bell, 1985. "Engineering Evaluation of Brick Veneer/Steel Stud Walls," *Proceedings, 3rd North American Masonry Conference,* The Masonry Society, Arlington, TX.

Historic Scotland, 1994. *Stonecleaning: A Guide for Practitioners,* Historic Scotland and the Robert Gordon University, Edinburgh, Scotland.

Isberner, A., 1974. *Specifications and Selection of Materials for Masonry Mortars and Grouts,* RD024.01M, Portland Cement Association, Skokie, IL.

Mattar, S., and T. Morstead, 1987. "Shrinkage in Concrete Masonry Walls: Case Study," *Journal of Performance of Constructed Facilities,* American Society of Civil Engineers, New York (May).

McKaig, T., 1962. *Building Failures,* McGraw-Hill, Inc., New York.

Sponseller, M., 1987. "In-Wall Bracing Bolsters Church Against Quakes," *Engineering News-Record,* McGraw-Hill, Inc., New York (January 8).

Sputo, T., 1993. "Sinkhole Damage to Masonry Structure," *Journal of Performance of Constructed Facilities,* American Society of Civil Engineers, New York (February).

Widhalm, C., E. Tschegg, and W. Eppensteiner, 1996. "Anisotropic Thermal Expansion Causes Deformation of Marble Claddings," *Journal of Performance of Constructed Facilities,* American Society of Civil Engineers, New York (February).

10

NONSTRUCTURAL
FAILURES

The majority of the content of the preceding chapters has centered on failures of structural materials and load-bearing engineered components of structures. In this chapter we review some problems associated with nonstructural components. Contemporary buildings are experiencing an increase in the incidence of complaints related to deficient mechanical systems, malfunctioning electrical equipment, poor interior air quality, annoying vibrations, and acoustical problems. Although not life-threatening or catastrophic situations, leaking building envelopes (facades and roofs) are nonetheless responsible for the vast majority of unhappy building owners and the resulting construction litigation.

There has been a proliferation of new synthetic materials in interior finishes and furnishings—materials whose properties are not yet fully understood, especially as related to health effects following long-term exposure in interior environments.

In the computer age, the contents of buildings are often more valuable than the buildings. Failure of nonstructural elements, such as fire sprinkler systems, can have serious economic consequences when these valuable contents are damaged. The random proliferation of computers and other electronic equipment in buildings has brought about new technical challenges for the design professions, such as the interesting phenomenon of harmonic currents.

Adding to the challenge is the fact that performance expectations regarding human comfort and use of energy are more demanding than in the past. Building owners and users expect instantaneous changes to the temperature and humidity conditions in interior spaces, with minimal use of energy resources.

The performance of nonstructural components and the performance of the structural system are not mutually independent subjects. It is, in fact, impossible to discuss the failure of nonstructural components, without considering the character of the structural system in which they are located. Many problems experienced by nonstructural components are directly related to excessive deformations or movements—deficiencies in the structural design. These include serviceability problems, cracking of partitions, and cladding distress. Conversely, there are cases where deficiencies in the nonstructural materials that comprise the building envelope have led to serious damage to the structure system. Some examples are found in Section 5.3, where structural timber members were weakened as a result of roof membrane deterioration.

A complete discussion of the wide variety of problems in constructed facilities related to nonstructural elements, mechanical systems, and electrical equipment is beyond the scope of this book. However, a few comments on some interesting contemporary problems and case studies will illustrate the need for technical attention to nonstructural, as well as structural, elements in buildings.

10.1   BUILDING ENVELOPE PROBLEMS: FACADES, CURTAIN WALLS, AND ROOFS

Chapter 9 contained reference to curtain-wall and facade problems, especially those associated with masonry walls and stone veneer. In this section that discussion is extended to include other materials and assemblies. Deficiencies in the performance of the building envelope account for more construction litigation than any other source. Despite the availability of technical expertise and improved materials and methods, leaking roofs and walls continue to plague the construction industry.

There is no better illustration of the need for competent overall project coordination than that given by a review of the dismal record of building envelope performance. Here, at the building envelope, is where the work of the various design specialists comes together. Errors in structural concept, mechanical system design, finish material selection, and other specialized decisions can each individually, or together, result in building envelope problems. There is a clear need for coordination, to assure that the decisions made by individual competent specialists are integrated into properly functioning roof and wall systems.

Following is an introduction to some of the reasons for roof problems and curtain-wall deficiencies. A few specific examples are given, but they are representative of many, many repetitive cases of building envelope failures. The very interesting case study of the John Hancock Tower in Boston, Massachusetts, is also presented here, since one of the many unfortunate events experienced by that building was failure of its beautiful reflective glass curtain wall.

10.1.1 Roofs

According to an editorial in *Architectural Technology* (Spring 1985), roofing—a mere 6 percent of construction costs—accounts for nearly 40 percent of liability claims made against architects. The editorial notes that quality design and well-written specifications are necessary but not sufficient to ensure satisfactory performance of a roof membrane. "It takes both knowledge and perseverance to get a roof down right." One developer said: "Anyone who hasn't had a major roofing problem is either lying or lucky" (ENR 1986b). "The roof must be treated as part of the building's thermal envelope—a living, breathing part of the structure that is affected by what happens inside and out. A good investment would be to put the effort into the details and monitoring that is (typically) wasted in litigation."

One source of problems associated with contemporary roofs is the immense and diverse variety of membrane and sealant materials and methods from which the architect and engineer can select. There has been an explosion of new materials, and until recently, useful statistical data on in-place performance history was lacking.

The use of single-ply roofing systems, employing a wide variety of new materials, was brought about by the energy "crisis" of the late 1970s (Fisher 1983, Heller 1985, Godfrey 1986). Radical change had not characterized the roofing industry until use of energy became a primary design consideration. Roofing materials and methods had evolved slowly. Of all the building technologies affected by the energy crisis, flat roofing technology was perhaps affected most significantly.

Only minor improvements had been made over several decades to the built-up-roofing process that uses alternating layers of bitumen and felt, surfaced with gravel. But as a result of the energy crisis, there was an increase in the cost and a limit to the availability of quality petroleum-based bitumen. Labor costs also escalated during the 1970s and 1980s, affecting adversely the competitiveness of relatively labor-intensive built-up roofs. At about the same time, concerns for the health and environmental effects of asbestos made it impossible to continue to use the dependable asbestos base sheets traditionally used for built-up roofs. In addition, the use of greater amounts of thermal insulation may have contributed to some built-up-roof failures, since they were exposed to greater temperature extremes. Because of these and other factors, the National Roofing Contractors Association (NRCA) estimates that over 40 percent of new commercial roof installations in the United States are now single membranes rather than built-up roofing.

Many new manufacturers entered the market in the 1970s and 1980s, supplying new products and proprietary roofing systems. Chemical companies developed a wide variety of new single-ply membrane roofing materials. These included modified bitumen materials (rubberized asphalt with chemical additives), EPDM (ethylene propylene diene monomer, a rubber), and PVC (polyvinyl chloride, or plastic). The introduction of so many new materials has generated confusion among designers, building owners, and roofing contractors.

Roofing industry organizations such as the National Roofing Contractors Association (NRCA) and the Asphalt Roofing Manufacturers Association (ARMA) responded by publishing technical manuals and information on performance characteristics of the various materials. NRCA publishes the *International Journal of Roofing Technology*. The Roofing Industry Educational Institute (RIEI) was established in 1979 and is quite active in publishing newsletters and conducting educational seminars. The roofing industry and independent standards-writing organizations have conducted a tremendous amount of research on roofing since the 1960s. However, the performance history of many of the new materials is still relatively short and the published data may be misleading with regard to long-term performance.

Polymer-based synthetic roofing was first used in Europe in the 1950s in the form of polyvinyl chloride in single-ply, tear-resistant sheets. Europeans also added polymers to bitumen to give it greater elasticity and weather resistance. This "modified bitumen" was fabricated in sheets, containing glass fiber or polyester reinforcement. In the 1960s the U.S. rubber industry began to produce synthetic rubber roofing in single-ply sheets. The waterproofing industry also began to market several new liquid-applied polymer roofs.

Initial experience with some of these new materials was accompanied by numerous technical problems, such as poor tear resistance and deterioration under exposure to sunlight. Some of the products were driven out of the market, but many were refined as the demand for synthetic roofing expanded.

Inexperienced applicators exacerbated the material problems with early single-ply roofs. As with built-up roofing, the performance of the newer single-ply roof membranes is highly dependent on quality of workmanship. Since many of the products are premanufactered, there is less labor required on the site, but single-ply systems are less forgiving of minor construction deficiencies than are multiple-ply built-up roofs.

Good roofing details are important no matter what material is selected for the membrane. This includes appropriate design, with slopes to eliminate ponding; good-quality flashing; and careful detailing around roof openings, wall intersections, and at parapets. Potential movements should be considered at all intersections and for all surfaces. Roof curbs should always be used to support mechanical equipment. The roof details must be planned and clearly communicated in advance of construction, not worked out at the site.

Sometimes the problems have been not with the roofing membrane itself but rather, with the adhesives used to hold it down. There has been more interest in design of roof membranes for the effects of wind uplift since Hurricane Andrew in 1992, as well as improved methods for fastening of roof sheathing materials. There is obviously a need for maintenance with any roofing material. All materials experience a degradation with time, even the best standing-seam metal roofs.

Some unique and "high-profile" structures have had more than their share of roofing problems. These include the air-supported fabric roofs. The Minneapolis Metrodome has experienced partial collapse on three occasions. The Pontiac, Michigan, Silverdome collapsed in March 1985, causing $8 million damage, when snow and ice accumulated on the 40,000-m^2 (10-acre) roof; total replacement of the fabric was required. The Sundome at the University of South Florida deflated under heavy rains on two occasions, in 1988 and 1989. And the Hoosier Dome in Indianapolis, Indiana, ripped in February 1988 under the accumulation of about 20 m^3 (over 5000 gallons) of rainwater. Ponding occurred because the roof had reportedly deflated due to failure of the instrumentation that monitored air pressure.

In 1980 the polyurethane covering and sprayed-on roof membrane of the Superdome in New Orleans, Louisiana, was damaged during a severe hailstorm. At first it was reported that the 10-year-old roof would require complete replacement at an estimated cost of $4.5 million. Later, in 1987, the decision was made to repair the roof for $2.7 million rather than total replacement.

For protection of the roof membrane from the effects of hailstorm damage, abrasion, and exposure to sunlight, one method receiving increased acceptance is the "inverted" or "protected membrane roof" (PMR) system. In this approach the rigid insulation and ballast is placed on top of the water-resistant membrane, creating a thermally protected environment for the roof and protecting the membrane from temperature extremes.

10.1.2 Facades

In Chapter 9, masonry exterior wall problems were reviewed, including leaking walls and falling facade materials. The need for quality design, construction, and maintenance discussed in that chapter applies to construction of any exterior wall assembly. Local climate and exposure conditions can make attention to specific details particularly important. The high winds and humid conditions in Florida are quite unforgiving of design and workmanship deficiencies. In 1982, facade problems in two public jail facilities were reported (ENR 1982b). Both projects were located in the Tampa Bay area. Facade sections were falling off both buildings, this after only two years and seven years, respectively, of exposure to the environment. In at least one of these cases, noncorrosive anchors specified in the design were not used to attach the stucco panels to the structural frame.

A third dramatic Florida stucco veneer failure was reviewed in an article in the *Journal of the National Academy of Forensic Engineers* (Ink 1990). The Lee County Justice Center in Ft. Myers, Florida, is a five- and six-story masonry and concrete building covering a city block. It was built in 1983–1985 for a construction cost of $33 million. A portion of the complex is used for a jail, so the concrete block exterior wall was left exposed on the interior surface throughout the project. The design incorporated thermal insulation outside the exterior walls. The exterior wall was comprised of stucco over foam insulation and metal furring strips that were attached to the concrete or masonry walls.

On Sunday morning, March 17, 1985, 16 Mg (18 tons) of the stucco veneer peeled from the building and fell into an unoccupied parking lot. The failure occurred during a thunderstorm with high winds. Debris included one entire five-story wall panel 16.8 m (55 ft) wide and 23.8 m (78 ft) high.

The veneer was originally attached to the structural walls by vertical galvanized Z-bar furring strips spaced at 610 mm (24 in.) on center. Attachment of the bars to the structure was with 13-mm (0.5-in.)-long ungalvanized T nails, power-driven through the legs of the Z-bars into the block or concrete. The foam insulation was held in position by the Z-bars and patches of adhesive. Outside the foam, a galvanized steel mesh was fastened to the Z-bars with screws. Stucco was then placed on the mesh [with control joints spaced at 1.2 to 2.4 m (4 to 8 ft) on center]. The stucco was painted.

According to the investigation by Ink (1990), the failure was caused by four observed phenomena: (1) nails pulling out of the masonry or concrete, (2) nails failing in shear, (3) shear failures in the Z-bar at the nail head, and (4) missing nails. Nail spacing varied from 355 mm (14 in.) on center to 1625 mm (64 in.) on center. In some areas no nails were found in 1 m^2 (10 ft^2) of wall.

Design specifications called for "hot-dip galvanized stub nails" that would probably not have been hand-driven, but nongalvanized power-driven nails were used. The investigator found through laboratory testing, however, that even the specified nailing system would have been deficient. A nail spacing of about 90 mm (3.5 in.) would have been required to provide an adequate factor of safety.

The deteriorated condition in which the nails were found was quite alarming; they were all severely corroded. Should the nailing pattern have been sufficient to resist the wind load initially, long-term durability would still have been questionable. Subsequent testing and inspection on all the rest of the project found severe corrosion, with many nails rusted completely through, even in areas where the veneer had not yet failed. These nails had been in place less than two years. Some were electroplated, but they, too, had been power-driven, which negated the plating protection.

Repairs involved providing additional noncorrosive plastic anchors with stainless steel bolts for all remaining stucco veneer surfaces. In all, 73,000 anchors were used to tie the mesh to the wall; then the stucco was patched. Where the stucco had to be replaced completely, improved materials and a better water-protective coating was applied. Repair costs were $600,000.

There is a need for specialized design consultants who truly understand curtain walls. A variety of trades are involved in constructing the building envelope assembly, and none of them takes full responsibility for performance. A coordinator is needed to integrate successfully the contributions of the various trades.

Wind and water penetration problems at the exterior wall comprise the second-largest area of construction litigation—second only to roofing deficiencies. Curtain wall construction has proven to be a bonanza for construction lawyers and waterproofing maintenance specialists. Metal and glass curtain walls are subject to all the forces causing trouble with masonry walls, sometimes to a greater degree. The coefficient of thermal expansion for aluminum is 2.5 times that of glass. When glass and aluminum are combined in an assembly, appreciable differential movement results, even in one-story panels.

Wind-pressure deformation and suction effects have broken the sealant joint between large glass panels and their metal frames, broken glass panes, and even pulled full panels from the structural frame. Expansion and contraction, lateral movement, and excessive wind pressures (negative and positive) all put added stress on the sealant. If the sealant material is prone to hardening, shrinkage, loss of solvent, or bond failure, a weathertight enclosure cannot be maintained.

Facade problems almost always involve moisture penetration. In many modern curtain walls the only defense against moisture penetration is the exterior surface sealant. A redundant line of defense against moisture is a wise expenditure. If possible, any repair project for such walls should include provision for an in-wall drainage system.

The bonding capability of the sealant to the materials in the wall must be sufficient if a good joint is to be achieved. The masonry, concrete, steel, or glass surfaces must be chemically clean and dry and at a proper temperature when the sealant is applied. Joint sealants must be compatible with the materials bounding the joint. The Johnson Wax Tower in Racine, Wisconsin, designed by Frank Lloyd Wright, has large areas glazed with 50-mm (2-in.)-diameter Pyrex tubing. When the structure was first enclosed in 1939, the sealant used was putty made of lead oxide and linseed oil. This substance did not adhere well to the Pyrex, and the building leaked extensively. A succession of newly developed sealants were then used without much improvement. Preformed polyvinyl chloride gaskets were replaced by a synthetic rubber compound that lasted about three years. In 1952 a specially designed two-part sealant was pulled away by thermal change, leaving appreciable separations along the tubes. Finally, in 1958, the 41,000 lineal meters (133,000 lineal feet) of glass jointing was filled with a one-part silicone rubber sealant using pressure injection. After about 20 years of trial and error a solution was found.

Taller and more flexible buildings experience movements that necessitated the development of better sealants. Chemical engineers have dramatically improved the properties of curtain-wall sealants in the past three decades. However, even with the availability of these new sealants, problems are escalating as designers push them to the limit (ENR 1987c). Sealants should not be asked to be the sole moisture barriers in buildings. It is always desirable to have a second line of defense, such as an in-wall drainage system.

Sealants are often selected and applied inappropriately. The Sealant, Waterproofing and Restoration Institute (SWRI), formed in 1976, conducts educational seminars and publishes guidelines for the industry to disseminate knowledge about the varieties of sealants now available.

Curtain walls that use manufactured panelized elements are improving, but early examples left much to be desired. Experience taught that good quality of the prefabricated elements alone was not enough to overcome problems with deficient connection details and poor workmanship in the field. In 1982 it was announced that 10,000 apartments in the United Kingdom had been condemned, to be demolished and replaced at an estimated cost of $400 million (ENR 1982a). The cladding system used in these public housing projects built in the 1960s and 1970s did not provide the durability that was promised. Problems caused by condensation and moisture penetration were so severe that demolition and replacement was the only economically feasible option. Most of these projects were built with precast concrete wall panels containing an inner layer of polystyrene insulation.

Again, in the United Kingdom in 1984 it was reported that $14 billion would be needed to modify and repair defects in 500,000 homes built using prefabricated panelized systems during the 1960s and 1970s. This form of construction had been encouraged during that time by government subsidies. Most of the problems were moisture induced and were not structural deficiencies (ENR 1984b).

Leaking walls often require costly repairs. Symptoms of leakage include corrosion, cladding fracture, material disintegration, rot, mildew, material distortion, and coating deterioration (Schwartz 1988). Sole reliance on a single moisture barrier is the root cause of most moisture-penetration problems. There exists no perfect sealant or gasket; intersections, corners, and parapets all present opportunities for moisture penetration into the walls even when the surface assembly is intact. Some incompatibilities among construction materials and sealants are nearly always present. A second line of defense for planned water collection and removal must be provided.

Laboratory mock-ups should be specified as part of curtain-wall design. These can help solve many problems, but they must be followed with tests in the field (Schwartz 1988). Performance of mock-ups in the laboratory is no assurance that in-place performance will be satisfactory. Laboratory mock-ups are often nonrepresentative of actual field conditions. Additionally, they do not address long-term performance and material degradation. Short-term adjustments may be effective in permitting the mock-up to pass the test, but these adjustments may not adequately address field conditions over the long term. Successful mock-up tests may simply provide a false sense of security. On-site testing is more indicative of performance than laboratory testing. Field water spray tests, randomly applied in the field, are a good check on construction quality.

One example of the cost of leaky facades, reported in 1990, is the case of a 23-story dormitory complex at Columbia University (ENR 1990a). The $28.7 million project was only eight years old but was troubled by falling and leaking ceramic tile cladding. The cladding was composed of approximately 1 million

100 by 200 mm (4 by 8 in.) tiles 9.5 mm (0.375 in.) thick, which resemble thin bricks. They were set in mortar on expanded metal lath over tarpaper waterproofing attached to concrete masonry units.

The facade materials separated from the wall, allowing moisture to penetrate and cause damage. In 1990, removal of the surface tile began, with plans announced to replace the construction with a conventional brick masonry cavity wall and new windows, at a projected cost of $20 million (ENR 1990b).

Despite the many failures of building facades, it is often difficult to assemble information about specific cases. Building owners are reluctant to discuss their problems openly, for a variety of reasons (ENR 1993). This has been a hindrance to those who seek to learn from the experiences of others. Some of the most prominent facade investigations and repairs have been shrouded in secrecy and speculation, including those of the Amoco Building in Chicago (see Section 9.6) and Boston's John Hancock tower (Section 10.1.3).

Curtain-wall systems are often, unfortunately, the object of cost-cutting maneuvers in the final stages of design (Schwartz 1988). When "value engineering" is focused exclusively on construction costs, the long-term results can prove to be far from cost-effective. For example, a 10-year-old building in Maryland required a $1 million reconstruction to replace deteriorated plastic through-wall flashing. The original design called for traditional metal flashing, but a misguided "value engineering" effort substituted plastic flashing, at an initial savings of only $7000.

Curtain-wall assemblies, with their diversity of materials, can fail due to incompatibilities of system materials. This is a general problem in complicated contemporary buildings (Cassady 1990). The proliferation of new materials and systems with unproven track records has understandably generated confusion and performance deficiencies. The case of the brick veneer/metal stud facade system discussed in Section 9.6 is a good example of problems that result when new systems are widely applied without the benefit of in-place performance information. Similarly, a number of buildings required complete facade replacement when weathering steel cladding failed to be the permanent, miraculous, noncorroding material that was promised by the product literature (Frauenhoffer 1987).

This confusion and lack of reliable performance information has increased the liability exposure for design professionals. Generally, the manufacturer of a specific sealant, cladding material, or even an entire proprietary system has not been held liable when the facade fails. The manufacturer always claims that the material or system has the properties claimed in the literature, but in this particular case the application departs in material respects from what is recommended.

A variety of proprietary "sandwich" cladding panels are now manufactured by various suppliers under a variety of trade names. These typically take the form of an aluminum, plastic, or stainless steel surface layer, and an insulating core of polyethylene foam plastic or other similar material. There has been an ongoing controversy over fire safety of these panelized systems. The controversy is fueled by charges and countercharges among manufacturers of competing systems.

The debate about the fire integrity of the many available exterior insulation and finish systems (EIFSs) was brought to the forefront by an unusual blaze in a New Hampshire office complex in 1985 (ENR 1990c). EIFS cladding systems, widely used throughout the United States on over 100,000 buildings, are competitive with brick and precast concrete facade products. The panels are similar in appearance to stucco or lightly etched concrete, limestone, or sandstone. Typically, they are composed of metal studs with fiberglass insulation, covered by gypsum board and expanded polystyrene beadboard or extruded foam, with a hand-troweled finish coat. In addition to new construction, EIFS products are often used in building renovations, applied over the existing exterior wall construction.

The controversy after the New Hampshire fire was over whether the material can be ignited by radiant heat from a fire as much as 6 m (20 ft) away. In New Hampshire, the fire began in an old four-story brick building, then spread across a 6-m (20-ft)-wide alley toward a new seven-story EIFS-clad office building. Some observers (firefighting personnel) saw the EIFS-clad building ignite without being directly exposed to flames. Subsequently, there has been much bitter debate in the literature over this issue. Further testing on these products has led to limitations in some of the model building codes on use of specific cladding panels. Prior to widespread application, many of these products had never been tested as to performance under radiant heat exposures.

Of course, there are a large variety of proprietary systems that fall under the name of EIFS. Many have shown excellent performance, but some have experienced cracking and leaking, for many of the same reasons that plague other wall assemblies. The sources of potential problems with EIFS assemblies are reviewed in an article by Kub, Cartwright, and Oppenheim (1993), along with identification of specific testing needs.

Manufacturers of new building materials and systems often complain about the length of time it takes to get code agency testing and approval of their products. Innovation is sometimes slowed by the review process. But materials with insufficient track records are often the subject of lengthy and widespread litigation when they do not perform as promised. Unfortunately, the first true objective laboratory testing of some new materials and systems has been conducted for presentation in the courtroom after the materials have failed to perform.

Even if construction materials have proven to be successful individually, new combinations of materials also fall into the category of unproven technology. The early costly "experiments" with brick veneer/steel stud systems illustrate this fact.

All of the problems associated with the exterior wall take on a greater level of complexity and economic importance as the building becomes taller and more flexible. In the tall building, untested materials and systems, convoluted project specifications, and confused approval and delivery systems for curtain walls can easily bring about economic catastrophe. When problems arise, they can often be traced to a lack of clear definition of responsibility for performance of the curtain-wall assembly.

Problems with metal and glass curtain walls in tall buildings led to dramatic developments in wind engineering since the 1970s. In addition to wind research, much has been learned about the technical properties of glass. It is now known that the strength of glass degrades considerably with time, especially in abrasive exposures.

In contemporary buildings, glass is being asked to perform more functions than in traditional buildings, and many new glass products have entered the market to respond to demands for thermal reflection and insulation properties. With the emergence of these new materials and nontraditional applications it became evident that reliable, independent research was needed to establish the technical properties of glass. Increasing liability exposure for design professionals convinced them that they could no longer rely exclusively on the old data and charts provided by the leading glass manufacturers. New standards for the manufacture and use of glass in construction were adopted by ASTM, ASCE, ANSI, and other organizations as a result of the updated research.

Glass fails from wind pressure, from wind-blown debris (such as gravel picked up from built-up roofs), thermal fracture due to differential heating of glass panels, failure of the seal in insulating (double-thickness) glass, and failure of the supporting system for curtain-wall construction. Sometimes manufacturing defects in the glass form the "critical flaw" where failure initiates. In several cases, for example, failure of tempered glass panels has been traced to minute particles of nickel sulfide trapped in the glass.

In Hurricane Alicia (1983), a number of tall office buildings in Houston, Texas, experienced glass breakage; approximately 3000 broken windows were reported. Subsequent studies showed that wind-driven gravel picked up from the roofs of adjacent buildings was a major contributing factor to the damage. Some have suggested that using a larger gravel size and higher parapets might mitigate this problem.

The boundary-layer wind tunnel has been established as an essential tool for understanding the character of fluctuating wind pressures distributed across the facade of tall buildings in the urban environment. The complex interaction between wind and the shape and dynamic properties of the structure can only be studied with a wind tunnel built specifically for such investigations. The effects of wind turbulence at the pedestrian level can also be determined and mitigated by using the boundary-layer wind tunnel as an interactive design tool. Of course, future development in the vicinity of the building can dramatically change the character of wind patterns on a building's curtain wall and increase the potential for wind-blown debris. A conservative design is warranted to prepare for such uncertainties.

From the foregoing discussion it can be seen that the curtain wall is required to respond to a highly complex set of technical performance criteria, including significant structural engineering criteria. Yet structural engineers are not typically paid to design curtain walls. As a result, the technical details sometimes fall through the cracks. In the engineering literature, a number of writers have decried the lack of engineering design as related to cladding of buildings (Becker and Robison 1985, Cohen 1991, Nicastro 1993). If there is to be a rational basis for cladding design, the participation of structural engineers is mandatory. One experienced curtain-wall forensic investigator notes that many curtain-wall failures can be traced to the fact that their design was not properly coordinated with the building structure. The "curtain" wall is not just a "drapery" hanging from the structure; it is composed of substantial materials that have specific technical properties and substantial demands placed on them.

10.1.3 Failure Case Study: John Hancock Mutual Life Insurance Company Building, Boston, Massachusetts

(*References:* Marlin 1977, Osman 1977, Campbell 1980, 1988)

The home office of the John Hancock Mutual Life Insurance Company is located in a prominent office tower in the Back Bay area of Boston, Massachusetts. The building is surely one of the most beautiful and controversial objects of Modern architecture (Figure 10.1). The many technical problems experienced by the building are legendary, even though the owner has taken extraordinary steps to stifle discussion in the technical literature. Much of the published information about the technical difficulties has been based in speculation or provided by knowledgeable persons who were involved only peripherally, not directly associated with sanctioned investigations. A 1981 settlement reached between the John Hancock Company and the numerous parties named in their lawsuit included an oath of secrecy about the terms of the settlement and forbade all technical consultants from ever discussing the results of their investigations.

The most public of the John Hancock Building's technical problems was the failure of its reflective glass curtain wall, and that is why the case study is included here (Figure 10.2). However, there were several other completely independent technical failures experienced by the building. What is known about the building, even though the information is incomplete, is sufficiently interesting and instructive to warrant a brief review. There are lessons in the John Hancock case for geotechnical and foundations consultants, curtain-wall designers, wind researchers, structural engineers, architects, and building owners.

The John Hancock Building was designed by one of the most prominent architectural firms in the United States, with a long list of prestigious award-winning projects. Despite this fact, the building suffered from a number of design and construction deficiencies, involving innovative but untested details,

Figure 10.1   John Hancock tower, Boston, Massachusetts.

Figure 10.2   Failed glass facade of the John Hancock tower was replaced temporarily by plywood panels. (Courtesy of R. Patton.)

a complicated plan configuration, foundation and excavation failures, excessive deformations and wind-induced accelerations, and problems with material performance. The record is illustrative of the immense scale of the technical challenges that can be encountered in the design of tall buildings with light-weight, flexible high-strength steel structural frames, especially when unusual plan configurations and large aspect ratios are involved.

The tower is 241 m (790 ft) tall, containing 60 stories of office space and two mechanical floors. The total floor area is 191,300 m² (2,059,100 ft²). At the time of its completion, the John Hancock Building was the tenth tallest building in the United States. The plan configuration of the tower is not rectangular; it is a rhomboid (a parallelogram in which the angles are oblique and adjacent sides are unequal).

The construction was started with a ground-breaking ceremony in August 1968. Because of the poor quality of the soil in the Back Bay (filled in the nineteenth century), the foundation system is extensive, as was the excavation contract. A reinforced concrete mat foundation 2.6 m (8.5 ft) thick is supported by 3000 steel piles driven to bedrock, 50 m (160 ft) below street level. Prior to excavating, 6000 m² (64,000 ft²) of interlocking steel sheeting was driven 17 m (56 ft) into the earth. The tower required the largest order of structural steel, 29,000 Mg (32,000 tons), ever used on any building in New England. The structure was topped out in September 1971, but numerous and diverse technical difficulties encountered during construction delayed occupancy until February 1976, three years later than anticipated. The original project budget was $95 million; final construction cost was more than $160 million.

From the initial announcement of intent to build the tower, this building has been marked by an extraordinary level of controversy. Much of the contro-versy—even today—is centered on the issue of context. The gigantic tower dwarfs the small-scale masonry structures that surround it and the adjacent Copley Plaza, many of them historically significant architectural works.

Concerns for preservation of the historic character of Boston's Back Bay neighborhoods and especially of certain buildings in the immediate vicinity of the proposed Hancock tower caused a great amount of opposition to the project, including vocal opposition from the Boston Society of Architects, the local chapter of the American Institute of Architects. The John Hancock Company expended considerable effort in eventually securing a variance to the local zoning ordinance so that the building could be built. The immense tower, with its reflective glass and metal curtain wall, presents a disquieting contrast in both scale and texture to the preexisting context. Important build-ings immediately adjacent to the site include the Clarendon Building and the Berkeley Building (both built by the John Hancock Company in 1920 and 1940, respectively); McKim, Mead, and White's neoclassic Boston Public Li-brary (1895), the Copley Plaza Hotel, and Henry Hobson Richardson's neo-Romanesque Trinity Church (1877).

The final architectural solution was heavily influenced by contextual consid-erations. Indeed, these considerations led directly to some of the technical difficulties experienced during and after construction. The aesthetic challenge of placing a project of this scale on such a sensitive urban site dictated the very shape of the building plan, which opened up views of Richardson's Trinity Church and later caused problems in the wind. In addition, the large dimensions and materials used for the reflective glass curtain-wall panels were selected so that the pedestrian would be less aware of the Hancock Building's presence. Glazing panels reflect the surrounding buildings, making the pedestrian more aware of the historic context (Figures 10.3 and 10.4). But these panels were of larger dimension than had ever been used in a curtain wall of this magnitude, and reflective glass was then a relatively new material, with little available data on in-place performance.

The reflective character and shape of the John Hancock tower gives it a magnificent sculptural quality when viewed from the distance, although opinions vary as to its success in achieving the designers' intentions at the pedestrian scale. It should be noted that since the Hancock tower was completed, many other historic buildings in the vicinity have been demolished to make room for new architecture, much of which shows less sensitivity for historic context than the John Hancock Building.

The John Hancock Building experienced four major technical failures, apparently unrelated:

1. Failure of the excavation system. This caused damage to adjacent buildings and utilities.
2. Unacceptable dynamic response in the wind. Excessive accelerations required the addition of tuned-mass dampers.
3. Discovery of a potential overturning problem with the primary structure system. Additional bracing was added in the direction of the major building axis.
4. Dramatic fractures of the glass facade. The facade was replaced entirely, after extensive studies, with a redesigned curtain wall. The curtain-wall revision necessitated substantial revision of the HVAC system throughout the building.

The excavation and foundation construction problems encountered during the early phases of construction are discussed in Chapter 11 (see Section 11.3 and Figure 11.2). The excavation disturbed the water table, causing damage to adjacent historic masonry structures supported on timber piles in the filled Back Bay soil. Collapse of the excavation support system also caused utility line damage. Two of the more severely affected buildings were the Copley Plaza Hotel and the Trinity Church. The John Hancock Company resolved the Copley Plaza Hotel dispute by purchasing the hotel; the Trinity Church matter was not so easily resolved and litigation on the irreparable damage done to the church continued for many years. An additional suit by Trinity Church against the Hancock Company claimed damage to the historic build-

Figure 10.3 The large reflective glass panels in the John Hancock tower's facade and the shape of the tower's plan were essential components of the building's aesthetic response to a difficult context.

Figure 10.4 The John Hancock tower glazing reflects its historic neighbors.

ing's leaded glass windows, allegedly caused by the increased heat load generated by the Hancock tower's reflective windows.

In 1984, a jury trial found the John Hancock Mutual Life Insurance Company responsible for cracks in the Trinity Church, caused during the foundation construction (ENR 1984a). The church was awarded $8.6 million. Attorneys representing the church in the dispute claimed that the Hancock tower's excavation retaining walls slipped 840 mm (33 in.), causing the adjacent street to sink 460 mm (18 in.) and moving the footings of the church. The foundations of the church settled and structural arches cracked. The case was appealed by the Hancock Company.

Three years later a Massachusetts Superior Judicial Court upheld the 1984 decision in favor of Trinity Church (ENR 1987a,b). The award was increased to $11.6 million. The 17-year litigation eventually was resolved using an innovative "take-down" theory for assessing damages (*Civil Engineering* 1987). The court awarded the costs required to compensate the church for demolition and complete reconstruction, since it was found economically unfeasible to repair the damage. The case was expected to set a precedent for assessing damage to "priceless" buildings. The *take-down cost* was defined as the cost of taking down and reconstructing a building. Eliminated from the claim would be all preexisting depreciation that would not have been affected by the defendants' actions.

A number of significant structural revisions were made during the four years the John Hancock Building sat unoccupied. Several windstorms passed through Boston during this period, and each one seemed to demonstrate some other unpredicted and undesirable behavioral characteristic of the tower. The belated addition of bracing to the structural core was deemed necessary after boundary layer wind tunnel tests were finally conducted at the University of Western Ontario, Canada. Neither the results of that testing nor the specific actions taken to correct deficiencies in the structure are available in the public domain. However, it is evident that stiffening of the primary structural system of a building after the construction is substantially complete is an extremely costly task. Several studies have hinted that the building would have been at risk of overturning in the direction of its major axis, had additional bracing not been added. Some reports suggest that over 1350 Mg (1500 tons) of steel was brought in to reinforce the elevator shafts and stair cores. In a seemingly unrelated wind design retrofit, tuned-mass dampers were added to reduce the unacceptable accelerations that would have caused discomfort to the users of the building.

The most visible problem at the John Hancock Building, however, was the failure and replacement of the facade, 75,000 m² (18.5 acres) of glass. The original glass panels were comprised of two sheets of glass separated by a 13-mm (0.5-in.) air space. After the panels started to break in late 1972 and early 1973, a year was invested in research, and the decision was made to replace the glass with 13-mm (0.5-in.)-thick single sheets of tempered safety glass. The reglazing began in May 1974 and was completed in August 1975, at a reported contract cost of $8.5 million. The new curtain wall contained 10,344 panes of coated tempered safety glass, framed by 900 Mg (1000 tons) of black duranodic aluminum. The surface dimensions of the glass panels remained the same as before, 1.4 by 3.5 m (4.5 by 11.5 ft), but the weight of each panel was increased from 225 kg (500 lb) to 360 kg (800 lb).

The plywood panels that had been used to replace the broken glass temporarily were reportedly sold for more than their original cost, since the price of plywood had risen during the year of research on the problem. This was perhaps the only economically profitable investment associated with this project.

There was much speculation in the public and professional literature regarding possible reasons for the glass failure. Some of the speculation centered on the larger spans inherent in the larger sheets of glass. A larger span generates higher bending stresses under wind pressure. A larger sheet of glass is also more susceptible to inclusion of a critical manufacturing defect. It was noted that more panels were broken near the base of the building than at the top, where designers had long assumed the higher stresses were located (Figure 10.2). This was blamed, in part, on the fact that shards of glass raining down from above would cause damage to the lower glass panels. Others speculated that the building was moving too much, or that the curtain-wall details were too tight.

Smaller panes were considered briefly at the time of the reglazing, but the idea was rejected on the basis of aesthetics. The change from double sheets of glass to single sheets increased the energy demand substantially. Several hundred thousand dollars was reportedly spent on upgrading the HVAC equipment, and projected energy usage over the life of the building was revised upward dramatically. The original glass was better in appearance, in that reflections were more consistent. It was smooth, perfectly flat, polished plate glass. The new glazing is not as smooth and the reflections are slightly distorted; tempered float glass cannot be manufactured to the same degree of flatness.

The owners announced that the new glass panels had solved the safety problem. Tests had indicated a pressure at failure of 10.56 kPa (220 lbf/ft²). The original glazing failed at 1.68 to 1.92 kPa (35 to 40 lbf/ft²) in laboratory tests. However, shortly after the reglazing was completed, some panels cracked. Continuing problems are expected, since a failure rate of 1 percent is an accepted standard for the tempered glass industry. This is due to minute particles of nickel sulfide stone which can become trapped in the glass during manufacture. The John Hancock Building, however, is quite unique. With over 10,300 windows, a 1 percent failure rate implies the fracture of over 100 glass panels, each weighing 360 kg (800 lb). Thus far, almost miraculously, no one has been injured by glass failures from the building. But the John Hancock Company is an insurance company, fully aware of potential liability problems.

When the unexpected failures began to reoccur, the John Hancock Company hired two security guards to watch the facade continuously with binoculars from 6:00 A.M. to midnight. Tempered glass loses its reflectivity when it cracks, about 5 to 10 minutes before falling. In June 1980, the "window watchers" were replaced by electronic sensors in every panel, connected to the mechanical system so that internal pressures can be reversed when a glass panel cracks. This will cause the glass to implode into the building rather than fall out onto the street or sidewalk (ENR 1981).

The John Hancock Building is not the only tall building that has experienced glass breakage in windstorms. In February 1988 wind gusts broke about 90 1.5 by 2.4 m (5 by 8 ft) windows in Chicago's 110-story Sears Tower. Panels between the fiftieth and eighty-eighth floors were broken. Wind-blown debris struck the glass on the eighty-eighth floor, causing many panels below to break, as airborne glass damaged the windows on lower floors, all the way down to the fiftieth level. Windows continued to break as much as $2\frac{1}{2}$ hours later. Fortunately, there were no injuries, since the damage occurred very early in the morning. Again, in April 1988, 97 windows were damaged in the Sears Tower in another windstorm, from the eightieth- to the thirtieth-floor levels. Wind-blown debris was again blamed for the incident.

Codes now recognize nonuniform, fluctuating wind pressures. New wind design tools, such as the sophisticated boundary-layer wind tunnel, are now used routinely by tall building designers. The problems experienced by the John Hancock Building led directly to greater use and refinement of boundary-layer wind tunnel testing and to other advances in wind engineering. The idea that wind pressures are highest at the top of a building has been proved incorrect. Boundary-layer wind tunnel tests have shown that depending on the building's configuration and that of adjacent buildings, the highest wind pressures may actually be at the street level.

Thus far, the most rational explanation to appear in the professional literature for the glass breakage in the John Hancock tower focuses on the unique character of the original double-glazed panel (Campbell 1988). A continuous lead spacer was used all around the panel edge, "soldered" with hot lead to both panes, leaving an air space between the panes. A reflective coating was applied to the inside surface of the outside pane. Thus the lead was "soldered" to glass on the inner surface of the inside pane and to the reflective coating on the outer pane. Reflective glass was a relatively new material at the time of the John Hancock curtain-wall installation.

Deformation studies on the building indicated some general problems, but they did not indicate that there was enough structural deformation to make this the sole cause of window breakage. One interesting clue was that the outside lite was always the first to break. This led to a study of the effects of cyclic wind and thermal stresses at the edge of the panel, and eventually to the conclusion that the lead edge assembly was failing in fatigue. Laboratory tests produced cracks in lead or solder; the cracks would then migrate into the glass because the bond between the materials was so effective.

Unfortunately, it has been nearly impossible—even to the present time—to access reliable information about the John Hancock case. As noted previously, a 1981 settlement included provision for an oath of secrecy on the terms of settlement. The results of all tests and technical analyses were sealed. The legal nondisclosure agreement among all the parties who were directly involved banned them "in perpetuity" from public comment on the Hancock's technical problems.

Fortunately, such a comprehensive ban on open discussion is quite rare. It is a disservice to the construction industry and it has been criticized quite justly and openly. In reference to Hancock's nondisclosure agreement, an *Engineering News-Record* editorial (August 27, 1981) stated: "The only remedy for building failures is information. The only way to avoid costly and dangerous blunders is to understand the mistakes of the past."

Paul Weidlinger, a prominent New York City structural engineer, has noted the advances in building wind engineering that resulted from the John Hancock case. He compares this to the impact on wind engineering for bridges that followed the Tacoma Narrows bridge failure. But there was a very significant difference. The technical research results on the Tacoma Narrows failure were widely disseminated and openly debated in both the public and professional literature, despite the existence of lawsuits (Ross 1984). This is perhaps the most positive outcome that can be derived from a tragic or costly failure.

One final note on the John Hancock Building: In 1977, one year after the building was finally occupied, it was the recipient of the AIA Honor Award, the highest design award given by the American Institute of Architects. Some interesting comments were made by the AIA Honor Award jury members: "This was a difficult building to premiate because of the adverse publicity it has received." "It is perhaps the most handsome reflective glass building . . . history may show it to be the last great example of the species" (Osman 1977).

10.2   HEATING, VENTILATING, AND AIR-CONDITIONING PROBLEMS

Structural problems can be caused by inadequate ventilation. Deterioration of timber structural elements in poorly ventilated environments was discussed in Section 5.3. Section 6.4 gives examples of corrosion of steel members in humid locations with minimal ventilation, including one fatal swimming pool ceiling collapse.

In addition to these problems, an increasing volume of construction litigation is specifically related to failures of heating, ventilating, and air-conditioning systems. Contributing to this increase are several factors: the higher level of comfort expectations among building users, the availability of sophisticated equipment that requires more sophisticated maintenance and operation skills, and increasing concerns about the potential health effects of poor-quality air in the interior environment. Several comprehensive books are available on the subjects of interior environmental health and unacceptable mechanical system performance, such as those by Singh (1994) and Schaffer (1991).

In this section we present only a very brief summary of some of the heating, ventilating, and air-conditioning (HVAC) problems experienced in contemporary buildings. Two case studies are discussed: the puzzling air quality problems at the new DuPage County Judicial and Office Facility near Chicago, and the costly mechanical system retrofit of the new State of Illinois Center building in downtown Chicago. At the time of this writing, both of these cases are so recent that complete information is not yet available; the information contained here is a summary of media reports that are in the public domain. These examples are included to illustrate the potential costs associated with deficiencies in the design of environmental control systems and the scale of litigation that may accompany the emerging controversies surrounding interior air quality and the *sick building syndrome.*

The poor quality of indoor air in contemporary buildings can, in many cases, be explained in terms of the narrow focus on energy conservation in the late 1970s and early 1980s, following the energy crisis. Ventilation requirements for new buildings were decreased during that period. Energy-efficient buildings were built tighter, usually with inoperable windows. A number of schools, as well as office and public buildings, began to experience ventilation problems. Some of these could be traced to underdesigned systems, some to poor maintenance and operation. In some cases the mechanical ventilation system was turned off altogether by building operations personnel in a misguided effort to save money and energy.

At the same time that designers were reducing ventilation, the interiors of buildings were finished and furnished with plastic and wood laminates, synthetic carpets, wall coverings, synthetic upholstery materials, and many other products provided by modern chemical engineering. These new materials have improved the comfort and aesthetics of building interiors, but they may have contributed to indoor air pollution, especially in poorly ventilated spaces.

Today, the general public has become quite concerned about measurable levels of contaminants in their living and working environments. Asbestos, radon, carbon dioxide, organic vapors, and formaldehyde are now common vocabulary associated with selecting materials and planning buildings. Indoor air quality concerns have brought together a diverse group of specialists in industrial hygiene, engineering, and microbiology. The science on the health effects of measurable contaminants is in its infancy, and there is currently much disagreement among the experts regarding acceptable levels. But one thing has become evident. There is a need for more ventilation than was recommended in the immediate post-energy-crisis standards. HVAC systems need increased fresh air intake, increased efficiency of air distribution, and improvements in filter efficiency.

In 1987 the American Society of Heating, Refrigerating, and Air-Conditioning Engineers (ASHRAE) issued new and increased ventilation standards to help address the sick building syndrome. This was a reversal of ASHRAE's reduced ventilation recommendations of the 1970s focused on energy conservation. The sick building syndrome is not yet well understood, but it is apparently brought about when some quantity or combination of harmful gases, fungi, bacteria, dusts, and other contaminants are trapped within a building.

The situation resulting has been blamed for employee fatigue, headaches, sinus infections, and in some cases, the development of severe and debilitating chemical sensitivities. The National Institute for Occupational Safety and Health (NIOSH) recognizes the concerns as genuine and is quite active in supporting research on the problem.

It should be noted that the concept of sick building syndrome is not intended to imply that the building is sick but that occupants become sick when in the building. Some researchers feel that most of the harmful gases dissipate over time and that buildings should not be occupied for about 90 days after new carpets, laminates, wall coverings, and furnishings are installed.

Thus far, sick building syndrome litigation has produced confusing and inconclusive results. One case that has received considerable publicity is that of the $53 million DuPage County Judicial and Office Facility near Chicago, Illinois (ENR 1995a). The building was completed in 1991. One year after occupancy, 124 county employees who worked in the building announced a lawsuit against the building's designers and contractor for negligence, alleging sick building syndrome. They also asked for a court order to close the facility until the condition in the building could be identified and corrected.

Employees complained of chronic headaches, dizziness, nausea, and respiratory irritation. Deficient ventilation was alleged in the four-story 33,450-m^2 (360,000-ft^2) building. In March 1992 a humidifier was involved in an incident that caused dozens of employees to be hospitalized and closed the building for several days.

Several different chemical and medical consultants were retained to study the building, as well as the NIOSH. None of the groups was able to determine the cause of the complaints conclusively. One consultant reported unacceptable levels of carbon dioxide and hydrocarbons. Corrective actions undertaken by the county included continuous operation of the ventilation systems using 100 percent outside air and implementation of a no-smoking policy. Employees continued to report the same symptoms.

A second major suit was announced. This action was by the county against the design firms. The suit claimed that the county would be ". . . subject to virtually unlimited liability if it were to operate the ventilation system in its designed and constructed condition." Subsequently, the county closed the building on September 21, 1992. Six months of employee complaints and consultant studies had not identified the problem or found a solution. The employees were relocated to the facilities they occupied prior to completion of the new building. At the same time, the county allocated approximately $3 million for major improvements to the HVAC system, including further separation of supply and exhaust air.

In response to the charges, the designers' position was that the structure was designed and built to meet or surpass all applicable building codes and air quality standards. Furthermore, they pointed to the fact that no conclusive evidence had been found by any of the many consultants to diagnose the problem or to assess blame.

One extensive medical study did at least find evidence of sick building syndrome. The study based its findings on 106 interviews with a random sample of the facility's 656 occupants in which the illnesses were confirmed. The blame was placed on energy conservation strategies that reduced ventilation beyond acceptable limits. The report recommended improving the ventilation with increased fresh air and diligent maintenance of the HVAC system. (By the time this particular report was received, $3.5 million had already been spent on enhancements to the HVAC systems.) Other recommendations in the report were to reduce the amount of carpeting, to limit the use of some cleaning chemicals, and to enforce bans on smoking, perfume, and use of aerosols.

In January 1995, a circuit court jury in Lake County, Illinois, found in favor of the designers. This decision was billed as a landmark decision for pending sick building syndrome litigation. The jury blamed the county for improperly operating and maintaining the building's HVAC system. Apparently, the jury accepted the designers' claim that county maintenance employees had damaged the air quality by pouring cleaning chemicals into a steam generator and accidentally dropping antifreeze into the building's air ducts. The two-year battle and month-long trial was over a claim by the county for $6 million in damages. (It was reported that the parties spent over $10 million in legal fees, more than the amount of the claim and the costs of repairs to the building.)

Subsequently, the personal injury lawsuit filed by the 124 employees against the designers was settled. The settlement found no fault on the part of the designers for the air quality problems in the building (ENR 1995b).

Another very interesting and publicly conspicuous HVAC failure was experienced by the State of Illinois Center building in Chicago (ENR 1986a, 1987d). This unique and controversial structure was completed in 1985. The building is organized around a 17-story atrium space of impressive scale, completely sheathed in glass.

The HVAC system for the State of Illinois Center was also unique and innovative. The cooling system was designed to use off-peak electricity rates to make ice during the night. During the daytime, air cooled by the melting ice was to circulate through the building to provide 40 percent of the day's cool air needs. But shortly after the building opened, it was discovered that the system capacity was woefully inadequate. Daytime temperatures inside the building were unacceptably high. Truck-mounted air-conditioning equipment was brought in as an emergency measure, and the state announced a $3 million allocation for repairs to the underdesigned system.

In April 1987, the state of Illinois filed a lawsuit against the designers for $20 million, citing an inadequate HVAC system. The $3 million allocated in 1985 had not addressed the problem, and an additional $7 million retrofit had been required to add further cooling capacity and fan power. A study commissioned by the state concluded that the building's original HVAC system design was far from adequate in delivering the amount of cool air needed to keep the building comfortable. It was alleged that the capacity of the ice-making system was inadequate, the fans were underdesigned, and the duct routing was convoluted.

Other knowledgeable observers held a contrary opinion that the sophisticated system controls were never properly installed or operated. They also blamed changing design standards that had shifted emphasis from energy conservation to higher comfort expectations.

It was acknowledged that part of the problem was that the ice-making operation was not able to work as many hours during the night as originally anticipated, so that there was insufficient volume of ice to supply the daytime demand. The hours of building use were extended past the normal working day, a condition that was not in the original project description.

This was an extremely challenging HVAC design problem, primarily because of the innovative character of the 17-story glazed atrium. It was a unique architectural design and an innovative HVAC concept. An ice-storage system on this scale had never been used before in a building. If there is a lesson to be learned in this example, it is that one must be extremely conservative when designing HVAC systems for innovative projects. The design should include strategies for easily accommodating corrective additions should they prove necessary.

Finally, a less interesting but nonetheless annoying problem associated with HVAC systems is noise and vibrations. These problems can be overcome much more easily and economically by design rather than after the equipment has been installed.

HVAC noise and vibration problems in buildings seem to be increasing, despite the availability of technical literature on the problem (Schaffer 1991, 1993). The underlying causes are typically traced to the following items: (1) the acoustical aspects of the HVAC system were not considered during design and construction; (2) the system components were selected properly but were not integrated into a quiet system; (3) postdesign cost cutting resulted in a system that although less expensive, generated more noise or vibration.

The common tendency to downgrade mechanical systems as the first source of cost reduction when estimates or project bids are beyond the budget is shortsighted. It is tempting to focus cost-cutting efforts on the hidden equipment and systems so that the quality of the aesthetics, building configuration, and finish materials is not affected. But reductions in the quality of the HVAC system can severely affect the ability of the facility to deliver the working conditions expected by today's building user. The examples reviewed here clearly show that restoring the user's satisfaction can be quite costly indeed.

10.3 SUSPENDED CEILINGS

One recurring problem involving nonstructural materials is that of suspended ceiling failures. Failures occur due to corrosion of the tie wires, deficiencies in the attachments at the ends of the wires (top or bottom), poor workmanship in the installation, and as the result of overload (loads not considered in the system design). When ceilings fail, there may be injuries or deaths even though there may be no impact whatsoever on the primary load-carrying structure system.

Beginning in 1961, the U.S. Army Corps of Engineers investigated a number of failures of metal-lath and gypsum-plaster suspended ceilings in army hospitals and classroom buildings. In all cases failure or hazardous conditions had occurred as the result of corrosion of galvanized tie wires. The ties secure the metal lath to furring channels and these furring channels, in turn, to runner channels. The most serious corrosion was found in the lathing tie wires, where they were embedded in the gypsum plaster, and in the furring tie wires, where they came in contact with the plaster. In a nationwide survey of army buildings by division and district offices of the Corps, the galvanized tie wires showed no signs of corrosion where the wires were not in contact with the gypsum plaster.

Failure of ceilings in a hospital under construction at Fort Lee, Virginia, alerted the Corps to the danger. Subsequently, serious corrosion of tie wires was found in every area of the country except the San Francisco and New England divisions. The Fort Lee ceilings failed within six months of plastering, but another ceiling in a classroom at Fort Belvoir, Virginia, had been in place more than 10 years before it failed. The Corps engineers found evidence of serious corrosion in a new hospital at Fort Eustis, Virginia, in tie wires that had been in place only three or four months. Corrosion was also uncovered in new hospitals at Fort Meade and Andrews Air Force Base in nearby Maryland.

To guard against tie-wire corrosion in new construction, the Corps changed its specifications to require stainless steel wire of nonmagnetic type. Army engineers warned that casual inspection may not reveal the extent of the corrosion. Small rust stains on the surface of the plaster may represent complete corrosion of the wires within the plaster.

Hadipriono (1988) discusses three more recent suspended ceiling collapses: (1) the failure at Walters Food Store in Columbus, Ohio, in July 1983; (2) the August 1983 incident at the Port Authority of New York and New Jersey (PATH) transit station in Jersey City, New Jersey; and (3) the accident at the Westin Hotel in Boston, Massachusetts, in August 1986. According to Hadipriono, all three ceiling systems were designed to accommodate dead load only, all failures were preceded by signs of distress, all involved questionable workmanship in spacing or attaching the tie wires, and all failures occurred while construction or maintenance workers were on the ceilings.

The Walters Food Store in Columbus, Ohio, was built in 1961. The owner decided in July 1983 to place additional insulation in the ceiling space, adding dead loads to the ceiling hanger assembly. During the installation, the weight of the workers was also present. On July 22, 1983, when 80 percent of the new insulation was in place, the ceiling assembly collapsed (Figure 10.5). A noticeable sag from the weight of the insulation was reported about 45 minutes before the failure.

Figure 10.5 Suspended ceiling collapse in food store. (Courtesy of F. Hadipriono.)

On August 8, 1983, a suspended ceiling above a pedestrian concourse at a Port Authority of New York and New Jersey (PATH) transit station in Jersey City, New Jersey, collapsed, killing two persons and injuring 10 others. Just before the collapse, a workman had stepped onto the ceiling to begin repairs to a large section that was sagging. The section of standard cement plaster ceiling that fell was approximately 24 by 46 m (80 by 150 ft) and weighed about 45 Mg (50 tons). An investigative report released in June 1984 blamed the incident on a number of contributing factors involving many project participants: faulty design; poor communication between architect, construction manager, and general contractor; inadequate supervision during construction; and poor maintenance, inspection, and repair procedures by the owner.

The PATH station ceiling was installed on metal lath on steel furring channels spaced at 300 mm (12 in.) and steel main runner channels spaced at 1220 mm (48 in.). The runner channels were hung by wires tied to prepunched tabs in the metal deck of the plaza floor slab, 2.6 m (8.5 ft) above the ceiling. The load on the metal tabs at the time of failure exceeded their specified load capacity by as much as 600 percent.

Notes on the specifications and the shop drawings for the steel floor deck clearly stated that the load on each prepunched tab was not to exceed 260 N (60 lb). However, the report found a lack of coordination during the shop drawing stage between the contract for the deck work and the contract for the suspended ceiling work. The 260-N (60-lb) restriction was overlooked when the ceiling was installed under the general construction contract.

Tests showed that the tabs could actually carry significantly more than the 260-N (60-lb) limit specified before failing. Their ultimate capacity was 1.7 to 2.5 kN (380 to 560 lb). However, even that capacity was not enough. As installed, the dead load was approximately 880 N (200 lb) per tab. When the repair worker stepped on the ceiling, an additional load of about 880 N (200 lb) per tab was sufficient to cause failure. A progressive collapse then followed, partly because some of the tabs had failed previously (indicated by a rusted fracture surface), and partly because the design did not include expansion joints to limit the area of potential failure.

More hangwires would have solved the problem. A better solution is to attach wires supporting heavy ceilings to inserts embedded in the concrete rather than tabs in the metal deck. There have been other occasions when cement ceilings attached to metal deck tabs have experienced similar failures.

The owners of the transit station had recommended repairs to the sagging ceiling on the incorrect assumption that a few metal tabs were corroded. Apparently, the owners did not undertake a full load analysis of the ceiling support system before sending workers onto the ceiling, nor was shoring provided. A load analysis would have clearly indicated the need for shoring, with so little redundancy in the construction. The forensic investigator's report warned that there are many similar ceilings in other office and public buildings that could have been constructed with similar deficiencies.

Another case where weight of workers contributed to a suspended ceiling collapse occurred at the Westin Hotel in Boston, Massachusetts, in August 1986. Thirty lineal meters (100 lineal feet) of plaster, concrete, and fiberglass canopy overhanging the hotel's main entrance collapsed, falling 5 m (16 ft) to the ground and injuring six persons. Two of the injured were plumbers, who were working on a renovation to the restaurant above the entrance overhang. Lack of communication between the owner and the architect was blamed for this collapse. The ceiling construction was designed for dead load only, but the owner may not have been aware of this. Metal straps supporting the ceiling had not been bolted to the ceiling runners, but had been fastened with wire. The city of Boston required reconstruction of the canopy with upgraded design to include both dead and live loads.

10.4 EQUIPMENT FAILURES

The modern building contains an inventory of increasingly sophisticated equipment. The scale and cost of this equipment, when it fails, can generate considerable litigation. Contemporary society has a great reliance on technology, and building users have become quite demanding of the level of performance of that technology. Problems with equipment in buildings are appearing more frequently in the literature. A few specific cases are presented here. It is likely that building designers and constructors will face some very interesting challenges in the future as equipment becomes even more technically sophisticated.

In some buildings, the electronic instrumentation is actually responsible for the integrity or performance of the load-bearing structure system. The 1988 failure of the air-supported fabric roof of the Hoosier Dome, discussed in Section 10.1.1, is one example. Faulty air-pressure-monitoring equipment allowed the roof to deflate and collect rainwater. Other examples are tuned-mass dampers and active energy dissipators used in base-isolation systems for improving wind and earthquake resistance.

As the scale and complexity of buildings advance, simple maintenance tasks take on new levels of sophistication, presenting technical challenges far beyond those of traditional buildings. Even the window-washing system design and operation becomes a major challenge for the irregularly shaped tall building (ENR 1989). It is almost impossible to conceive of a safe and reliable method of washing the windows in some of today's high-rise buildings, with their setbacks, unconventional roof lines, and inclined glass curtain walls. There are many one-of-a-kind curtain-wall configurations with nonvertical glazed surfaces where it is impossible to use the conventional continuous window-washing equipment track along the window mullions. In such cases, the equipment is unique, project specific, and very expensive; and there has been costly litigation on some occasions when it fails to operate or when accidents occur. Window-washing systems should be considered early in the conceptual design phase; this has become a new area of specialization for design consultants.

Specialized equipment that fails to work properly can be responsible for substantial delays in construction completion schedules. Two examples are the case of mobile library shelving equipment at the British Library Building in London (ENR 1992b) and the incredibly costly problems experienced at the new Denver (Colorado) International Airport with its sophisticated automated baggage-handling system.

The British Library Building was designed to accommodate a compact movable steel shelving system involving over 300 lineal kilometers (186 lineal miles) of floor-to-ceiling units. The mobile shelving in the new $865 million building was selected for its ability to store books at high density. The shelving contract was for $16.3 million. The units are accessed by their ability to move laterally on rails. However, faulty drive mechanisms and corrosion of metal components made it necessary for workers to dismantle over 96 lineal kilometers (60 lineal miles) of the shelving, delaying the transfer of 12 million books by nearly two years, according to the U.K. Department of National Heritage in London. New paint specifications were ordered to mitigate the corrosion problems. The final cost for the shelving was reported to be over $19.1 million, including design modifications to the drive mechanism.

The controversial Denver International Airport project, completed in 1995, was plagued by serious problems with its $193 million state-of-the-art automated baggage handling system. At the time originally scheduled for opening the facility, with all other construction substantially complete, the baggage system was experiencing both hardware and software problems. Bags were being damaged by the equipment, which was reportedly throwing them more violently around the facility than is typically done by human baggage handlers. Completion of the $4 billion airport was delayed 16 months while the baggage system was redesigned. Some reports estimated the cost of the delay at over $1 million per day.

The first year of operation of the Denver International Airport, however, was apparently a great success, and the baggage system is operating admirably (ENR 1996). There was not a single air traffic control delay at the new airport in December 1995 as compared with an average over the seven previous Decembers of 1100 delays. For the year, there was an 80 percent reduction in delays over those at Denver's previous Stapleton International Airport. Revenues were higher than predicted, resulting in a first-year profit, contrary to many predictions. And despite brief problems during the Christmas holidays, the automated baggage system damaged just 367 bags in December, or only one per 3200 handled. This statistic is a 40 percent improvement over the industry standard.

One final example of nonstructural equipment problems in modern buildings is given here. The dilemma of "harmonic" electrical currents in buildings is a problem that was most certainly not considered by the designers of past generations. This serves as an example of the potential emergence of surprising new problems in the future as the contents of buildings become more and more technically and electronically sophisticated.

New electrical problems are occurring in buildings due to the random proliferation of electronic equipment: computers, printers, FAX machines, and related hardware. Electronic ballasts in lighting systems are another source of these currents. *Harmonic currents* are produced by electronic equipment because of the way that the equipment switches alternating current (ac) to direct current (dc) for use within electronic circuits. The resulting harmonics invade the three-phase conductors of (ac) circuits, distorting the waveform and reducing the quality of available power. The neutral conductor, which is supposed to carry limited currents, can become overloaded.

Harmonic currents can cause severe damage to electric transformers and mechanical systems, can destroy computers, and can cause fires. This problem is forcing a review of existing electrical and fire codes and standards (ENR 1992a). New buildings will be designed to these higher standards when they are available, but older buildings may not be able to make the necessary adjustments economically. One interesting aspect of the problem is that all electrical-system equipment and components are by their very nature interconnected. This means that harmonic currents can reach the building's transformers and return to the utility power grid, thereby gaining access to other buildings and doing similar damage there.

In a random survey, overloaded neutral conductors were found in many buildings that had not yet reported any problems. The extent of the problem is not yet known, but one estimate is that about one-fourth of all the buildings in the United States have the potential for future problems with harmonic currents. More difficulties will undoubtedly occur in the future as more electronic equipment is integrated into buildings. Some solutions are already available. These include equipping computer equipment rooms with special transformers, filters, and specially designed components. The design would have to be based on knowledge of future computer usage in the building; equipment could not continue to be added randomly to the system. Perhaps the best solution is to require the manufacturers of electronic office equipment to address the problem at the equipment level. It may be possible to manufacture equipment that does not produce the harmonics in the first place.

10.5 NONSTRUCTURAL COMPONENT REPAIR COSTS FOLLOWING RECENT SEISMIC EVENTS

Nonstructural components can contribute substantially to the total damage experienced in an earthquake (Lagorio 1990, AIA/ACSA 1992). Heavy nonstructural components add to a building's mass, generating larger seismic shear forces and becoming a threat to life safety when they fall. From past experience we know that falling nonstructural components (precast concrete panels, parapets, and masonry curtain walls) are potentially as hazardous to human life as failure of load-bearing structural components.

In the 1989, Loma Prieta, California, earthquake and the 1994 Northridge, California, event, nonstructural cladding panels and masonry infill panels unintentionally participated in resisting lateral loads. This behavior becomes a life-safety issue as well as a source of costly repairs. A number of observers have called for a more sophisticated rational design approach to consider the effects of cladding on seismic performance (Cohen 1995).

Masonry infill panels combined with a flexible, moment-resistant reinforced concrete or steel frame have always presented serious problems in large earthquakes, as have poorly attached precast concrete cladding panels (Figure 10.6; see also Figures 8.9 and 9.12). When these panels are not carefully considered in the design, they may interfere with the planned performance of the structural frame. If improperly detailed, the panels can restrict frame deformations, leading to total collapse of the structural system in a large earthquake.

Unsymmetrical placement of rigid nonstructural elements, such as masonry partitions or exit stair enclosures, can shift the location of the center of resistance to lateral forces, leading to unanticipated torsional effects. Partial height masonry infill walls can lead to damage from the "short column" effect, where some column deformations are restricted, causing them to attract more than their fair share of the lateral loads.

These are all examples of cases where nonstructural elements can influence the response of the structure system. In addition to these problems, the damage to nonstructural elements and contents within buildings has become a major concern in the past decade. The extent of such damage in the Loma Prieta

Figure 10.6   Masonry infill failure in earthquake. (Courtesy of the Federal Emergency Management Agency.)

earthquake, and especially in the Northridge earthquake, was a shock to the media, the public, building owners, and the insurance industry. Over 112,000 structures were damaged in the relatively moderate Northridge earthquake in an area that was not densely populated. The Northridge earthquake was the most costly natural disaster to date in the United States as measured by property damage; estimated losses are $30 billion. The 1989 Loma Prieta event caused damage amounting to about $6 billion. Some dramatic structural collapses in both earthquakes have been discussed previously. But much of the costly damage in engineered buildings was to nonstructural components, including interior damage and damage to building facades, typically caused by excessive deformations. It is expected that the surprising extent of this damage will lead to consideration of more restrictive drift limits in future designs.

A flexible structural system provides the ability to store energy temporarily through deformations, thus preventing total collapse in a major earthquake. The desirability of a ductile, flexible structural system in a large seismic event is discussed in Section 7.7. Seismic design codes in the United States and abroad have encouraged the use of such systems, because structures cannot economically be designed to resist large earthquakes with elastic range stresses. Inelastic deformations are needed to absorb or dissipate the energy demands of a great seismic event. However, the recent earthquakes in California caused so much damage to nonstructural building components and building contents that this generally accepted approach is now the subject of considerable discussion among engineers, architects, insurance professionals, and their clients, the building owners.

Although such damage had been predicted by structural engineers and was assumed in the commentary of the relevant building codes, the insurance industry and the public were surprised by its extent. Designers are facing pressure from building owners and users to upgrade stiffness requirements (drift control) so that facilities will be economically repairable following a moderate earthquake. The Northridge earthquake was not the "Big One" predicted for California, and the insurance industry is asking: How many of these $30 billion incidents can we afford?

It is extremely important that all architects and structural engineers understand this basic conflict in seismic engineering: the need for ductility and flexibility in the major event, and the need for stiffness in the moderate event. If flexibility and ductility are not present, the structure is in danger of total collapse in a major earthquake. Such structures may survive moderate earthquakes very well, and this may give owners and occupants a false sense of security. Unfortunately, excellent performance in a small event is absolutely no guarantee of survival in a large earthquake. The well-publicized damage to poorly detailed (nonductile) concrete buildings and highway structures in the Loma Prieta and Northridge earthquakes, and to the Hanshin Expressway in Kobe, Japan, serve as dramatic examples of the need for ductility (see Figures 7.8, 7.9, and 7.10).

Conversely, if drift control is insufficient, there will be substantial damage in the more frequent moderate earthquakes (Figure 10.7). In windstorms, excessive flexibility can also cause damage to contents and discomfort to building occupants. Resolving this fundamental conflict between the need for both flexibility and drift control should be a prime consideration early in the schematic design, when the structural bracing system is being selected. There are several means whereby both criteria can be satisfied at reasonable cost. The *Uniform Building Code* has encouraged use of "dual systems" that include both stiff shear walls or shear core systems along with continuous moment-resistant frame elements in the same structure. The *eccentric braced frame* (EBF) system also provides an effective response to these conflicting requirements.

Christopher Arnold, a California architect and researcher who has long been interested in seismic effects on buildings, was once asked by an engineering colleague: "Why are you—an *architect*—interested in earthquakes? Isn't this a *structural engineering* problem?" Arnold replied that he was interested in "the other 80 percent of the building." In the past, Arnold has urged seismic engineers to be "honest" and to refer to the "ductility factor" as the "*damage factor*," for a ductile response inevitably results in damage to nonstructural components. The experience in recent California earthquakes has given credibility to Arnold's position.

Figure 10.7 Earthquake damage to building contents. (Courtesy of the Federal Emergency Management Agency.)

When designers rely on large inelastic deformations to dissipate energy in a moderate seismic event, the price is paid in damage to architectural materials and building systems. Building contents are destroyed when this nonstructural damage occurs. An example is water damage from broken fire sprinkler lines. With the proliferation of costly computer systems and related equipment, the contents of contemporary buildings are quite often more valuable than the buildings themselves.

Since the Northridge earthquake, a number of individuals and organizations are promoting an entirely new approach to considering the potential for nonstructural component damage in earthquakes. The move toward "performance-based design" is expected to be reflected in future editions of the *Uniform Building Code* and other model seismic design codes. There will probably be increased concern for damage control. This implies more attention to drift limits or control of deformations. Anticipated cost of repairs following the design-level earthquake will be part of the preliminary planning for future new buildings.

Sophisticated building owners now recognize that the higher cost of construction required by upgraded seismic engineering may be justified if repair costs following the earthquake are reduced. Anticipated cost of repairs in the design-level earthquake will be a programming consideration discussed between architect, engineer, and client. Since the Northridge earthquake, with over \$20 billion of direct losses, it has become more difficult to obtain earthquake insurance in California. In the past, owners of facilities valued at \$200 million simply purchased \$200 million of earthquake insurance coverage. Now they are unable to obtain full coverage, and the reduced level of coverage now available may cost more than was previously paid for full coverage. In this insurance climate it becomes economically attractive to building owners to expend more funds on construction quality and hazard mitigation rather than on insurance. Reducing the expected damage has become more economically feasible than simply buying insurance to cover the expected loss.

In the past, the approach of U.S. codes has implied that if the occupants can walk out of the building alive after a major earthquake, the design was acceptable. This attitude is now being questioned. The prospect has arisen to offer the client the option of several upgraded levels of seismic performance—each, of course, with a higher cost. The Federal Emergency Management Agency (FEMA) and the Building Seismic Safety Council (BSSC) have proposed the development of guidelines that include minimum standards, along with upgraded options, or *performance-based levels.*

Buildings may eventually be given specific seismic-safety-level ratings. The current code is designed to provide safe egress. Postearthquake damage can be extensive, yet the design satisfy current code requirements. The underlying problem is that while codes are intended to be minimum standards of performance, they typically become the norm due to economic pressures. There is a need for specified performance objectives: life safety protection *plus* damage control.

The Structural Engineers Association of California (SEAOC), supported by FEMA funding, is writing recommendations for performance-based seismic engineering of buildings. In these new guidelines, deformations are expected to replace forces as the key design parameter. Several levels of performance will be defined, all achieved by controlling story drifts in the design-level earthquake. These will help to clarify realistic expectations of performance and will establish a direct relationship between initial construction cost and performance expectations.

10.6 REFERENCES

AIA/ACSA, 1992. *Buildings at Risk: Seismic Design Basics for Practicing Architects,* American Institute of Architects/Association of Collegiate Schools of Architecture, Council on Architectural Research, Washington, DC.

Becker, R., and R. Robison, 1985. "Get Involved with Cladding Design," *Civil Engineering,* American Society of Civil Engineers, New York (June).

Campbell, R., 1980. "Evaluation: Boston's John Hancock Tower in Context," *AIA Journal,* American Institute of Architects, Washington, DC (December).

Campbell, R., 1988. "Learning from the Hancock," *Architecture,* American Institute of Architects, Washington, DC (March).

Cassady, L., 1990. "Incompatibility of Building Components," *Journal of Performance of Constructed Facilities,* American Society of Civil Engineers, New York (February).

Civil Engineering, 1987. American Society of Civil Engineers, New York (April).

Cohen, J., 1991. "Cladding Design: Whose Responsibility?" *Journal of Performance of Constructed Facilities,* American Society of Civil Engineers, New York (August).

Cohen, J., 1995. "Seismic Performance of Cladding: Responsibility Revisited," *Journal of Performance of Constructed Facilities,* American Society of Civil Engineers, New York (November).

Engineering News-Record, 1981. McGraw-Hill, Inc., New York (February 12).

Engineering News-Record, 1982a. McGraw-Hill, Inc., New York (May 13).

Engineering News-Record, 1982b. McGraw-Hill, Inc., New York (December 16).

Engineering News-Record, 1984a. McGraw-Hill, Inc., New York (February 2).

Engineering News-Record, 1984b. McGraw-Hill, Inc., New York (April 19).

Engineering News-Record, 1986a. McGraw-Hill, Inc., New York (August 14).

Engineering News-Record, 1986b. "Roofing Reaches for Quality," McGraw-Hill, Inc., New York (November 6).

Engineering News-Record, 1987a. McGraw-Hill, Inc., New York, (January 22).

Engineering News-Record, 1987b. McGraw-Hill, Inc., New York (February 5).

Engineering News-Record, 1987c. "Stretching the Watertight Seal," McGraw-Hill, Inc., New York (November 12).

Engineering News-Record, 1987d. McGraw-Hill, Inc., New York (December 3).

Engineering News-Record, 1989. "Scaling the Heights to Wash the Windows," McGraw-Hill, Inc., New York (May 4).

Engineering News-Record, 1990a. McGraw-Hill, Inc., New York (January 18).

Engineering News-Record, 1990b. McGraw-Hill, Inc., New York (June 14).

Engineering News-Record, 1990c. McGraw-Hill, Inc., New York (August 2).

Engineering News-Record, 1992a. McGraw-Hill, Inc., New York (May 18).

Engineering News-Record, 1992b. McGraw-Hill, Inc., New York (September 28).

Engineering News-Record, 1993. "Big Curtain Wall Woes," McGraw-Hill, Inc., New York (November 15).

Engineering News-Record, 1995a. McGraw-Hill, Inc., New York (January 16).

Engineering News-Record, 1995b. McGraw-Hill, Inc., New York (January 30).

Engineering News-Record, 1996. McGraw-Hill, Inc., New York (February 26).

Fisher, T., 1983. "Radical Roofing," *Progressive Architecture,* Reinhold Publishing Corporation, New York.

Frauenhoffer, J., 1987. "Weathering Steel Cladding Failure," *Journal of Performance of Constructed Facilities,* American Society of Civil Engineers, New York (May).

Godfrey, K., 1986. "Roof Membranes: New Systems, New Problems," *Civil Engineering,* American Society of Civil Engineers, New York (March).

Hadipriono, F., 1988. "Investigative Studies of Ceiling Collapses," *Journal of Performance of Constructed Facilities,* American Society of Civil Engineers, New York (February). (See also Discussion by D. Dusenberry in May 1989 issue.)

Heller, B., 1985. "Rudiments of Roofing," *Architectural Technology*, American Institute of Architects, Washington, DC (Spring).

Ink, S., 1990. " 'And the Wall Came Tumbling Down'—Analysis of Stucco Veneer Failure at Lee County Justice Center," *Journal of the National Academy of Forensic Engineers*, NAFE, Hawthorne, NY (December).

Kub, E., II, L. Cartwright, and I. Oppenheim, 1993. "Cracking in Exterior Insulation and Finish Systems," *Journal of Performance of Constructed Facilities*, American Society of Civil Engineers, New York (February).

Lagorio, H., 1990. *Earthquakes: An Architect's Guide to Nonstructural Seismic Hazards*, John Wiley & Sons, Inc., New York.

Marlin, W., 1977. "Some Reflections on the John Hancock Tower," *Architectural Record*, McGraw-Hill, Inc., New York (June).

Nicastro, D., 1993. "Can Engineers Cut Curtain Wall Failures?" *Civil Engineering*, American Society of Civil Engineers, New York (November).

Osman, M., 1977. "The 1977 AIA Honor Awards," *AIA Journal*, American Institute of Architects, Washington, DC (May).

Ross, S., 1984. *Construction Disasters: Design Failures, Causes, and Prevention*, Engineering News-Record, McGraw-Hill, Inc., New York.

Schaffer, M., 1991. *A Practical Guide to Noise and Vibration Control for HVAC Systems*, American Society of Heating, Refrigerating, and Air-Conditioning Engineers (ASHRAE), Atlanta, GA (June).

Schaffer, M., 1993. "Controlling HVAC System Noise and Vibration," *ASHRAE Journal*, American Society of Heating, Refrigerating, and Air-Conditioning Engineers, Atlanta, GA (June).

Schwartz, T., 1988. "Drawing Up Waterproof Curtains," *Civil Engineering*, American Society of Civil Engineers, New York (March).

Singh, J., ed., 1994. *Building Mycology: Management of Decay and Health in Buildings*, E & FN Spon, London.

11

CONSTRUCTION SAFETY AND FAILURES DURING CONSTRUCTION

Failures occurring during construction have been discussed in earlier chapters. Some of these failures were due to design errors; others were caused by construction mistakes or construction equipment accidents. In this chapter we focus on the subject of construction safety and structural failures related to the construction process itself.

The construction industry in the United States has a poor safety record. Although construction is an inherently dangerous occupation, many accidents are avoidable. The safety record can be improved through education, legislation, and vigorous administrative enforcement of safety regulations. Careful inspection, from initial site work through completion, can also contribute to safety. In this chapter the reasons why so many failures occur during construction are reviewed, and efforts aimed at improving safety on the construction site are discussed.

11.1 CONSTRUCTION: A DANGEROUS OCCUPATION

Failures of structures frequently occur during construction (Carper 1987). These failures may involve components, assemblies, or large portions of the incomplete structure. Failures during construction are costly and in the extreme case may result in injury or death.

Public interest in construction safety has become more pronounced as a result of media coverage of statistics released by the National Safety Council in 1985 (NSC 1985). Over 2200 deaths were reported for the construction industry during 1984, the largest total for the eight major industries surveyed. In that year there were 39 deaths per 100,000 workers in construction; only mining and agriculture experienced higher rates. Also recorded were 220,000 disabling injuries (ENR 1985g).

A 1987 National Institute for Occupational Safety and Health (NIOSH) report reviewed records collected during 1980–1984. Out of a total of 7000 annual occupationally related deaths, 20.4 percent were in construction, again the highest total for any industry. The death rate was 23 deaths per 100,000 workers, a rate second only to mining, with 30 deaths per 100,000 workers (ENR 1987b). In 1988, the National Safety Council (NSC) reported that the construction industry was still averaging 2200 fatalities, 500,000 disabling injuries, and 34 deaths per 100,000 workers annually (ENR 1989a).

Many factors contribute to the risks associated with construction; some of these have little to do with the performance of the structure itself. Leading causes of injury and death include falls, being struck by falling objects or moving vehicles, electrocutions, equipment accidents, and excavation failures. NIOSH statistics indicate that falling is the leading cause of deaths, accounting for 26 percent of construction fatalities (NIOSH 1987). Numerous crane accidents due to manufacturing defects, maintenance problems, or operator errors are recorded each year (Allen and Schriever 1973). The fatality rate for trenching accidents is 112 percent higher than for construction accidents overall, according to Occupational Safety and Health Administration (OSHA) statistics. NIOSH has noted that trenching accidents account for 1 percent of all work-related deaths in the United States and for at least 1000 disabling injuries annually (ENR 1985d). Other hazards inherent to construction processes are fire, explosions, toxic chemicals, dust, and ventilation problems in confined work spaces.

Large failures that cause multiple deaths, such as the L'Ambiance Plaza collapse in 1987 (Sections 11.6 and 8.5) and the Willow Island Cooling Tower failure in 1978 (Section 11.5), receive much media attention. But the vast majority of accidents affect only one or two construction workers at a time (Preziosi 1989). Small projects are not immune from serious accidents. Half of all deaths from falling are from heights less than 9 m (30 ft); 80 percent of trench cave-in fatalities occur in trenches less than 5 m (15 ft) deep (Korman, Setzer, and Bradford 1990).

The economic consequences of construction accidents are also disturbing. Research conducted in 1982 by the Stanford University Department of Civil Engineering for the Business Roundtable's Construction Industry Cost Effectiveness Project showed that over $1.6 billion annually is lost in the United States directly due to construction accidents. This cost is on the order of 3 percent of the total annual costs of construction (ENR 1982a). The same study suggested that when indirect losses are considered, the overall costs approach $8.9 billion (CECS 1990).

Other countries have noted the dangers of construction, although the accident rates appear to be less than in the United States. In the United Kingdom,

2300 construction workers were killed or seriously injured in 1985, and the degree of enforcement of regulations was increased as a result (ENR 1987e). In the European Community, the construction industry employs 10 percent of the workforce but accounts for 15 percent of workplace accidents and 30 percent of occupational fatalities (ENR 1992f). A 1988 report issued by the National Safe Workplace Institute of the International Labor Office, a United Nations agency, gave the annual rates of construction deaths per 100,000 workers in several countries. The United States led the list with 39 deaths per 100,000 workers, followed by 30 in France, 24 in Spain, 18 each in New Zealand and Finland, 16 in Greece, and 15 in the United Kingdom (Dinges 1991).

It is not clear from the statistics available what proportion of worker injuries or deaths result from failure of structural assemblies or components. The statistics are not complete, since many minor or partial failures are not reported. Statistics regarding economic consequences are also incomplete and should be adjusted to include the related costs of workers' compensation insurance, dispute resolution, litigation, and so on.

What is clear, however, is that the safety record in the construction industry can be improved on many fronts. Most construction accidents are due to unsafe actions. These are avoidable. Many construction accidents are simply the result of ignorance, carelessness, or greed. Efforts by federal, state, and local organizations to improve construction practices have achieved some degree of success and should be continued.

James Lapping, director of safety and health of the Building and Construction Trades Department of the AFL/CIO Labor Union, testified in 1988 before the U.S. House Subcommittee on Health and Safety. He noted that construction workers constitute only 5 percent of the workforce, yet account for 26 percent of all occupational fatalities. Lapping called for more aggressive safety enforcement by OSHA (Preziosi 1989).

A new office charged with specific attention to the construction industry was established within OSHA in 1988 because of the alarming safety record. OSHA activities include both enforcement and education. Assisting the education effort are several projects, such as the Architecture and Engineering Performance Information Center (AEPIC) at the University of Maryland and the Center for Excellence in Construction Safety (CECS) at West Virginia University (Eck 1987, Loss 1987).

Some encouraging statistics reflect the positive influence of education and enforcement strategies. For example, the alarming death rate of 33 per 100,000 workers in 1987 compares favorably with the even more alarming rate of 71 deaths per 100,000 workers recorded in 1973. Some of this reduction is likely due to increased safety awareness and more vigorous enforcement of safety regulations (ENR 1988a). California's Division of Occupational Safety and Health Administration (CalOSHA) is credited with contributing to a significant reduction in the construction accident rate in that state. Construction in California was targeted by special legislation in 1973 as a hazardous industry. A comparison of safety performance in 1967–1972 with the performance in

1973–1978 shows a 38.1 percent reduction in the total number of deaths (ENR 1980b). CalOSHA noted that this improvement was achieved despite a 20 percent increase in the number of construction workers. During that five-year period, construction experienced the most significant reduction in occupational-related deaths of any of the industries monitored by CalOSHA.

Vigorous enforcement is evidenced by an increase in the number of citations and the size of penalties levied on offenders, including prison sentences for criminal actions that endanger human life. Most citations have been aimed at construction companies, since they have control of the site, but there are an increasing number of specialty contractors and even design professionals who have been prosecuted for unsafe construction site conditions. Most citations are for deficient scaffolds, failure to provide fall protection, and unsafe excavations, particularly unbraced trenches. The regulations for such activities are clear and there is much information available to prevent these common accidents, but unsafe practices continue to occur due to carelessness and other marketplace factors.

Although aggressive enforcement is producing measurable improvements, there are some who argue against expanding safety legislation. While the number of inspectors is barely sufficient to enforce existing regulations, some are demanding a general reduction in the size of government. The challenge is to improve safety without inordinate costs to the taxpayer and without placing unnecessary burdens on conscientious constructors.

Much of the safety problem is an attitudinal one. Despite educational efforts, many construction workers in the United States still consider construction to be a "man's world," one in which risks are to be taken and flaunted. To some, the safety meetings are a waste of time, and conforming to safety regulations is for "wimps." Such persons may wear their safety belts if absolutely required to do so, but refuse to admit vulnerability by tying them to a secure support. In some states, consideration is being given to citing individual workers for unsafe practices rather than their employers. This concept is patterned after the success of similar provisions in Canadian legislation and is intended to place responsibility for certain regulations directly on the person involved (ENR 1991a). Thus enforcement will go hand in hand with education and awareness programs to bring about attitude adjustments within the industry.

Safe construction practices include monitoring the performance of the structure under construction. In several respects a structure is most vulnerable to failure during the construction phase. In this chapter we review the reasons for the large number of failures that occur in incomplete structures. Efforts to reduce the potential for structural failure during the construction phase will reduce some of the risk of injury and of unforeseen costs and delays.

11.2 CONSTRUCTION FAILURES DUE TO DESIGN ERRORS

Failures may occur at any time during the expected useful life of a structure. For example, catastrophic collapses of bridges and other structures have occurred due to lack of maintenance after many years of satisfactory performance. Similarly, an error made during construction may go undiscovered until failure occurs, years after occupancy of the facility.

The first test of a structure's adequacy, however, takes place during the construction phase. Thus a collapse during construction may not necessarily imply a construction error. It may be the result of an error made during the design phase. Indeed, for many structures built in past eras, failures during construction must have served as valuable tests of incomplete or inaccurate designs. Masonry construction failures in Europe during the Gothic period, for example, undoubtedly played a role in improving engineering theory as builders conducted trial-and-error and trial-and-success experiments.

The 1981 Harbour Cay Condominium collapse in Cocoa Beach, Florida, occurred during construction, but the most probable cause was a design deficiency (USDC/NBS 1982a). Eleven construction workers were killed in this collapse (see Section 7.6). Some may argue that this test of a structural design during construction is desirable. Certainly, many more lives could have been lost at Harbour Cay or in the great Gothic cathedrals had all failures occurred after occupancy. Such an argument, however, is of little comfort to those who spend their lives working on the construction site.

Design errors sometimes are discovered and corrected by competent observers of structural performance during construction. It has been suggested that the catastrophic collapse of the Hartford, Connecticut, Civic Center Coliseum roof (see Section 6.2.2.c) could have been prevented. Certainly, the structure gave sufficient warning during construction of impending collapse, but no one competent to interpret the implications of inadequate performance was retained for inspection services during the construction phase (Committee 1978). This failure, five years after completion, could have caused injury or death to 12,800 occupants. Fortunately, the building was not occupied the night of the collapse. The risk of this type of failure is compounded when field inspection procedures are compromised or poorly administered with minimal financial support (Carper 1984).

Whatever the cause of structural failure, it is clear that a structure is extremely vulnerable to failure while it is under construction. Not only is this the first test of the adequacy of design, but the construction process itself provides several opportunities for failure.

11.3 EXCAVATION, TRENCHING, AND FOUNDATION CONSTRUCTION ACCIDENTS

In no phase of construction is a keen eye and the skill that comes from experience more important than in work below the ground surface. Just as in the art of surgery, success does not permit much deviation from carefully planned and executed sequential actions in which all potential mishaps have been anticipated. There is considerable merit in the analogy between surgery and below-surface construction. Both expose the unknown and must expect to remedy unforeseeable conditions. Both require preparation to counteract operational shock to existing physical conditions, conditions that are often much weaker than assumed. Every logical step should be preconsidered before starting the sequence, with alternative protective moves in readiness to correct for a miscalculation. Unfortunately, not all work is so well planned. Failures in subsurface work are common. They are expensive—in money, in time, and too often in loss of life.

Under the theory that a calculated risk is permissible in temporary excavations, as if temporary structures were safe at a standard insufficient for permanent work, sheeting and shoring for subsurface construction are prone to accident and consequent damage. Sheeting and bracing are seldom installed to the dimensioned tolerances required of permanent framing, and many critical details and structural redundancies built into permanent work are missing in hastily contrived excavation support systems. These deficiencies have been responsible for catastrophic excavation collapse incidents throughout the world (Lin and Hadipriono 1990).

Deep excavation failures continue to occur in the United States, as illustrated by the collapse of an earth retention system for a large 14.3-m (47-ft)-deep foundation excavation in Washington, D.C., in November 1990 (ENR 1991c). Foundation excavations in dense urban areas present special problems because of site access restrictions and congested work areas as well as the existence of adjacent structures and utilities.

The most common causes of cave-in accidents are inadequate or nonexistent shoring, use of defective shoring materials, inability to accurately assess soil conditions, failure to consider changing weather conditions, failure to properly locate heavy loads away from the trench/excavation, poor planning, and improper or lack of training (CECS 1989).

Small unplanned events can trigger catastrophic collapse in even well-designed excavation shoring systems. Such an accident occurred in a properly sheathed and braced excavation for a three-tier garage below street level in Washington, D.C., in the early 1960s. With machine excavation completed, the final earth removal at subgrade was being done by a very small back-hoe scratching the soft disintegrated rock away from the soldier beams. The bucket teeth on the back-hoe engaged the flange of a soldier beam and in the attempt to free the contact, pulled the beam inward, causing a local collapse. The soil poured into the excavated area and the added forces crippled the two lower sets of crossbracing, the top frame remaining intact (Figure 11.1). Street traffic was closed off on one of the major arterials. Corrective work consisted of a new line of soldier beams surrounding the original sheathing, some 5 m (15 ft) away, with enclosures to tie to undisturbed portions. Fortunately, a small park area adjacent to the point of failure was available for the corrective enclosure.

The purpose of sheeting is not only to protect the enclosed work but also to prevent loss of support for adjacent structures. In the foundation work for

Figure 11.1 Sheeting failure caused by backhoe bucket hooking a soldier beam.

the Cleveland, Ohio, Federal Building in 1966, the adjacent street and utility lines suffered damage due to subsidence because the shoring did not prevent movement of the sheeting. In such vulnerable areas where sheeting positions must be maintained, a system of jacking facilities may be necessary, so that movement can be neutralized by rapid adjustment of reactions induced into the supporting members.

One of the many technical problems encountered by the John Hancock Tower in Boston, Massachusetts, was damage to adjacent buildings and infrastructure during the foundation construction. Excavation disturbed the water table and modified earth pressures around the site, causing damage to adjacent structures that were built on timber piles in filled land. Several of these structures were important historic treasures, including the Copley Plaza Hotel and the Boston Trinity Church, designed by H. H. Richardson (Figure 11.2). According to court records, excavation retaining walls slipped 840 mm (33 in.) as the office building's foundation was under construction in 1969, allowing the adjacent street to sink 460 mm (18 in.) and moving the footings of Trinity Church. Parts of the church building shifted, the central tower tilted 125 mm (5 in.), and structural arches cracked. A lengthy lawsuit was initiated in 1975 and settled 17 years later, in 1984, when the John Hancock Mutual Life Insurance Company agreed to pay the Trinity Church $11.6 million in damages. The controversial settlement recognized that the damage to the

Figure 11.2 Foundation construction for Boston's John Hancock tower damaged adjacent structures including the historic Trinity Church.

church building was irreparable. The compensation amount was based on the cost of completely demolishing and reconstructing the historic masonry building (ASCE 1987, ENR 1987a).

Control of movement, which can lead to total collapse, requires careful attention to details, even assuming a proper design of the sheeting system. Unwanted surcharge loadings must be prevented, such as unforeseen equipment operation at the edge of the excavation and, more often, stockpiling of construction materials at the convenient but hazardous position at the edge of the sheeting. Existing surcharge loadings such as neighboring buildings with shallow foundations must be considered in the design, as well as traffic loading and vibration effects on the disturbed soil or backfill. Groundwater control must be positive and continuous, or else the sheeting must be designed for hydrostatic pressure in addition to soil pressure.

Reactions imposed on wales change with excavation depth and the concentrations at rakers almost always require welded stiffeners on the webs of economically designed wales. Diagonal rakers impose upward vertical loading on the sheet piles or on the soldier beams. If the soil friction, modified by water and vibration, cannot resist such reactions, the embedded lower ends must take the full upward pull. Where steel sheeting rests on rock, anchorage may be necessary in the form of grouted rods welded to the sheeting. In the case of soldier beams, either sockets may be provided in the rock for grouted embedment or rod anchors welded to the beams may be grouted into the rock. A buckled wale or a bowed brace is a sign of trouble and must not be tolerated. Heel blocks for the spur brace reactions must be stable and large enough to transmit the load to the soil. Disregard of any of these small details can generate a large failure. Each of the previous items has been noted as the cause of excessive sheeting movements, with accompanying damage to adjacent structures.

Interior bracing, whether horizontal or sloped, interferes with excavation equipment and usually requires cutouts in cellar walls with incipient leakage in the finished walls. Cable or rod tiebacks to hold the sheeting to the exterior soil or rock are therefore preferable where practical, so that exposed bracing is eliminated and simple stressing can be accomplished to develop the desired reactions. It should be noted that even for temporary use, tiebacks entering private property are considered illegal trespass and may not be permitted. Even if public or private land is available, tiebacks are an acceptable solution only if the anchorage is reliable.

Failure of an entire tiebacked 1:5 sloped earth cut 10.7 m (35 ft) high with anchors all tested for 89 kN (10 tons) at the Seattle, Washington, City Hall construction in 1961 was caused by the internal weakening of the fissured clay. Expanding-type anchors had been inserted into 200-mm (8-in.)-diameter holes drilled 4 to 6 m (12 to 20 ft) into the embankment. The surface was covered with wire mesh and gunite and 89 by 292 mm (4 by 12 in.) timbers set in a grid to connect the anchor rods. The slide was 46 m (150 ft) long and sheared at 5 m (15 ft) back from the face of the bank.

Cofferdams, designed to resist water pressure for construction of bridge piers and dam projects, are susceptible to failure under unusual loads such as impact loads, ice buildup, or flooding. These hazards are in addition to those already discussed for general excavation projects.

Soil failures can be induced by vibrations. Earth masses that are not fully consolidated will change volume when exposed to vibration impulses. One source of vibrations is construction equipment, especially pile drivers (Dowding 1996). Traffic on a rough pavement and blasting shock can also cause damaging vibrations.

In 1963–1965, the insertion of steel H-piles by heavy impact hammers and by vibrating hammers for a tall government building in New York City severely damaged adjacent buildings. The buildings were resting on a medium-dense sand of uniform grain size. All of the adjacent buildings, more than 120 m (400 ft) of frontage, were condemned as unsafe and had to be demolished, including one 19-story building. The adjacent street pavement settled up to 400 mm (16 in.) when only a fourth of the piles had been installed, before the vibrating hammers were brought to the job. As a precaution against further street subsidence, a line of steel sheeting was driven and covered with a sand berm on the excavation side. This was successful in protecting the street against further settlement and damage to the utility lines buried therein. In a similar building construction in 1960, some 300 m (1000 ft) away, when the first pile driving indicated a serious effect on the adjacent building, work was stopped and the existing footings were underpinned by piles carried to the expected tip elevation of the new building piles. Thereafter, pile driving had no effect since the sand layers, which were consolidated, no longer carried any building load.

Methods to mitigate the vibration effects of pile-driving operations have been developed, as have sophisticated instruments for monitoring the effects of such vibrations. The delay interval in impact sequence can be controlled to minimize dynamic response of structures on adjacent sites or of earth masses (Fairweather 1990).

Digging trenches for defense, for drainage, for waterways, and for the insertion of conduits and foundations has been done for many centuries. Even without any theoretical developments, knowledge gained from the unfortunate experience of others should by now prevent any trenching failures. Theoretical studies, combined with experience, have led to guidelines and standards for safe side slopes in various soils, and for sheeting the sides of excavations where adequate side slopes are impractical. Full compliance with these existing standards would almost certainly eliminate fatalities related to trench excavations. Yet trench collapses continue to occur at an alarming, unwarranted, and inexcusable rate, often with dramatic news stories of fatal or near-fatal results. Trench excavations for utility lines near existing structures continue to undermine foundations, leading to failures of entire building structures.

It is a fact that trenches can be dug in any type of soil in a way that will not harm the workers, the public, or adjacent existing facilities. However, the reported incidents of open-trench failures are so numerous and so depressingly similar that specific examples are not necessary here. Many, many fatalities occur in trenching accidents, frequently in shallow utility line excavations only 2 to 3 m (7 to 10 ft) deep. Many of the fatalities result from successive collapses of unbraced trenches, burying would-be rescuers of earlier victims.

In 1985, a special conference on trench safety was sponsored by the Architecture and Engineering Performance Information Center (AEPIC 1985) at the University of Maryland. A clear consensus was reached at the conference that the knowledge is already available to eliminate most trenching accidents. Most occur simply due to greed and carelessness. The prevalent attitude is: "If we work fast enough, we won't have to brace it; nothing is likely to happen." Modern equipment that can excavate deeper and faster has contributed to the problem. Support must be installed quickly to keep pace with trench excavation.

In recognition of the problem, a major education and enforcement initiative was announced by OSHA in 1985 (ENR 1985d). OSHA administrators were alarmed at the statistics, including the fact that more than 100 workers are killed every year in preventable trench accidents. Substantial increases in the number of OSHA inspections and citations have been matched by new state and municipality regulations.

Since 1986, much higher economic penalties have been levied for violations of safety regulations. Criminal homicide and manslaughter charges have been prosecuted against those who have been found guilty of repetitive willful violations. Jail sentences have become commonplace. Texas, for example, has focused on this specific safety problem as a result of over 60 fatalities in trench accidents in that state between 1980 and 1986 (ENR 1986b). Some local jurisdictions are now requiring construction firms to be certified in trench safety prior to receiving building permits.

Tunneling projects present unique safety hazards. In addition to the normal hazards associated with construction, work must be carried out in confined spaces, sometimes in a pressurized environment, and the potential for cave-ins is present. In Section 11.8 we discuss some of the hazards associated with confined spaces. In January 1990 a dramatic tunneling accident occurred in Tokyo, Japan, where workers were drilling a 12.5-m (41-ft)-diameter tunnel for a bullet train route. Air leaking from a pneumatic-shield tunneling machine burst upward, spewing a large volume of water, soil, and rocks onto a busy Tokyo street. The accident created a 13 by 10 by 5 m (42 by 32 by 17 ft) deep hole into which four cars disappeared. Ten people were injured, but there were no fatalities (ENR 1990c).

The U.S. Environmental Protection Agency (EPA) in 1987 conducted a substantial investigation into the safety record of the $3.2 billion Chicago sewer project, partially in response to 10 fatal accidents on the 80-km (50-mi) tunneling project. The accidents were caused by asphyxiations, falls, and equipment malfunctions. One of the owner's representatives on the project noted that because of the use of large boring machines rather than the more hazardous drill-and-blast methods, the project's accident rate was actually far better than on most tunneling projects. He claimed that tunneling work generally produces an average death rate of two workers per mile of tunnel (ENR 1987c). Projecting this alarming statistic to the 80-km (50-mi) project predicts an expected death rate of 100 workers on this one project! Surely, such a rate is unacceptable and should not be used as a measure of success. If the average accident rate is of this magnitude, the industry must focus on tunneling safety to improve the record.

11.4 CONSTRUCTION LOADS MAY EXCEED DESIGN LOADS

One recurring cause of structural failures during construction is excessive construction loading. Often the loads applied to structural members while construction is taking place are in excess of service loads anticipated by the designer.

Construction loads may include transportation and erection loads in the case of prefabricated components. Frequently, the loads exerted on a precast concrete member during erection are the greatest loads ever experienced by the member. The designer should anticipate these loads and communicate the lifting points and erection sequences clearly to the construction workers.

Other construction loads may be unforeseeable by the designer. These must be monitored carefully by the construction superintendent. The stockpiling of heavy construction materials, such as masonry units, has caused numerous catastrophic failures (Figure 11.3).

Three workers were killed in an accident in December 1985 when 73 Mg (80 tons) of structural steel was stockpiled on one bay on the fifth floor of a 21-story building under construction in Los Angeles, California. The bay was loaded to twice its capacity, causing three beams to fail at bolted connections, and then 10 bays were destroyed in a progressive collapse down to the ground floor (ENR 1986a). Those responsible were prosecuted for negligence under criminal statutes.

Roofing contractors are often cited by OSHA for roofing gravel overloads during construction of flat built-up roofs. Collapse due to such overloads can be sudden, accompanied by serious injury to workers. In one such failure in Florida in 1983, a 14 by 12 m (45 by 40 ft) section of roof over a shopping center collapsed, killing one worker and injuring two when the roof was overloaded with 2.1 Mg (4800 lb) of rock. The OSHA report noted that the accident could have been averted if the load had been distributed carefully on the roof (ENR 1983a).

A similar roof collapse in the Pacific northwest was due to stockpiling of roofing repair materials. A minor local failure had occurred as a result of an inadequate bearing detail for parallel chord timber roof trusses. The roofing contractor who was hired to repair the detail on the remaining structure stockpiled all of his materials on the roof. This immediately precipitated the

Figure 11.3 Masonry contractors have carefully distributed construction materials on this project to reduce the potential for overloading during construction. (Courtesy of the Brick Institute of America.)

collapse of the rest of the roof as failure occurred in exactly the same manner, involving the same deficient connection the roofing contractor had originally been retained to repair.

Building failures have occurred due to concrete mixing operations on the roofs of buildings under construction. While locating the mixing operation at the roof may simplify placement procedures, the delivery and storage of mixing materials must be carefully monitored, or the loads can easily exceed the capacity of structural members supporting the roof.

Other failures have been recorded when lifting cranes were supported by structural members not designed for such loads. Columns and walls have been used as supports for jacking procedures completely unanticipated by the designer (McKaig 1962). Such practices must be avoided. Again, the importance of field inspection, preferably by a designer's representative, cannot be overemphasized.

Even the weight of construction workers themselves has caused the collapse of certain structures that were designed only to carry dead loads, such as suspended architectural ceiling systems (Hadipriono 1988). Six people were injured when a suspended ceiling collapsed under the weight of plumbers working in the space above the ceiling in a Boston, Massachusetts, hotel in 1986 (ENR 1986c). When such ceilings have been deficiently installed in the first place, the weight of construction workers can be just enough to trigger collapse, with catastrophic results. The failure of a transit station ceiling in New Jersey killed two persons when an improperly installed suspended ceiling grid gave way under the weight of a worker who stepped onto the grid to make repairs to a sagging area (ENR 1983b).

11.5 MATERIALS AND ASSEMBLIES NOT YET AT DESIGN STRENGTH

Another common cause of structural failure during construction is the presence of materials and structural assemblies that are working at less than anticipated design strength.

For certain types of construction, especially cast-in-place reinforced concrete, this may be the principal source of construction failure. Premature removal of formwork or shoring is responsible for many injuries and deaths each year, as is the application of construction loads on insufficiently cured concrete. These events are well documented and widely disseminated. Yet such accidents continue to occur. Insufficiently cured concrete has little structural capacity. The emphasis on reducing construction time, however, is a constant threat to cautious scheduling. Despite the existence of standards for formwork and shoring removal time developed and enforced by OSHA and other organizations, this type of failure occurs over and over again. Those responsible for scheduling must understand and respect the initial characteristics of this material or costly accidents are inevitable. Three significant failures from the 1970s illustrate the potential for catastrophic failures due to premature loading.

On January 25, 1970, the 17-story Commonwealth Avenue building in Boston, Massachusetts, collapsed during construction. The progressive failure caused 11,000 m^2 (120,000 ft^2) of cast-in-place reinforced concrete floor slabs to fall to the basement level. Approximately 7300 Mg (8000 tons) of construction material was involved in the collapse. Four construction workers were killed and 20 others were injured. Numerous design and construction deficiencies were uncovered during the subsequent investigation, including violations of a number of standards (Kaminetzky 1991). The triggering cause of the collapse was punching shear in the slabs due to premature removal of forms. Low temperatures during construction contributed to the very low compression strength of the concrete at the time of the collapse, only about 4.8 MPa (700 psi).

The 1973 Skyline Plaza collapse at Bailey's Crossroads in Fairfax County, Virginia, is another example of catastrophic failure due to premature shoring removal (USDC/NBS 1977). Fourteen construction workers were killed, and 35 others injured, in a progressive collapse involving 27 stories (see Chapter 7).

The most costly construction accident in the history of the United States was due to premature loading of insufficiently cured concrete. On April 27,

1978, a hyperbolic reinforced concrete cooling tower at a power plant in Willow Island, West Virginia, collapsed during construction, killing 51 workers. The top 1.5 m (5 ft) of the outer shell around the entire circumference of the cooling tower fell inward, carrying with it the four-level erection system, work platforms, and safety nets. All 51 construction workers on the platforms fell over 50 m (168 ft) with the erection system and construction equipment.

The jack-up erection system, designed to be supported on the cooling tower as it was being constructed, was a patented system that had been used successfully on many previous projects (Figure 11.4). The system was not adequately secured to the construction, however. The thorough investigation conducted by the National Bureau of Standards came to the disappointing conclusion that inadequately cured concrete and lack of a few bolts in the formwork–tower connections caused the bolts in the most recently placed concrete to pull out of the concrete shell. Human judgment errors placed construction loads on the shell before the concrete of the uppermost lift— placed the preceding day—had gained adequate strength (Lew 1980, USDC/ NBS 1982b). Apparently, this unsophisticated mode of failure is possible even in the most technically sophisticated construction projects.

Some important lessons were learned from the Willow Island cooling tower collapse. A similar failure at a cooling tower under construction in Washington State in 1981 killed only two workers. The accident was a local collapse of part of a jack-up form system. Unlike the continuity of the Willow Island system, which contributed to the totality of that collapse, the erection system used in the Washington State project involved 60 gang forms that were each independently attached to the shell, with no connections between forms (ENR 1981a).

The requirement of temporary support for cast-in-place concrete is, in itself, a source of construction accidents. Formwork and shoring system collapses are all too common. All types of formwork systems—timber, steel, and even air-supported forms for thin shell concrete dome construction—have failed. Usually these failures are sudden, causing injuries and deaths to construction workers.

A system of formwork for the reception of wet concrete at a height above the previously constructed floor is not the most stable structure unless carefully braced. The weight is almost entirely at the top and is supported by a array of posts that are not rigidly connected to the form at the top or to the floor at the bottom. Considerable resistance against lateral sway can be provided if the columns are placed at least a day ahead of the floor. The unit labor cost of placing the concrete may be somewhat increased, but the small expenditure is inexpensive insurance.

Shoring posts supported on a lower completed floor can be assumed to have equal and uniform bearing. However, forms for the lowest slab level are often supported on "mud sills" that are not on solid ground, usually on backfills recently placed with great probability of softening from the flow of water, either natural runoff or wash water from forms or from truck mixers. Unequal

Figure 11.4 Failure of the scaffold and work platforms due to overloading of insufficiently cured concrete killed 51 workers on a cooling tower construction project in Willow Island, West Virginia, in 1978. (Courtesy of the Occupational Safety and Health Administration.)

settlement of the sills seriously disarranges the designed equality of post reactions, with a good possibility of overloading the posts that do not settle. High posts without lateral bracing will buckle and break apart. Of course, the use of bent shores or out-of-plumb installation of shoring can contribute to the tendency toward buckling.

Vibration of the concrete and movement of construction equipment can easily impart sufficient lateral load to cause stability failures in unbraced shoring. Such stability failures often lead to progressive collapse of large areas of shoring. In addition, the wind exposure of many projects during construction may introduce lateral and uplift forces.

Fatalities involving formwork and shoring collapses are found throughout the engineering literature since the early days of concrete construction. As projects have gotten larger, so have the failures. The record of cast-in-place concrete construction during the past decade shows that lessons from the past have not been adequately learned. A chronological review of just a few of these cases will illustrate the need for greater attention to critical temporary structures.

In 1981, a steel form for a retaining wall at a Ohio River hydroelectric plant failed, dumping 19 m³ (25 yd³) of concrete on construction workers from a height of 4.3 m (14 ft). Two workers were killed and 10 injured (ENR 1981b).

On April 15, 1982, 12 construction workers and a state inspector were killed when the Riley Road Interchange Freeway Ramp in East Chicago, Indiana, collapsed (USDC/NBS 1982c). An additional 18 workers were injured in this failure, which involved collapse of an entire 55-m (180-ft) span, all but 6 m (20 ft) of another 55-m (180-ft) span, and 41 m (135 ft) of a partially completed span (Figure 11.5). The failure occurred prior to post-tensioning of the cast-in-place superstructure, during which time all the construction loads were being carried by the falsework system. According to the investigation conducted by the National Bureau of Standards, the falsework system was designed improperly. Reused shoring materials were cited, along with inadequate bracing at the top of shoring towers. The collapse was triggered by cracking of concrete pads that supported a shoring tower. Workers had earlier reported cracks in the 1500 by 1500 by 300 mm (60 by 60 by 12 in.) pads that were reinforced only with nominal wire mesh reinforcement (ENR 1982b, Ratay 1987).

In November 1982 the timber falsework supporting one span of a reinforced concrete overpass under construction at Elwood, Kansas, collapsed, killing a woman state inspector and injuring eight workers (ENR 1982d).

One worker was killed and six injured in Tampa, Florida, in 1984 when 370 m² (4000 ft²) of the top floor of a retail arcade collapsed. The wood floor formwork failed while the concrete was being placed. The dead worker was crushed under 540 Mg (600 tons) of wet concrete. Her body was pulled from the rubble by her husband, who was a construction supervisor on the project. A subsequent investigation by OSHA cited an unstable shoring system due

Figure 11.5 Twelve construction workers were killed and 18 injured in the 1982 collapse of this freeway ramp in East Chicago, Indiana, during construction. (Courtesy of Raths, Raths & Johnson, Inc.)

to inadequate bracing of shores and installation of shores that were not plumb (ENR 1984, 1985a).

In October of the same year in the same state, one worker was killed and seven injured when shoring collapsed at a building under construction for Florida Atlantic University, causing the workers to fall 12 m (40 ft) (ENR 1985c).

Thirteen workers were injured in a 1985 failure in El Paso, Texas, at a reinforced concrete garage under construction. Inadequate bracing of shoring led to a progressive collapse of a 7.6 by 18.3 m (25 by 60 ft) section (ENR 1985b). Also in El Paso, a bridge under construction collapsed in June 1988 due to faulty scaffolding or defective forms, killing one worker and injuring seven (ENR 1988c).

In 1989, a 15 by 23 m (50 by 75 ft) section of plywood forms for a garage deck in Los Angeles, California, collapsed. One worker was killed and 13 injured. The cause was inadequate bracing of shoring (Korman, Ichniowski, and Rosta 1989). That same year in Los Angeles, five workers were injured when a 360-Mg (400-ton) cast-in-place concrete bridge girder collapsed while falsework was being dismantled to lower the girder into its final position (ENR 1989b).

In August 1989 a 30-m (100-ft)-long overpass over the Baltimore–Washington Parkway near Laurel, Maryland, collapsed during construction. Nine workers and five motorists were injured in this dramatic falsework failure. The bridge had not yet been post-tensioned. Screw jacks at the top of the shoring were not of the capacity specified. In addition, other shoring details and imbalanced loads created complex load paths during construction (ENR 1989c, 1990a).

A 61-m (200-ft)-long temporary truss supporting falsework for a bridge over the Mississippi River near St. Paul, Minnesota, collapsed and fell 28 m (90 ft) in April 1990, killing one worker and injuring several others. The falsework was for the crown of a concrete box-section arch, one of two such twin arches, each spanning 169 m (555 ft). The temporary truss supported the falsework over a navigation channel (ENR 1990e).

Six construction workers were injured in 1991 when formwork collapsed at a three-story science building in Buffalo, New York on the Amherst campus of the State University of New York. A 100-m^2 (1100-ft^2) section of the roof slab collapsed due to an inadequate shoring and form system (ENR 1992b).

In September 1991 a 38-m (125-ft)-long section of timber falsework for a Los Angeles freeway interchange collapsed, falling 23 m (75 ft) onto an active six-lane freeway. Fortunately, the freeway was not occupied at the time, and there were no injuries. The falsework had supported a concrete span as the concrete was placed and for six weeks after it had been post-tensioned. The falsework was scheduled for dismantling. Apparently, an unexpected load imbalance contributed to the collapse (ENR 1991d, 1992a).

The United States is not alone in its experience with failures of temporary structures. In 1985 a 350-Mg (385-ton) steel box girder bridge fell in Germany, injuring several workers when a temporary pier collapsed (ENR 1985f). Numerous failures of formwork and temporary structures occur in developing countries, such as Far Eastern and Southeast Asian countries (Hadipriono and Sierakowski 1988). Even in Japan, where such accidents on large construction projects are rare, a recent formwork collapse killed seven workers and injured 13 others. In February 1992 this accident in Japan's Kanagawa Prefecture occurred during concrete placement on the second floor of a gymnasium when 726 Mg (800 tons) of material fell into a swimming pool on the building's ground floor (ENR 1992c).

These collapses of temporary structures usually involve very unsophisticated modes of failure. Nothing new is hardly ever learned from the subsequent investigations; there are nearly always many precedents.

While formwork may represent over half the cost of construction of a reinforced concrete structure, there is a tendency to construct it quickly and with minimal care. Formwork and shoring is "temporary," and standard practice is to accept a higher risk for temporary structures. This attitude has contributed to the many costly failures involving temporary structures used to protect or support a permanent work during construction (Ratay 1987). Such failures involve not only concrete formwork and shoring, but also cofferdams, sheeting and bracing for excavations, underpinning, roadway decking, construction ramps, runways and platforms, and scaffolding. The safe design and construction of falsework requires the same degree of care as the design of the finished structure. Construction safety is compromised when temporary supporting structures are designed without guidance by qualified technical professionals.

Similar problems arise when steel frames, timber structures, and precast concrete facilities are erected with temporary field connections. These temporary connections may have less capacity, especially under unusual loads, than the designed final connections. Thus while erection is taking place, the assembly may have less strength than the finished assembly. This condition is unfortunate, particularly when the construction process often introduces loads that are larger than those anticipated by the designer, as discussed in Section 11.4.

Missing bolts along with other erection procedural errors contributed to the August 1979 collapse of the timber arch roof over the Rosemont Horizon Stadium in Chicago, Illinois (see Section 5.2). Five construction workers were killed and 16 injured. Stockpiled decking on the roof and general overall stability also were cited by the OSHA investigators. But the principal triggering cause was the temporary nature of the connections. Over half of the specified bolts were missing. Of the 966 bolts needed for the connections already installed, only 444 were in place. Of these, 338 had no nuts. Those nuts that were installed were only fingertight. The roofing subcontractor indicated that bolts were intentionally omitted to facilitate subsequent construction (ENR 1979). The next year, concrete stands under construction collapsed at the same Rosemont Horizon Stadium, dumping 31 Mg (34 tons) of concrete to the ground. The failure again involved inadequate temporary connections, this time poor-quality welds (ENR 1980a).

The 1986 steel frame failure in Los Angeles, California, discussed in Section 11.4, involved temporary connections. The principal cause of this collapse, which killed three construction workers, was excessive loading from stockpiling of structural steel. However, a contributing factor was that the erected portion of the frame was held together by temporary connections with less strength than that of the designed final connections (ENR 1986a).

In May 1990 three ironworkers were killed by collapsing structural steel at a building project in Columbus, Ohio. Erection bolts were overloaded while the workers were pulling a beam into position. The bolts involved were not those that were to be used in the final designed connection. When the erection bolts sheared, 14 steel members failed in a progressive collapse (Korman and Giannone 1990).

Two ironworkers were killed at the Denver International Airport project in February 1992, when a column fell due to a failed temporary weld at the column base. The column was being pulled into a plumb position when temporary tack welds failed. Both workers were tied to the structural steel, one to a beam and one to the column that fell. When the column began to fail, the construction worker on the beam grabbed for his colleague, and his own belt slipped off the end of the beam. Both workers then fell 7.6 m (25 ft) with the steel column (ENR 1992d).

11.6 STABILITY PROBLEMS WITH INCOMPLETE STRUCTURES

Possibly the most dramatic structural failures during construction are those resulting from a lack of stability. The designer conceives of the structure as a completed entity, with all elements interacting to resist the loads. Stability of the completed structure depends on the presence of all structural members and connections, including floor and roof diaphragms. During construction, the absence of floors and walls reduces the resistance to lateral forces, and, in the case of compression members, even for vertical loads.

During the process of construction, the configuration of the incomplete structure is constantly changing and stability often relies on temporary bracing. The safe assembly of a well-designed framework of timber, steel, or precast concrete requires awareness that the components by themselves are not necessarily stable. At no stage in the assembly is the factor of safety equal to that of the completed structure unless temporary support is designed carefully to ensure safety.

Analysis of the stability requirements for these irregular, incomplete, and constantly changing assemblies presents a challenging problem to the most capable structural engineers (Figure 11.6). Unfortunately, the provision of temporary bracing is often left to construction workers who may not have the necessary expertise.

Figure 11.6   Erection of a precast concrete building requires constant attention to temporary bracing to maintain stability of the incomplete structure. (Courtesy of C. Prussack.)

Many one-story buildings fail during erection, where columns are slender and erection is so rapid that temporary sway bracing is omitted. As is often the case for catastrophic trench cave-ins, the fatal attitude is: "If we work fast enough, bracing is not necessary."

Masonry walls under construction are not stable until they are tied to roof and floor diaphragms. Additional stability problems in masonry wall construction exist when work platforms are supported on one side of the wall with no exterior bracing. In downtown Edmonds, Washington, a 9-m (30-ft)-high masonry wall under construction collapsed in 1984. It was only one wythe thick and had no bracing. No one was injured, and the masonry contractor explained the collapse by saying, "We just got some unexpected wind gusts."

It is quite revealing that such events are nearly always characterized as "freak" or "unexpected" circumstances. Certainly, the frequency of these failures would be reduced if predictable natural hazards were expected and accommodated by design.

In 1986, two construction workers were killed in Atlanta, Georgia, when an inadequately braced concrete block wall collapsed on them in a windstorm (ENR 1986d). Two workers were killed and 14 injured in New York City in 1990 when an unbraced concrete block wall fell. The 4.6-m (15-ft)-tall by 30-m (100-ft)-long wall was being constructed atop an existing one-story warehouse. No building permit had been obtained for the vertical expansion of the building (ENR 1990d).

Catastrophic stability failures frequently occur with little warning and are usually the result of imbalanced or lateral loading. Many roof structures fail prior to the installation of bracing or sheathing in the plane of the roof. Erection failures of timber trusses and arches are fairly common. Lack of temporary bracing to replace the completed roof deck in holding the long, tall, slim members in equilibrium or an accidental blow from erection equipment are the usual causes.

Seven concrete girders tumbled over like dominos on a highway construction project near Seattle, Washington, in 1988. The 26-m (85-ft)-long cast-in-place girders each weighed 35 Mg (38.5 tons). Diaphragms that would have provided stability were not yet in place (ENR 1988b). In 1991, a temporarily braced steel box girder slipped from supporting concrete tees and fell onto a busy street in Hiroshima, Japan. The girder, part of an elevated rail system, was 65 m (213 ft) long and weighed 54 Mg (60 tons). When it fell it flattened a number of vehicles and killed 14 people, including five construction workers (ENR 1991b).

Figure 11.7 shows a stability failure of a radial arch dome structure under construction in Alexandria, Louisiana, in 1964. The 73-m (240-ft) span dome was framed with 36 glued-laminated timber arches, supported by a tension ring on 9-m (30-ft)-high columns around the perimeter. All the arches were in place with purlins attached, ready for the installation of a wood fiber concrete deck. With the failure of two of the four erection cables connecting the tension ring to the 3-m (10-ft)-diameter compression ring at the top, the compression ring rotated and the entire roof collapsed within the arena. A temporary pipe shore for the compression ring had been removed an hour before the collapse.

In 1958, an 11-story welded steel frame in Toronto, Canada, collapsed due to inadequate bracing (Allen and Schriever 1973). The building, 66 m (215 ft) long and 20 m (65 ft) wide, was framed with columns spaced at 6 m (20 ft) and girders spanning the entire width. Connections had been welded through the ninth floor, and all of the 1680-Mg (1850-ton) frame was erected with temporary cable bracing. Wind bracing across the building was provided with girder-to-column moment connections. In the longitudinal direction, deep concrete spandrels were to be provided with the floor construction, which had not yet been started. With no one at the site, some wind action caused the top two stories at one corner to fall and the entire frame then collapsed in the longitudinal direction (Figure 11.8). Not a single piece of steel in the wreckage could be salvaged, although only one of the several hundred welds had failed. Without the concrete spandrels, the very light longitudinal tie beams at each floor level in the face of the walls did not have enough rigidity to brace the 11-story columns.

In 1987 the north stands addition of the University of Washington football stadium in Seattle collapsed during construction. The addition is a steel-framed

(a)

(b)

Figure 11.7 Collapse of a radial timber arch dome in Louisiana in 1964. The collapse occurred prior to installation of the roof diaphragm. (Courtesy of the Architecture and Engineering Performance Information Center.)

Figure 11.8 Failure of an inadequately braced steel frame during erection in 1958, Toronto, Ontario, Canada. (Courtesy of Federal Newsphotos, Ltd., Canada.)

structure providing seating for 20,000 people. The structure is over 52 m (170 ft) high and includes nine bents of steel framework, each supported by four 711-mm (28-in.)-diameter steel pipe columns filled with concrete on cast-in-place concrete piles. At the time of the collapse, the pipe columns were not yet filled with concrete. Lateral loads in the completed structure are resisted by braced frame action in one direction, and by crossbracing and diaphragm action of the seat plates and metal roof deck.

The sudden collapse sequence of this structure is documented in Figure 11.9. Inadequate temporary support was the most probable cause of failure. In this case, lateral loads were not necessary to cause failure. There was no indication of either wind or seismic disturbance. Without bracing, the cantilevered structure was unstable under gravity loads alone.

The design for the completed structure was found to be adequate. An incomplete system of temporary guying cables was cited by investigators as the critical deficiency (WJE 1987). Some guy cables had reportedly been removed on the morning of the collapse to facilitate progress of the steel erection. Fortunately, there was some warning that this collapse was imminent. An alert construction superintendent cleared the construction site and prevented loss of life and injury.

In Brattleboro, Vermont, 4600 m² (50,000 ft²) of a 28,000-m² (305,000-ft²) warehouse under construction collapsed in a 1980 windstorm, killing one

(c)

(b)

(a)

Figure 11.9 Collapse of a steel stadium project during erection due to inadequate temporary bracing. Seattle, Washington, February 1987. (© John Stamets, Seattle, Washington.)

worker and injuring four others. Workers were in the process of installing guy cables to brace the steel skeleton, but erection was proceeding at a faster pace than bracing installation (ENR 1980c).

A 610-m (2000-ft)-tall guyed television transmission tower collapsed in 1988 in Colony, Missouri, while three construction workers were replacing the steel structure's crossbracing members. All three were killed. The tower had stood for over a year, but collapsed when members were removed for replacement (ENR 1988e).

Preengineered metal building systems are susceptible to collapse during construction when erection procedures developed by the industry are not implemented. Usually, these failures involve insufficient or nonexistent bracing (Sputo and Ellifritt 1991).

Minor structural failures take place every day because of insufficient or nonexistent temporary bracing. A small building in a town in the Pacific northwest is shown in Figure 11.10 immediately following such a failure. The building was a church, built with donated materials and volunteer labor. The roof trusses were toe-nailed to the top plates of the supporting walls, and all workers went home for the weekend without installing any form of lateral bracing. The inevitable failure was not newsworthy; it did not even appear in the local newspaper. But the economic loss to the parties involved was relatively substantial. The cumulative effect of recurring minor failures of this kind has an impact on overall costs of construction.

Figure 11.10 Collapse of unbraced roof trusses.

When construction workers are present, an otherwise insignificant stability failure can become very serious. In January 1993, three construction workers were killed at the site of a small hotel expansion project in Post Falls, Idaho (Figure 11.11). The two-story light-frame addition was located on a site with severe wind exposure. When a strong wind began, the workers went onto the roof to attempt to secure it. A section of about 185 m² (2000 ft²), complete with trusses and sheathing, was lifted off the wall framing along with the three workers. The entire section was carried intact through the air over 60 m (200 ft) and crushed the workers when it came to rest in the parking area. A contributing factor may have been that the exterior wall sheathing was not yet in place, so internal upward wind pressure was able to combine with uplift on the exterior roof surface to overcome the dead load. The existence of the roof sheathing was detrimental, since it simply presented the wind with a large uninterrupted surface area.

Extreme natural hazards (wind and earthquake) always cause a disproportionate degree of damage to projects under construction. Some of these failures are stability problems due to incomplete diaphragms or bracing. Other failures result from the complex distribution of unanticipated torsional stresses in irregular structural configurations, as the uncompleted structure responds to lateral loading from windstorms or seismic events.

Construction sequencing is extremely important in preserving the stability of incomplete structures. The catastrophic collapse of a Denver viaduct in 1985 occurred when eight 50-Mg (55-ton) girders were placed on incomplete pier tables (ENR 1985e). One worker was killed and four were seriously injured. This failure was apparently the result of the contractor's misunderstanding of the proper construction sequence.

The responsibility for developing a construction sequence is currently the subject of some controversy. Some engineers believe that the designer should be responsible for developing and supervising the construction sequence; others advocate making this the responsibility of the contractor or construction manager. Most contractors prefer to retain the freedom to be creative in the development of a construction sequence. The contractor traditionally maintains control of the means and methods of construction. Sequencing has a major influence on construction costs. Thus innovative approaches to sequencing may produce profits.

Recent failures, however, suggest that competent technical advice is required for the safe sequencing of construction. A possible compromise may be found in requiring the designer to develop at least one acceptable sequence, in consultation with a construction professional. This sequence should be communicated clearly to the construction superintendent for the project. If the contractor wishes to propose an alternative sequence, it should be prepared by qualified engineering professionals and should be submitted to the designer for approval. Critical stages in the approved sequence should be accompanied by adequate field inspection, preferably by the design engineer's representative.

(*a*)

(*b*)

Figure 11.11 Three construction workers were killed when a portion of this incomplete roof was lifted off the structure in a wind storm in Post Falls, Idaho, 1993.

Figure 11.12   L'Ambiance Plaza, Bridgeport, Connecticut, 1987. Failure of this lift-slab project killed 28 construction workers. (Courtesy of the National Institute for Standards and Technology.)

Frame instability was a definite factor contributing to the severity of the 1987 collapse of the L'Ambiance Plaza lift-slab project in Bridgeport, Connecticut (Figure 11.12). Construction of the shear walls was lagging behind the erection of the columns and lifting of the slabs. Several theories have been advanced regarding the triggering cause of this catastrophe, which killed 28 construction workers. Most investigators agree, however, that, whatever the triggering cause, the failure would have been limited to a local area, had the structure been more stable at the time (Moncarz et al. 1992).

The L'Ambiance Plaza collapse is discussed in considerable detail in Section 8.5. A review of this case illustrates the many interrelated conditions existing during the construction phase of a project that contribute to the potential for failure.

11.7   RENOVATIONS, ALTERATIONS, AND DEMOLITIONS

Stability problems occur frequently in alteration work. When portions of an existing structure are removed, the structure may become unstable. This is particularly true when floor or roof diaphragms are removed. Some historic conservation projects involve the preservation of only the street facade of tall nonreinforced masonry buildings (Figure 11.13). This type of project requires experienced professional judgment. Until building owners become aware of the potential hazards associated with such projects, stability failures are likely to continue.

The actual load distribution paths in existing aging buildings are difficult to predict prior to the commencement of alterations. Uneven foundation settlements over years of use can cause redistribution of loads, to the point that "nonstructural" partitions may, in fact, be carrying substantial dead and live loads. An experienced forensic engineer once noted that many older buildings are standing today, despite deteriorating conditions, simply because they have ". . . gotten in the habit of doing so." It can be disastrous to unwittingly disturb the mechanism that the structure is using to enable it to stand.

Changing times often require alteration of physical conditions and change in occupancy of buildings. Building codes carefully spell out the necessity for building permits based on application to remove or modify any structural element, or to introduce new loads on existing structures. It is remarkable, however, how little awareness exists of the importance and care needed when altering an apparently safe structural assembly.

In 1987 and 1988, two aging buildings in New York City collapsed while they were undergoing illegal alterations—"repairs"—executed by incompetent construction workers. In each case, neither building permits nor inspections nor engineering expertise was retained by the building owners. Both projects involved work on the foundations of the buildings. Both failures resulted in fatalities, when load-bearing walls were undermined. One collapse, a six-story building in Manhattan's garment district, killed the building's owner and injured 10 others. In the other, the owner's 13-month-old baby was killed in the collapse of a five-story townhouse (ENR 1987d, 1988d).

Knowledgeable expertise is essential to establish the location and function of critical structural members before alterations are initiated. The Los Angeles, California, Elks Club was being altered in 1929, and a door opening was cut through a "partition." The partition actually turned out to be a 4.6-m (15-ft)-deep reinforced concrete girder spanning 18 m (60 ft). The entire floor area collapsed. Two steel trusses were added to replace the lost support over the room below.

Brownstone residences in New York City and similar buildings in other cities have been altered to form separate apartments. One such project in 1963 involved combining two adjacent houses into four modern apartments on each of the four floors. During the renovation, without advising the architect, it was decided to lower the first floor, originally at stoop level, to match the sidewalk and reduce the cellar height to a minimum. The 200-mm (8-in.) brick party wall at each floor had a 38 by 89 mm (2 by 4 in.) flat timber sill under each level of floor beams. In effect, there was a 38-mm (2-in.) timber separating the masonry at each story. The outside walls had been chased out to receive the beams at the lower level. A four-beam section of floor was cut loose and

Figure 11.13 Historic conservation project in Washington, D.C. Temporary support for the unreinforced masonry wall is required when floor and roof diaphragms are removed for renovations.

the operation of prying out the beams, which did not have fire cuts in the party wall, acted like a 6-m (20-ft)-long crowbar, twisting the wall. The entire building collapsed, providing space for a much-needed neighborhood parking lot.

Unique structures may present unusual vulnerabilities to failure during maintenance or alteration work. The Lake Washington Floating Bridge failure in Washington State in November 1990 is discussed in detail in Section 4.2.2.b. During construction, a turbulent storm caused water to enter the concrete pontoons through cracks and holes. Water from the hydroblasting process used to remove the old concrete deck had also been allowed to accumulate temporarily in the pontoons, and holes made by the contractor were not covered sufficiently to prevent stormwater accumulation. As a result, several of the pontoons sank and much of the deck was lost. The cause of this failure is obvious: Concrete pontoons do not float when full of water. But this example illustrates the extraordinary vulnerability of certain structures during alteration work—structures that have provided satisfactory service for many years. Those performing the work must appreciate the increased exposure to natural hazards during construction and take extraordinary caution to provide insurance against such potential failures.

Demolitions involve work that is particularly risky. In 1982, three workers were killed and three injured when a 10.4-m (34-ft)-high brick wall fell. The collapse was associated with the demolition of a two-story wood frame building in Sanford, Florida. At the time of collapse the building was just a shell, with both the floors and the roof removed. A sudden wind gust collapsed the unbraced wall (ENR 1982c).

When explosives are used, accidental damage to adjacent structures from falling debris or vibrations is common. Bystanders can be injured by falling or dislodged debris. Predicting the exact time, sequence, and direction of collapse relies on the judgment of experienced demolition experts. Minor errors in calculation, involving inaccurate assumptions of strength or stiffness of the existing structure, can bring about serious accidents or cause damage to adjacent structures. In one recent demolition project in Pittsburgh, Pennsylvania, a steel mill collapsed 30 minutes ahead of schedule, as workers were cutting out the last row of columns. Fortunately, there were no serious injuries, even though a considerable amount of debris fell into a busy street (ENR 1993b). Such accidents serve as regular reminders of the hazards inherent in renovation and demolition activity.

Prestressed concrete structures require special expertise when undergoing alterations or demolitions. A construction worker was crushed by falling floor slabs in a nine-story prestressed concrete parking garage demolition project in Washington, D.C., in December 1989. At the time of the collapse, prestressing tendons were being cut (ENR 1990b). Owners and operators of post-tensioned prestressed concrete buildings have unknowingly caused distress in facilities by cutting through tendons for installation of electrical floor boxes. It is imperative that the operators of such facilities appreciate the stresses that are locked into the structures—stresses that can be life threatening if they are released at inopportune times. Several references are available, provided by the prestressed concrete industry, to guide those who are undertaking renovation or demolition work in prestressed concrete structures. Usually, an engineered shoring system is required to support the structure temporarily while detensioning is taking place. When accurate as-built construction documents are not available, the response of the structure is unpredictable. Explosive failures are possible, especially in unbonded post-tensioned concrete structures.

11.8   OTHER CONSTRUCTION HAZARDS

Construction accidents involving equipment failure and procedural errors are common. Sometimes these accidents result in failure of the structure or components of the structure. Examples include improperly executed connections, improper placement of prefabricated elements, and erection procedural errors, including crane accidents. Also common are falls and hazards associated with confined spaces, such as suffocations, explosions, and construction fires. Many construction workers are electrocuted each year, and many suffer less catastrophic injuries, such as back injuries from heavy lifting.

As noted previously, falls are the leading cause of disabling injury and death in the construction industry. Falls represent 23 percent of all construction injuries and one-third of all fatalities (Potts 1991). As is the case for accidents in general, most falls are preventable. Most are due to carelessness, ineffective or nonexistent protection, and defective scaffolding.

Transportation projects such as roadway or bridge construction projects in hazardous work zones result in many deaths and injuries each year. Workers in highway and heavy construction projects are killed and injured at a greater rate than building construction workers, partly as a result of exposure to unsafe traffic conditions (Korman 1992).

Accidents involving moving construction vehicles cause construction site injuries and fatalities each year. In June 1992, for example, a dump truck driver was killed at a highway project in Clearwater, Florida, when he drove his truck beneath a partially completed overpass. The bed of his truck was still in the raised position; he had just dumped a load of fill. After delivering his load, the driver then attempted to exit the site by a different route, forgetting to lower the dump box. As he drove under the overpass, the raised truck snagged a 41-Mg (45-ton) precast concrete girder. The girder had been placed on pier caps but had not yet been secured by bolts. When the truck was approaching the overpass, inspectors on the site began yelling at the driver, but he apparently did not hear their warnings. The girder toppled onto the cab of the truck and crushed the driver (ENR 1992e).

Certain facilities, such as grain storage elevators, are especially susceptible to explosions due to accumulations of dust or other materials. Construction work in these environments must be done with care, as the work often involves potential sources of combustion, such as welding and electrical construction equipment. Air compressor explosions on construction sites are also common and have injured workers and pedestrians.

Most electrocutions and explosions injure or kill only one person, so they do not receive national media coverage. Multiple deaths are widely reported, however, and nearly always lead to better safety standards for hazardous construction activities. Such was the case when three workers were killed in an explosion at a Milwaukee, Wisconsin, tunneling project in November 1988 due to a buildup of methane gas (Tarricone 1990).

Crane accidents are reported regularly in the construction literature. Nearly every major city in the United States has experienced fatalities and serious injuries due to construction accidents related to lifting operations. Bystanders, either motorists or pedestrians, are often among the victims, as construction materials or portions of the collapsing equipment fall to the street level.

A detailed listing of the crane accidents of the past decade is beyond the scope of this book, but the list would include many, many accidents causing one or two fatalities each, about as many serious injuries, and considerable damage to partially completed construction. There are also many stories of "close calls," where quick action by operators or fortuitous circumstances prevented catastrophe. Among these is an accident in Seattle, Washington, in 1985, when a 61-m (200-ft)-long boom section of a tower crane being raised at the twenty-fifth-floor level of a high-rise building broke loose and dropped. The crane operator rode the cab 100 m (325 ft) down to the ground without serious injury, although damage amounting to $1 million was done to the construction below.

Numerous fatal crane accidents in New York City led to the adoption of city regulations requiring special certification of crane operators, increased protection over pedestrian paths and city streets, and limitations on the hours of lifting operations over streets. Other major cities have adopted similar regulations following accidents in their jurisdictions. San Francisco, California, experienced a tragic accident in November 1989 when a tower crane toppled in the financial district, killing five people and injuring 21 others.

Crane accidents may involve mechanical equipment failure or operator errors. Common operator errors include overloading and errors in setting up or dismantling the equipment. Overload indicators, although helpful, are not infallible. The operator's expertise is critical, especially when unusual conditions, such as wind or loads of unique configuration, are present. The erection, jumping, and dismantling of tower cranes is a risky operation, testing the skill of crane operators and their knowledge of their equipment. The 1989 San Francisco accident described earlier occurred during a jumping operation. In April 1987, three construction workers were killed in Dallas, Texas, when a 12-story tower crane collapsed. The crane was being dismantled and failed due to an imbalanced load; the counterweights had not been removed to balance removal of the boom sections.

There are also many accidents involving incompetent use of mobile cranes and small hoists. Usually, these result from failure to level the equipment or failure to ensure adequate support. Overturning often occurs simply because the operator did not extend the outriggers properly for sufficient stability. In many of these cases there are serious injuries, and fatalities are not uncommon. Casualties almost always include the operator of the equipment.

11.9   STRATEGIES TO IMPROVE CONSTRUCTION SAFETY

As long as structures are constructed by human beings using imperfect equipment, materials, and procedures, failures are likely to continue. Many of these failures will occur during the process of construction, endangering the lives of construction workers. Construction is an inherently dangerous occupation.

Improvements can be made in the safety record of the construction industry, however (Levitt and Samelson 1993). Federal, state, and local ordinances have achieved documented reductions in avoidable construction accidents. Data collection and educational awareness programs are also expanding. The collection and dissemination of information will assist in reducing the incidence of recurring construction accidents (Khachaturian 1985).

The Occupational Safety and Health Administration (OSHA) and the American National Standards Institute (ANSI) are continually refining construction site safety standards and regulations for improved work zone safety. OSHA has recently released a new standard to protect workers in confined spaces, for example. *Confined spaces* are defined as work areas that have limited means of egress, are subject to accumulation of toxic or flammable contaminants, or have an oxygen-deficient atmosphere. This new standard is expected to prevent 54 fatalities and 5000 serious injuries annually from accidents involving suffocation, burns, and explosions.

Economic pressures and time constraints sometimes take precedence over safety considerations, with disastrous results. Yet just a well-timed simple reminder of the danger may bring about renewed diligence. OSHA's popular "Fatal Facts" information program has been a successful way to get the message out into the workplace. These one-page handouts review actual incidents of construction site fatalities, complete with a drawing of how the accident occurred. The simple message is clearly communicated at site safety sessions (ENR 1993a).

Some of the quality assurance/quality control (QA/QC) strategies currently under discussion have the potential to reduce structural failures during construction. Most of these strategies encourage greater involvement of the designer in the construction process and more involvement of construction professionals during the design phase. The goal is to anticipate problems before they occur.

Promising QA/QC proposals include constructability reviews during the design phase and a renewed emphasis on inspection procedures (Hinckley

1986). Constructability reviews are a joint effort by the designer and a construction consultant. Many failures result from the cursory review of change orders necessitated by unbuildable design details. Constructability reviews seek to uncover these problems prior to the construction phase. Such problems as site access, spatial constraints, conflicts in contract documents, construction sequence problems, temporary support requirements, and component incompatibilities can be resolved during these reviews. Bringing a team together early in a project can reduce some of the confusion that may later lead to accidents. The "partnering" approach to project delivery shows some promise, as each party is encouraged to contribute to the successful performance of all other parties.

The renewed emphasis on construction inspection by qualified independent professionals is encouraging. Experiments with self-inspection by the contractor have not been generally successful. Quality and safety are improved when the designer is involved or represented in on-site inspection. A representative of the design firm is most qualified to interpret initial performance of the structure, to check for conformance with design intent, and to verify field conditions. There is no question that vigorous inspection contributes to reduction of accidents and pays economic dividends. Jacob Feld often stated: "The love of money is the root of all evil; but the fear of losing it is the root of good inspection." (Feld 1964)

As the global community becomes more interrelated, opportunities will arise to learn from other countries regarding construction safety. In some countries the design engineer plays a more active role on the construction site. This fact was illustrated in a report of a failure of a bridge in Latvia in 1986. A load test was being conducted on the 40-year-old steel arch bridge. The load test called for 14 gravel-loaded trucks to be driven onto the bridge. Just as the last truck drove on the bridge, it collapsed, killing 10 workers and engineers, including the engineer who designed the load test. The design engineer was deliberately standing under the bridge during the test. Standing under the construction is a time-honored local engineering custom, intended to indicate the engineer's confidence in his work and his responsibility for it (ENR 1986e). Adopting such a policy in the United States might result in improved construction site safety as well as a general trend toward conservative design!

In France, Belgium, and Holland, technical control bureaus review and approve all phases of construction for major projects. These technical bureaus, such as the French SOCOTEC, are financed by insurance companies that provide coverage to the owner for partial or total collapse for a period of 10 years after completion. The clear focus of responsibility makes members of the control board more accountable for safety during construction and for safety of the completed project. The board represents the insurance company, and that company will suffer direct economic loss if failure occurs. The technical control boards are independent of governmental supervisory agencies, thus operating with greater efficiency, and are responsible only to the private insurance companies. This process has been quite successful. In France, SOCOTEC took on 12,000 projects in 1983, representing one-third of the total volume of construction (Ross 1984).

Japan has achieved an enviable safety record, according to some sources, especially on larger construction projects. Part of this record may be due to an entirely different organization of the construction industry, which is much more centralized and heavily regulated than in the United States. A few large construction companies divide up the large projects in a procedure that would be considered unfair trade practice in the United States. These companies are comprehensive organizations, with employees who represent all design and construction professions. The entire project is organized from the start with close participation of all parties that will be involved in the design and construction. Thus opportunities for miscommunications are reduced. The project team is assembled prior to any major decisions, and the team has a "family" attitude regarding successful delivery of the project. Safety is viewed with great diligence. Highway construction projects may assign more workers to direct traffic and watch for hazards than are actually doing the work. Major construction sites are immaculate. A serious accident is probably a career-ending incident for supervisory personnel. An injury or death is treated as if the victim were a member of the family, and an accident is viewed as the responsibility of the entire "family," a great sorrow and an embarrassment to all involved. This feeling of sorrow and shared responsibility is stronger than the desire to shift blame to other parties that is so evident in the frag-mented U.S. construction industry.

Although there may be benefits to the Japanese approach, there are also disadvantages. The U.S. organizational system provides a much greater opportunity for innovation and the freedom to approach each new project with an entirely reorganized team. The open competition in the United States and low-bid approach generally produces substantially lower construc-tion costs. These benefits of the U.S. system are envied by the Japanese, and despite the regulation and centralization, not all construction accidents have been eliminated in Japan, as evidenced by several examples already included in this chapter. In fact, when small projects are included, the overall safety record of Japanese construction may be no better than the U.S. record, according to some sources. A recent journal article notes a crisis in the Japanese construction industry, an inability to attract young people into the industry (Peterson 1992). Although the industry employs 10 percent of the Japanese workforce, only 4 to 6 percent of high school and college graduates are selecting jobs in construction. Undesirable working conditions are often cited as the major factor. Construction workers have to work longer hours for less pay than workers in other sectors of the economy. And according to this source, casualties are high, with over 1000 Japanese construction workers killed every year since 1989. Such statistics are difficult to verify, as failures and accidents are not discussed as openly in Japan as they are elsewhere. The point is that we have much to learn from each other, and the expanding internationalism of the construction industry will provide valuable opportunities to share information that may improve safety.

The U.S. construction industry is confronting the challenge of improving safety on several fronts. In 1991 the Construction Industry Institute (CII) announced a coordinated program to achieve a 25 percent reduction in accidents by the year 2000. The CII is an organization that includes 45 large owners of constructed facilities and 47 major contractors. Data collection is under way so that educational efforts can be prioritized. Efforts such as this from within the industry are a promising development. They may be even more successful than regulations imposed and enforced by governmental agencies.

11.10 REFERENCES

AEPIC, 1985. *Excavation Failures Conference,* Architecture and Engineering Performance Information Center, College Park, MD.

Allen, D., and W. Schriever, 1973. "Progressive Collapse, Abnormal Loads, and Building Codes," in *Structural Failures: Modes, Causes, Responsibilities,* American Society of Civil Engineers, New York.

ASCE, 1987. *Civil Engineering,* American Society of Civil Engineers, New York (April).

Carper, K., 1984. "Limited Field Inspection Versus Public Safety," *Civil Engineering,* American Society of Civil Engineers, New York (May).

Carper, K., 1987. "Structural Failures During Construction," *Journal of Performance of Constructed Facilities,* American Society of Civil Engineers, New York (August).

CECS, 1989. *Excel,* Center for Excellence in Construction Safety, West Virginia University, Morgantown, WV (January).

CECS, 1990. "The Challenge of Education and Training in Construction Safety," *Excel,* Center for Excellence in Construction Safety, West Virginia University, Morgantown, WV (May).

Committee to Investigate the Coliseum Roof Failure, 1978. *Report of Committee to Investigate the Coliseum Roof Failure,* City of Hartford, Hartford, CT.

Dinges, C., 1991. "Construction Site Safety Back in Focus," *Civil Engineering,* American Society of Civil Engineers, New York (July).

Dowding, C., 1996. *Construction Vibrations,* Prentice Hall, Upper Saddle River, NJ.

Eck, R., 1987. "Center for Excellence in Construction Safety," *Journal of Performance of Constructed Facilities,* American Society of Civil Engineers, New York (August).

Engineering News-Record, 1979. McGraw-Hill, Inc., New York (November 22).

Engineering News-Record, 1980a. McGraw-Hill, Inc., New York (February 14).

Engineering News-Record, 1980b. McGraw-Hill, Inc., New York (February 28).

Engineering News-Record, 1980c. McGraw-Hill, Inc., New York (September 18 and October 30).

Engineering News-Record, 1981a. McGraw-Hill, Inc., New York (March 19).

Engineering News-Record, 1981b. McGraw-Hill, Inc., New York (October 29).

Engineering News-Record, 1982a. McGraw-Hill, Inc., New York (February 4).

Engineering News-Record, 1982b. McGraw-Hill, Inc., New York (April 22).

Engineering News-Record, 1982c. McGraw-Hill, Inc., New York (October 14).

Engineering News-Record, 1982d. McGraw-Hill, Inc., New York (November 25).

Engineering News-Record, 1983a. McGraw-Hill, Inc., New York (October 27).

Engineering News-Record, 1983b. McGraw-Hill, Inc., New York (November 10).

Engineering News-Record, 1984. McGraw-Hill, Inc., New York (August 2).

Engineering News-Record, 1985a. McGraw-Hill, Inc., New York (January 3 and January 17).

Engineering News-Record, 1985b. McGraw-Hill, Inc., New York (May 2).

Engineering News-Record, 1985c. McGraw-Hill, Inc., New York (June 20).

Engineering News-Record, 1985d. McGraw-Hill, Inc., New York (October 17).

Engineering News-Record, 1985e. McGraw-Hill, Inc., New York (November 28).

Engineering News-Record, 1985f. McGraw-Hill, Inc., New York (December 12).

Engineering News-Record, 1985g. McGraw-Hill, Inc., New York (December 19).

Engineering News-Record, 1986a. McGraw-Hill, Inc., New York (January 2 and March 27).

Engineering News-Record, 1986b. McGraw-Hill, Inc., New York (May 1).

Engineering News-Record, 1986c. McGraw-Hill, Inc., New York (September 18).

Engineering News-Record, 1986d. McGraw-Hill, Inc., New York (August 21).

Engineering News-Record, 1986e. McGraw-Hill, Inc., New York (October 23).

Engineering News-Record, 1987a. McGraw-Hill, Inc., New York (February 5).

Engineering News-Record, 1987b. McGraw-Hill, Inc., New York (August 6).

Engineering News-Record, 1987c. McGraw-Hill, Inc., New York (August 13).

Engineering News-Record, 1987d. McGraw-Hill, Inc., New York (November 19).

Engineering News-Record, 1987e. McGraw-Hill, Inc., New York (December 24).

Engineering News-Record, 1988a. McGraw-Hill, Inc., New York (April 28).

Engineering News-Record, 1988b. McGraw-Hill, Inc., New York (June 2).

Engineering News-Record, 1988c. McGraw-Hill, Inc., New York (June 23).

Engineering News-Record, 1988d. McGraw-Hill, Inc., New York (November 3).

Engineering News-Record, 1988e. McGraw-Hill, Inc., New York (December 8).

Engineering News-Record, 1989a. McGraw-Hill, Inc., New York (May 4).

Engineering News-Record, 1989b. McGraw-Hill, Inc., New York (May 18).

Engineering News-Record, 1989c. McGraw-Hill, Inc., New York (September 7).

Engineering News-Record, 1990a. McGraw-Hill, Inc., New York (January 4 and February 15).

Engineering News-Record, 1990b. McGraw-Hill, Inc., New York (January 11 and March 1).

Engineering News-Record, 1990c. McGraw-Hill, Inc., New York (February 1).

Engineering News-Record, 1990d. McGraw-Hill, Inc., New York (April 5).

Engineering News-Record, 1990e. McGraw-Hill, Inc., New York (May 3).

Engineering News-Record, 1991a. McGraw-Hill, Inc., New York (March 18).

Engineering News-Record, 1991b. McGraw-Hill, Inc., New York (March 25).

Engineering News-Record, 1991c. McGraw-Hill, Inc., New York (August 5).

Engineering News-Record, 1991d. McGraw-Hill, Inc., New York (September 30).

Engineering News-Record, 1992a. McGraw-Hill, Inc., New York (January 6).

Engineering News-Record, 1992b. McGraw-Hill, Inc., New York (January 27).

Engineering News-Record, 1992c. McGraw-Hill, Inc., New York (March 2).

Engineering News-Record, 1992d. McGraw-Hill, Inc., New York (March 9).

Engineering News-Record, 1992e. McGraw-Hill, Inc., New York (July 6).

Engineering News-Record, 1992f. McGraw-Hill, Inc., New York (July 20).

Engineering News-Record, 1993a. McGraw-Hill, Inc., New York (February 1).

Engineering News-Record, 1993b. McGraw-Hill, Inc., New York (February 8).

Fairweather, V., 1990. "Monitoring Vibration," *Civil Engineering,* American Society of Civil Engineers, New York (January).

Feld, J., 1964. *Lessons from Failures of Concrete Structures,* American Concrete Institute, Detroit, MI.

Hadipriono, F., 1988. "Investigative Studies of Ceiling Collapses," *Journal of Performance of Constructed Facilities,* American Society of Civil Engineers, New York (February). (See also Discussions in May 1989 issue.)

Hadipriono, F., and R. Sierakowski, 1988. "Concrete Construction Problems in Far East and Southeast Asian Countries," *Journal of Performance of Constructed Facilities,* American Society of Civil Engineers, New York (August).

Hinckley, J., 1986. "Reviewing for Potential Failure," *Civil Engineering,* American Society of Civil Engineers, New York (July).

Kaminetzky, D., 1991. *Design and Construction Failures,* McGraw-Hill, Inc., New York.

Khachaturian, N., ed., 1985. *Reducing Failures of Engineered Facilities,* American Society of Civil Engineers, New York.

Korman, R., 1992. "Taking Danger Out of Work Zones," *Engineering News-Record,* McGraw-Hill, Inc., New York (May 25).

Korman, R., and M. Giannone, 1990. "Leveling Method Cited in Fatal Steel Collapse," *Engineering News Record,* McGraw-Hill, Inc., New York (November 8).

Korman, R., T. Ichniowski, and P. Rosta, 1989. "Shoring Details Eyed in Accidents," *Engineering News-Record,* McGraw-Hill, Inc., New York (September 14).

Korman, R., S. Setzer, and H. Bradford, 1990. "Jobsite Dangers Defy Worker Protection Drive," *Engineering News-Record,* McGraw-Hill, Inc., New York (November 1).

Levitt, R., and N. Samelson, 1993. *Construction Safety Management,* second edition, John Wiley & Sons, Inc., New York.

Lew, H., 1980. "West Virginia Cooling Tower Collapse Caused by Inadequate Concrete Strength," *Civil Engineering,* American Society of Civil Engineers, New York (February).

Lin, H., and F. Hadipriono, 1990. "Problems in Deep Foundation Construction in Taiwan," *Journal of Performance of Constructed Facilities,* American Society of Civil Engineers, New York (November).

Loss, J., 1987. "AEPIC Project: Update," *Journal of Performance of Constructed Facilities,* American Society of Civil Engineers, New York (February).

McKaig, T., 1962. *Building Failures: Case Studies in Construction and Design,* McGraw-Hill, Inc., New York.

Moncarz, P., R. Hooley, J. Osteraas, and B. Lahnert, 1992. "Analysis of Stability of L'Ambiance Plaza Lift-Slab Towers," *Journal of Performance of Constructed Facilities,* American Society of Civil Engineers, New York (November).

NIOSH, 1987. *National Traumatic Occupational Fatalities,* Division of Safety Research, National Institute for Occupational Safety and Health, U.S. Department of Health and Human Services, Cincinnati, OH.

NSC, 1985. *Accident Facts,* National Safety Council, Chicago.

Peterson, A., 1992. "Japan Survey: Time for Reflection," *International Construction,* Maclean Hunter, Ltd., Barnet Hertfordshire, UK (November).

Potts, D., 1991. "Fall Prevention and Protection," *Excel,* Center for Excellence in Construction Safety, West Virginia University, Morgantown, WV (Spring).

Preziosi, D., 1989. "Setting Sights on Safety," *Civil Engineering,* American Society of Civil Engineers, New York (January).

Ratay, R., 1987. "Building Around a Building," *Civil Engineering,* American Society of Civil Engineers, New York (April).

Ross, S., 1984. *Construction Disasters: Design Failures, Causes, and Prevention,* Engineering News-Record, McGraw-Hill, Inc., New York.

Sputo, T., and D. Ellifritt, 1991. "Collapse of Metal Building System During Erection," *Journal of Performance of Constructed Facilities,* American Society of Civil Engineers, New York (November).

Tarricone, P., 1990. "Tales from Milwaukee's Underground," *Civil Engineering,* American Society of Civil Engineers, New York (January).

USDC/NBS, 1977. *Investigation of Skyline Plaza Collapse in Fairfax County, Virginia,* U.S. Department of Commerce, National Bureau of Standards, NBS Bldg. Sci. Ser. 94, U.S. Government Printing Office, Washington, DC.

USDC/NBS, 1982a. *Investigation of Construction Failure of Harbour Cay Condominium in Cocoa Beach, Florida,* U.S. Department of Commerce, National Bureau of Standards, NBS Bldg. Sci. Ser. 145, U.S. Government Printing Office, Washington, DC.

USDC/NBS, 1982b. *Investigation of Construction Failure of Reinforced Concrete Cooling Tower at Willow Island, West Virginia,* U.S. Department of Commerce, National Bureau of Standards, NBS Bldg. Sci. Ser. 148, U.S. Government Printing Office, Washington, DC.

USDC/NBS, 1982c. *Investigation of Construction Failure of the Riley Road Interchange Ramp, East Chicago, Indiana,* U.S. Department of Commerce, National Bureau of Standards, NBSIR 82-2593, U.S. Government Printing Office, Washington, DC.

WJE, 1987. *Investigation of the Collapse of the North Stands Addition to the University of Washington Husky Stadium,* Wiss, Janney, Elstner Associates, Inc., Northbrook, IL.

12

RESPONSIBILITY FOR FAILURES: LITIGATION AND ADR TECHNIQUES

The past few decades have been marked by a growing litigiousness in American society. The increase in litigation has affected nearly all segments of society. Within the design/construction industry, adversarial relationships among parties were responsible for an increase in the number and size of claims, contributing to the insurance crisis of the mid-1980s. While many of these claims involved failures, many more were founded on procedural problems: construction delays, change orders, and unclear responsibilities. Design professionals were brought into more disputes—the result of evolving theories of liability that held them more accountable for their actions and accountable to a broader circle of parties.

From 1985–1989, membership in the Forum Committee on the Construction Industry of the American Bar Association doubled to over 4000 attorneys. "Everybody wants to be a construction lawyer," according to Robert A. Rubin, of Postner & Rubin, a New York City law firm; "there's a lot of money in it" (Schriener, McManamy, and Setzer 1989). At the present time, more than 90 percent of all construction disputes are settled before they go to court, but they still are very costly and the cost is rising.

Design and construction professionals have responded to the increasing threat of litigation in several ways that seem to be producing positive results, as reflected by stabilizing insurance premiums. Some of the response has been aimed at reducing accountability for risk. Such efforts have been consistent with activities supported by other professionals, such as medical practitioners. This response, in itself, may not reduce the risk of failure or improve the quality of the constructed project, any more than it has a direct relationship with quality of health care in the medical arena, but it has relieved some untenable stress in the industry.

This category of response is seen in tighter contracts that limit or transfer liability to other parties. More contracts are including limitation of liability clauses, indemnification clauses, or disclaimers intended to shift risks to others. This is especially true for environmentally sensitive or hazardous work. Along with changes in contract language, there have been efforts to reduce the number of frivolous lawsuits. The tort reform movement, active in most states, seeks to place reasonable and consistent limits on awards for noneconomic damages (such as pain and suffering).

On another front, increasing litigation has produced changes in the industry that actually have the potential to improve the quality of the constructed project. Risk management programs have arisen with a greater focus on partnering, quality assurance, and quality control. Projects are managed better, with more complete documentation. Effective alternative dispute resolution (ADR) techniques have been introduced that promise to be less adversarial, less costly, and more expeditious than traditional litigation. Most of these ADR techniques encourage parties to focus on the project and ongoing business relationships rather than preparation for the courtroom. New insurance concepts include risk management programs and incentives that encourage teamwork. Construction lawyers with a better understanding of the industry are helping their clients write clear contracts, avoid disputes, and settle them quickly and equitably when they do arise.

The preface to an excellent text on construction litigation gives this review of the past decade:

> Perhaps the 1980s can be characterized as the era of hardball construction litigation, conducted by pitbull attorneys, and encouraged by more-than-willing clients. Now it appears there is a quiet revolution underway, shifting the focus in the 1990s to alternative dispute resolution of construction disputes.
>
> Dispute review boards have been used in 100 projects worth over $6 billion. The success rate of the DRBs has been impressive. On 21 completed projects, 63 recommendations for settlements have been made and accepted with no litigation. Minitrials and mediations, once textbook concepts, are becoming commonplace.
>
> Environmental laws proliferated rapidly in the last decade and interpretations and amendments of those laws will continue. . . . Delivery systems have also changed in the last ten years. Construction and program management, fast track construction, and design-build gained widespread acceptance. Other methods are surfacing in the attempt to streamline the [construction] process. . . ." (Rubin et al. 1992)

In this chapter the subject of responsibility for failures is discussed. Traditional litigation of construction disputes and evolving alternative dispute resolution techniques are reviewed. This is a brief and incomplete discussion, since many of these techniques are still being tested and revised. A list of references at the end of the chapter provides many case studies and more extensive information on the topics included here.

12.1 RESPONSIBILITY: IGNORANCE, NEGLIGENCE, AND THE STANDARD OF CARE

The difficulty of assigning responsibility for construction failures was discussed in Chapter 1. Most failures are the result of interrelated multiple causes. Even if a single technical cause can be identified and agreed upon, the responsible parties may be difficult to sort out given the complex project delivery system.

Standard forms of agreement in which the limits of responsibility are fixed, separately prepared by associations of architects, engineers, construction managers, contractors, subcontractors, and vendors, are seldom sufficiently coordinated to eliminate gaps or overlaps. In the typical construction project responsibility is assigned, subdivided, and endlessly fragmented. With the fragmentation of accountability, the possibility of oversights, gaps, and errors becomes greater. Unless the industry develops a workable arrangement of control and responsibility, the incidence of failure will increase and greater governmental control will result. This will not solve the problem. It will merely interfere with the work of each member of the construction team and increase construction costs.

In the recent past, efforts have intensified to reach industry-wide agreement on the division of responsibility. The latest product of this effort is the first edition of a comprehensive publication, *Quality in the Constructed Project* (ASCE 1988). In the many joint committee meetings, conferences, symposia, and roundtable discussions on this subject, there is general agreement that a clear-cut division of responsibility to the owner and the public is desirable and as yet nonexistent.

Whatever the division of responsibility, four factors must be satisfied for proper performance and acceptable results:

1. A proper design with competent specifications to provide safety and suitability to the desired use of the structure. A proper design is not necessarily a scientific solution of the problem, but a technical determination of requirements based on geographical and climatological conditions.
2. The design in all phases of architectural and engineering detail must be buildable, by available materials, labor, equipment, and local experience.
3. The contractor, and all subcontractors and vendors, must have the capability of reading the plans, must read the specifications before they

take on the work, and must have the willingness to follow the contract requirements and the desire to do a good job.

4. An inspector to watch that the contractor performs according to the contract is essential, but the inspector cannot be expected to do the impossible. The inspector cannot correct for incompetence or lack of commitment of others who make up the team.

If each of the team members does his or her job well and avoids trying to do the work of anyone else, the performance will be successful and the product will be satisfactory. Attempts to bypass this division of responsibility by design–construct contracts, or contracts that assign self-inspection to the contractor, have not eliminated construction failures. Quite the contrary is indicated by the record.

In addition to the legal confusion present in the traditional project delivery system, further confusion is resulting from nontraditional approaches to project delivery, and evolution in case law. The introduction of single-party contracts for turnkey project, design–build projects, and systems buildings has tended to blur the traditional legal distinction between construction law and consumer product law.

A new definition of "failure" has emerged, where failure is defined as improper administration or deficient management of procedures. Falling into this category are costly and complex claims for delays, and disputes involving unexpected or changed conditions. As new approaches to project delivery are introduced, there will be a corresponding evolution in theories of liability.

The confusion of liabilities is extended further when one recognizes that designers and builders have obligations to parties beyond those included in the project contracts. When physical failures occur, claims for injuries and property damage are often submitted by third parties. The doctrine of strict privity of contract, according to which architects and engineers had a legal relationship only with the owner who contracted with them to provide services, has been eroded by decades of case law. It is clear that the legal responsibility of design professionals extends beyond the party with whom they have contracted—to any other party who may be injured by their negligence. These are the cases that lead to the most peculiar lawsuits, and these are the disputes that are most difficult to resolve outside the traditional judicial system.

Assignment of culpability for a failure or object of dispute often is based on issues of ignorance, incompetence, and negligence. One is *negligent* if he or she is likely to omit what ought to be done. *Ignorance,* the state of not being informed of what is required, usually entails *incompetence,* the inability to do what is required whether one knows or does not know. When ignorance or incompetence cannot be proven, the cause of the failure may be explained by negligence.

Negligence is often found in the preparation of contract documents, especially in lack of coordination between plans and site conditions and in discrepancies between plans and specifications. Proven negligence on the part of a design professional may result in the temporary or permanent suspension of his or her license to practice. A license is revoked both as a punishment and to protect the public. The state has discretion to determine whether the design professional is properly qualified to continue in the profession. One widely publicized case in this regard was the administrative decision following the 1981 pedestrian walkway failure in the Hyatt Regency Hotel in Kansas City, Missouri (see Chapter 6). The state of Missouri revoked the licenses of the structural engineers responsible for the project design (Rubin and Banick 1987).

Neither absolute nor strict liability has generally been applied by the courts to design professionals, except in cases of extremely hazardous activities such as blasting. Design professionals are suppliers of services, not products. The design professional—architect or engineer—is required to possess and use the same degree of skill, knowledge, and ability possessed and used by other members of his or her profession. Furthermore, the design professional is charged with the exercise of ordinary care and his or her best judgment in carrying out the assignment. However, the designer does not guarantee a perfect set of plans nor warrantee a satisfactory result. The liability of the architectural and engineering professions in this regard is not at all equivalent to that applying to the producers of consumer products.

The doctrine of *reasonable standard of care* implies that one who undertakes to render services in the practice of a profession or trade is required to exercise the skill and knowledge normally possessed by members of that profession or trade in good standing in similar communities. Recent decisions also imply that the design professional has a duty to stay informed of advances in his or her discipline.

In his paper "Professional Negligence of Architects and Engineers," George M. Bell states:

An architect or engineer does not warrant the perfection of his plans nor the safety or durability of the structure any more than a physician or surgeon warrants a cure or a lawyer guarantees the winning of a case. All that is expected is the exercise of ordinary skill and care in the light of the current knowledge in these professions. When an architect or engineer possesses the requisite skill and knowledge common to his profession and exercises that skill and knowledge in a reasonable manner, he has done all that the law requires. He is held to that degree of care and skill and that judgment which is common to the profession. (Roady and Andersen 1960)

The architect or engineer is neither a manufacturer nor vendor of any product. Like the doctor or accountant or lawyer, the design professional renders purely professional services. Hence, unless the designer specifically guarantees a particular result, his or her liability in negligence (or malpractice) is similar to that of other professionals. There is no professional negligence connected with an unfortunate event unless the design professional has failed to follow the normal, accepted practices used by other members of the designer's profession.

Much of the judicial foundation for defining the liability of architects and engineers was already developed by the end of the nineteenth century (Sapers and Merliss 1988). The law recognized architects as professionals and began to apply standards of care for architects similar to those applied to doctors and lawyers. The architect was not charged with the requirement to guarantee a defect-free building. Courts determined instead that the legal test of an architect's performance would be a test of professional judgment. To defend successfully against a claim of negligence, the architect did not have to prove that he or she had designed a perfect building, but merely that he or she had acted with reasonable skill, knowledge, and judgment equivalent to that generally exhibited by other members of the architectural profession.

Sapers and Merliss (1988) note that the professional status of the architect and the architect's consequent duties were stated definitively at the end of the nineteenth century in *Coombes v. Beede,* 89 Me. 187,36 A. 104 (1896). The court summarized the issues as follows:

> The responsibility resting on an architect is essentially the same as that which rests upon the lawyer to his client, or upon the physician to his patient, or which rests upon anyone to another where such person pretends to possess some skill and ability in some special employment, and offers his services to the public on account of his fitness to act in the line of business for which he may be employed. The undertaking of an architect implies that he possesses skill and ability, including taste, sufficient to enable him to perform the required services at least ordinarily and reasonably well; and that he will exercise and apply in the given case his skill and ability, his judgment and taste, reasonably and without neglect. But the undertaking does not imply or warrant a satisfactory result.

Sapers and Merliss further observe that the *Coombes* decision placed the architect in the company of learned professionals who, in the words of a twentieth-century court, ". . . deal in somewhat inexact sciences and are continually called upon to exercise their skilled judgment in order to anticipate and provide for random factors which are incapable of precise measurement" (*City of Mounds View v. Walijarvi,* 263 N.W.2d 420, 424; Minn. 1978). "In such circumstances, certainty as to the exact result to be obtained by relying on an architect's plans or supervision is impossible, and perfection is to be neither anticipated nor expected."

Part of the justification for not applying product liability law to constructed projects is that the design professional does not have control over the entire construction process and all aspects of the complex project delivery system. Nevertheless, inability to reach the proper defendants and application of a "deep pockets" theory brings the design professional into many curious lawsuits. For example, injured construction workers who are prevented from suing their employers under workers' compensation laws may seek compensation from the designer even though the design professional does not generally have control over means and methods used on the construction site.

Specialization of technical disciplines has contributed to a confusion of accountability for design errors. In a simpler age an architect and a contractor constituted the entire construction team. Today, the erection of a structure has become so complex that one cannot expect an architect to possess such detailed engineering knowledge as to be capable of preparing complete plans and specifications for the structural work, the electrical work, the environmental control systems, and so on. These designs must be prepared by specialists in specific engineering professions. The architect's function is usually to prepare the overall design for the building, subcontract the engineering work to competent consulting engineers, and then coordinate these engineering designs with the architect's own design so as to produce a building that is both functional and aesthetically appropriate.

A problem that frequently arises deals with the responsibility assumed by an architect when he or she hires competent engineers to prepare the various engineering designs. If the engineering designs prove to be inadequate, the owner or a third party may seek to hold the architect responsible on the basis of negligent design.

A typical source of design professional liability claims is found in questions regarding responsibility for "supervision," "observation," or "inspection" during construction. The meaning of *supervision* in the common language differs somewhat from the accepted meaning in construction. When supervision is defined as having a general oversight of the work, or as inspection of the work, there is no conflict or disagreement in meaning. When defined as equivalent to superintendence, however, which implies that the supervisor is in charge of the work, giving direction as a manager, there is a distinct conflict. Having charge or managing the work entails a type of control that would not be agreed to by the contractor or accepted by the professional inspector. The contractor has traditionally maintained control over means and methods of construction. The architect and engineer do not typically tell the contractor how to do the work, except for those operations where a particular sequence or procedure is critical.

To avoid the potential conflict of authority, whether intended or not, it would be wise to eliminate the use of the word *supervision* from the designer's responsibility, and call the service by its correct names, *inspection* or *observation*. There is then no confusion regarding the necessary single control of means and methods of construction; supervision is the province and duty of the contractor. Inspection is the province and duty of the designer, usually delegated to a representative, as a service to protect the economic rights of the owner and the welfare of the public. The inspector is on the job to see that the owner receives what the owner has bought. Because much of the work is eventually covered from view, inspection must be provided at each step of construction.

There is another facet of field service that can best be defined as technical coordination, or even technical control. Included is the continual need for clarification of the construction documents, the explanation of meanings, and the fitting together of details involved in separately specified items, sometimes performed or supplied by different subcontractors. The field coordination of installations for the various services, such as plumbing lines, ventilation ducts, and electrical systems in modern buildings, entails a major expenditure of technical assistance.

Many of the examples of structural failure described in earlier chapters could have been avoided if proper field inspection by qualified design professionals had been provided. Unfortunately, such inspection service is usually not provided at the very projects where the need is greatest, that is, where appropriate contractor control is minimal or even nonexistent, and where funding is inadequate from the start. In work based on documents with incomplete details, when minimum planning fees are combined with lack of sufficient field control, the stage is always set for something to go wrong.

Reliance by many owners on the inspection service of local building inspectors, usually on the argument that such service has been paid for by the building permit fee, is of questionable value. This is especially true in smaller jurisdictions. The cry in the technical press is for the designer to inspect his or her own work, and such a cry is loudest immediately following a significant construction failure. Building codes and standards are gradually adding requirements for certification by the designer that the finished project complies with the intent of the approved drawings. If provided honestly, such inspection can significantly reduce the number of construction failures (Carper 1984). The implementation of such certification requirements, however, must recognize that the designer is not in control of all aspects of the project delivery system.

One area of increasing concern for design professionals, both in the United States and abroad, is exposure to claims involving responsibility for accidents on the construction site. (See Chapter 11 for many examples of failures during construction due to unsafe construction practices.) A recent directive in the European Community (EC) makes architects and engineers in Europe responsible for construction site safety. The Architects' Council of Europe has gone on record stating that architects are not educated to recognize unsafe conditions, nor are they under contract to provide safety programs. The council's position is that responsibility for health and safety on sites must remain with contractors, who have control of the construction site and of means and methods of construction. The Architects' Council has begun a lobbying effort to convince EC member nations to rectify the misconceptions that led to this directive.

Frequently, similar claims are brought against design professionals in the U.S. Usually, these claims are based on unsafe construction practices that were existing on the site during an inspection visit by the design professional. There is some evidence to suggest that should an architect or engineer bring an unsafe condition to the attention of the contractor, the design professional may increase his or her exposure to liability for all unsafe conditions on the site throughout the course of the project. Although such increased liability exposure is clearly unfair, given that the designer does not maintain control of the site, such claims have caused design professionals to be wary of giving advice to contractors, even when human lives are at stake. During the 1980s, some professional liability insurers actually advised their clients to stay away from construction sites, simply because of the potential exposure to liability for construction site safety. This is not in the interest of public safety, nor of construction quality. The licensed design professional has an ethical obligation to express concern for safety of construction procedures, but the legal responsibility for maintaining safety must continue to rest with the contractor or construction manager who controls the site.

For some time the general contractor has been responsible for the methods and procedures used to erect the project, albeit in conformity with the architect's plans and specifications. The general contractor is typically responsible for construction methods, techniques, practices, and procedures. He or she hires the various subcontractors and assumes (by contract) legal responsibility for the work of these subcontractors.

In many current projects, construction managers have assumed this role, and the role of contract administration, scheduling, sequencing, and so on. Construction management is a new profession. The liabilities of the construction manager have not been fully established but are evolving through legal precedence and the introduction of formal contracts developed by the Construction Management Association of America (CMAA), the American Institute of Architects (AIA), and other professional organizations.

Failures in completed structures, although still too frequent, are much less common than when steel bridges were being sold complete with design by nontechnical salespeople in open price competition. Sometimes those who promote the design–build single contract forget this historical fact. The very existence of an independent licensed engineering profession is based in the potential conflict of interest inherent to single contracts. When separate accountability does not exist, additional safeguards must be provided in the interest of public safety.

The problem with the design–build contract is simply stated: Who is ultimately responsible for the health, safety, and welfare of the public? And has that person been judged competent by the state through appropriate licensing procedures? The construction client is not the only party to whom the design profession maintains a responsibility. Professional licensing and the professional society's code of ethics obligate the designer to serve in the public interest. The architect's and engineer's licenses allow them the right to practice in exchange for protection of the public welfare. This concern is particularly valid when speculative projects are constructed. The owner or developer may have no interest whatever in long-term performance or safety of the project.

In such cases the design professional may be the only party who is ethically motivated to consider these critical issues.

The vendor or material supplier also bears some liability for successful performance. If materials or products delivered to the project prove to be defective or not in accordance with the specifications upon which they were sold, the supplier is liable. Similarly, if a manufacturer or vendor warrants that a product is fit for the purpose for which the architect has indicated it will be used, the vendor is liable for any resultant damages if the product fails to measure up to the warranties (in the absence of a valid limitation of liability clause). However, an architect or engineer may not safely rely on the manufacturer's or vendor's representations or brochures in specifying a new, untested material. Numerous cases have established that the architect-engineer is liable for the failure of such material if it is not adequate for the particular application in question. In such an event the architect-engineer may look to the vendor for indemnification. Needless to say, the designer who specifies a new material and does not test it thoroughly should require written warranties as to the sufficiency of the material for the purpose intended. In addition, a written indemnification should be obtained from the vendor. This is the only way the architect-engineer can protect himself or herself in the event of the failure of the material.

The owner's responsibility for construction failures, deficiencies, and disputes should also be mentioned. The owner/developer is often the real but undisclosed villain. This is especially true when the owner is planning a speculative-type building, as noted above, so that virtually the entire cost will be paid for out of the mortgage money. If when bids are received, the cost exceeds the owner's tight budget, there is an immediate squeeze placed on the architects, the engineers, and the contractors to find a way of reducing the cost. This price squeeze is often reflected in a completed project that represents something less than high-quality work and material. The deficiencies are most often observed in inadequately performing heating systems, deficient mechanical installations, and high maintenance costs. Even structural collapse is a possibility.

The evils brought about by excessive price cutting should be ascribed to the owner. The legal responsibility, however, never shifts from the contractors or the architects or the engineers, who undertake what are often difficult if not impossible projects, hoping somehow that they will work out satisfactorily. When such a building fails, causing either property damage or personal injuries, it is the architect-engineer or the contractor, not the owner, who ultimately bears the legal liability for the failure and who will be required to respond in damages.

In a formal report following a recent structural collapse in Canada, the investigating commission commented on the owner's contribution to the catastrophe. The report noted that ". . . the role played by the owners in the construction process places them in a key position to influence the quality of construction and thus the safety of the public."

Although the report further outlined numerous specific defects in the owner's performance, it directed its recommendations to the design and construction community rather than the owners. This was done because, in the words of the report, "owners, as such, are not governed by standards of ethical practice as are architects, engineers and other professionals" (Government of B.C. 1988).

It clearly behooves all parties to be wary of contracting for work with owners who are not committed to quality or who are not willing or able to provide sufficient funding to achieve expected results.

12.2 RISK MANAGEMENT AND DISPUTE AVOIDANCE

The best approach to dealing with disputes is to avoid them altogether. Although this may not be entirely possible, the early identification of potential problems enables the participants to manage the risks associated with construction more successfully. Much is currently being written about avoiding liability. Certainly, design and construction professionals must accept responsibility for their actions, errors, and omissions. Accountability is inseparable from professionalism. But there are proven techniques for avoiding unnecessary liability exposure.

Many of the companies that insure design professionals and construction firms have developed loss-prevention programs. Their clients are encouraged to participate in these programs through premium rebates. These rebates are a form of profit sharing, made possible through reductions in claims.

In response to the growing number of professional liability claims against medical professionals in the 1980s, at least 12 states have adopted legislation that requires at least one colleague (a professional peer) to vouch for the validity of a malpractice lawsuit. Some of these laws have been challenged successfully in the courts as limiting access to justice, but there is no question that such provisions would tend to discourage frivolous or misdirected lawsuits.

More lawyers are now trained in construction, with specialized skills for anticipating potential disputes. "In-house" counsel is now being employed even by small construction, construction management, and design firms. This reflects a recognition that legal advice at the outset of a project is important to the prevention of disputes. A lawyer can bring to the table an understanding of the proper allocation of risk and can help the various parties express their intentions clearly.

Most risk management techniques focus on contractual relationships. Risks cannot be avoided entirely, but the ideal contract allocates risks clearly, assigning the sharing of risk in such a way that teamwork is encouraged rather than confrontation.

Contractual attempts simply to shift risks to other parties have not been successful. What is needed is a clear definition of responsibility, such as that suggested in the manual *Quality in the Constructed Project* (ASCE 1988).

Each party must be willing to share equitably in the risk, but with a clear understanding of their individual accountability.

Some design firms have recently begun to include *limitation of liability* clauses in their contracts. These clauses are composed of special contract language that limits liability for errors and omissions to a negotiated amount such as the value of professional liability insurance coverage or the amount of the fee paid for professional services. When properly drafted, such clauses have been enforceable in some court cases. However, since these clauses are part of the contractual arrangement, they are only enforceable against the client. They have no limiting effect whatever on third-party claims such as those involving bodily injury.

Limitation of liability clauses in contracts imply that clients are willing to share the risk for professional errors and omissions. Construction organizations have protested that these clauses could have the unfair effect of making the contractor the "deep pocket" party in the project. Some engineers and architects also object to this practice since it has the potential to erode their authority further. Limitation of liability clauses are particularly unfair when the design professional carries an inadequate level of insurance so that the limit is set low. Because these contract clauses are relatively new, the law is unclear in many jurisdictions as to their enforceability. The design professional should be wary of relying on these clauses exclusively for protection, although their use is encouraged by some professional liability insurers.

Indemnification clauses, although prohibited in some jurisdictions, have been used successfully for projects on which insurance is not available. These clauses are especially appropriate for environmentally sensitive or hazardous work, such as toxic waste site cleanup and asbestos removal projects. It is important to remember that indemnification clauses are no better than the assets of the indemnitor or the contractual liability insurance provided to cover such indemnification clause. Unfortunately, professional liability is excluded from most contractual liability insurance coverage.

In addition to contractual provisions, risk management techniques used by design professionals apply internal and external quality assurance/quality control (QA/QC) strategies. These strategies include formal peer review of organizational operations and peer review of individual project designs. Constructability reviews may bring in construction consultants during the design phase. These reviews are generally conducted by a third party, such as an independent experienced construction management firm. The constructability review seeks for unresolved details, errors, inconsistencies, and omissions. The intent is to resolve potential problems before they escalate into costly disputes.

Another risk management approach taken by an increasing number of design firms is incorporation. Architecture and engineering firms maintain assets that are exposed to liabilities arising from their errors or omissions. A substantial claim may exceed the net worth of the firm. If the firm has insufficient insurance or assets to pay a claim, the responsible architects or engineers may discover that their personal assets are exposed. One advantage to corporate organization is the ability to write contract language that requires all claims to be made only against the assets of the corporation. This will limit the exposure of individual members of the organization.

It should be noted, however, that incorporation as a *business* corporation is not available to engineers and architects in many states. They are limited to *professional* corporations, which in many instances do not enjoy the same limited liability as do business corporations.

During the construction phase, risk management techniques include contractual provisions for continued involvement of the design professional and better management of the construction process. Contractor hiring based on merit or prequalification, rather than low bid price, is clearly another important way to reduce risks to all parties.

Preconstruction meetings can clarify technical issues and define roles and responsibilities. Every attempt should be made throughout the construction phase to foster cooperation in avoiding disputes and resolving them promptly when they do occur. Of course, willingness and capability to negotiate is critical. All parties must be committed to the project. This entails an understanding that timely resolution of disputes and a continuing harmonious relationship is to the benefit of the project. The principal advantage of successful internal negotiation is that conflicts may be resolved prior to development of a fixed adversarial position.

Several successful projects have been constructed using the concept of *partnering*. This concept is aimed at producing a "win–win" environment through owner–contractor relations that emphasize team building (Stephenson 1996). Key personnel at the executive levels meet each other prior to the development of any problems. Trusting personal relationships are established and cultivated. Workshops are scheduled to resolve problems early. The focus is on preventing problems and disputes. The rewards include cost-effectiveness, through reducing the money normally spent on dispute resolution, and the potential for a long-term commitment between owner and construction firm.

Even in the public sector, partnering can establish a team-building atmosphere immediately upon the contract award. This environment will enhance communication and minimize disputes. Standard agreements are often ambiguous and subject to interpretation. Partnering meetings can clarify responsibilities and resolve discrepancies.

The U.S. Army Corps of Engineers has completed a number of complex projects successfully using this approach. The Construction Industry Institute (CII) notes that expected benefits of partnering include improved efficiency and cost-effectiveness, increased opportunity for innovation, and the continuous improvement of quality products and services.

Partnering encourages trust, long-term commitment and shared vision. It calls for a more personal relationship among parties. While the specific arrangement may be flexible, partnering always involves a formalized approach to building trust and opening lines of communication. It is a cooperative, nonadversarial approach to contract management.

Of course, it is not possible to avoid completely all disputes within the construction industry. However, there are many ways in which the construction environment can be made less adversarial. One group that is committed to developing and encouraging effective risk management strategies is the Dispute Avoidance and Resolution Task Force (DART). Established in 1991, DART is an industry-wide partnership formed by owners, designers, contractors, and even some lawyers, aimed at reducing unnecessary disputes and litigation (McManamy 1991).

12.3   TRADITIONAL LITIGATION OF CONSTRUCTION DISPUTES

Traditional litigation in the courts is the most structured approach to construction dispute resolution. There are strict rules of evidence and procedures for reports, discovery, interrogatories, depositions, direct and cross examination, and redirect and re-cross examination.

The late Joseph S. Ward, a practicing forensic engineer and geotechnical consultant, has briefly described the process of traditional litigation:

> Traditional litigation begins with the filing of a claim. Such filing may take place well after the event has occurred. The legal profession is involved from the outset as attorneys are retained as adversaries on behalf of the plaintiff and the one or more defendants.
>
> In the typical case, the following scenario then develops. The attorneys for the defendants file an answer to the complaint and may also file a counterclaim. The plaintiff's attorney will answer the counterclaim and may also produce an amended complaint. Third parties may be brought into the litigation by the plaintiff or the defendants and claims will be made against these third parties, with the ensuing answers. There will be a demand by all parties for the production of documents, and interrogatories will be filed, with the expected answers to such interrogatories. Expert and fact witnesses will be retained by attorneys for the various litigants who will conduct their investigations. The experts will produce reports that will be distributed to all parties involved in the conflict. As part of the discovery proceedings, witnesses may be subpoenaed for deposition testimony. Motions will be made to the court by the attorneys, such as to dismiss the complaints, and the courts will rule on these motions.
>
> Finally, a court date will be set. However, with the current crowded court calendars in most jurisdictions throughout this country, a year or more can elapse before the court date is fixed, and this may be further extended by the courts.
>
> At any time during the foregoing process the parties may agree to settle their differences, and a settlement agreement may be reached. But even when the parties agree to settle, a long time usually has elapsed since the occurrence of

the event that initiated the conflict. If a settlement is not reached before the trial date, the court proceedings commence, and if settlement is not attained prior to the end of the trial, a protracted period will ensue because of the testimony of witnesses and the pleadings of the attorneys. (Ward 1989)

Traditional litigation is costly, lengthy, and complex. In some cases the cost of resolving the dispute exceeds the amount of the initial claim. Such expenditures add absolutely nothing to the value of the project, nor do they compensate the party that has suffered loss.

The cost of litigation is escalating and it is not unusual for a dispute litigated in the courts to take years before it is resolved. The average life of a design professional liability claim file is between three and four years, according to Sandra L. Nelson, an insurance professional (Nelson 1987).

Nelson notes that these costs are accrued by all parties to the dispute, primarily through legal fees and the time expended by the employees of the plaintiff and the defendants. The final award made to the aggrieved party will be a long time in coming. Even when damages are assessed and an award is made, it has been reported that an average of 70 percent of that award goes to the legal costs incurred during this entire procedure.

No one group in particular can be blamed for this alarming situation. Professionals, clients, third parties, insurance companies, and lawyers have contributed to the problem in one way or another, yet each has pointed the finger at the other, almost as if to perpetuate the conspiracy born of an authoritarian system which is, by its nature, complex and adversarial, but which is the only one we have known: litigation. (Nelson 1987)

Even attorneys have noted dissatisfaction with the current state of traditional civil litigation. A 1992 survey conducted by the Defense Research Institute, composed of 18,000 lawyers who represent defendants, reported that most defense lawyers—as well as plaintiff's lawyers and judges—support reforming the civil justice system. Of those responding to the survey, 59 percent said the civil justice system is working only "somewhat well," compared with 18 percent who felt that the present system works "very well." Major criticisms related to delays, heavy case loads, frivolous claims, the jury selection system, excessive use of pretrial discovery procedures, excessive attorney fees, and high costs for expert witnesses.

These concerns have provided impetus to the development of a number of alternative dispute resolution (ADR) techniques, discussed in the following section. Yet some cases simply must proceed through traditional litigation. Unusual cases involving untested theories of liability are useful to establish legal precedents. Third-party suits may require resolution by the courts when ADR techniques are inappropriate.

One party to the dispute may insist on litigation. Litigation may be preferred by one party in cases where legal precedents are favorable to that party, when emergency relief is sought, when the amount of money at stake is large, or when information from third parties through discovery rules is required (Pavalon 1987). Sometimes the structured rules of litigation are essential to a case. Attorneys who favor traditional litigation remind their clients that the procedures used in the courtroom, although cumbersome, include important safeguards. These procedures have evolved over centuries and are intended to promote the cause of justice.

12.4 ALTERNATIVE DISPUTE RESOLUTION TECHNIQUES

Dissatisfaction with traditional civil litigation and the adversarial atmosphere it encourages has led to the development of a number of alternative dispute resolution (ADR) techniques for the construction industry. ADR methods involve negotiation or mediation to settle a disagreement without going to court. ADR techniques can result in expeditious settlements, to the satisfaction of all parties involved in the dispute.

The premise governing ADR techniques is that a conflict can be resolved most easily immediately after the conflict has arisen rather than through the long and costly route of traditional litigation. A growing variety of ADR alternatives are available. These generally fall into the categories of arbitration, mediation, or minitrials.

Some jurisdictions are requiring mandatory ADR because of crowded court schedules. These courts will not accept a case until ADR methods have failed to produce a settlement. Bar associations and some jurisdictions, through ethical guidelines and statutes, are beginning to require attorneys to inform clients of the existence of ADR methods. Many construction attorneys now even favor the use of mediation in most cases and minitrials in at least some cases (McManamy 1991). ADR was officially recognized by the Federal government in the Administrative Dispute Resolution Act of November 1990. The law strongly recommends ADR forums where appropriate.

Although opinions of lawyers vary on ADR, nearly all attorneys advise caution in selecting ADR methods over traditional litigation. Careful consideration should be given to selecting the method that is appropriate for the particular situation. Most attorneys favor those ADR methods that are voluntary and nonbinding.

Goals and objectives should be clearly understood prior to committing to a particular method of ADR (Marcus 1988) As noted in the preceding section, some disputes simply require formal litigation because of the issues, the interplay of parties, the intransigent attitude of one or more parties, or because it would be inappropriate to compromise the particular dispute in the manner offered by ADR methods.

Several of the most common ADR methods are reviewed briefly in this section. Further discussion and case studies are presented in the references at the end of the chapter (ASCE 1991, Rubin et al. 1992).

12.4.1 Arbitration

For construction disputes, binding arbitration is by far the most often used alternative to litigation. Arbitrated construction hearings usually involve two parties who are having a contract dispute and an arbitrator they jointly choose to resolve the dispute. The arbitrator is usually someone familiar with the construction industry. Most large claims involve three arbitrators instead of one.

Accepted rules and procedures are followed to ensure that hearings are run fairly. Arbitration is more structured than some of the newer ADR techniques, but arbitration procedures are more relaxed than traditional litigation. Formal courtroom procedures may be incorporated, but many time-consuming rules associated with traditional litigation are simplified or bypassed altogether.

Hearings can be scheduled by the participants rather than waiting for unpredictable court schedules. The average time to process a construction arbitration case from filing to award was 192 days in 1986, and 164 days in cases settled without the need to complete arbitration. In some jurisdictions, court trials entail a two- or three-year wait before the case is even scheduled.

The American Arbitration Association (AAA), a nonprofit organization founded in 1926, operates from more than 30 offices nationwide to help resolve more than 46,000 disputes annually. Arbitration is used to resolve disputes in many fields, but AAA reports that the construction industry has for some time been the largest user of arbitration. It is quite common for a clause to be included in construction contracts to require dispute settlement by arbitration.

More than 32,000 arbitrators serve on AAA's construction panel. AAA holds training sessions for arbitrators, who are selected on the basis of their impartiality and expertise. It should be noted that in addition to arbitration, AAA promotes out-of-court settlements of all types, including dispute review boards and mediation.

Among other relaxed procedures, discovery is not normally employed in arbitration. This can accelerate the process of dispute settlement. Each party can only speculate as to the probable testimony of opposing witnesses. The elimination of discovery has both advantages and disadvantages. The principal advantage is the reduction in time.

Rules of evidence are also much more informal than in traditional litigation. In the courtroom, the rules governing admissibility of evidence, and the way it is introduced, are quite stringent. In arbitration, application of the rules of evidence is left to the discretion of the particular arbitrators. This may result in reduced legal costs because less time is needed to prepare and present a case.

In binding arbitration, the decision of the arbitrator is final. It can be appealed only under very unusual circumstances—where the procedures were clearly flawed, fraud is alleged, or conflict of interest on the part of an arbitrator can be proved. This aspect of arbitration also has some disadvantages, but it definitely shortens the duration of most disputes, since the case cannot be retried after arbitration has concluded. The arbitrator is usually required to submit a decision within 30 days of the close of hearings.

All this informality makes for a less adversarial environment than the courtroom. Arbitration helps to preserve ongoing business relationships. The proceedings are usually private, so the reputations of involved parties are not damaged. The final decisions and awards are confidential, known only to the arbitrators and the parties at dispute. A transcript of the hearings is made only when requested by one of the parties, and files can be obtained only by subpoena.

Although arbitration can be very useful for solving complex disputes quickly and equitably, there are also a number of disadvantages to this approach. If there is no contract clause requiring the parties to go to arbitration, the case may end up in court anyway, even if one party desires arbitration. Of course, a contract clause committing all parties to use arbitration may preclude the use of other, more appropriate ADR methods unless all parties agree to explore other methods.

One party can use delay tactics to prolong arbitration hearings to the point where the time-saving advantage is lost. Even after an arbitration award is made, there is no way to force payment without going to court. When arbitration is selected, certain rights present in litigation are given up. There is no jury, and the decision cannot be appealed. There is no right to broad discovery. The "structureless" character of the hearings may allow for the introduction of duplicative, irrelevant, or prejudicial evidence that would not be permitted in court. These rights are given away in the interest of saving time and money. Traditionally, arbitrators were not required to give any reasons for their decisions, but new procedures encourage them to do so, especially for large complex cases.

Edward A. Hannan, an attorney and partner in the Milwaukee firm Godfrey, Trump & Hayes, has warned of the potential disadvantages of arbitration:

> Arbitration has been widely praised as a speedy, inexpensive alternative to litigation for resolving a wide range of disputes, including disputes arising from construction failures. However, the *quid-pro-quo* for reducing litigation costs is a waiver of substantial safeguards available to litigants in civil actions. The waiver of procedural safeguards must be taken into account when assessing the "costs" of arbitration. The true cost of arbitration may be imposition of an unjust award which cannot be remedied thereafter. . . .

> While arbitration may be appropriate for resolving minor claims not involving substantial dollars, it is questionable whether arbitration is a sound alternative to litigation for resolving claims arising from building failures. . . . (Hannan 1986)

Hannan further notes that there are significant procedural limitations to the arbitration remedy. These pose many significant risks that may exceed the benefits, including:

1. *The risk of the inability to join third parties.* In civil litigation, the rules of procedure typically grant the parties a broad right to implead other parties whose conduct may have contributed to a loss. In the context of a claim arising from a building failure, there are typically many potentially culpable parties: contractors, subcontractors, product manufacturers, product suppliers, construction managers, owners and design professionals. Since potential damages in building failure cases are typically high and usually beyond the capability of any one defendant, the right to join other parties and their insurers becomes significant. Unless all parties to the design and construction process have mutually agreed to submit all disputes to arbitration, there is a significant risk that the truly culpable party may not be required to join in the proceeding.

2. *No right to discover the opponent's position.* Discovery is a method of compelling an adverse party to disclose facts within its own knowledge, information or belief or to disclose and produce documents within its possession to enable a party to defend an action. Federal Rules of Civil Procedure provide that parties may obtain discovery regarding any matter, not privileged, which is relevant to the subject matter involved in the pending action. In submitting a claim to arbitration the parties may lose the litigation right to fully discover the basis of the opponent's claim. In arbitration, parties not only lack an absolute right to discovery, but also discovery is typically limited to depositions. This is a significant loss of the right of access to information available in civil proceedings: an arbitration hearing can become a "game of blind man's bluff" rather than a fair contest with basic issues and facts disclosed ahead of time.

3. *Absence of formal rules of evidence.* The fundamental purpose of the evidentiary code is to ensure that reliable, trustworthy facts are presented to the finder of facts. Civil actions are always governed by an evidentiary code which defines not only what evidence may be received but also the manner in which such evidence is to be introduced. Arbitration proceedings, however, are typically not governed by rules of evidence. The AAA rules provide that the arbitrator shall be the judge of the relevancy and materiality of offered evidence, and that a conformity to legal rules of evidence shall not be necessary. There is a significant risk that irrelevant, inflammatory or otherwise untrustworthy evidence (such as hearsay) may be admitted. Arbitrators usually err on the side of admitting any and all evidence, regardless of its trustworthiness to avoid charges of excluding relevant material evidence. The effect of this is that otherwise inadmissible evidence may become the basis for the arbitration award. The award may be determined on the basis of inaccurate and untrustworthy evidence.

4. *Lack of explanation of the basis for the arbitration award.* Upon conclusion of the evidence presented by both sides, the arbitration hearing is declared closed and under AAA rules, the arbitrator has 30 days in which to make an award. The award is the final decision determining the submitted dispute. AAA merely requires that the award be in writing and signed by the arbitrator or a majority of the arbitrators. The general rule is that arbitrators are not required to give reasons for making the awards. The lack of explanation for the basis for the award significantly hampers a party who may wish to challenge an award in the courts.

5. *The risk of a limited scope of judicial review of arbitration decisions and awards.* As a general rule, arbitration awards may only be vacated on very

narrow grounds. The modern tendency of the courts has been to insulate an arbitration award from judicial scrutiny, provided that the arbitrators appear to be impartial and that the hearing itself appears to have afforded a fair opportunity to present evidence. Absent a transcript of the proceedings, a reviewing court must presume that the evidence was adequate to support the award. Also, conflicts of interest are very difficult to establish. Ironically, avoidance of the cost of a transcript is sometimes touted as a cost-saving benefit favoring arbitration over a civil trial. Yet arbitrators do not even need to state their reasons for the award they have made. The practical effect of the rules strictly limiting the scope of judicial review of arbitration awards and the rules imposing a near insuperable burden of proof upon the challenging party, is to render the arbitrator's decision final in most cases. In cases such as building failure cases, where the damages are high, the risks of loss can be unfairly shifted to nonculpable parties without further redress.

6. *The direct costs of a complex arbitration dispute can be substantial— approaching or even exceeding the costs of litigation.* These costs include an AAA administrative fee, arbitrator's fees, rental of conference rooms, etc. In complex construction cases, the direct costs of an arbitration proceeding may equal or exceed the direct costs incurred in civil litigation.

Although arbitration is widely touted as a "speedy" and "inexpensive" method for dispute resolution, there are substantial risks to be assessed in considering whether to agree to employ the method. Not only can arbitration proceedings in a complex case be lengthy, but also delays may be encountered by way of disagreements over the arbitration panel or the filing of amended claims. The hearings themselves can be protracted if a party elects to submit volumes of evidence and many witnesses; arbitrators tend to receive all evidence. (Hannan 1986)

In summary, Hannan notes that parties to arbitration may save costs through avoiding the expense of depositions. However, they also face the risk of surprise that is usually diminished by prelitigation discovery. Parties to arbitration may avoid the cost of transcribing the hearing but at the risk of sacrificing the right to judicial review of an unfair or partial award. Parties to arbitration may avoid some cost through participation in hearings not encumbered by formal rules of evidence. However, they must also bear the cost of a decision predicated upon otherwise inadmissible evidence.

Finally, although arbitration litigants present their disputes to persons purportedly having "technical" experience, frequently arbitrators have little or no "judicial" experience. However, these judicially-inexperienced arbitrators will often be called upon to decide difficult legal issues concerning the meaning of contracts, the standards of conduct, the reception of evidence and the scope or measure of damages. Moreover, their decisions will often be final. (Hannan 1986)

Eugene I. Pavalon, past president of the Association of Trial Lawyers of America, also cautions against the blind acceptance of mandatory arbitration as preferable to litigation, for many of the same reasons given by Hannan. Pavalon further notes that:

> [C]ommercial arbitrators are discouraged from lengthy written opinions, largely to withhold grounds on which a losing party can seek a reversal in court. While arbitrators typically do not disclose the reasons for their decisions, it is precisely through written appellate court opinions, accompanied often by lengthy explanations, and through publicly reported lower-court opinions that standards and norms to guide private and commercial behavior are promulgated. (Pavalon 1987)

The advice from these two attorneys is worthy of consideration. Their concerns have been extensively noted here, because the relaxation of safeguards to which they refer is present to some degree in all of the ADR methods. However, it is also evident that arbitration and the other ADR methods are growing in popularity. Historically, the courts have frowned on nonjudicial procedures, but now many jurisdictions prefer arbitration because courts are so overloaded and judges are unfamiliar with the construction industry. Between 1979 and 1989 the number of arbitration cases nearly tripled because of the desire to avoid lengthy and costly legal action. Binding arbitration remains by far the most widely used form of alternative dispute resolution in the construction industry, and many users report satisfaction with the results (Riggs and Schenk 1990).

According to most attorneys, parties with complex disputes involving large amounts of money are probably best advised to seek satisfaction within the shelter of the judicial system if their contract permits (Rubin et al. 1992). The advantages of arbitration over litigation diminish as the size of the claim rises. Some of the other ADR methods may be more useful in complex cases, since the judgments are nonbinding.

12.4.2 Mediation

Mediation is facilitated negotiation. It is the least formal of the methods available after negotiation has failed. Mediation is a structured negotiation that should probably be the next step above internal negotiations.

In mediation, an objective third party may render a nonbinding decision based on his or her opinion of how the opponents would do in court. However, it should be noted that a decision by the mediator is not absolutely required. Unlike an arbitrator, the mediator has no authority. The mediator can only suggest, persuade, and recommend. The mediator's influence is in the fact that this is the disputing parties' last chance to retain control over the decision-making process before the case escalates to binding arbitration or the courts.

Mediation is a flexible system that takes dispute resolution away from the lawyers and the courts. This process diminishes the importance of legal knowledge, although the mediator may indicate what the parties may expect in arbitration or litigation. In mediation, the emphasis is on solving the problem, not on the formalities of judicial review.

Among the advantages of mediation is that it is the least costly method of resolving disputes. Reasonable people who approach the resolution of disputes in good faith can settle even emotionally charged issues through a structured negotiation, which is far less costly than litigation. Another advantage is that the parties at dispute find out early whether or not mediation is going to work. Not much time is wasted in mediation even if the decision is made to carry the dispute forward into another forum. Even if no final agreement is reached in mediation, it is probable that many issues will be disposed of, so that the resultant litigation or arbitration will address a simplified list of unresolved problems (Goodkind 1988).

Associations of architects, contractors, and engineers have endorsed mediation and encouraged their members to mediate disputes. Major insurance companies are also pleased with the track record of mediation and are encouraging their clients to use this method. Many construction lawyers who are opposed to mandatory binding arbitration support mediation, according to a 1992 survey conducted by the American Bar Association's Forum on the Construction Industry.

The unique aspect of mediation is that the parties may negotiate directly rather than through an advocate representing them in an adversarial environment (Goodkind 1988). Sometimes, lawyers may serve as advocates in mediation, however. When the parties are genuinely committed to finding reasonable solutions to their dispute, the result can be a "win–win" agreement. The agreement will resolve immediate problems while enabling both sides to express their position and reach a mutually acceptable compromise. One major advantage is the confidentiality of the mediation forum.

Mediation should be considered as a first step in most construction disputes. All other methods involve obtaining a decision by a third party. Most other methods involve one party seeking to win by causing the other party to lose. Mediation allows the disputing parties to uncover their own problems and devise their own solutions.

Of course, the skill of the mediator is crucial to the success of this process. The mediator acts as a facilitator, providing a forum that helps the parties craft a mutual agreement. The mediator works to stimulate discussion but should not be overly concerned with formalities. When evidence and witnesses are brought into the proceedings, the purpose is to persuade the other party, not the mediator. The mediator has no enforcement power and imposes no decisions. Mediation is and must be voluntary. The mediator's power rests on continuing consent, not authority (Nelson 1987).

The ideal mediator is one who can clearly articulate issues and focus discussion. The mediator is a catalyst for settlement as trust is built among the parties at dispute and agreements are forged. To be successful the mediator should have some technical knowledge of the topics involved in the dispute as well as conflict resolution skills. According to some experienced mediators, the method seems to work more effectively when no attorneys are present

(Ward 1989). This is because attorneys are typically trained to be advocates in an adversarial forum, and most attorneys are more comfortable in such an environment. Mediators use conciliatory skills. In general, attorneys can be a hindrance to the forging of compromise agreements. Of course, there are exceptions to this rule; some attorneys have conciliatory skills. In addition, attorneys are helpful in drafting the settlement documents after the principals themselves have agreed to a resolution of the problem.

One disadvantage of both arbitration and mediation is a shortage of qualified arbitrators and mediators. The American Arbitration Association (AAA) has a construction industry advisory committee with nine association members who have helped to draw up guidelines for mediation, and the associations are currently promoting the committee's use and training mediators, so this situation will improve. Another disadvantage is that the parties are required to present their best case in this nonbinding forum, perhaps to the detriment of their position in later formal litigation.

While the nonbinding aspect of mediation facilitates open negotiation, there always exists the potential that one party will reject the mediator's recommendations—leading to litigation or arbitration—even after the parties have reached an agreement. Sometimes the agreement reached in mediation is contingent on other events that may not materialize, so the dispute may resurface and require further mediation or litigation at a later date.

Mediation does not work well when one party feels that its case is extremely strong or when resolution is beyond the financial capacity of one of the parties. Litigation may be the only workable avenue.

On the other hand, mediation has been used successfully for some very large and complex claims, including multiparty claims. One such case was the construction collapse of L'Ambiance Plaza in Bridgeport, Connecticut (see Section 8.5). This was a remarkably complex case involving multiple fatalities. All claims, however, were resolved through a creative and controversial global mediated settlement (Felsen 1989).

12.4.3 Mediation/Arbitration

Mediation/arbitration is a formal contract agreement to structure dispute resolution by attempting mediation first. If an acceptable settlement cannot be reached, the contract calls for subsequent binding arbitration. This agreement establishes that the first dispute resolution forum will be nonbinding and nonadversarial. The parties are given an incentive to settle through mediation, since they know that they will be subjected to a mandated decision by a third party should the dispute escalate to arbitration.

A further variation on this concept is the "100-day-document" agreement discussed in Section 12.4.7.

12.4.4 Minitrials

The minitrial is a voluntary, structured, nonbinding alternative dispute resolution process that takes place in an arena that usually resembles a courtroom, although other forums may be used. Attorneys for both sides prepare informal briefs in simple, straightforward language, then make their case before senior business representatives or neutral advisors, who then decide the case. This is the unique aspect of the minitrial. The presentations are not made before judges, juries, or arbitrators; they are made before the disputing parties themselves.

The executives who represent each party must have the authority to make settlement decisions. These management representatives need not have been close to the dispute. In fact, the method seems to work best if they have not been intimately involved with the specific dispute in question.

During the course of the minitrial, the corporate executives hear both sides of the dispute and develop a feeling for the probable outcome should the dispute proceed to litigation. Immediately following the presentations, these executives meet to discuss settlement. The minitrial is not bound by traditional rules of evidence or legally accepted means of compensation. Settlement options are far less restrictive. The disputing parties may discover a creative resolution that satisfies each party's business objectives—a "win–win" solution. One extremely important benefit of the minitrial is the preservation of ongoing business relationships that might be destroyed in the adversarial environment of litigation or arbitration.

The minitrial atmosphere results in a business-based solution rather than an imposed judicial decision. The dispute is treated more like a business problem than a legal problem. The procedure is much quicker and less formal than traditional litigation (Green 1981).

Presiding over the entire proceeding is a neutral advisor selected by both parties to the dispute. This advisor may provide guidance to the parties after the minitrial as to his or her perception of the strengths and weaknesses of each side's position.

If successful, the minitrial is clearly a way to reduce legal expenses. Some advocates of this method estimate costs at one-tenth that of litigation. The procedure has produced some rapid settlements of complex cases that would have required years in litigation. Because of the informality, the content of the deliberations relies more on facts than on points of law. Minitrials are flexible; the parties can design a procedure and schedule with which they are comfortable.

At the very least, the minitrial gives a good indication of where the parties agree and disagree and where liability is likely to lie if the case goes to trial. If the parties desire confidentiality, the proceedings can be private.

Minitrials have been very successful in resolving disputes that involve mixed questions of law and fact, where the disagreement centers on how the facts of a particular case may apply to existing settled law. Construction disputes usually involve numerous issues of fact and are therefore excellent candidates for settlement by minitrials. These complex fact-oriented cases are the very cases for which traditional litigation can be particularly expensive and lengthy. Minitrials have been used to settle complex multiparty cases, international commercial disputes, and cases involving the U.S. government (Green 1981, Henry 1988).

12.4.5 Private Judges

Private judges now serve in various capacities to provide a private forum for resolution of civil disputes through mediation, arbitration, or any other format found acceptable to the disputing parties. In some jurisdictions, crowded courts have been referring cases to this alternative forum, sometimes known as "rent-a-judge."

Usually, the company is founded by a judge or judges who have left the bench to establish an alternative to civil litigation. Individual judges or panels of hearing officers comprised of retired judges preside over the hearings. The company provides all facilities, usually in the form of conference rooms rather than courtrooms.

Formats vary from very informal mediation to minitrial, to advisory (non-binding) arbitration, or even binding decisions, depending on the desires of the participants. The judge serves as a neutral party with legal expertise. The decision or advice of this respected neutral party usually is sufficient to serve as a catalyst for settlement (Knight 1988).

12.4.6 Dispute Review Boards

This method uses a dispute review board (DRB) that is established at the outset and maintained throughout the project to serve as a settlement forum for disputes arising between owner and contractor (ASCE 1991). The typical DRB is composed of three members: one chosen by the owner, one chosen by the contractor, and a third chosen by the first two members. The third member usually serves as chair of the group. Each board member is an experienced construction professional, and all three must be acceptable to both the owner and contractor. DRB members are selected for their knowledge and technical expertise in the type of project to be constructed. Prestige of the board members selected is important to the success of this method.

The American Arbitration Association has prepared standard procedures for setting up DRBs. These procedures and case histories of successful DRB projects are available from AAA.

The first meeting of the DRB occurs very early in the project and is usually held at the project site. Board members are supplied with copies of all progress reports. The board holds regular meetings at the site to follow construction progress and keep aware of potential areas of conflict.

When a dispute arises, the parties are first encouraged to settle it themselves through internal negotiations. If the dispute is not resolved by negotiation, it is referred to the DRB, which holds a hearing on site. Hearings are informal and use flexible procedures. Following the hearings, the board meets privately and makes recommendations to the owner and contractor. These recommendations are not binding. If one party is not satisfied, that party can request the board to hold another hearing and review its first decision. If the party is still in disagreement with the recommendations, arbitration or litigation can be pursued.

While the board's recommendations are not binding, it is highly unlikely that they will be rejected. This is because the DRB recommendations and all supporting material will usually be made available to a subsequent court or arbitration panel unless the contract specifies otherwise. Neither body would be likely to ignore the conclusions of a board of recognized experts, particularly when that conclusion was reached at a time when the dispute was fresh and the evidence was immediately available. The very existence of the DRB motivates the owner and contractor to try harder to settle disputes promptly and equitably.

The DRB approach works very well for disputes that arise during a project, especially those that would otherwise lead to project delays. A timely decision by the DRB enables the project to go forward. Litigation and arbitration often occur long after the dispute. These methods require that evidence must be reconstructed from old records, some of which may be incomplete or missing altogether. The DRB is a more timely solution that deals with disputes at the time they occur, when the facts and the people involved are readily available and their memories are fresh. The parties at dispute may not yet be intransigent, making it more likely that an early agreement can be reached. Within the DRB environment, project work continues and relationships are maintained, despite the existence of disagreements over specific problems.

One disadvantage of the DRB alternative may be its initial cost. Provision for a DRB must be included in the initial contract. The functions of the board are specified in contract language, so it can be organized and set in operation before any disputes arise. Retaining appropriate board members and providing for their expenses is an unavoidable cost of this technique. The dispute review board alternative has been called "real-time" dispute resolution. Complex disputes are resolved as they occur, without litigation. On larger projects this capability is well worth the cost of maintaining the board.

The history of the DRB approach was in underground construction projects. The idea was first promoted by the Underground Technology Research Council sponsored by the American Society of Civil Engineers and the American Institute of Mining, Metallurgical and Petroleum Engineers. To date, the method has been used very successfully on a number of large, heavy construction projects, such as tunneling, highways, water projects, and transportation. This type of project has traditionally been marked by costly delays due to disputes, often involving unexpected circumstances. Qualified, experienced experts on a DRB can serve to resolve such disputes in a very short time.

The typical cost of a DRB is generally about $20,000 to $30,000 throughout the life of a project. This method is now being recommended for projects greater than $5 million (Shanley 1989).

12.4.7 Other ADR Techniques

100-Day Document. This alternative was pioneered by the Deep Foundations Institute Construction Round Table (DFICR). It was especially formulated for the type of project in which unexpected or changed conditions may be encountered, as in excavation, tunneling, and other below-grade construction.

The 100-day document provides by contract that either party can demand a discussion between the parties. This discussion must take place within seven days of the event that is the topic of dispute. The contractor can terminate work if the requested discussion does not occur within the seven days.

If the meeting does not produce an acceptable resolution, a mediator is retained. The mediator's involvement is agreed to beforehand as part of the initial contract language. If the mediator cannot forge an agreement between the parties within 30 days, the dispute is carried forward into either arbitration or litigation.

In the case of litigation, the issues at dispute may now be clearly stated while the facts are fresh and the relevant documents are readily available. In the case of binding arbitration, the same documents are introduced, and limited discoveries and reviews are performed. The arbitrator is required to submit a binding conclusion by the end of the 100-day period.

By stipulating time limits on the various activities, this method hastens the dispute resolution process. It is a mandatory commitment to discussion, mediation, and arbitration—in that order. This agreement is somewhat more structured than the mediation/arbitration concept discussed in Section 12.4.3 (Loftus 1988).

Baseball Arbitration. This unique method borrows from the sports world, where the technique has been used to negotiate player salaries (Rubin et al. 1992). Each party to the dispute puts its "last best offer" in writing. After a presentation of positions in arbitration, the arbitrator is not permitted the freedom to make a compromise award. Instead, the arbitrator much choose one of the two "last best offers."

This places the burden on each party to suggest a realistic settlement, for if the offer submitted by one party is considered unfair by the arbitrator, the other party's submittal will probably be selected. This method has not been used extensively, but it may be useful in cases where a fair compromise is desired. In practice, this method is not unlike dividing a candy bar between children by asking one to make the division and giving the other the opportunity to choose first.

Resolution by State Agency Panels. This concept was proposed first by medical professionals to ease the medical malpractice crisis. The idea is to remove malpractice claims from the civil court system. Instead, they would be heard by a state agency that would settle the claim. The agency would also discipline incompetent or negligent physicians. The method was expected to be faster and more consistent than traditional litigation. This approach has not been used sufficiently to judge its effectiveness. The Association of Trial Lawyers of America and other consumer advocacy groups have vigorously opposed its implementation on the grounds that a patient's right to a jury trial is eliminated.

Nevertheless, some states, including California, have proposed the creation of similar statewide dispute resolution boards to oversee all public construction contracts. If the decisions of such boards are binding, the right to a trial by jury may be denied.

Federal Government Disputes Resolution. This process for disputes on federal projects has been available for several decades. The method uses contract appeals boards made up of government employees known as administrative law judges. These judges are assumed to be knowledgeable and impartial as well as experts in construction. Decisions by the appeals board are generally considered to be binding as to findings of fact, if supported by substantial evidence, but are not binding as to conclusions of law. The decisions will carry significant weight if the dispute goes on to arbitration or litigation (Rubin et al. 1992).

12.5   TRENDS IN DISPUTE AVOIDANCE AND DISPUTE RESOLUTION

The tort reform movement is expected to result in changes in the civil justice system in the next few years. There are a number of powerful vested interests standing in the way of these reforms, but nearly all states have begun to respond to the public opinion against inconsistent and unreasonable awards. The changes are coming in a variety of forms, but the goals are consistent: curbing the number of frivolous suits, reducing the cost of the dispute resolution process, and placing equitable limits on damage awards, especially for noneconomic damages such as pain and suffering. Many of these reforms are opposed by the Association of Trial Lawyers of America. Trial attorneys and some consumer advocacy groups see tort reform as an attack against the protection of the rights of victims.

In an effort to reduce the number of frivolous suits, more jurisdictions will introduce new legal hurdles, including certificates of merit and mandatory reviews by panels of conciliators. New sanctions are now being considered for lawyers who bring unfounded suits to federal courts, and other jurisdictions are expected to follow this example. Some attorneys oppose these sanctions as misguided—an unconstitutional limit on access to justice.

The tort reform movement was given additional momentum in the early 1990s by the *Council on Competitiveness Report.* This report, completed during the Bush administration under the direction of Vice President Dan Quayle, included a number of recommendations to reform the legal system. The report called for reducing the number of frivolous suits, placing limits on noneconomic damage awards, encouraging ADR, limiting punitive damage awards, simplifying discovery processes, and limiting expert testimony.

Implementation of some of these reforms, along with the acceptance of ADR methods and other risk management techniques, has begun to be reflected in construction industry statistics. The July 1992 issue of *Liability Update,* published by Schinnerer Management Services, Inc., notes that claims against architects and engineers increased significantly in the 1980s. In 1983 there were 42 claims filed per 100 professionals insured by Schinnerer. This statistic declined dramatically thereafter and has stabilized for the years 1988–1991, down to 26 or 27 per 100 insured. However, the severity of paid claims was up significantly, to an average of $211,000 per claim.

Contracts will similarly evolve, experimenting with limitations on liability and disclaimers that shift risks. It is expected that the courts will continue to insist that these clauses be equitable, precise, and consistent; if not, they will probably be unenforceable.

Design and construction firms will adopt more effective loss-prevention programs. More design organizations will probably incorporate, attempting to avoid personal liability. Quality assurance/quality control programs will continue to be helpful in reducing losses throughout the industry. The team-building concept of partnering shows great promise for reducing the adversarial atmosphere that has been counterproductive to the industry. New alternative dispute resolution techniques will emerge, and those already in existence will be refined to streamline the process of settling disagreements between parties.

Finally, creative new ideas about construction insurance may reduce the necessity for unproductive conflict. Contractors' risk insurance and professional liability insurance premium costs skyrocketed in the mid-1980s, with lower coverage limits, more exclusions, and higher deductibles. The result was a growing number of firms going without insurance (22 percent of engineering design firms in 1990) and a movement of these firms out of markets that required insurance. Captive insurance companies were formed by engineers in some disciplines, starting with an insurance company founded by the Association of Soil and Foundation Engineers (ASFE). These captive insurance companies developed mandatory loss-prevention programs for their insured members. The insurance situation has improved recently, especially for those firms that have implemented loss-prevention programs and adopted risk management practices. As noted earlier, insurance premiums have stabilized since 1987 and are even going down as result of reduced claims.

A number of construction industry organizations are investigating the unified risk insurance concept. C. Roy Vince, an insurance professional and advocate for this concept, has noted that construction disputes are very expensive and that the funds spent to resolve them add nothing of value to the project or to society. *Unified risk insurance* is a project-based single insurance policy that provides incentives for all parties to work together to resolve disputes. The policy includes umbrella coverage that insures the entire construction team for design and construction deficiencies: owner, construction manager, architect, engineers, material suppliers, contractor, subcontractors, and subconsultants. This encourages a partnership approach. Ideally, the insurance money will be spent on solving the problems rather than on settling the disputes (Vince 1989).

Of course, the current adversarial nature of the construction industry will not improve unless attitudes are changed. James W. Poirot, past national president of the American Society of Civil Engineers, has written:

> Our nation is embroiled in a litigious society where everyone feels someone else is to blame. We must be responsible for our own actions and when our actions have created additional expense, delay or harm to other individuals, corporations, or parties, we must be responsible for rectifying the harm we created. As each party and member of a construction team recognizes that each is responsible for his own actions and is willing to develop an attitude of compromise and understanding, resolutions to disputes will be more rapid and less costly. It is all in our attitudes. Proper attitudes among all team members will mean more successful projects with fewer disputes, less legal fees and a greater respect for the members of the design-construct industry. (Poirot 1987)

12.6 THE ARCHITECT OR ENGINEER AS EXPERT WITNESS

Every incident where there is damage to life or property is followed by an attempt to pin down the responsibility. Unless the cause of failure is determined first, the search for the responsible person is a fruitless effort, accompanied by incriminations and unethical claims and counterclaims. Even with the cause determined, there is often complete disagreement about assigning responsibility.

Forensic engineers and architects contribute important factual and interpretive information as expert witnesses in dispute resolution. The activities of forensic experts and the ethics of forensic practice are not presented here; relevant information is available in several references (ASFE 1985; TCFE 1989; Carper 1989, 1990). Often, forensic experts have been instrumental in providing information that results in an out-of-court settlement. One prominent geotechnical forensic expert has noted that over a 38-year period, fewer than 10 percent of his cases reached the trial or arbitration stage (Ward 1989). It can be assumed that his contributions were at least partially responsible for achieving timely and equitable settlements.

Expert witnesses can play an important role in traditional litigation or in any of the alternative dispute resolution techniques. Indeed, forensic engineers have been at the forefront in supporting the development of ADR methods. As noted in Section 12.4, some of the ADR methods focus on technical issues—issues of fact—rather than points of law. Thus the engineering investigations become paramount to the settlement negotiations. Also the decision makers in the ADR processes may be more acquainted with construction and engineering technology than are the judges and juries comprising traditional litigation. The ethical and competent forensic expert is quite comfortable in such an environment.

12.7 REFERENCES

ASCE, 1991. *Avoiding and Resolving Disputes During Construction*, Technical Committee on Contracting Practices, Underground Technology Research Council, American Society of Civil Engineers, New York.

ASCE, 1988. *Quality in the Constructed Project: A Guideline for Owners, Designers and Constructors*, American Society of Civil Engineers, New York.

ASFE, 1985. *Expert: A Guide to Forensic Engineering and Service as an Expert Witness*, Association of Soil and Foundation Engineers, Silver Springs, MD.

Carper, K., 1984. "Limited Field Inspection Versus Public Safety," *Civil Engineering*, American Society of Civil Engineers, New York (May).

Carper, K., ed., 1989. *Forensic Engineering*, Elsevier Science Publishing Co., Inc., New York.

Carper, K., 1990. "Ethical Considerations for the Forensic Engineer Serving as an Expert Witness," *Business and Professional Ethics Journal*, Vol. 9, No. 1, Rensselaer Polytechnic Institute, Troy, NY.

Felsen, M., 1989. "Mediation That Worked: Role of OSHA in L'Ambiance Plaza Settlement," *Journal of Performance of Constructed Facilities*, American Society of Civil Engineers, New York (November).

Goodkind, D., 1988. "Mediation of Construction Disputes," *Journal of Performance of Constructed Facilities*, American Society of Civil Engineers, New York (February).

Government of B.C., 1988. *Report of the Commissioner Inquiry: Station Square Development*, Government of British Columbia, Victoria, British Columbia, Canada.

Green, E., 1981. "Resolution of Business Disputes Outside the Courts," *Corporate Counsel Review*, Texas State Bar Journal, Corporate Counsel Section, Houston, TX.

Hannan, E., 1986. "Arbitration: A Risky Method for Resolving Disputes," in *Forensic Engineering: Learning from Failures*, K. Carper, ed., American Society of Civil Engineers, New York.

Henry, J., 1988. "ADR and Construction Disputes: The Minitrial," *Journal of Performance of Constructed Facilities*, American Society of Civil Engineers, New York (February).

Knight, H., 1988. "Use of Private Judges in Alternate Dispute Resolution," *Journal of Performance of Constructed Facilities*, American Society of Civil Engineers, New York (February).

Loftus, W., 1988. "The 100-Day Document: A Better Way to Solve Construction Disputes," *ASTM Standardization News*, American Society for Testing and Materials, Philadelphia (February).

Marcus, S., 1988. "Goals and Objectives for Alternative Dispute Resolution," *Journal of Performance of Constructed Facilities,* American Society of Civil Engineers, New York (February).

McManamy, R., 1991. "Quiet Revolution Brews for Settling Disputes," *Engineering News Record,* McGraw-Hill, Inc., New York (August 26).

Nelson, S., 1987. "ADR—A Different Ritual: An Insurer's Perspective," *Journal of Performance of Constructed Facilities,* American Society of Civil Engineers, New York (November).

Pavalon, E., 1987. "ADR: Trial Lawyer's Perspective," *Journal of Performance of Constructed Facilities,* American Society of Civil Engineers, New York (November).

Poirot, J., 1987. "Alternative Dispute Resolution Techniques: Design Professional's Perspective," *Journal of Performance of Constructed Facilities,* American Society of Civil Engineers, New York (November).

Riggs, L., and R. Schenk, 1990. "Arbitration: Survey on User Satisfaction," *Journal of Performance of Constructed Facilities,* American Society of Civil Engineers, New York (May).

Roady, T., and W. Andersen, eds., 1960. *Professional Negligence,* Vanderbilt University Press, Nashville, TN.

Rubin, R., and L. Banick, 1987. "The Hyatt Regency Decision: One View," *Journal of Performance of Constructed Facilities,* American Society of Civil Engineers, New York (August).

Rubin, R., V. Fairweather, S. Guy, and A. Maevis, 1992. *Construction Claims: Prevention and Resolution,* Van Nostrand Reinhold, New York.

Sapers, C., and P. Merliss, 1988. "The Liability of Architects and Engineers in Nineteenth-Century America," *Journal of Architectural Education,* Association of Collegiate Schools of Architecture, Washington, DC (Winter).

Schriener, J., R. McManamy, and S. Setzer, 1989. "Lawyers: Whose Side Are They On?" *Engineering News Record,* McGraw-Hill, Inc., New York (March 16).

Shanley, E., 1989. "A Better Way," *Civil Engineering,* American Society of Civil Engineers, New York (December).

Stephenson, R., 1996. *Partnering and Alternative Dispute Resolution in the Planning, Design, and Construction Business,* John Wiley & Sons, Inc., New York.

TCFE, 1989. *Guidelines for Failure Investigation,* Technical Council on Forensic Engineering, American Society of Civil Engineers, New York.

Vince, C., 1989. "Unified Risk Insurance: An Update," *Journal of Performance of Constructed Facilities,* American Society of Civil Engineers, New York (November).

Ward, J., 1989. "Role of the Forensic Expert in Dispute Resolution," in *Forensic Engineering,* K. Carper, ed., Elsevier Science Publishing Co., Inc., New York.

13

LEARNING FROM FAILURES

Studying the failures of the past can be useful in mitigating the potential for future failures. Reduction of failures is particularly critical in our modern technological society. In many ways, society has become increasingly vulnerable to failures and accidents even since the first edition of this book was published.

Dense concentration of the world's population in urban centers places more lives at risk from the effects of industrial accidents. This is true not only in the highly developed industrial nations, but also in the developing countries, as evidenced by the thousands of casualties in Bhopal, India, following the methyl isocyanate gas leak of December 1984. Hazardous wastes and the transportation of hazardous materials through populated areas are problems new to contemporary society.

The technical characteristics of facilities and industrial processes have become increasingly complex, while the potential for human error has not decreased. An error made by a single person can have catastrophic implications. Human error in the operation of a modern large aircraft can cause hundreds of deaths in a single accident. A single connection failure in a modern long-span sports stadium can threaten the safety of thousands of spectators. Practices simply must be monitored and improved to ensure a greater degree of integrity in facility design and operation if the risk of catastrophic failures is to be reduced. Quality assurance and quality control strategies must also be implemented to reduce the risk of human error.

Failure examples from the past suggest specific technical and procedural improvements. Perhaps even more important, these case studies provide motivation for the acceptance of effective revisions to building codes and standards of practice. All of the decision makers (designers, builders, owners, and users) need to be convinced of the importance of failure-reduction strategies in order for the desired results to be realized. Education is the key to achieving this goal. Failure case studies are a significant component of this education.

13.1   WHY ARE THERE SO MANY FAILURES?

In January 1985, the Technical Council on Forensic Engineering of the American Society of Civil Engineers organized a workshop in Clearwater, Florida: *Reducing Failures in Engineered Facilities* (Khachaturian 1985). Professionals representing a wide variety of construction industry interests attended the workshop, which was conducted to identify reasons for the apparent increase in failures and performance deficiencies in engineered facilities and to recommend strategies for improving the quality of performance.

The general consensus among those who attended the workshop was that both the frequency and severity of failures were on the increase, as measured by the number of claims being filed against design professionals and construction companies. There was not, however, general agreement as to the reasons for the increase in performance deficiencies. Several contributing factors were suggested by the participants; these are discussed here.

A number of practicing design professionals noted concerns about loss of control over execution of the design in the field. This loss of control appears to be the result of an undue emphasis on reduction of construction time and cost. Nontraditional "fast-track" approaches to project delivery became prevalent in the 1970s and 1980s. These presented new opportunities for misunderstood communications and unclear lines of responsibility.

Competing against the costs of time is not a new challenge for designers, but the era of unprecedented double-digit inflation in the 1970s placed extraordinary pressures on design and construction schedules. Society began to define the "best" project as the one that could be completed in the shortest possible time and for the least initial cost.

During this period, a new profession came into existence—construction management. This new profession promised to deliver projects faster and for less cost. Although few question the need for better management of the fragmented and complex project delivery system, the construction management profession was not well defined in its formative years. Construction managers now play an important role in contributing to scheduling, cost control, *and* quality assurance, but in the late 1970s, many persons were practicing "construction management" who were trained only as developers or business managers. Services were not clearly delineated, and when something went wrong, these "professionals" often were neither willing nor able to accept accountability for their errors and omissions. The recognized and established professionals, the architect and the engineer of record who were licensed or registered by the jurisdiction in which the project was located, continued to bear the burden of professional liability. Architects and engineers found that their legal liabilities remained, even for those responsibilities that had purportedly been "transferred" by contract to the construction manager, and even when the architects and engineers had neither been retained nor compensated to provide the relevant services.

Competitive bidding for design services and the "low-bid-is-the-best-bid" psychology became firmly entrenched in the construction industry during the 1970s. The ethical conflicts arising from "economical" design–build package services were also identified by workshop participants as contributors to escalating failure rates.

Along with this emphasis on reducing costs and construction time came society's desire for projects of ever-increasing scale and complexity and with more demanding performance expectations. Yet the users and owners of these facilities often have little understanding of the technologies required to deliver these expectations. Building systems have become increasingly complex, so much so that it is absolutely essential to retain technically competent personnel to maintain and operate modern facilities properly. When owners are not sufficiently sophisticated to recognize this need, operational failures are inevitable.

Still other workshop participants noted that the process of designing and constructing complex modern buildings has necessitated the development of a large number of engineering design specialties. Coordinating the work of the specialists has become an increasingly difficult task and communication problems have escalated with the number of participants. Problems are developing at the interface, where the work performed by the various specialists must be successfully integrated for adequate performance. Incompatible systems result in deficient performance.

In addition, new materials and construction methods are appearing at an unprecedented rate. These new materials and assemblies are often heavily promoted to idealistic designers who are all too willing to believe the unsubstantiated claims of product manufacturers. Proprietary systems and new materials are applied simultaneously in a variety of environmental exposures without the benefit of long-term trial-and-error experience. When the new "miraculous" materials do not achieve the expected performance, the result is costly litigation, almost always at the designer's expense.

A large number of workshop attendees also expressed concerns about potential misuse of computers and engineering software. The computer is an essential tool, but it has been introduced to the construction industry only recently. The computer clearly has great potential to contribute toward resolving some of the current problems within the industry. Nevertheless, misapplication of this new tool has already caused new problems to surface.

For example, the computer can process information very rapidly. However, the designer may lose critical time for reflection as a result. Also, basic errors in concept may not be uncovered by this tool when it is used carelessly or by an incompetent designer.

The computer can be used to explore a wide range of alternatives and can relieve the designer of repetitive tasks, allowing more time for quality control, including increased participation by the designer during the construction phase. Unfortunately, the computer is instead being advertised as a tool that can merely speed up the design process, allowing for the completion of more jobs in less time.

The proliferation of specialized engineering software is also of concern. The probability exists that unqualified designers, lacking skills for appropriate verification, will misuse the software. Despite the availability of this useful computational tool, the necessity remains to consult the relevant specialized professionals who are qualified to make experience-based engineering judgments. It should always be remembered that computers do not lie awake nights worrying about the possibility of failure; computers could not care less whether people live or die. Professional judgment exercised by competent human beings is a critical link in failure mitigation.

The "litigation explosion" was also cited by some workshop attendees as a contributing factor to the increase in construction deficiencies. On the surface, one would expect that the threat of litigation would cause all participants to be more careful in executing their responsibilities. The *uncertainty of the outcome* of litigation, however, has caused many to see litigation as simply a "cost of doing business," an inevitable aspect of design and construction practice. An inordinate amount of time and effort now goes into the writing of contracts that limit or shift liability, leaving less time to devote to engineering design, site inspection, and quality control. The resulting "tighter" contracts, unfortunately, may leave certain responsibilities unassigned. Rather than exercising greater attention to quality assurance and quality control, some designers have simply bought more insurance. This practice brought about severe stress during the insurance crisis of the mid-1980s, when insurance for design professionals became unavailable or unaffordable.

13.2 WHAT CAN WE DO TO REDUCE THE FREQUENCY AND SEVERITY OF CONSTRUCTION FAILURES?

At the present time, many projects are under way to respond to the negative effects of the foregoing trends. These failure-reduction strategies are supported by a broad base of individuals and professional societies.

A number of interdisciplinary conferences on the subject of construction failure mitigation have been conducted, in addition to the 1985 ASCE workshop on *Reducing Failures of Engineered Facilities* referred to in Section 13.1. These include:

- November 1983: *Building Structural Failures: Their Cause and Prevention,* a conference held in Santa Barbara, California, sponsored by the Engineering Foundation (Gross 1986)

- March 1986: *Construction Industry Roundtable Meeting,* Kansas City, Missouri—site of the Hyatt Regency Hotel pedestrian walkway collapse (1981) and the Kemper Memorial Arena roof collapse (1979)
- December, 1987: *Structural Failures II,* a conference held in Palm Coast, Florida, sponsored by the Engineering Foundation (Bell, Kan, and Wright 1989)

These conferences identified several specific activities that could lead to an improvement in the construction industry environment. Several industry-wide projects were initiated to respond to the remarkably similar recommendations emerging from these meetings. The events were attended by professionals representing all disciplines and interests in the construction industry, including architects, engineers, contractors, developers, attorneys, and insurance professionals. As such, the conferences presented unique opportunities for an exchange of information. The interdisciplinary discussion was quite enlightening, causing each participant to become more aware of the viewpoints of colleagues from other disciplines.

As noted earlier, the convoluted project delivery system and the technical complexity of modern construction has necessitated the development of technical specialists. Some of the procedural deficiencies resulting from specialization may be resolved through enhanced communication among disciplines. In this regard, the conferences themselves were a step in the right direction.

Following are excerpts from the most specific recommendations emerging from these conferences:

- *The Need for Improved Structural Integrity.* Design concepts and details must provide ductility, continuity, and redundancy. Higher safety factors should be required for nonredundant structures. More attention should also be given to the construction process, providing structural integrity throughout this phase.
- *The Design Professional to Be Responsible for Ensuring That the Completed Building Is Safe for Occupancy.* This includes safety of the basic design, structural details, checking of shop drawings, field inspection, and verification of all structural safety related matters. The design professionals should certify that to the best of their knowledge, all safety requirements have been met, before a certificate of occupancy is issued.
- *Project Peer Review.* All buildings above a certain threshold level should have a comprehensive review of the structural design and details performed by an independent professional. Organizational peer review (where an organization is reviewed by a representative from a peer organization) should also be encouraged.
- *Better Definition and Assignment of Responsibility.* A document should be developed that sets forth the many areas of responsibility necessary for the design and construction of buildings. The document should be used

as a checklist to assure that responsibilities for all safety considerations are assigned and understood by all parties.

- *Unified Risk Insurance.* The industry should develop a blanket, all-purpose risk policy that will insure all members of the project team simultaneously under a single umbrella. Such a combined policy would help restore cooperation and teamwork in the overall construction process.
- *Better Code Enforcement.* There is a need to strengthen building departments in many jurisdictions in order to carry out the necessary regulatory functions.
- *Competitive Bidding for Design Professional Services to Be Discouraged.*
- *Improved Education.* There is a need for improved dissemination of failure information. The industry should encourage the collection and dissemination of information to learn from failures, to minimize the repetition of failures, and to improve the practice of architecture, engineering, and construction. The industry should assist in educational efforts through specialty conferences, newsletters, journals, monographs, and through the development of curriculum materials for undergraduate, graduate, and continuing education.
- *Establishment of a Repository for Failure Information.* A central repository for data collection about the technical and procedural causes of failures should be established and supported by the industry.
- *Establishment of a Journal.* There is a need for an interdisciplinary journal on case studies and on the causes and costs of failures. It should be refereed and open for discussion, so that controversial issues can be discussed in the professional arena.
- *Quality Assurance/Quality Control.* Design professionals should take the lead in promoting the use of QA/QC procedures for construction projects, from inception to completion.

These goals are ambitious, but the list indicates that there is much that can be done to improve the performance of constructed facilities. Some of the goals have been at least partially addressed in the decade following the first Engineering Foundation conference, and these efforts have met with varying degrees of success.

In October 1993, a day-long symposium on these topics was held at the ASCE national convention in Dallas, Texas. The symposium, sponsored by the ASCE Technical Council on Forensic Engineering, was a retrospective look at *The Hyatt Regency Walkways Failure: Progress in the Construction Industry over the Last Decade.* Industry experts reviewed each of the foregoing recommendations from the conferences held in the 1980s and commented on progress toward implementing these objectives.

The consensus of the speakers was that progress has been made in several areas—while in others there has been little progress, either due to pressures from outside the industry or opposition to change from within the industry. With regard to some of the concerns expressed at the conferences, the situation has improved in the past decade. In other cases, the construction environment has gotten worse.

On the positive side, the ASCE and other organizations have expended a great deal of effort on the task of establishing a clearer definition of roles and responsibilities of the various participants who contribute to a construction project. In addition, there is greater industry recognition of the need for increased attention to structural integrity issues. Considerable work has been done toward implementing project peer review for important projects, and tangible advances have been made in the area of failure information dissemination. The progress in these four areas is briefly reviewed here.

13.2.1 Definition and Assignment of Responsibilities

In response to the need for greater clarity in the definition and assignment of responsibilities, the American Society of Civil Engineers invested considerable financial resources and volunteer time in the development and publication of a manual, *Quality in the Constructed Project: A Guideline for Owners, Designers and Constructors* (ASCE 1988). While the ASCE took the lead role in preparing this publication, many other societies and individuals have contributed to the project. The publication attempts to list all design and construction tasks and suggests several models for assigning responsibilities and the attendant risks. Unclear lines of responsibility have in the past contributed to many failures and accidents.

While there is general agreement that such a document is needed, there has been resistance from a few organizations to the adoption of this particular document. The disagreements center on specific language and concerns regarding the legal implications of the document. Still, widespread interest in this highly controversial and ambitious project is an indication that discussions may proceed to an acceptable conclusion. It is hoped that eventually all parties within the industry can reach agreement on a better definition of responsibilities. The alternative is the inevitable imposition of restrictive governmental regulations.

13.2.2 Structural Integrity

The topic of structural integrity has been addressed through new requirements in some of the model building codes. Concerns regarding the prevention of catastrophic progressive structural collapse have led to provisions that encourage greater redundancy and continuity in structural systems and details. Some of these provisions are modeled after those adopted by the United Kingdom, following the Ronan Point apartment tower collapse in 1968. The U.S. Department of State began to retrofit State Department facilities to guard against progressive collapse as a result of the 1983 terrorist bombings of two buildings in Beirut, Lebanon, and all new State Department facilities are designed in accordance with structural integrity provisions. It is expected that all the model building codes will adopt these concepts and encourage their implementation by requiring higher safety factors for nonredundant designs. Experience indicates that creative solutions can provide redundancy and continuity without excessive cost. These concerns have become even more prominent in response to recent incidents of domestic terrorism within the U.S.

Committees of design and construction professionals are also working on the equally important topic of structural integrity for temporary structures. Greater structural redundancy in the design of shoring and bracing systems for temporary structures will most certainly reduce the frequency and severity of failures during the construction phase.

13.2.3   Project Peer Review

The concept of project peer review has been the subject of considerable discussion in the literature (Bell, Kan, and Wright 1989; Zallen 1990). Project peer review is a comprehensive review of the structural concept and details by an independent professional retained for this specific purpose. A few jurisdictions, including the city of Boston, Massachusetts, now require project peer review for certain projects. A joint Task Force on Project Peer Review, sponsored by ASCE and the American Consulting Engineers Council (ACEC), published a guideline for implementing the concept (ACEC/ASCE 1990). Project peer review can be an important component of quality assurance/quality control (QA/QC) for the construction industry. Other components of QA/QC include enhanced communication techniques, constructability reviews, and greater participation by the designer in the construction phase. Organizational peer reviews, in which a representative of a peer organization conducts a formal review for the purpose of recommending improvements in the organization, are also endorsed by several professional societies, including the American Consulting Engineers Council, the American Society of Civil Engineers, and the Association of Soil and Foundation Engineers.

13.2.4   Dissemination of Failure Information

There have been significant advances in dissemination of failure information since the early 1980s. There is more discussion in the current engineering literature about failures, accidents, and performance deficiencies. This exchange of information has helped to bring about improvements in performance. A number of books have been published on failures (see Section 1.3). One of the more tangible achievements was the establishment of the ASCE *Journal of Performance of Constructed Facilities* in 1987. This journal on failure case studies and the causes and costs of performance deficiencies has received widespread acceptance by practicing professionals and academicians, both in the United States and abroad.

The principal purpose of failure information dissemination is education, so that errors will not be repeated. The important role of education in the reduction of failures is discussed further in Section 13.3.

13.2.5 Other Failure Mitigation Developments

There are other positive developments that have the potential to reduce the number of construction failures. The computer is being used successfully to assist in the management of procedures and to cross-check the coordination of design and construction. Also, computer methods are now used to better coordinate the collection, collation, and dissemination of performance information through such projects as the Architecture and Engineering Performance Information Center (AEPIC).

Unfortunately, the AEPIC project has not yet become a useful resource for design professionals, at least in the way it was envisioned originally. The AEPIC repository for performance information was established in 1982 at the University of Maryland at College Park. Its establishment was marked by considerable fanfare and declarations of support from all the significant design and construction industry organizations. However, the ambitious project was never adequately funded, nor was it adequately staffed. Design professionals did not contribute significant numbers of case studies to the repository as originally expected. As a result, the database is limited and the record of tangible AEPIC achievements is not impressive. Nevertheless, the concept is valid, and such a database will certainly be desired in the future. Advances in database management, expert systems data processing technology, and efficient information retrieval software can only help to enhance the eventual usefulness of a performance information library. It should also be noted that AEPIC served as a catalyst for many other developments in failure mitigation. For example, the interdisciplinary Editorial Board of the ASCE *Journal of Performance of Constructed Facilities* included many AEPIC members. AEPIC, along with the National Society of Professional Engineers (NSPE), continues to be a cosponsoring organization for the journal.

Another positive development has emerged from the liability crisis. This is the evolution of several new methods of dispute settlement and dispute avoidance (see Chapter 12). These new strategies are proving to be more successful than traditional litigation in certain cases involving construction failures and deficiencies. The teamwork emphasis of these new problem-solving methods can contribute to an improved project atmosphere. The product resulting from this less confrontational atmosphere is likely to be more defect-free, since the goal is to anticipate and avoid problems before they become disputes.

Innovative approaches to insurance, which encourage cooperation rather than confrontation, are also available. Many of these insurance programs are coupled with loss-prevention incentives. Recently, insurance premiums for design professionals have been declining. These reductions may be a reflection of the success of loss-prevention strategies. However, there is some evidence to suggest that increased competition, by way of new insurers entering the market, may be more responsible for the reduction in premiums.

The industry-wide attention and effort focused on mitigation of failures in the past decade is quite impressive. These activities are founded in the conviction that failures are not inevitable and that the industry can improve its performance. Opportunities exist for an expansion of these and other efforts in the interest of reducing the frequency and severity of failures.

13.3   THE CRITICAL ROLE OF EDUCATION

Education is an essential component of any failure mitigation strategy. This fact has been emphasized by the inclusion of vigorous discussions regarding the importance of education in nearly every forum on the subject of construction failure. All of the ongoing failure information dissemination activities discussed in the previous section are intended for educational purposes.

The rationale for bringing failure case studies into the classroom has been established by a number of writers (Carper 1990; Kaminetzky 1991, 1994; Bosela 1993; Rendon-Herrero 1993a,b). Learning from the actual performance of engineered facilities can be a valuable supplement to the basic undergraduate curriculum for professional design and construction education. Case histories from the field of forensic engineering, integrated into the design and construction curriculum, provide motivation for learning technical concepts. Failure case histories also encourage classroom discussion of important nontechnical issues, including ethics and other professional practice topics.

The discussion of failure case histories need not be disillusioning to students, nor need it diminish their creativity and enthusiasm for their chosen profession. In fact, such discussion usually serves to enhance enthusiasm and commitment to professionalism. Students become enlightened regarding the consequences of decision errors. They are motivated to learn the technical and procedural skills for preventing failures. Students entering the design and construction professions are best equipped if they have an appreciation for the possibility of failure and an understanding of each participant's role in mitigating risk.

Graduate-level courses on forensic engineering also should be encouraged, along with continuing education courses, seminars, specialty conferences, and convention sessions for practicing design and construction professionals. Resource materials are available to support all these activities and more are being produced by organizations such as the Committee on Education of the ASCE Technical Council on Forensic Engineering. One very useful publication by this committee is a monograph prepared for use in undergraduate education, *Failures in Civil Engineering.* This monograph presents brief summaries of about 50 historically important case studies from all fields of civil engineering and gives a reference list for each case to encourage students to conduct literature searches for further information (Shepherd and Frost 1995).

Numerous case studies are also published in the ASCE *Journal of Performance of Constructed Facilities.*

At least one author has taken the position that education is the *only* effective failure mitigation strategy, since education is the key to improving the design and construction workforce. This position may be a bit simplistic, however, since it presumes that those doing the educating know all there is to know. There is also a need for research and experience, the foundations of useful knowledge. Also, while it is clear that competence in every phase of the production segment of the construction industry is needed if failures are to be avoided, and that education is the source of competence, it must be remembered that the causes of failure involve other human failings besides lack of knowledge.

McKaig (1962) listed the underlying causes of failure as ". . . ignorance, carelessness, negligence, and greed." While education can address the ignorance factor, a comprehensive failure mitigation program must also include legislation and punitive enforcement for willful acts of carelessness, negligence, and greed. Education alone cannot produce ethical and moral behavior. This is not intended to diminish the important role of education; rather, it demonstrates that education, research, experience, and diligent code enforcement are all critical elements in the mitigation of construction failures.

13.4 CONCLUDING REMARKS

The preparation of this book was undertaken with the belief that design and construction professionals and students have both the responsibility and desire to learn from failures. Case histories from the field of forensic engineering can enhance their understanding of technical concepts and provide opportunities for interpersonal and professional skill development. These opportunities, and the attitudes they encourage, may prove to be as important to professional survival as the technical skills.

As long as structures are designed and built by human beings using imperfect materials and procedures, failures will be experienced along with successes. The tools and theories of modern technology are available to enable the designer and builder to improve the quality of the constructed project. Unfortunately, there are trends in modern society that may lead to an increase in both the frequency and the severity of failures. Designers and construction professionals have recognized these trends and are implementing strategies to mitigate their effects.

Some of these strategies involve the sharing of experiences, both failures and successes. The tradition of learning lessons from experience, including failure, continues to be important to the future success of the design and construction professions. Just as the pathologist contributes to the education of medical professionals, the forensic consultant contributes information essential to advances in the design and construction professions. Lessons learned from experience can supplement modern mathematical theories and computational tools to predict the behavior of engineered facilities with a greater degree of confidence than ever before.

13.5 REFERENCES

ACEC/ASCE, 1990. *Project Peer Review: Guidelines,* American Consulting Engineers Council and American Society of Civil Engineers, New York.

ASCE, 1988. *Quality in the Constructed Project: A Guideline for Owners, Designers and Constructors,* American Society of Civil Engineers, New York.

Bell, G., F. Kan, and D. Wright, 1989. "Project Peer Review: Results of the Structural Failures II Conference," *Journal of Performance of Constructed Facilities,* American Society of Civil Engineers, New York (November).

Bosela, P., 1993. "Failure of Engineered Facilities: Academia Responds to the Challenge," *Journal of Performance of Constructed Facilities,* American Society of Civil Engineers, New York (May).

Carper, K., 1990. "Engineering Pathology: Case Studies in the Classroom," *Education and Continuing Development for the Civil Engineer: Setting the Agenda for the 90's and Beyond, Proceedings of the National Forum,* American Society of Civil Engineers, New York.

Gross, J., 1986. "Building Structural Failures: Their Cause and Prevention," *Journal of Professional Issues in Engineering,* American Society of Civil Engineers, New York (October).

Kaminetzky, D., 1991. *Design and Construction Failures: Lessons from Forensic Investigations,* McGraw-Hill, Inc., New York.

Kaminetzky, D., 1994. "Guest Editorial," *Journal of Performance of Constructed Facilities,* American Society of Civil Engineers, New York (February).

Khachaturian, N., ed., 1985. *Reducing Failures of Engineered Facilities,* American Society of Civil Engineers, New York.

McKaig, T. H., 1962. *Building Failures: Case Studies in Construction and Design,* McGraw-Hill, Inc., New York.

Rendon-Herrero, O., 1993a. "Too Many Failures: What Can Education Do?" *Journal of Performance of Constructed Facilities,* American Society of Civil Engineers, New York (May).

Rendon-Herrero, O., 1993b. "Including Failure Case Studies in Civil Engineering Courses," *Journal of Performance of Constructed Facilities,* American Society of Civil Engineers, New York (August).

Shepherd, R., and J. Frost, eds., 1995. *Failures in Civil Engineering: Structural, Foundation and Geoenvironmental Case Studies,* Committee on Education, Technical Council on Forensic Engineering, American Society of Civil Engineers, New York.

Zallen, R., 1990. "Proposal for Structural Design Peer Review," *Journal of Performance of Constructed Facilities,* American Society of Civil Engineers, New York (November). (See also Discussions in August 1992 issue).

INDEX

Abrasion, 111, 262–265, 375, 381
Abutments, 105–108
Acoustical problems, 371, 396
Addis, W., 1
Addleson, L., 11, 173
Administrative Dispute Resolution Act, 466
Admixtures, 235
Aggregate, concrete, 20, 235, 246–248, 260, 262, 271, 285, 293, 295
Aging facilities, 24, 52, 72, 88, 92, 143–145, 179, 228, 255, 263, 281, 332–335, 439
Air Force base, 92, 255–259, 309, 397
Airplane accidents, 52, 158, 187
Airport, 66, 69, 247, 294, 429
Air quality, 21, 371, 392–396. *See also* Environmental control systems.
Air-supported structures, 42, 375, 400
Alabama, 247
Alabama Department of Transportation, 301
Alaska, 207, 229, 295
Albany, New York, 84, 97, 345
Alexander, Louisiana, 183, 431–432
Alkali aggregate reaction, 247, 293
Allen, D., 304, 410, 431
Al-Mandil, M., 286
Alterations, 21, 48, 84, 107, 120, 165, 290, 438–442. *See also* Renovation.
Alternate dispute resolution (ADR) techniques, 13, 451–482, 491
    arbitration, 467–471
    baseball arbitration, 477
    dispute review boards, 475–476
    federal government disputes resolution, 478
    mediation, 471–473
    mediation/arbitration, 473
    minitrials, 473–475
    one hundred-day document, 477
    private judges, 475
    state agency panels, 477–478
American Arbitration Association, 467, 469–470, 473, 475
American Bar Association, 451, 472
American Concrete Institute, 8, 11, 246, 315, 332

American Consulting Engineers Council, 490
AFL/CIO Labor Union, 411
American Institute of Architects, 10–11, 171, 203, 211, 257–260, 285, 320, 385, 392, 450
American Institute of Mining, Metallurgical and Petroleum Engineers, 476
American Institute of Steel Construction, 213, 227
American National Standards Institute, 381, 444
American Plywood Association, 169
American Society of Civil Engineers, 2, 9, 12–13, 22, 44, 163, 174, 178, 224, 274, 302, 313, 331, 381, 417, 453, 461, 466, 475–476, 480, 484, 486, 488–493
American Society of Heating, Refrigeration and Air Conditioning Engineers, 393
American Society for Testing and Materials, 247, 332, 381
American Welding Society, 206
Ammann, O., 141
Amoco Corporation, 357–360
Amrhein, J., 30
Amsterdam, the Netherlands, 52
Amtrak railroad accident, 133
Anchorage, Alaska, 31–32, 35, 56, 95, 97, 275, 316–317
Anchorages, 290, 292, 295–296, 298, 315–316, 319, 322–324
Andersen, W., 455
Annapolis, Maryland, 267
Apartment buildings, 8–9, 20, 52, 60, 83–84, 89, 91, 101–102, 163, 188, 243, 249, 267–268, 303–306, 310–315, 340, 343–344, 368, 378, 439, 489
Aqueduct, 72, 327
Arbitration, 466–478, 480
Arch, 170–172, 174–178, 183, 310, 320–321, 332, 416, 431–432
Architects Council of Europe, 458
Architecture and Engineering Performance Information Center, 12, 411, 419, 432, 491
Arena. *See* Stadium.

Arizona, 306
Arkansas, 116
Arlington, Massachusetts, 313
Arnold, C., 404
Asbestos, 373, 393, 462
Ashland Petroleum oil tank, 208
Asphalt, 97–98
Asphalt Roofing Manufacturers Association, 374
Asphyxiation, 419, 442, 444
Association of Soil and Foundation Engineers, 11, 60, 108, 479–480, 490
Association of Trial Lawyers of America, 470, 478
Athena, Oregon, 182
Atlanta, Georgia, 45, 48–49, 71, 84, 190, 309, 430
Atlantic City, New Jersey, 248
Auditorium, 10, 17, 42, 189, 191, 204, 240
Azad, A., 286

Babel, Tower of, 14
Backfilling, 71, 73, 81, 95–96, 417, 423
Bailey's Crossroads, Virginia, 9, 242–245, 422. See also Buildings, Skyline Plaza.
Baker, Sir J. Fowler Benjamin, 135
Baltimore, Maryland, 88
Baltimore–Washington Parkway, 427
Baluch, M., 286
Banick, L., 224–225, 455
Bangkok, Thailand, 104
Barge collisions, 51, 133
Baseball arbitration, 477
Base isolation, 38–39, 366, 400
Basement walls, 95
Bayside, New York, 191
Bearing, 290, 292, 295–299, 308, 328
Beasley, K., 332, 349, 353–354, 359
Beaver, Pennsylvania, 190
Bech-Andersen, J., 174
Becker, R., 382
Beirut, Lebanon, 50, 490
Belgium, 207, 445
Bell, G. M., 455
Bell, G. R., 12, 27, 197, 351, 487, 490
Below-grade construction, 60, 65–83, 236, 413–420, 476–477
Bennett, J., 142, 144–145
Bennett, L., 64, 66
Benton Harbor, Michigan, 249
Beri, P., 188
Berry, E., 246
Bessemer, Alabama, 190
Bhopal, India, 483
Big Rock Mesa slide, California, 100–101
Big Sur area, California, 100
Bijlaard, P., 207
Birkeland, H., 298, 315
Birkeland, P., 298, 315
Birmingham, Alabama, 17, 190

Blast, 48–52, 83, 95, 418, 420, 455. See also Explosion; Terrorism.
Boeing Company, 94
Boehmig, R., 224
Bolivar, Tennessee, 20, 27–28, 197–199
Bombay, New York, 189
Bond, 237–238, 271, 273, 306, 312, 315, 322–324, 337, 377, 391
Borescope, 178–179
Bosela, P., 492
Boston, Massachusetts, 8, 13, 27, 38, 46, 65, 87, 147, 150–151, 172, 327, 367, 372, 379, 382–392, 397, 399, 416, 421, 490
Boston Society of Architects, 385
Bouch, Sir Thomas, 133, 135
Boundary layer wind tunnel, 43, 357, 381, 389, 391
Bracing, 43, 88, 150, 166–167, 171, 178, 182–183, 187–204, 308, 311, 335, 386, 389, 417, 419, 422, 425, 427–431, 433–435, 441, 490
Bradford, H., 410
Brattleboro, Vermont, 433–435
Bray, J., 108
Brazil, 60, 87, 91, 267
Brick Institute of America, 332, 421
Brick veneer/metal stud wall. See Metal stud/brick veneer systems.
Bridges, 27, 29, 36, 38–41, 49, 51–52, 70, 79, 85, 107, 111–112, 130–158, 187–188, 207–208, 228–231, 250, 275–276, 283, 285, 295–296, 306, 319, 365, 392, 412, 418, 427, 442, 445, 459
Ashtabula Bridge, Ohio, 136–137
bascule bridge, 137
Bissell Bridge, Windsor, Connecticut, 323–324
Bronx–Whitestone Bridge, New York City, 139
Brooklyn Bridge, New York City, 111–112, 138, 141
Carquinez Bridge, California, 207
Chickasawbogue Bridge, Alabama, 149
covered bridges, 163
Duplessis Bridge, Quebec, Canada, 208
Eads Bridge, St. Louis, Missouri, 141
Firth of Forth Bridge, Scotland, 135–136, 152
Firth of Tay Bridge, Scotland, 40–41, 133–135
floating bridges, 153–156, 441
Galloping Gertie. See Bridges, Tacoma Narrows Bridge.
George Washington Bridge, New York City, 138, 141
Golden Gate Bridge, California, 111, 138–139
Hackensack River Bascule Bridge, New Jersey, 137
Harvard Bridge, Boston, 147

Hatchie River Bridge, Tennessee, 149
Hood Canal Bridge, Washington State, 51, 154
Hutt Bridge, New Zealand, 307
iron bridges, 103
James River Bridge, Newport News, Virginia, 263–264
Kenai River Bridge, Alaska, 295
Kings Bridge, Melbourne, Australia, 207–208
Lake Maracaibo Bridge, Venezuela, 51, 230
Lake Washington Bridge. *See* Bridges, Murrow Floating Bridge, Seattle
Menai Straits Bridge, 138
Mianus River Bridge, Connecticut, 13, 21, 52, 145–148, 158, 228
Murrow Floating Bridge, Seattle, Washington, 153–156
Point Pleasant Bridge, West Virginia, 142–145, 148, 228
Quebec Bridge, Canada, 9, 135, 152–153
railroad bridges, 5, 130–131, 133–137
Raritan River Bridge, New Jersey, 51
San Francisco–Oakland Bay Bridge, California, 139
Schoharie Creek Bridge, New York, 13, 21–22, 148–150
Severn and Wye Bridge, United Kingdom, 158
Silver Bridge, West Virginia. *See* Bridges, Point Pleasant Bridge.
Sunshine Skyway Bridge, Florida, 51, 133
suspension bridges, 133, 138–145, 206, 230
Tacoma Narrows Bridge, Washington, 41, 138–141, 155, 392
Tappan Zee Bridge, New York, 132
Wheeling Bridge, West Virginia, 138
Yankee Doodle Bridge, Connecticut, 147
Yns-y-Gwas Bridge, Wales, 323
British Board of Trade, 135
British House of Commons, 308
British Institution of Civil Engineers, 7, 253
Brittle fracture, 38, 136–137, 187, 207–209, 225–227, 280
Brittle materials, 88, 213, 235, 239, 275, 330, 347, 350, 352
Brooklyn, New York, 66, 77, 93–94, 132, 191–192, 248, 345
Brookville, New York, 204
Brown, D., 133, 135
Brown, J. G., 114
Browne, M., 145, 158
Brussels, Belgium, 84
Buckling, 18, 27–28, 189, 191, 193–198, 201, 257, 294
Buckner, C., 165, 228
Buffalo, New York, 84, 93, 191, 249, 427
Building codes. *See* Codes, design.

Building and Construction Trades Department, AFL/CIO, 411
Building envelope, 331, 336, 348–360, 371–373. *See also* Facades; Roofing.
Buildings:
  Amoco Tower, Chicago, 13, 357–360, 379
  Atlantic Richfield Towers, Los Angeles, California, 209
  British Library, London, 400
  Commonwealth Avenue Building, Boston, 8, 242, 268, 422
  Denver International Airport, Colorado, 400–401, 429
  DuPage County Judicial and Office Building, Illinois, 393–396
  El Paso, Texas Civic Center, 209
  Empire State Building, New York City, 52
  Federal Building, Cleveland, Ohio, 416
  Foothills Communities Law and Justice Center, California, 39
  Ford Theater, Washington, DC, 84
  Harbour Cay Condominium, Cocoa Beach, Florida, 9, 18, 271–274, 413
  Hartford Civic Center Coliseum, Hartford, Connecticut, 9–10, 18, 25, 198, 200–204, 413
  Hoosier Dome, Indianapolis, Indiana, 375, 400
  Hyatt Regency Hotel walkways, Kansas City, Missouri, 10, 12, 18, 25, 210, 214–225, 455, 487–488
  Ireland Building, New York City, 90
  J. C. Penney Building, Anchorage, Alaska, 316–317
  John Hancock Building, Boston, 13, 209, 372, 379, 382–392, 416
  Johnson Wax Tower, Racine, Wisconsin, 377
  Justice Building, Little Rock, Arkansas, 93
  Kemper, Crosby, Memorial Arena, Kansas City, Missouri, 10–11, 18, 209–214, 487
  Lafayette Hotel, Buffalo, New York, 84
  L'Ambiance Plaza, Bridgeport, Connecticut, 9, 189, 206, 310–315, 410, 438, 473
  Lee County Justice Center, Ft. Meyers, Florida, 375–376
  Lincoln First Tower, Rochester, New York, 357
  Magic Mart Store, Boliver, Tennessee, 197–199
  Miami Metropolitan Detention Center, Florida, 241
  Minneapolis Metrodome, Minnesota, 375
  Murrah, Alfred P., Federal Building, Oklahoma City, 50
  New Haven, Connecticut Coliseum parking structure, 231
  New York City Convention Center, 206
  Old Criminal Building, New York City, 93

Buildings (*continued*)
  Olive View Medical Center, California, 36
  Orpheum Theater, New York City, 191
  Pittsburgh, Pennsylvania Airport, 294
  Port Authority (PATH) Transit Station,
    New Jersey, 397–399, 422
  Post, C. W., College Auditorium,
    Brookville, New York, 204
  Palace of Fine Arts, Mexico City, 88
  Polo Grounds, New York, 253–254
  Pravda Building, Moscow, Russia, 303
  Robins Air Force Base, Macon, Georgia,
    255–259
  Ronan Point Housing Project, United
    Kingdom, 8, 50, 303–306, 489
  Rosemont Horizon Arena, Illinois, 9,
    170–172, 428–429
  Salt Lake City and County Building, Utah,
    39, 366
  Sears Tower, Chicago, 391
  Silverdome, Pontiac, Michigan, 375
  Skyline Plaza, Bailey's Crossroads,
    Virginia, 242–245, 268, 422
  Stamford, Connecticut Train Station, 241
  Standard Oil of Indiana Building. *See*
    Buildings, AmocoTower, Chicago.
  State of Illinois Center, Chicago, 393–396
  Station Square Shopping Center, Burnaby,
    British Columbia, Canada, 193–197
  Strand Theater, Pittsburgh, Pennsylvania,
    173
  Suffolk County Courthouse, Boston,
    Massachusetts, 367
  Sundome, University of South Florida, 375
  Superdome, New Orleans, Louisiana, 375
  Telč Castle, Czech Republic, 163, 179
  Texas State Fair Exhibition Building,
    Dallas, 254
  Trinity Church, Boston, Massachusetts,
    385–389, 416
  U.S. Post Office, Chicago, 206
  University of Illinois Memorial Stadium,
    53, 230
  University of Washington Stadium
    expansion, Seattle, 431–435
  Walters Food Store, Columbus, Ohio, 397
  Washington State Convention Center,
    Seattle, 209
  West Berlin Congress Hall, Germany,
    320–321
  Westin Hotel, Boston, Massachusetts, 397,
    421
  Wilkins Air Force Base, Shelby, Ohio,
    255–259
  Wolftrap Performing Arts Center, Fairfax,
    Virginia, 208
  World Trade Center, New York City, 47,
    50, 187
  Yonkers Raceway parking structure, New
    York, 263–266

Building Research Station, Great Britain, 282
Building Seismic Safety Council, 38, 406
Building Standard Law, Japan, 276
Business Roundtable Construction Industry
  Cost Effectiveness Project, 410

Cable structures, 43, 112, 229–230
Cairo, Egypt, 84
Caissons. *See* Deep foundations, caissons.
California, 100–101, 227, 310, 402–406, 411,
  478
California Dam Safety Law, 116
California Seismic Code, 35
California State University, 172, 316, 318
Camber, 27, 132, 297–298, 301
Campanile, 332
Campbell, R., 382, 391
Canada, 193–197, 248, 460
Canals, 102–103, 118
Carbonation, 262, 284
Carper, K., 11, 22, 38, 52, 55, 183, 188, 409,
  413, 458, 480, 492
Cartwright, L., 380
Cassady, L., 379
Cast iron, 91
Cathodic protection, 231, 285, 287, 320
Cavity walls, 349–350
Ceiling collapse, 53, 180, 229, 242, 392,
  396–399, 421–422
Celestial Globe Company, 240
Center for Excellence in Construction Safety,
  9, 410–411, 414
Challis, Idaho, 36
Chambersburg, Pennsylvania, 107
Champaign, Illinois, 174, 294
Change orders, 197, 445, 451
Charleston, South Carolina, 30
Charlotte, North Carolina, 42
Chemical reactions, 247, 260, 262, 321
Chemical treatment:
  of soils, 62, 64, 78
  of timber products, 174, 228
Chen, F., 62
Cheyenne Mountain, Colorado, 74, 78
Chicago, 9, 13, 26, 44–47, 69–70, 77, 85, 93,
  107, 132, 170, 179, 206, 355, 357–360,
  379, 391, 393–396, 419, 428
Chicago River, 44, 65
Chiles, J., 5
Chimneys, 338–339
Christie, S., 148
Cincinnati, Ohio, 45, 344
Cladding, 348–360, 372, 378–380, 382, 402.
  See also Curtain walls; Facades.
Clearwater, Florida, 12, 442, 484
Cleveland, Ohio, 77, 83, 87, 93, 267, 311, 362,
  416
Coalinga, California, 36
Coastal construction, 100–101, 104

Coatings, 229–230, 284, 287, 321. *See also* Finishes.
Cocoa Beach, Florida, 9, 18, 271–274, 413
Codes, design, 26, 32, 34–36, 38, 41, 46, 48, 50, 222, 226, 239, 243, 248, 256–258, 267, 274–276, 315, 318, 380, 391, 394, 404–406, 439, 458, 483, 489–490, 493
  enforcement of, 26, 47–48, 248, 274, 318, 458, 488, 493
Cofferdam, 418, 428
Cohen, J., 316, 357, 382, 402
Cold regions, 16, 61, 64, 248
Cold storage facility, 107
Colin, D., 262
Coliseum, 9–10, 18, 25, 42, 198, 200–204, 413
Collin, A., 102
Collision, 48–52, 130, 133, 191
Colombia, 102
Colony, Missouri, 435
Columbia, South Carolina, 103
Columbus, Ohio, 49, 343, 355, 367, 397, 429
Columns, 254–255, 265, 267, 318, 344, 346–347
Comerio, M., 366
Comfort expectations, 371, 392, 396, 404, 406, 485
Communication, 17–18, 293, 398, 420
Composite materials, 157, 237–238
Compression:
  in concrete, 236, 240, 265, 267, 291
  in masonry, 327–328, 330, 346, 351, 355
  in steel, 189, 193–197, 201–203
  in timber, 166
Computer-aided design, 115, 200–204, 280, 485–486, 491
Computer software, misuse of, 17–18, 201, 204, 485–486
Concrete, precast, 20, 32, 36, 38, 43, 50–51, 154, 156, 170, 172, 225, 276, 278, 290–326, 335–336, 378, 402, 420, 428–430, 442
  connections, 298–302
  erection, 306–315
  performance in wind and earthquakes, 315–319
  production techniques, 292–298
  properties of, 290–292
  quality control, 302–306
Concrete, prestressed, 9, 20, 156–157, 166, 189, 228, 283, 285, 290–326, 425, 441, 442
  connections, 298–302
  corrosion of tendons, 319–324
  erection, 306–315
  performance in wind and earthquakes, 315–319
  production techniques, 292–298
  properties of, 290–292
  quality control, 302–306
Concrete, reinforced (cast-in-place), 6–9, 17–18, 20, 29, 33–39, 93, 113–114, 225, 228, 235–289, 343, 345–346, 350, 355, 368, 402, 421–425, 427, 431
  abrasion, deterioration, 262–265
  corrosion, 281–287
  ductility, seismic performance, 274–281
  formwork, temporary support, 242–245
  historic examples, 239–242
  movements, 251–262
  properties of, 235–239
  quality control, 243–250
  shear failures, 265–274
Concrete Reinforcing Steel Institute, 332
Condensation, 165, 174, 330, 348, 352, 378
Condominium, 9, 18, 271–274, 413
Conemaugh River, Pennsylvania, 118–119
Connecticut Department of Transportation, 147
Connecticut Turnpike, 107, 145–148
Connections, 18, 21, 27, 34–36, 38–39, 42, 50, 130, 140, 143–148, 150, 153, 333, 335, 360, 378, 421, 423, 428–429, 442, 483
  concrete connections, 251–252, 255, 290, 292–293, 298–302, 306, 315–316, 318–319, 355–356, 363
  steel connections, 187–188, 200–201, 205–228, 280–281, 315
  timber connections, 165–172, 178, 183, 228
Constructability, 13, 223, 291, 444–445, 462
Construction errors, 18–20, 170–172, 183, 188–193, 213, 240, 243, 250, 260, 267, 273–274, 286, 290, 302–306, 310–316, 321, 382, 409, 413, 422, 480
Construction Industry Institute, 447, 463
Construction loads, 150, 171, 191, 206, 239, 243, 260, 308, 310–315, 414, 417, 420–423, 425, 428–429
Construction management, 200–204, 436, 452, 459, 461, 463, 484–485
Construction Management Association of America, 459
Construction safety, 8–9, 20, 77, 239, 243, 310–315, 409–450, 458–459, 487
Cook, W., 319
Cooling tower, 9, 20, 242, 311, 410, 423–424
Coombe, J., 53, 230, 281
Cooper, T., 152–153
Corley, W. G., 227
Corning, New York, 90
Corrosion, 21, 52–53, 67, 72–73, 111–112, 143–148, 165–167, 187–188, 228–231, 236, 246, 251, 260, 262, 281–287, 291, 293, 319–324, 337, 352, 355, 357, 360–363, 376, 378, 392, 396–397, 399–400
Council on Competitiveness Report, 478
Coyne, A., 121
Cracking, 238, 247, 249–250, 252–254, 258–262, 265, 267, 274, 280, 283, 291–295, 299–302, 319, 321–324, 330, 333, 335, 338, 340–342, 344–353, 355, 361, 367, 372, 380, 389, 391, 416, 441

Craftsmanship, 330, 348. *See also* Workmanship.
Crane accidents, 152, 189, 191, 243, 307, 310, 410, 421, 442–444
Creep, 239, 251–252, 265, 291, 293, 298, 319, 321, 343, 346
Crescent City, California, 32
Crestwood, New York, 73
Cross, Hardy, 137, 241
Cuba, 302
Culver, C., 313
Culverts, 73
Cuocco, D., 313
Curtain walls, 39, 42–43, 228, 254, 283, 348–360, 362, 372–373, 376–392, 400, 402. *See also* Cladding; Facades.
Cyclic loading, 112, 137, 143–144, 156, 187, 213, 340, 355, 391. *See also* Stress reversals.
Cypress Viaduct, California, 276
Czech Republic, 163, 179

Dallaire, G., 205
Dallas, Texas, 106, 355, 443, 488
Damping, 139–140, 281, 386, 389, 400
Dams, 13, 43–44, 74, 80, 102, 104, 111–130, 157–159, 250, 263, 418
  Alexander Dam, Hawaii, 129
  Ambursen Buttress Dam, Massachusetts, 117
  Buffalo Creek, West Virginia, 124, 157
  Calavaras Dam, California, 128
  Fontanelle Dam, Wyoming, 117
  Fort Peck Dam, Montana, 124, 129
  Indian Lake Dam, Ohio, 250
  Lafayette Dam, California, 116
  Malpasset Dam, France, 114, 117, 120–121
  Maltilija Dam, California, 117
  Mercer Dam, Pennsylvania, 117
  Oahe Dam, South Dakota, 80, 129
  St. Francis Dam, California, 115, 124
  Spokane River Dam, Washington, 127–128
  Teton Dam, Idaho, 13, 117, 124–127, 157
  Toccoa Dam, Georgia, 127, 157
  Vaiont Dam, Italy, 44, 113, 117, 121–123
  Waco Dam, Texas, 129
  West Branch Dam, Ohio, 130
  Wheeler Lock, Tennessee River, 114, 123
Dam safety programs, 124, 126
Daniels, G., 118, 124
Davies, M., 228
Day, R., 262, 265
Debris flows, 60, 103, 119
Decay. *See* Deterioration.
Deep foundations, 63–71
  caissons, 65–71
  piles, 65, 79, 91–94, 107–108, 163
    concrete, 66–71, 88, 263–264
    steel, 66–69, 385, 418
    timber, 65, 85, 94, 386, 416

Deep Foundations Institute Construction Round Table, 477
Defense Research Institute, 465
Deformations, excessive, 27, 60, 138, 140, 187–188, 201–202, 227, 236, 239, 242, 252, 275–276, 306, 316, 318, 331, 338, 344–345, 351, 372, 377, 385, 391, 402–403, 405–406
Degenkolb, H., 28, 30
Demolition, 87, 150, 155, 290, 323–324, 378, 389, 438–442
Denmark, 173–174
Denver, Colorado, 20, 83, 362–363, 400–401, 429, 436
Department store, 20, 27–28, 193–199
Derbalian, G., 72
Design-build, 135, 452, 454, 459, 485
Design errors, 9, 17–18, 125, 127, 135, 137, 152–153, 156, 193–198, 201–203, 213, 222, 224, 230–231, 240–241, 260, 272, 274, 283, 286, 290–291, 293–306, 308, 312–313, 319, 321, 375, 382, 393, 395–396, 398–399, 409, 412–413, 422, 455, 480
Design services, competitive bidding for, 12, 195, 485, 488
Deterioration, 52, 113, 144, 286, 372, 374, 376, 378, 439
  in concrete, 246–247, 250, 262–265
  in masonry, 333, 335, 349, 352, 357–360
  in steel, 228
  in timber, 165, 167, 173–178, 392
Detroit, Michigan, 355
Developing countries, 236–237, 246, 286, 428
Dewatering, 60, 71, 92–94, 105
Dias, W., 293
Diaz, C., 95
Dimensional stability, 165–167, 186, 239, 246, 251–262, 337–344, 346, 355–360, 367
Dinges, C., 411
Disaster preparedness, 38
Dispute Avoidance and Resolution Task Force, 464
Dispute review boards, 452, 475–476, 478
Dome:
  concrete, 42, 309
  steel, 10, 191–192, 204
  timber, 183, 431–432
Dormitory. *See* Housing.
Dowding, C., 418
Downdrag and heave of foundations, 91–94
Drafting errors, 18–19
Ductility, 36, 38, 165, 187–188, 207, 213–214, 239, 274–281, 315–316, 318, 328, 330, 352, 363, 367, 403–404, 487
Dunker, K., 145, 158
Durability, 17, 21, 52–53, 165, 173, 230, 236, 238, 246, 282–283, 323–324, 330, 337–338, 352, 356–357, 376, 378
Dusenberry, D., 153, 156

Dynamic loads, 130–131, 137, 170, 213, 218, 227, 239, 306, 308, 315, 330, 335, 367, 418

Eads, J., 152
Earthquakes, 24, 28–39, 44, 52, 54, 72, 80, 95–97, 99–100, 102, 104, 106, 111, 113, 122, 124, 157, 163, 165–167, 178, 186–187, 210, 225–227, 236, 238–239, 261, 274–281, 293, 315–319, 328, 330, 333, 363–367, 400, 402–406, 436
   Anchorage, Alaska, 31–32, 35, 56, 95, 97, 275, 316–317
   Challis, Idaho, 36
   Charleston, South Carolina, 30
   Coalinga, California, 36
   Great Hanshin Earthquake, Japan. *See* Earthquakes, Kobe, Japan.
   Kobe, Japan, 34, 38, 276, 279–280, 318, 404
   Loma Prieta, California, 32, 37–38, 157, 276, 402–404
   Long Beach, California, 35, 275
   Mexico City, Mexico, 28–29, 33–34, 238, 275
   New Madrid, Missouri, 30
   Niigata, Japan, 95
   Northridge, California, 32, 34, 37–38, 40, 157, 210, 225–227, 276–278, 280, 316, 318, 402–405
   San Fernando, California, 35–37, 157, 275
   San Francisco, California, 31, 35, 275
   Santa Barbara, California, 275
   Tokyo, Japan, 35
   Whittier, California, 36, 316
Earthquakes, effects of, 30–32, 62
Earthquake Engineering Research Center, 277
Earthworks, 59–110
East Chicago, Indiana, 9, 242, 425–426
Ebeling, D., 178
Eberhardt, A., 72
Eccentric loading, 66, 73, 201, 240, 265, 297, 308, 310, 328, 353
Eck, R., 411
Edison, New Jersey, 27
Edmonds, Washington, 430
Education, 13, 377–378, 409, 411–412, 419, 444, 447, 484, 488, 490–493
Egyptian pyramids, 327
El-Eswani, H., 298
Electrical equipment and systems, 127, 371–372, 401–402, 441, 443
Electrocutions, 410, 442–443
Ellifritt, D., 188, 435
El Paso, Texas, 209, 427
Elwood, Kansas, 425
Energy efficiency, 371, 373, 389–390, 393, 395–396
Engineering Foundation, 12, 486–488

Environmental control systems, 21, 268, 330, 392–396, 457. *See also* Air quality; HVAC systems.
Environmental hazards, 60, 208, 228–229, 357, 373, 462, 483
Environmental regulations, 16, 155, 452
Epoxy coatings, 284
Eppell, F., 330
Eppensteiner, .W., 356
Epstein, H., 203
Equipment, 18, 48, 51, 95, 98, 127, 191, 246, 268, 290, 308, 371, 374, 392, 399–402, 405, 409–410, 414–415, 417–419, 423, 425, 431, 442
Erosion, 85, 97–98, 100–101, 104, 113, 122, 126–127, 148–149
Estenssoro, L., 180
Ethics, 135, 207, 224, 459–461, 480–481, 485, 492–493
European Community, 411, 458
Evanston, Illinois, 249
Everett, Washington, 191
Excavation, 18, 62, 68–69, 74–78, 80–81, 83–85, 91, 94, 97, 99–100, 102–103, 106, 129, 385–386, 389, 410, 412–420, 428, 477
Expansion, 251–262, 293, 298–299, 302, 335, 337–344, 346, 350, 356–361, 367, 377
Experience, 1, 32, 60, 108, 162, 173, 193, 204, 246, 290, 316, 353, 413, 439, 441, 486, 493–494
Experience, trial-and-error, 1, 30, 34–35, 138, 141, 227, 239, 363, 377, 413, 485
Expert witness, 22, 363, 464, 479–481
Explosion, 48–52, 70–72, 77, 115, 187, 410, 441–444, 489. *See also* Blast; Terrorism.
   gas explosion, 8, 49–50, 77, 303–306
Eyre, D., 319

Fabrication errors, 293, 312
Fabric roofs, 42, 375, 400
Facades, 13, 20–21, 231, 316–317, 329, 331, 342, 348–363, 372–373, 375–392, 403, 438–440. *See also* Cladding; Curtain walls.
Factor of safety, 114, 125, 203, 267, 331–332, 336, 376, 487, 490
Failure, causes of, 13–21
   concept errors, 14–16
   construction errors, 18–20
   design errors, 17–18
   material deficiencies, 20–21
   operational errors, 21
   programming deficiencies, 16–17
   site selection, 16
Failure, definition of, 2, 454
Failure, economic consequences of, 410, 435, 445–446
Failure, political implications of, 75–77, 207
Fairweather, V., 418, 466, 471, 477–478
Falls, 410, 412, 419, 442

Falsework, 9, 238, 425, 427. *See also* Formwork; Temporary support.
Fargo, North Dakota, 89
Farquharson, F., 41, 139
Fast-track construction, 200, 203, 206, 349, 452, 484
Fatigue, 52, 112, 136–137, 143–144, 156, 178, 187, 213–214
Federal Emergency Management Agency, 36, 43, 56, 157, 317, 364–365, 403, 405–406
Federal Energy Regulatory Commission, 127
Federal Government Disputes Resolution, 478
Federal Highway Administration, 158
Federal Housing Administration, 255
Federal Power Commission, 49
Feld, J., 8, 445
Feldmann, G., 313, 315
Felsen, M., 313, 473
Fields, M., 136, 187
Finishes, 230, 371, 393–396. *See also* Coatings.
Finland, 411
Fire, 45–48, 100–101, 119, 136, 172, 180, 187, 236, 261, 379–380, 401, 405, 410, 442
  Beverly Hills Supper Club, 46
  Chicago Civic Center, 45
  Coconut Grove Nightclub, 46
  Dupont Plaza Hotel, San Juan, Puerto Rico, 47
  First Interstate Bank Building, Los Angeles, California, 48
  General Motors Transmission Plant, Livonia, Michigan, 48
  Iroquois Theater, Chicago, 46
  McCormick Place Exhibition Hall, Chicago, 46
  MGM Grand Hotel, Las Vegas, Nevada, 47–48
  One Meridian Plaza, Philadelphia, 48
  Our Lady of the Angels School, 47
  Stauffer's Inn, New York, 47
  Triangle Shirt Waist Company, New York, 46
Fire protection, 46, 187, 349
Fire-resistive plywood, 13
Fire sprinkler systems, 371
Firth, C., 153
Fisher, J., 157, 187
Fisher, T., 373
Flashing, 333, 336–337, 349–350, 352, 374, 379
Flat-plate construction, 243, 267–274, 310–315
Floods, 24, 43–44, 54, 60, 62, 93, 102, 104–105, 113, 115–116, 118–121, 148, 155–156, 180, 405, 418
Flood control projects, 43–44, 105, 122
Florida, 167, 274, 301, 315, 375, 420
Florida Atlantic University, 427

Florida Department of Transportation, 51, 252
Florida Sunshine State Parkway, 97
Flotation of foundations, 93–95
Fonseka, M., 293
Forensic engineering, 21–22, 166, 172, 182, 187, 332, 382, 399, 439, 464, 480–481, 492–493
Formwork, 103, 162, 236, 238, 242–245, 248, 250–252, 274, 422–423, 425, 427–428
Ft. Collins, Colorado, 357
Fort Lee, Virginia, 397
Ft. Meyers, Florida, 375–376
Fort Worth, Texas, 231
Forum Committee on the Construction Industry, ABA, 451, 472
Foundations, 59–110, 112–116, 120–121, 127, 135, 167–168, 262, 303–304, 342, 382, 385–386, 389, 413–420, 439
Fowler, Sir John, 253
Fracture mechanics, 130, 206, 208–209, 224
France, 120–121, 188–189, 263, 411, 445–446
Frank, W., 118
Frauenhoffer, J., 174, 229, 231, 349, 379
Freedonia, New York, 93
Frost damage, 93, 103–104, 174, 246, 248–251, 283, 285, 296, 303, 324, 335, 337–338, 340, 344, 349–350, 361
Frost, J. D., 13, 108, 492
Frost, Robert, 25
Fungi, 65, 173–174, 393
Furnishings, 371, 393–396

Galambos, T., 188
Galvanizing, 229, 284, 319
Gasparini, D., 136, 187
General Accounting Office, 158
Geology, 30, 56, 61, 74, 78, 99, 101, 104, 115, 123, 125, 129
Georgia, 247
Geosynthetic materials, 64, 104, 108
Geotechnical engineering, 59–62, 65, 74, 104, 107–108, 115, 121, 382, 464, 480
Gerber beams, 195
Germany, 87, 320–321, 427
Giannone, M., 429
Glass, 372, 377, 381–392, 400
Glazed brick, 347, 350
Glued-laminated timber, 163, 166, 170, 178, 183, 431–432
Gnaedinger, J., 11
Godfrey, E., 7
Godfrey, K., 373
Goodkind, D., 472
Gordon, J., 54
Gorman, T., 174–177
Gould, J., 60
Grain storage elevators, 50, 87–90, 442–443
Gravity, 25–28, 99, 120, 130–131, 170, 218, 433

Great Salt Lake, Utah, 105
Great Wall of China, 327
Greece, 411
Green, E., 474–475
Green, P., 287
Greenfield, S., 61–63, 65, 91, 100
Greenvale, New York, 10
Grimm, C. T., 344, 346
Gross, J., 208, 486
Ground freezing, 78
Ground modification techniques, 31, 82, 89, 91. *See also* Soil modification.
Ground water, 61–62, 65, 67, 80, 85, 89–90, 92–95, 99–101, 104–105, 262, 386, 416–417
Grout, 296, 303, 320–324
Guadalajara, Mexico, 49
Gumpertz, W., 351
Gurfinkel, G., 208, 281, 294
Guy, S., 466, 471, 477–478
Gymnasium, 182, 189–190, 306

Hadipriono, F., 71, 95, 246, 397–399, 414, 421, 428
Hailstorm, 375
Hamilton, Ontario, 49
Hammond, R., 7
Hammurabi, Code of, 3–4
Hangers, aircraft, 163, 309
Hannan, E., 468–471
Hansen, K., 157
Hanson, T., 298, 315
Hanshin Expressway, Japan, 276, 280, 404
Harmonic electrical currents, 371, 401–402
Hartford, Connecticut, 9–10, 18, 25, 95, 116, 192, 198, 200–204, 413
Hawaii, 32, 75–77, 129
Hazardous wastes, 60, 62, 483
Health, 371, 373, 392–396
Heger, F., 314
Heller, B., 373
Hennebique, 240
Henry, J., 475
High Point, North Carolina, 361
High-rise buildings, 42, 70, 90, 93, 188–189, 243, 254–255, 267–268, 276, 303–306, 314, 316, 330, 336, 340, 344–346, 348, 353, 357–360, 377, 380–392, 400, 422, 431, 433, 443, 489
Highway structures, 35, 37, 52, 100–101, 107, 132, 142–157, 242, 247, 251, 275–277, 280, 291, 295, 301, 306, 323–324, 425–427, 431, 436, 442, 446, 476
Hinckley, J., 444
Hiroshima, Japan, 431
Historic structures, 163–164, 332–335, 385–386, 413, 416–417, 438–440
History of failures, 3, 239–242, 413, 492
Hokenson, R., 127
Holland, 445

Hollywood, California, 106
Hooley, R., 189, 313, 438
Hospitals, 36, 54, 69, 90, 93, 338, 345, 397
Hotel, 10, 12, 18, 25, 45, 47–48, 84, 210, 214–225, 240, 249, 267, 329, 340, 397, 436–437, 455, 487
Housing, 38, 50, 167, 250, 286, 303–306, 346, 349, 362, 366, 378. *See also* Residential construction.
Houston, Texas, 104, 333, 381
Hsieh, H., 62
Huber, F., 149
Huck, R., 165, 228
Huckelbridge, A., 298
Human error, 14, 44, 225, 236, 241, 423, 483
Humidity, 239, 248, 283, 321, 375, 392. *See also* Moisture.
Hurricane, 24, 39–40, 54–55, 62, 138, 165, 167, 170, 315–316, 374, 381
HVAC systems, 330, 390, 392–396, 460. *See also* Airquality; Electrical systems; Environmental control systems; Equipment; Mechanical systems.
Hydrogen embrittlement, 285, 319–321

Ice, 117, 204, 249–250, 263, 375, 418
Ichniowski, T., 427
Impact loads, 187, 239, 418, 431
Incompatible materials, 17, 331, 343–347, 351–352, 355, 365, 368, 377–379, 445, 485
India, 98, 286, 329, 483
Indianapolis, Indiana, 375
Indian Concrete Journal, 260–262
Industrial accidents, 483
Information dissemination, 1, 8, 11–12, 22, 292, 314, 392, 444, 488–491, 492
Infrastructure, 71–72, 112, 157, 281, 416
Ink, S., 375–376
Insects, 173
Inspection, 7, 12–13, 18, 21, 34, 52–53, 61, 66–70, 108, 113–114, 117, 124, 143–150, 152, 157, 167, 177, 183, 197, 202–204, 206–207, 225, 228, 238, 242–243, 248, 254, 259, 274, 281, 283, 295–296, 302, 318, 323–324, 331, 336–337, 355, 373, 398, 409, 412–413, 419, 425, 436, 439, 444–445, 454, 457–458, 486–487. *See also* Supervision.
Instrumentation, 74, 80, 104, 115–116, 122, 126, 156, 214, 375, 400, 418
Insulation, thermal, 373, 375–376, 378, 381, 397
Insurance, 12–13, 197, 390, 403–404, 406, 445–446, 451–452, 459, 461–462, 465, 472, 479–480, 486–488, 491–492
International Association of Concrete Repair Specialists, 286
International Commission on Large Dams, 114, 123

International Decade for National Disaster Reduction, 34, 54
International Labor Office, 411
Irwin, M., 51
Isberner, A., 337
Italy, 104, 113, 117, 121–123, 157, 267, 332

Jackson, Michigan, 268
Jackson, Mississippi, 87
Jamaica, 191–192
Jamaica Bay sewage treatment plant, New York, 73–74
Jansen, R., 114
Japan, 32, 34, 38, 74, 77, 95, 101–103, 164, 206, 276, 279–280, 318, 404, 419, 428, 431, 446
Jayanandana, A., 293
Jennings, D., 41
Jennings, J., 80
Jersey City, New Jersey, 397–399
Johannesburg, South Africa, 80
John Hancock Mutual Life Insurance Company, 382–392, 416
Johnson, S., 11, 59, 62
Johnson, W., 114
Johnstown, Pennsylvania flood, 44, 115, 118–120
Jones, C., 197
Jones, D., 64
Jones, K., 64
Joseph, L., 148–149
Journal of the Construction Division, ASCE, 302
Journal of Performance of Constructed Facilities, ASCE, 12, 314, 490–491, 493
Judgment, 14, 21, 30, 60–61, 80, 116, 126, 141, 183, 423, 439, 441, 456, 486

Kagan, H., 182–183
Kalamazoo, Michigan tornado, 41
Kaminetzky, D., 11, 71, 187–188, 208, 252, 298, 422, 492
Kan, F., 12, 153, 156, 487, 490
Kansas, 116
Kansas City, Missouri, 10–12, 18, 25, 92, 192, 209–225, 357, 455, 487
Karp, L., 63
Karter, M., 45
Keith, E., 167
Kellogg, R., 173
Khachaturian, N., 12, 444, 484
Khan, M., 315
Kingston, Jamaica, 191–192
Kircher, C., 280
Knight, H., 475
Kobe, Japan, 34–38, 276, 279–280, 318
Korman, R., 410, 427, 429, 442
Kovacs, G., 88
Krakatoa volcano, 32
Kub, E., 380

Kuesel, T., 53
Kulicki, J., 148
Kyoto, Japan, 164

Laboratory mockup tests, 378, 390–391
Lagorio, H., 32, 316, 402
LaGuardia, New York airport, 69
Lahnert, B., 189, 313, 438
Lamellar tearing, 209
Landgren, J., 319
Landslides, 31, 35, 56, 60, 80, 95, 99–104, 113, 116–117, 122, 129
Land-use planning, 30–31, 35, 45, 54–57
Lankersham, California, 106
Lapping, J., 411
Las Vegas, Nevada, 47–48
Latvia, 445
Laurel, Maryland, 427
Lavon, B., 178
Lebanon, 75
Legislation, 409, 412, 493
Leonards, G., 2, 108
LePatner, B., 11, 59, 62
Levitt, R., 444
Lev Zetlin Associates, 201–203
Levy, M., 120, 133, 138, 145, 148, 204, 304
Lew, H. S., 423
Liability, 100–101, 108, 120, 203–204, 373, 379, 381, 390, 394, 451–482, 491
Liberty ships, 208
Lichtenstein, A., 142, 187
Liepens, A., 153, 156
Lift-slab construction, 9, 188–189, 206, 310–315, 410, 438
Light-frame construction, 39, 61, 63, 167, 170, 182, 316, 436–437
Lightning, 127
Lightweight structures, 27, 41–43, 138–139, 165, 167, 186, 291, 295, 315–316, 367, 385
Lim, K., 187
Lin, H., 71, 414
Liquefaction, 31, 37, 62, 80, 95–96, 113
Litigation, 2, 13, 17, 21, 50, 55, 60, 70, 75–77, 92, 94, 100–101, 105, 108, 116, 121, 130, 204, 210, 215, 223, 229, 243, 310, 331, 335, 348–351, 361–363, 372–373, 377, 380, 382, 386, 389, 392–395, 399–400, 416, 451–482, 486, 489, 491
Little Rock, Arkansas, 93
Livonia, Michigan, 48
Load transfer foundation failures, 85–88, 92
Loftus, W., 477
Loma Prieta, California, 32, 37–38, 157, 275, 402–404
London, Ontario, Canada, 357
London, United Kingdom, 50, 92, 97, 104, 116, 303–306, 400
Long Beach, California, 35, 104, 172, 189, 192
Long Island, New York, 349
Long Island Expressway, New York, 66, 72

Longspan structures, 10–11, 27, 41, 43, 163, 174–177, 180, 189–190, 198, 200–204, 209–210, 213, 239, 252, 291–292, 483
Loomis, R., 203
Los Angeles, California, 48, 100, 105, 115, 117, 123, 189–190, 280, 366, 420, 427, 429, 439
Loss, J., 411
Lossier, H., 7
Louisville, Kentucky, 49
Luft, R., 153, 156
Lynch, M., 187

MacDonald, C., 136–137
Maevis, A., 466, 471, 477–478
Maintenance, 13, 17, 21, 52–53, 71, 94, 103, 108, 111, 115, 127, 130, 132, 144–149, 156–158, 165, 173–178, 187, 228–230, 236–237, 252, 254, 265, 283, 285, 287, 290–292, 333–334, 348, 374–375, 377, 392–393, 395, 397–398, 400, 410, 413, 441, 460
Malhotra, V., 246
Malibu, California, 100–101
Malpractice, 455, 461, 477
Management errors, 127, 207, 241, 243, 454
Manhasset, New York, 18
Manhattan, New York, 340
Manitoba, Canada, 89
Manufacturing defects, 381, 390, 410
Marble facade panels, 355–361
Marcello, C., 123
Marcus, S., 466
Marlin, W., 382
Marshall, R., 222
Masonry, 27, 32, 36, 43, 66, 86–88, 113, 228, 252–253, 255, 327–370, 372, 375, 386, 402–403, 413, 416–417, 421, 430, 439
  aging masonry, 332–335
  cladding, curtain walls, facades, 348–360
  corrosion, 360–363
  incompatibilities with other materials, 343–347
  interior partitions, 367
  properties of, 327–332
  seismic and wind performance, 363–367
  settlements and other movements, 337–343
  unreinforced masonry, 15, 33, 35, 38, 42, 275–276, 280, 328, 363–367, 438–440
  workmanship, 335–337
Masonry Conservation Group of Historic Scotland, 333
Masonry Institute of America, 30, 332, 350
Masonry Research Foundation, 351
Masonry Society, The, 332
Massachusetts, 309, 389
Material deficiencies, 20–21, 72, 111, 213, 230–231. See also Incompatible materials.

Mattar, S., 331, 338
Maurel, G., 319
McCullough, D., 118
McFadden, T., 64, 66
McGuire, W., 313–314
McKaig, T., 8, 328, 421, 493
McLean, F., 157
McManamy, R., 451, 464, 466
Mechanical equipment and systems, 127, 268, 330, 371–372, 374, 391–396, 443, 460. See also Air quality; Environmental control systems; Equipment; HVAC systems.
Mediation, 21, 156, 313, 452, 466, 471–473, 475, 477
Mediation/arbitration, 473, 477
Meehan, R., 63
Merliss, P., 456
Mertz, D., 157
Metal roofing, 42
Metal stud/brick veneer systems, 13, 350–353, 379–380
Mexico, 49
Mexico City, 28–29, 33–34, 49, 88, 104
Meyer, R., 173
Miami, Florida, 241, 310
Microbiologically-induced corrosion, 228
Milwaukee, Wisconsin, 77, 443
Mindlin, H., 142
Mines, 79–80, 85, 102, 124, 157
Minitrials, 452, 466, 473–475
Minneapolis, Minnesota, 362, 375
Mississippi River, 43, 51, 427
Missouri, 223
Missouri River, 43
Misuse of facilities, 27. See also Operational errors.
Mitchell, D., 319
Modeling, 137, 139, 141, 256–259
Moisseiff, L., 139
Moisture, 52, 60, 64, 87, 93–95, 98, 101, 103, 165–167, 173–174, 178, 228–229, 231, 236–238, 250, 252–253, 260–263, 265, 282–284, 319–320, 322, 324, 331–332, 335–338, 340, 342–345, 348–360, 362, 367, 375–380
Moncarz, P., 53, 72, 189, 313, 438
Monongahela River, 209
Monteiro, P., 357
Montreal, Quebec, Canada, 67, 308
Moore, R., 172
Morley, J., 89
Morris, M. D., 302–303
Morris, W., 135
Morstead, T., 331, 338
Mortar, 13, 332–333, 335–338, 340, 347–348, 350, 352, 361–363, 379
Moscow, Russia, 303
Moses, F., 298

Movements, 112, 115, 121, 132, 137, 213, 239, 251–262, 280, 291–292, 295, 298–299, 301–302, 321, 337–343, 372, 374, 377, 390, 425
  earth, soil and rock movements, 59, 63, 73–74, 79–81, 83–84, 87–90, 92, 95, 100–102, 106–107, 116, 129, 416–417. *See also* Stability, soil, rock stability.
Movement joints, 112, 132, 206, 251–262, 291–292, 295, 298–299, 301–302, 315, 336–347, 349–350, 352, 355, 360, 365
Mt. Laurel, New Jersey, 173
Mud slides, 24, 60, 75, 100–101
Mullick, A., 260
Multistory buildings. *See* High-rise buildings.
Murillo, J., 149
Murrow, E., 155
Murrow, L., 155
Mycology, 173–174, 228

Napoleonic Code, 5
Natchez, Mississippi, 99
Natchitoches, Louisiana, 49
Nathan, N., 197
National Aeronautics and Space Administration, 274
National Association of Corrosion Engineers, 287
National Bridge Inspection Standards, 144, 148
National Bureau of Standards, 144, 210, 215–218, 222, 243–245, 272, 274, 305, 311–314, 423, 425. *See also* National Institute of Standards and Technology.
National Concrete Masonry Association, 332
National Earthquake Hazards Reduction Program, 38
National Fire Protection Association, 45
National Geophysical Data Center, 29
National Institute of Standards and Technology, 313, 438. *See also* National Bureau of Standards.
National Institute for Occupational Safety and Health, 9, 394, 410
National Oceanic and Atmospheric Administration, 31
National Research Council Committee on Ship Structural Design, 208
National Research Council of Canada, 7, 102
National Roofing Contractors Association, 373–374
National Safe Workplace Institute, 411
National Safety Council, 9, 409–410
National Science Foundation, 12
National Society of Professional Engineers, 12, 491
National Transportation Safety Board, 142, 144, 147, 149
Natural hazards, 16, 24–58, 165, 170, 406, 430, 436, 441

Naval Research Laboratory, 15
Negligence, 77, 121, 215, 224, 229, 240, 243, 267, 394, 420, 453–455, 457, 477, 493
Nelson, S., 465, 472
New Brunswick, New Jersey, 368
New Hampshire, 380
New Madrid, Missouri, 30
New Orleans, Louisiana, 104, 375
New York, 42, 46–47, 83, 99, 148, 241, 296, 333, 362
New York City, 38, 47–48, 50, 79–80, 90, 105, 111, 138–139, 173, 251, 269, 298, 345, 355, 418, 430, 439, 443, 451
New York Housing Authority, 250, 345
New York State Thruway, 97, 103, 148–149
New Zealand, 38, 307, 411
Newcastle, Indiana, 189
Newman, A., 205
Niagara Power Authority, 81, 102
Nicastro, D., 382
Niigata, Japan, 95, 102
Nondestructive evaluation, 66, 145, 149, 157, 178, 206–207, 242, 287, 331
Nonstructural components, 32, 38, 157, 162, 316, 348, 363, 365, 367–368, 371–408, 439
Norfolk, Virginia, 367
North American Air Defense (NORAD), 74, 78
Northridge, California, 32, 34, 37–38, 40, 157, 210, 225–227, 276–278, 280, 316, 318, 402–405
Norway, 173
Nuclear Power Plant, 16
Nuclear Regulatory Commission, 40

Oakland, California, 37, 45
Occupational Safety and Health Administration, 9, 171–172, 410–411, 419–420, 422, 424–425, 428, 444
O'Connor, J., 286
Office buildings, 47–50, 67, 70, 82, 87, 91, 348–363, 381–396, 399, 416
Ohio River, 209, 425
Ojdrovic, R., 227
Oklahoma City, Oklahoma, 50
Omaha, Nebraska, 293, 307
One-hundred-Day Document, 473, 477
Operational errors, 21, 72, 111, 127, 158, 174, 230, 290, 392–393, 396, 410, 414–415, 441, 443–444, 485. *See also* Misuse of facilities.
Oppenheim, I., 380
Optimization, 314, 319
Oregon, 182
Orlando, Florida, 190, 335
Osman, M., 382, 392
Osteraas, J., 53, 189, 313, 438
Ottawa, Canada, 84, 92, 191

Overloading, 21, 27, 29, 52, 69, 84–85, 92, 102, 111, 143, 180, 188, 201, 261, 267–268, 397–399, 420–421, 423–424, 443
Overtopping of dams, 113, 115, 117, 122, 157

Pacific Coast Highway, 100–101
Padgett, T., 183
Palm Coast, Florida, 12, 487
Palos Verdes Peninsula, California, 100
Panama Canal, 103
Parapet, 340–343, 346, 349–350, 355, 374, 378, 402
Paris, France, 188–189, 240
Parker, J., 27, 197
Parking structures, 53, 83, 225, 231, 252, 254, 263–266, 269, 283–285, 293–294, 299, 309, 311, 316, 318, 321, 414, 427, 441
Partnering, 445, 463–464, 479
Pasternack, S., 95
Pathology, 22, 493
Patton, R., 384
Pavalon, E., 466, 470–471
Pavements, 97–99, 247, 252–253, 263, 265, 442
Pearson-Kirk, D., 286
Pedestrian walkways, 10, 12, 18, 25, 210, 214–225, 455, 487–488
Peer review, 12–13, 197, 203, 223, 241, 462, 487, 489–490
Pennsylvania, 208
Peoria, Illinois, 349
Peraza, D., 313–314
Perenchio, W., 319
Perera, A., 293
Performance-based design, 405–406
Permafrost, 16, 61, 64, 66
Peru, 55
Peterson, A., 446
Petroski, H., 11, 22, 130, 138, 141–142
Pfeifer, D., 319
Pfrang, E., 222
Philadelphia, Pennsylvania, 48, 77, 85, 297, 355
Philippine Islands, 282
Pile-driving, 50, 418
Pile foundations. *See* Deep foundations, piles.
Pipelines, 49, 71–73, 207, 228–229, 285, 302
Piping failures in dams, 113, 126
Pisa, Tower of, 88
Pittsburgh, Pennsylvania, 106, 173, 294, 333, 441
Plainfield, New Jersey, 69
Plate tectonics, 30
Platteville, Colorado, 250
Pneumatic structures, 42, 375
Point Loma, California, 73
Poirot, J., 480
Ponding roof failures, 26–28, 178–180, 197–199, 213, 375, 400

Pontchartrain Causeway, Louisiana, 51
Pontiac, Michigan, 375
Port Authority of New York and New Jersey, 397–399
Port Washington, New York, 178
Portland Cement Association, 37, 258, 260, 293, 332
Portland, Oregon, 88, 106, 254
Post, C.W., Center Auditorium, 10
Post Falls, Idaho, 436–437
Post Tensioning Institute, 323
Poston, R., 313, 315
Potts, D., 442
Power plant, 72, 81, 102, 113, 122, 127–128, 425
Powers, J., 60
Precast concrete. *See* Concrete, precast.
Precast/Prestressed Concrete Institute, 293, 301, 315
Precipitation, 62
Pre-engineered products, 166–167, 180–183, 188, 293, 309, 378, 435
Prendergast, J., 73
Prestressed concrete. *See* Concrete, prestressed.
Preventive engineering, 158, 405
Preziosi, D., 410–411
Pride, R., 262
Priestley, M., 227
Private Judges, 475
Procedural deficiencies, 194–195, 223–224, 292, 442, 451, 487
Product Liability Law, 456
Programming errors, 16–17, 405–406
Progressive collapse, 8–9, 18, 50, 135, 137, 139, 156, 167, 182, 191, 203, 213, 223, 239, 243, 271, 303–306, 310–315, 399, 420, 422, 425, 427, 429, 489
Proprietary systems, 165–167, 180–183, 373, 378–380, 485
Providence, Rhode Island, 307
Prussack, C., 430
Pullar-Strecker, P., 285–286
Purdue University, 126
Puri, S., 138, 187

Quality assurance, quality control, 70, 72, 108, 114, 118, 135, 165, 167, 172, 207, 223, 236, 238, 240, 243–250, 280, 290, 292, 302–306, 318, 321, 336, 353, 375, 378, 444–445, 452, 462, 479, 483–484, 486, 488, 490
Quayle, D., 478
Quebec, Canada, 102

Rabbat, B., 145, 158
Racine, Wisconsin, 377
Radio telescope, 15–16
Raghu, D., 62
Raikar, R., 286

Railroad ties, 293
Rain, 10, 26–28, 99, 101, 103, 105, 115,
    118–122, 124, 155, 165, 178–180, 197,
    210, 213, 306, 330, 335–337, 348, 367,
    375, 400
Rajkumar, C., 260
Raleigh, North Carolina, 294
Ramabhushanam, E., 187
Rancho Cucamonga, California, 39
Ratay, R., 428
Raths, Raths & Johnson, Inc., 284, 426
Redundancy. *See* Structural redundancy.
Reese, D., 174
Rehabilitation, 38, 72, 78, 156, 158, 276, 319.
    See also Renovation; Repair; Retrofit;
    Upgrade.
Reinforced concrete. *See* Concrete,
    reinforced.
Reinforced earth, 64. *See also* Ground
    modification techniques.
Reliability, 165, 236, 302
Relief joints. *See* Movement joints.
Rendon-Herrero, O., 492
Renovation, 53, 127, 177, 193, 290, 380, 399,
    430–431, 438–442. *See also*
    Rehabilitation; Repair; Retrofit;
    Upgrade.
Renton, Washington, 94
Repair, 52–53, 66, 72–73, 79, 90–91, 97, 107,
    115, 127, 129–130, 132, 138, 147, 154,
    156–157, 176–178, 207–209, 211, 214,
    225, 227, 231, 247–248, 250, 252–253,
    259, 263–266, 269, 280–281, 285–287,
    294–297, 301, 303, 321–324, 332–333,
    336–337, 344, 348–349, 353, 355, 357,
    362, 365, 368, 375–376, 378, 389,
    398–399, 402–406, 420–422, 439. *See also*
    Rehabilitation; Renovation; Retrofit;
    Upgrade.
Research Council on Performance of
    Structures, 8
Reservoirs, 83, 113, 115, 117–125, 127, 253,
    263, 267
    Baldwin Hills Reservoir, Los Angeles,
        California, 117, 123–124
    Colombo, Ceylon, 253
Residential construction, 62, 65, 83, 85, 87,
    95, 97, 100–101, 103, 108, 124, 163, 168,
    243, 439. *See also* Housing.
Ressler, M., 224
Retaining walls, 18–19, 95–96, 99, 103,
    105–108, 237, 389, 414, 416, 425
Retrofit, 38–39, 52, 147–148, 157, 276, 280,
    294, 332–333, 365–366, 389, 393. *See also*
    Rehabilitation; Renovation; Repair;
    Upgrade.
Revie, R., 281
Riggs, L., 471
Riley Road Interchange Freeway Ramp, East
    Chicago, Indiana, 9, 242, 425–426

Rio de Janeiro, Brazil, 87
Risk management, 461–464, 479
Roady, T., 455
Robert Gordon University, Scotland, 334
Roberts, M., 262
Robison, R., 112, 157, 205, 230–231, 283, 382
Rochester, New York, 357
Rock, 78–83
Rock bolting, 74, 78–81, 107, 157
Roebling, J., 112, 138
Rollings, M., 97
Rollings, R., 97, 253
Rome, Italy, 89
Roofing, 21, 174, 333, 348, 372–375, 377, 420
Roofing Industry Educational Institute, 374
Roof trusses, 26, 167, 169–170, 172–182, 189,
    191, 212, 420, 431, 435–437
Rose, J., 167
Rosemont, Illinois, 9, 170–172, 428–429
Ross, S., 11, 35, 48, 120–124, 130, 133, 137,
    142, 209, 392, 446
Rosta, P., 427
Rubin, R., 224–225, 451–452, 455, 466, 471,
    477–478
Ruff, B., 118
Russia, 191
Rzonca, G., 262

Sabnis, G., 286
Safety regulations, enforcement of, 409,
    411–412, 419–420, 443–447, 493
St. Lawrence River, 70, 152–153
St. Paul, Minnesota, 65, 427
Salinas, California, 71
Salt Lake City, Utah, 39, 336
Salvadori, M., 120, 133, 138, 145, 148, 204,
    304
Samelson, N., 444
San Andreas fault, 31
San Diego, California, 73
San Fernando, California, 36–37, 157
Sanford, Florida, 441
San Francisco, California, 20, 31, 35, 37–38,
    66, 111, 138–139, 247, 397, 443
San Jose, California, 105
San Juan, Puerto Rico, 47, 73
San Mateo, California, 310
Santa Ana River, California, 44
Santa Barbara, California, 12, 35, 486
Sao Paulo, Brazil, 87, 91
Sapers, C., 456
Scaffold, 20, 335, 412, 423–424, 427–428, 442
Scarangello, T., 313
Schaffer, M., 392, 396
Schenk, R., 471
Schlaich, J., 321
Schoellkopf power plant, New York, 81, 102
Schools, 26, 35, 42, 47, 54, 85, 103, 163,
    179–180, 182, 190–191, 240–241, 248,

250, 269, 301, 306, 333, 335, 349, 353, 367, 393, 397
Schriener, J., 451
Schriever, W., 102, 304, 410, 431
Schupack, M., 320–324
Schutt, C., 319
Schwartz, T., 378–379
Scotland, 333–334
Scott, P., 228
Scour, 13, 21, 115–116, 125, 127, 148–150
Scribner, C., 313
Sealants, 265, 331, 340, 342, 349, 373, 377–379
Sealant, Waterproofing and Restoration Institute, 377
Searcy, Arkansas, 49
Seattle, Washington, 38, 102–103, 153–156, 307, 417, 431–435, 443
Secondary stresses, 166, 202, 241, 252, 337
Seepage, 113, 119, 124–126
Seismic events. *See* Earthquakes.
Selfridge Air Force Base, Michigan, 92
Sequencing of construction, 20, 150, 170, 255, 293, 303, 309, 313–314, 420, 436, 445, 459
Serviceability problems, 60–61, 132, 372
Settlement of foundations, 31, 35, 61–62, 67–68, 70, 72, 79, 83–95, 97, 107, 113, 119, 124, 260–261, 337–343, 418, 439
Setzer, S., 410, 451
Sewer lines, 71, 73, 77, 84–85, 93–94, 101, 419. *See also* Pipelines.
Shakespeare, W., 173
Shanghai, China, 104
Shanley, E., 476
Sharif, A., 286
Shear:
  in concrete, 235, 238, 241, 243, 258, 265–274, 297, 299, 316, 422
  in timber, 166–167
Shear, punching, 18, 243, 265–274, 422
Shen, C., 61–63, 65, 91, 100
Shepherd, R., 13, 492
Shinnerer Management Services, Inc., 479
Shop drawings, 12, 210, 223–224, 281, 398, 487
Shoring, 20, 84, 103, 162, 238–239, 243, 249, 267, 269, 273, 414, 416, 422–423, 425–428, 442, 490. *See also* Bracing; Formwork; Temporary support.
Shrinkage, 239, 251–262, 265, 291, 293, 296, 298–300, 302, 308, 319, 337, 343–344, 346–347, 368
Shuirman, G., 60–61, 101, 105
Shyne, J., 72
Sicily, 101
Sick building syndrome, 393–396
Sierakowski, R., 246, 428
Simpson, Gumpertz & Heger Inc., 28, 198–199
Singh, J., 173, 392

Sinkholes, 61, 64, 79
Sioux Falls, South Dakota, 361
Site drainage, 59, 64, 97–98
Site rehabilitation, 60, 62. *See also* Soil modification.
Site selection, Site development errors, 16, 54–56
Skokie, Illinois, 258
Slabs-on-grade, 97–99
Slope stability, 61. *See also* Landslides.
Slosson, J., 60–61, 101, 105
Smeaton, John, 40, 135
Smith, E., 203
Smith, J., 208
Snow, 10, 25–26, 136, 178–181, 198, 200, 204, 250, 321, 375
SOCOTEC, 445–446
Soil modification, 62–63, 78. *See also* Ground modification techniques.
Soil nailing, 64, 108
Soils, 59–110, 413–420
  bearing capacity of, 61
  characteristics of, 59
  compaction, 61–62, 64, 73, 114
  fills, 61–63, 88, 98, 105–106, 113–115, 123, 135, 250, 385–386, 416
  moisture content of, 61–64
  problem soils, 61–65, 109
    contaminated, 61–62
    expansive, 16, 61, 63–64, 83, 93, 95, 98
    organic, 61–62, 64
  testing of, 59–60, 63, 67, 78, 82–83, 100, 114–115, 123, 135
  zone of influence, 59
Soils report, 61
Somayaji, S., 163, 173
Somes, N., 304
South Africa, 98
Southern Building Code Congress, 190
Southgate, Kentucky, 46
Soviet Institute of Construction and Architecture, 303
Space truss, 9, 18, 198, 200–204, 206, 211
Spain, 113, 116, 411
Specifications, 17, 72, 152–153, 246, 281, 285, 309, 337, 373, 376, 381, 398, 453, 455, 457, 460
Sponseller, M., 366
Springfield, Illinois, 85
Sputo, T., 64, 188, 337, 435
Stability, 20, 27, 36, 43, 66, 120, 150, 166, 171–172, 187–205, 230, 236, 238, 275, 280, 308, 313, 315–316, 335, 353, 423, 425, 428–439
  soil, rock stability, 59, 78, 80, 83–85, 89, 91–92, 94, 99–101, 105, 108, 114, 116, 122, 157. *See also* Movements, earth, soil and rock.
Stadium, 53, 163, 170–172, 174–177, 188, 190, 209–214, 230, 253, 265, 431–435, 483, 487

Stafford, D., 112
Stamets, J., 434
Stamford, Connecticut, 241
Standard of Practice, Standard of Care, 223, 453, 455
Stanford University, 410
State Agency Panels, 477–478
State University of New York, 427
Statue of Liberty, 228
Steel, 17–18, 20, 38, 43, 45, 93, 135, 141–145, 152–153, 155, 166, 186–234, 330, 344–345, 350, 362–363, 385, 389, 392, 402, 420, 428–429, 431–435, 459
    connections, 205–227
    corrosion, 228–231
    properties of, 186–188
    stability problems, 188–204
Steel, corrosion-resistant, 13, 230–231, 379
Steel joists, 27, 193–199, 229
Steinman, D. B., 137, 141
Stephenson, R., 463
Stevenson, R., 7
Stiffness, 27, 138–141, 165, 404
Stone panels, 336, 342, 344–345, 354–360, 372
Stone, W., 50
Stratta, J., 212–213
Stress concentrations, 187, 207–208, 227, 261, 292, 295–297, 328, 336, 342, 346
Stress reversals, 187, 189, 275, 308, 321. *See also* Cyclic loading.
Structural Clay Products Institute, 350
Structural engineering, definition of, 30
Structural Engineers Association of California, 35, 227, 406
Structural integrity, 8, 46, 50, 53, 69, 139, 148, 167, 172, 180, 197, 228, 243, 303–306, 315, 333, 360, 365, 367, 400, 483, 487, 489–490
Structural redundancy, 8, 17–18, 32, 50–52, 143–144, 147–149, 166, 182, 200, 203, 208, 213–214, 222–223, 243, 303–306, 315, 322, 399, 414, 487, 489–490
Stucco, 375–376
Stuttgart Institute for Massive Construction, 321
Suarez, M., 313, 315
Suaris, W., 315
Subsidence, 60, 64, 80, 104–105, 116, 124, 416, 418
Subway, 78, 82–83, 93–94
Sulfate, 246, 262, 293, 338
Supervision, 172, 237–238, 241–242, 267, 269, 303, 318, 398, 420, 456–457. *See also* Inspection.
Surface treatments. *See* Finishes; Sealants.
Surpula lacrymans, 173
Suspended ceilings. *See* Ceiling collapse.
Swimming pool, 53, 88, 100, 173, 229–230, 283, 392, 428

Switzerland, 53, 229–230, 240, 267
Syracuse, New York, 94, 263

Tall buildings. *See* High-rise buildings.
Tamaro, G., 60
Tampa, Florida, 51, 72, 133, 247, 353, 375, 425
Tanks, 73–74, 99, 208, 344
Tarricone, P., 443
Technical Council on Forensic Engineering, ASCE, 2, 12–13, 313, 480, 484, 488, 492
Telford, T., 138
Temperature effects, 251, 253–255, 261, 265, 294, 298, 301–302, 321, 335, 338, 342–346, 350, 355, 357, 368, 373, 377. *See also* Thermal stresses.
Temporary support, 20, 84, 150, 182–183, 188–193, 236, 238, 242–245, 248, 335, 414, 417, 419, 423, 427–440, 442, 445, 490. *See also* Bracing; Formwork; Shoring.
Tents, 43
Tera River, Spain, 116
Termites, 173
Terpening, T., 51
Terrorism, 50–51, 187, 489–490
Terzaghi, K., 114, 116, 123
Testing of materials and components, 240, 247–248, 380, 445
Texas, 419
Texas Tech University, 168
Thames River Tunnel, 328
Thermal stresses, 111, 130, 132, 144, 150, 260, 301, 321, 330, 340, 342, 381, 391. *See also* Temperature effects.
Thornton, C., 148–149
Tidal action, 92, 105, 262–263
Tighe, M., 285
Tilt-up construction, 316, 318
Timber structures, 9, 27, 34, 162–185, 316, 428–429, 431–432
    connections, 167–172, 228
    maintenance, 173–177
    properties of, 162–167
    proprietary systems, 180–183
    repairs, 177–178
Truss Plate Institute, 183
Tokyo, Japan, 35, 77, 101, 104, 419
Tomasetti, R., 148–149
Topeka, Kansas, 51
Tornado, 24, 39, 40–41, 62, 167
Toronto, Ontario, Canada, 188, 431, 433
Torsion, 139–140, 166, 187, 189, 193–198, 203, 252, 257, 267, 353, 365, 402, 436
Tort reform, 452, 478
Tower of Babel. *See* Babel, Tower of
Tower of Pisa. *See* Pisa, Tower of
Trans-Alaska Oil Pipeline, 207, 229
Transcona Grain Elevator, 89–90
Transmission tower, 435

Transportation accidents, 72, 306–315
Tree roots, 62, 92, 94, 80, 103
Trench accidents, 71, 84, 410, 412–420, 430
Tschegg, E., 356
Tsunami, 30, 32, 35, 54
Tunisia, 89
Tunnels, 60, 64, 71, 74–78, 80, 82, 85, 94, 102, 285, 328, 419–420, 443, 476–477
  Mont-Blanc Tunnel, France, 81
  Mt. Washington Tunnel, Pennsylvania, 78
  Shimizu Railway Tunnel, Japan, 74
  Thames River Tunnel, United Kingdom, 328
  Wilson Tunnel, Hawaii, 75–77
Turbulence, 39, 43
Turek, W., 200
Turk, A., 183
Typhoon, 24, 39

Uhlig, H., 281
Underground Technology Research Council, 476
Undermining of foundations, 83–85, 115, 418, 439
Unequal foundation support, 88–91
Unified Risk Insurance, 479–480, 488
Uniform Building Code, 34–35, 276, 280, 404–405
United Kingdom, 173, 303–306, 319, 323, 378, 410–411, 489
U.K. Department of National Heritage, 400
U.K. Department of Transportation, 323
United Nations, 54, 411
U.S. Army Corps of Engineers, 44, 143, 397, 463
U.S. Bureau of Reclamation, 125
U.S. Bureau of Public Works, 282
U.S. Committee on Large Dams, 121, 123
U.S. Congress, 10, 15–16, 144
U.S. Department of Defense, 258
U.S. Department of State, 50, 489–490
U.S. Environmental Protection Agency, 419
U.S. Geological Survey, 97
U.S. House Subcommittee on Health and Safety, 411
U.S. Navy "Big Dish" project, 15, 16
Union of Soviet Socialist Republics, 101, 303
University of California, 227, 277
University of Idaho, 174–177
University of Illinois, 53, 230, 258
University of Maryland, 12, 411, 419, 491
University of Michigan, 309
University of South Florida, 375
University of Washington, 139, 188, 190, 431
University of Western Ontario, Canada, 389
Unreinforced masonry. See Masonry, unreinforced.

Upgrade, 36–38, 53, 157–158, 276, 280, 365–366. See also Rehabilitation; Renovation; Repair; Retrofit.
Urban construction, 60, 71–72, 109, 307, 357, 386, 414

Vaccoro, G., 263
Value engineering, 379
Van Volinburg, D., 285
Veneer, 330–331, 343–347, 349–353, 355, 361, 375–376
Venice, Italy, 121, 332
Ventilation, 77, 173, 229, 283, 330, 350, 392–395, 410. See also Air quality.
Vibrations, 32, 35, 48–52, 62, 64, 77, 95–97, 130, 137, 139–140, 178, 218, 230, 236, 238, 246, 255, 261, 330, 333, 367, 371, 396, 417–418, 425, 441
Vinalhaven Island, Maine, 80
Vince, C. R., 479–480
Virginia, 242–245, 295, 308, 367, 422
Volcano, 32
Von Karman, T., 141

Wales, 103, 115, 323
Waltham, Massachusetts, 336
Ward, J., 464–465, 473
Washington, DC, 84, 332–333, 414, 440–441
Washington (State), 27, 41, 51, 138, 153–156, 423, 441
Washington State Department of Transportation, 29, 154, 156
Washington (State) Public Power Supply System, 16
Water. See Moisture.
Watertown, New York, 249
Waterville, Maine, 26, 179–181
Watson, S., 112
Weather, 239, 242, 246, 248–250, 260, 263, 286, 330–331, 335–337, 348–360, 377, 414
Weathering steel. See Steel, corrosion-resistant.
Weidlinger, P., 392
Welding, 130, 203, 206–209, 225–227, 275, 280–281, 301, 313, 429, 431, 443
West, A., 187
Westchester County, New York, 78, 338
West Virginia Highway Department, 144
West Virginia University, 9, 411
Wheaton, Illinois, 83
White, A., 265
White, Sir Bruce, 7
Whittier, California, 36
Whittle, C., 165, 228
Widhalm, C., 356
Wiggins, J., 63
Wilborn & Associates, 11
Wildfire, 24, 45, 100–101
Wilkinson, E., 53, 230, 281

Willow Island, West Virginia, cooling tower, 9, 20, 242, 311, 410, 423–424
Wind, 10, 39–43, 54, 99, 111, 130, 133–136, 138–141, 155, 165–168, 170, 178, 180, 187–189, 191, 200, 210, 213–214, 261, 263, 293, 308–309, 311, 315–319, 321, 328, 330, 338, 348, 351, 353, 357, 363–368, 374–377, 381–392, 404, 425, 430–431, 433, 436–437, 441. *See also* Hurricane; Tornado; Typhoon.
Wind engineering, 391–392
Window-washing equipment, 400
Windsor, Connecticut, 323
Wind tunnel testing, 43, 141, 357, 381, 389, 391
Wiss, Janney, Elstner Associates, Inc., 433
Wolf, J., 53
Wood borers, 65
Woodruff, G., 141

Workers Compensation Laws, 457
Workmanship, 165, 238, 281, 335–337, 342, 349, 352, 374–375, 378, 396–397. *See also* Craftsmanship.
Wright, D., 12, 487, 490
Wright, Frank Lloyd, 377
Wright, R., 50, 208

Yokel, F., 50
Yonkers, New York, 17, 240, 263–266, 306, 346

Zallen, R., 26, 179–181, 490
Zarghamee, M., 153, 156, 227
Zetlin, L., 158
Ziraba, Y., 286
Zschokke, B., 282
Zurich Institute for Testing Materials, 282
Zwerneman, F., 187